32.50

2

Pests of Field Crops

PESTS OF FIELD CROPS

Third Edition

F. G. W. Jones
M.A., Sc.D.(Cantab.), F.I.Biol. formerly Deputy Director and Head of
Nematology Department, Rothamsted Experimental Station and Visit-
ing Professor in the Department of Zoology and Applied Entomology,
Imperial College, London

and the late
Margaret G. Jones
M.A., Ph.D.(Cantab.), F.R.E.S., Entomology Department, Rothamsted
Experimental Station

EDWARD ARNOLD

First published in Great Britain 1964
by Edward Arnold (Publishers) Ltd
41 Bedford Square
London WC1B 3DQ

Edward Arnold (Australia) Pty Ltd
80 Waverley Road
Caulfield East 3145
PO Box 234
Melbourne

First published in United States of America 1984
by Edward Arnold
300 North Charles Street
Baltimore
Maryland 21201

British Library Cataloguing in Publication Data
  Jones, F. G. W.
    Pests of field crops.——3rd ed.
    1. Field crops——Diseases and pests——
    Great Britain
  I. Title   II. Jones, Margaret G.
    633'.0896'0941   SB605.G7

    ISBN 0-7131-2881-X

Reprinted with corrections 1966
Second edition 1974
Reprinted with corrections 1977
Reprinted 1980
Third edition 1984

Text set in 10 on 11 pt Times New Roman
Printed in Great Britain by
Thomson Litho Ltd, East Kilbride, Scotland

Preface to the Third Edition

In this edition, to accommodate new developments and to effect economies, Chapter 1 has been greatly shortened. The earlier controversies about the role of population density in regulating numbers have been resolved or have been replaced, and the subject is now well covered by numerous publications on population ecology. The reference list has also been shortened by deleting the section on reference books and by removing from the main list all but important references dating from earlier than about 1970. As the government sponsored schemes for pesticide registration, safety precautions and agricultural chemicals approved are well established and supported by legislation, an attempt has been made to mention only those products approved at the time of going to press. For historical and other reasons other products are included when necessary. Other changes and additions seek to bring the book up to date in other respects. I regret the necessity of pruning rather than extending the reference list but missing references may be found by consulting earlier editions in libraries.

It is a pleasure to acknowledge assistance in bringing the book up to date from: R. A. Dunning and colleagues, Broom's Barn, G. A. Wheatley and colleagues, NVRS, Roger Cook, WPBS, J. Bowden, D. Griffiths and numerous other colleagues at Rothamsted, J. A. MacFarlane, TDRI, Richard Brown, Imperial College, Tom Mabbott, DAFS and A. W. Colling, formerly of the Nature Conservancy. Members of the British Museum (Natural History) and of CIE checked some scientific names. The following members of MAFF/ADAS were consulted: P. Aitkenhead, R. Gair, J. F. Southey and A. L. Winfield. Appropriate parts of the script were read by staff in the *MAFF* Agricultural Science Service at Slough (Storage Pests and Pest Control Chemistry), Tolworth (Mammals) and Worplesdon (Birds). Sybil Clark and Pamela Evans produced Plates XVII and XIX and Janet Cowland assisted with the script and with proof reading. I am most grateful for the help all have given. Any errors, omissions or mistakes in interpretation are mine.

F. G. W. Jones

1984

Preface to the Second Edition

Since the first edition was written the harmful effects of the powerful and persistent organochlorine pesticides have been realized increasingly. In Britain, by voluntary agreement, they have been phased out and are now retained for a few approved uses only. The voluntary scheme for the approval of products for farmers and growers agreed between the manufacturers of agricultural chemicals and the Agricultural Departments of the United Kingdom and various Acts affecting poisonous substances have gone far to regulate pesticide usage in a practical and sensible manner. Rachel Carson's book *Silent Spring*, despite its inaccuracies and emotional tone, did much to arouse public awareness of the potential dangers from the misuse of pesticides. Consequently, during the last ten years much has been written about pesticides and much new work stimulated. In this second edition we have tried to cover new developments and to bring the book up to date. To counterbalance the increase in length necessitated by new material we have eliminated some passages that referred to foreign pests and to pests of fruit.

We thank those who have helped us to make these changes, notably G. A. Wheatley and A. J. Dunn, National Vegetable Research Station, R. Howe, D. C. Drummond, Gwyneth C. Williams, P. S. Tyler, H. K. Heseltine of the M.A.F.F. Pest Infestation Control Laboratory, Gwenyth R. Raw, M.A.F.F. Plant Pathology Laboratory and J. Mason, Plant Health Branch, M.A.F.F.

F. G. W. Jones
M. G. Jones

1973

Preface to the First Edition

This book chiefly describes pests of British field crops and is intended primarily for students in Agricultural Colleges and Universities, but some pests of garden and orchard are mentioned, as are a few pests from abroad. Balancing the contents of chapters has been difficult, because we do not have specialist knowledge in every group and because some groups of pests, such as mites and slugs, have been less thoroughly studied than others. Inevitably, greatest emphasis falls on insects, for, although individually they may be no more important than some vertebrate or nematode pests, the number of species known is so much greater.

After an introductory chapter, the book is divided according to zoological grouping, but only brief outlines of classification are given and sufficient about structure to enable students to make preliminary identifications. Chapters on crops, methods of control and pesticides follow to coordinate matters that cannot be fully covered in the early chapters.

The authors thank Mrs Janet A. Cowland for help in preparing the script and some of the line diagrams, Mr V. Stansfield, Mr F. D. Cowland and Mr W. E. Dant for their help with photographs, Mr H. W. Janson for providing us with specimens and helpful criticisms, and to Mr H. L. G. Stroyan for the use of his aphid slides. The following specialists kindly read and criticised portions of the script: Mr P. Aitkenhead, Dr E. W. Bentley, Mr E. B. Brown, Dr J. A. Dunn, Dr J. A. Freeman, Dr H. C. Gough, Mr H. G. Lloyd, Dr E. J. Miller, Dr R. K. Murton, Dr F. Raw, Mr J. H. Stapley, Mr J. W. Stephenson, Dr H. V. Thompson, Dr M. A. Watson, Mr A. L. Winfield, Mr D. W. Wright and Mr E. N. Wright. We gratefully acknowledge their help but we are responsible for any errors or omissions.

Most of the line diagrams are original, some are redrawn or adapted from the work of others. Due acknowledgement is made in the text. Photographs acknowledged as 'Crown copyright' are published by permission of the Controller of Her Majesty's Stationery Office.

F. G. W. Jones
Margaret G. Jones

Rothamsted
May 1964

Preface to the First Edition

Contents

Contents

1

The origin and nature of pest problems

Most animal populations are influenced by man in various ways, usually as a consequence of agriculture, which may be defined as 'interfering with nature'. Before crops can be grown or livestock raised, the natural vegetation must be removed or radically changed. This process necessarily destroys most of the natural fauna and disrupts the complex interrelationships that have grown up over long periods between various species of animals and plants. Destruction of the vegetation and cultivation of the soil drastically change the habitat, which becomes harsher because the soil surface is exposed directly to the weather. Few animals survive the change. The herbivorous species that do are those able to withstand the new environment and to feed on the crops planted or the weeds that colonize the exposed soil. Where a succession of different crops is grown, polyphagous species which feed on many kinds of plants are favoured; where monoculture is practised, species that previously fed on wild plants related to the crop grown may survive. Relicts of the old fauna remain in woodlands, hedgerows, ditch sides, road verges and waste land, but, where agriculture is intense and permanent, habitats are mostly man-made. In Britain the prehistoric forests and grasslands have mostly disappeared, and natural habitats are found only on the very tops of mountains, on barren, inaccessible heaths and bogs or in salt marshes.

To provide more food for the expanding human population, the process of alteration and destruction of natural vegetation goes on at an increasing pace. Where areas of virgin land are brought under cultivation, the surviving species of animal generally fail to fill all the available niches provided by crops. Nature abhors a vacuum, physical or biological, and alien species soon move in from distant habitats. The old geographic barriers such as seas, great rivers, high mountains and areas of desert or tundra have become less formidable obstacles to animal dispersal because of the development of human transport, which intentionally or unintentionally carries harmless and potentially harmful species over much of the world. Quarantine services set up to check dispersal do no more than delay it, because sampling and inspection methods are inadequate, and it is impossible to watch all possible points of entry. It is therefore only a matter of time before a potentially harmful species from one place finds a vacant niche well suited to its way of life in another. What applies to animal pests applies equally to organisms causing plant diseases, and to noxious weeds.

The crops grown in newly developed areas are themselves mostly aliens. Potential pests are often introduced with them in seeds, foliage, roots or soil, so that the distribution of pest and host often coincides from the start. When this is not so, other means of dispersal bring about the same result in time, provided climate and soil in the new environment favour the pest.

One notable feature of natural communities of animals is an apparent stability. Although numbers of plant-feeding species fluctuate, they are rarely so many that the host plants are destroyed, and, for most of the time, numbers appear very much below the carrying capacity of the hosts. The artificial communities of animals that spring up in agriculture appear far less stable, and epidemics of one pest species or another are common. One contributory factor is monoculture, or at least covering large areas annually with a single type of plant which has been bred for yield or succulence rather than for resistance to pest attack. Apart from this,

Table 1.1 Pests introduced into the U.S.A. (see Wardle, 1929).

1. *From N.W. Europe*
   Hessian fly (*Mayetiola destructor* (Say.))   Wheat
   Gipsy moth (*Lymantria dispar* (L.))   Tree foliage
   Small White Butterfly (*Pieris rapae* (L.))   Brassicas
   Carrot fly (*Psila rosae* (F.))   Carrots
   Netted slug (*Deroceras reticulatum* (Meull.))   Many crops

2. *From the Mediterranean*
   Pink boll worm (*Pectinophora gossypiella* (Saunders)) Cotton
   European corn borer (*Ostrinia nubilalis* (Hbn.))   Maize
   Codling moth (*Cydia pomonella* (L.))   Apples
   Grain weevil (*Sitophilus granarius* (L.))   Stored grain

3. *From S.E. Asia*
   San José scale (*Comstockaspis perniciosa* (Comst.))   Citrus
   Cottony cushion scale (*Icerya purchasi* (Mask.))   Citrus
   Japanese beetle (*Popillia japonica* (Newm.))   Tree foliage, field crops

4. *From South and Central America*
   Potato tuber moth (*Phthorimaea opercullela* (Zell.))   Potato
   Potato cyst-nematode (*Globodera rostochiensis* (Woll.)) Potato, via Europe
   Sugar-cane borer (*Diatraea saccharalis* (F.))   Sugar-cane

however, the artificial communities seem inherently less stable. A factor contributing to instability and sometimes leading to partial or complete destruction of the host plant is the lack of a range of enemies that would prey on the pest and prevent it multiplying excessively.

The history of agriculture in the United States of America illustrates some of the points made above. Agriculture has developed in some regions there within living memory and few crops are native. A few of the numerous pests introduced from abroad are shown in Table 1.1. The same situation exists in Canada, Australia and New Zealand and in other areas recently cleared for cultivation.

In Europe and other areas cultivated for centuries this phase has passed, and the artificial populations dependent on agriculture show some evidence of secondary balance (i.e. of less violent fluctuation), although the artificial circumstances of agriculture leave many old pests that develop excessive populations and seriously menace their host crops. Also, important new introductions still occur, of which Colorado beetle and potato cyst-nematode may be cited as examples. Population studies suggest that relative constancy of food supply such as is provided by stable European agriculture, where the same crops are grown in approximately the same proportions for many years, does not of itself produce stability of animal numbers, although it may decide the average level over a period. Probably the main factors affecting numbers are climatic, acting differentially upon pests, their host plants and their enemies. The importance of enemies, especially specific enemies, is best seen where the numbers of an alien pest have been drastically curtailed by an enemy brought in for the purpose (p. 256) or where enemies are killed by pesticides (p. 166).

In pest problems, the population density of the animal concerned is important for there is usually a relationship between crop injury and numbers, and thus an 'economic threshold' above which the organism is harmful and below which it is of no importance. The population level at which serious economic injury occurs is not constant from year to year or from field to field (Fig. 1.1), but partly depends on the skill of the grower and partly on edaphic, topographic and climatic factors beyond his control. Another critical level especially important is that population density which, under a particular set of circumstances, will repay treatment by chemical or other means (Strickland, 1962). The criterion of what is a pest is purely economic.

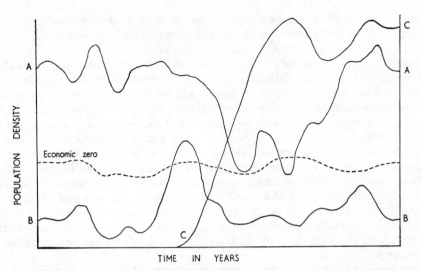

Fig. 1.1 Types of pest. A, one that fluctuates above the economic zero most of the time; B, one that fluctuates below the economic zero most of the time; and C, a successful new introduction that multiplies rapidly and establishes itself for the time being at an epidemic level. (Economic zero = economic threshold.)

Biologically injury may sometimes be trifling, such as a few wireworm holes in potatoes or capsid blemishes on fruits, but damage of this kind may greatly lessen market value. Canning, freezing and packaging have placed a new emphasis on freedom from blemishes caused by pests and have lowered critical levels. Whereas some damage was permissible in unprocessed or unwashed produce, almost none is tolerated when food is washed processed or packaged. These developments have forced farmers to apply pesticides where they would otherwise have been unnecessary.

Types of pest

Pests can be classified as follows:

1. *Regular*. A pest that rarely falls below the economic threshold, e.g. frit fly (Fig. 1.1A).
2. *Sporadic*. A pest that only occasionally rises above the economic threshold, e.g. antler moth, silver Y moth, millepedes (Fig. 1.1B).
3. *Potential*. Organisms that might be highly noxious if allowed to become established, e.g. Colorado beetle, Japanese beetle.
4. *Pests of special circumstances*. e.g. Wireworms after ploughed up grass. Frit fly attacking wheat after rye-grass ley.
5. *Extinct pests*. e.g. The wheat gall nematode not reported since 1956; turnip sawfly reappeared in 1947 after an absence of more than 40 years, now rarely recorded; beet carrion beetle, virtually disappeared since modern pesticides were introduced.

The status of a pest may differ in different parts of its range. Thus the beet fly in Britain used to occur regularly in the coastal regions of East Anglia but was sporadic inland.

Some pests tend to be cyclic in appearance, but the so-called cycles are so irregular that it becomes difficult to distinguish this class of pest from (1) or (2) above. More definite cycles sometimes occur in insects with long life histories such as the cockchafer in Europe and the periodic cicada of the U.S.A. These insects exist in several more or less independent streams some of which are far more abundant than others. Swarms of adults from the more abundant streams appear in the years when that stream completes its life cycle, whereas adults from less abundant streams are inconspicuous in intervening years.

The applied zoologist concerned with pests deals mainly with herbivorous animals, and the plants on which they feed. The relationship between the animal and plant ranges from browsing or grazing when the animal is large (e.g. rabbit, pigeon, slugs and some large, gross-feeding insects), through wounding (e.g. wireworms, pigmy mangold beetle, flea beetles), to ectoparasitism (e.g. aphids, mites, thrips, many soil nematodes) and endoparasitism (e.g. leaf and stem miners, seed beetles, stem nematodes, root-invading nematodes), when the animal is small relative to its host plant. The applied zoologist strives or should strive to limit species that cause economic injury by lessening their numbers per unit area until injury is insignificant. Eradication, though desirable, is usually impossible. The aim of the farmer facing the loss of his crop is often rather different. His object is to destroy as many of a pest species as possible for the least immediate cost. Indiscriminate use of pesticides can produce new problems by destroying predators which may lead to an increase in pests feeding on plants, by the development of strains of pests resistant to particular pesticides, or by the accumulation of toxic residues in the soil (Chapter 16).

It is impossible to destroy harmful animals without also influencing the plant-animal community of which they are a part. Although the artificial communities based in crops are perhaps less stable and less complex than those in natural habitats, enemies nevertheless still play a part in the limitation of numbers.

Population balance

The manner in which natural populations are regulated is the subject of much controversy. Chapman (1931) thought that *biotic potential* (power of increase under optimum conditions) of a species was opposed by *environmental resistance*. But all factors opposing increase are not in the environment, for some are created by the species or depend on the genetic constitution of its individuals. Environmental factors can be divided descriptively into physical (climate, structure of the environment) or biotic (other organisms of the same or different species). Another classification is based on the way in which factors work, that is whether they are *dependent* or *independent* of the population density of the animal in question (Solomon, 1949). Density-independent factors include climate and quality of food which kill or reduce fecundity in the same proportion, regardless of population density. Density-dependent factors are quantity of food, availability of nesting sites, territory or available space for which competition increases as numbers rise and enemies (parasites, predators, organisms of disease), which find victims more readily when numbers are large. Density-independent factors alone cannot limit numbers, for if these factors are not adverse enough to cause extinction, numbers would continue to rise exponentially even though the multiplication factor was less than the maximum potentially possible. Density-dependent factors, however, act as a brake that becomes progressively more effective as numbers rise and check unlimited increase (Fig. 1.2).

Often the equilibrium achieved is not a steady state as depicted by the continuous line in Fig. 1.2 but a series of oscillations, their amplitude and form being determined by the interaction between rate of increase, damage to the host which diminishes the food supply, initial population density and other factors (May, 1981). Animals with great powers of multiplication and mobility are able to find and exploit new territory rapidly (e.g. aphids): their numbers fluctuate greatly and they are referred to as *r*-strategists (Southwood, 1981). Other animals with lesser powers of increase and perhaps a lesser degree of mobility have a more conservative strategy. These so-called *k*-strategists tend to persist in places they have colonized and may have mechanisms that protect their hosts from extreme injury from their feeding (e.g. cyst-

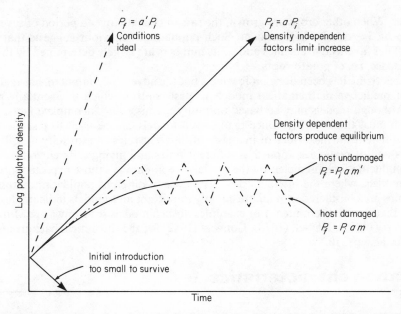

Fig. 1.2 Graphical illustration of the hypothesis of population control by density-dependent factors. Pi, population density initially; Pf, population density finally after an arbitrary interval; a', maximum possible increase rate; a, real increase rate; m', host plant unaffected; m, host plant (food supply) affected.

nematodes; Jones, Parrott & Perry, 1981). Many animal species do not fall neatly into these two categories. Further, species that have the attributes of k-strategists in a natural setting may exhibit some of the attributes of r-strategists when attacking farm crops (e.g. stem nematodes; Green, 1981) largely because farming greatly increases mobility and provides dense contiguous stands of host plants.

Many population experiments are done in chambers or vessels which eliminate all possibility of migration. In nature, although populations are often distributed in aggregates or islands, there is always the possibility of interchange. Events that lead to the extinction of individual islands or groups of islands do not usually affect all islands and re-population may therefore occur. Because nematodes are relatively immobile, it is sometimes possible to do experiments with them in field plots that, with insects, can only be done in closed laboratory systems. Figure 1.3 summarizes information on the potato cyst-nematode derived from observations on many fields. When potatoes are grown, population increase is density dependent. If preplanting numbers are large, numbers after harvest may decrease because juveniles compete for space and food in the potato roots and also diminish the supply because heavily invaded root systems

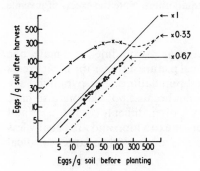

Fig. 1.3 The relationship between the population densities of the potato cyst-nematode before planting and after harvest. ×, susceptible potato crops; ·, fallow and other crops; —·—·—, hatching effect of resistant potatoes. Derived from data in Winfield (1965).

are smaller. When other crops are grown, the rate of decrease over a period of several years is almost linear, i.e. density independent. Such population behaviour suggests that this introduced pest has no effective enemies and so its numbers are largely determined by the supply of food and space, i.e. of potato roots.

Much theorizing has been done on few facts but the advent of computers has made possible the use of predictive mathematical models in pest control, which is essentially population control. Although models may be based on faulty premises and incomplete facts, those who make them must think deeply. Once made, such models can be used to test the theoretical effects of different intrinsic rates of increase, of different rates of mortality at different times and stages and of pesticides applied at different times and giving different rates of kill. The population fluctuations predicted by the models are not always those expected intuitively. In complex models, where the interaction of a number of different population parameters and relationships are considered, not all are equally important and models indicate those that are and those that can be discounted. For examples of such models see Conway & Murdie (1972), Jones & Perry (1978), Hassell (1978), Conway (1983) for the theoretical background see May (1981). (See also p. 310.)

Population characteristics

Two attributes of animal populations are of import in relation to pest control: their structure in terms of adult and immature stages and their mobility, which governs the speed with which they occupy new territory or re-populate old territory. A stable hypothetical population composed of equal numbers of males and females reproducing sexually, and with females laying an average of 100 eggs each, would have a structure of the kind shown in Fig. 1.4. The

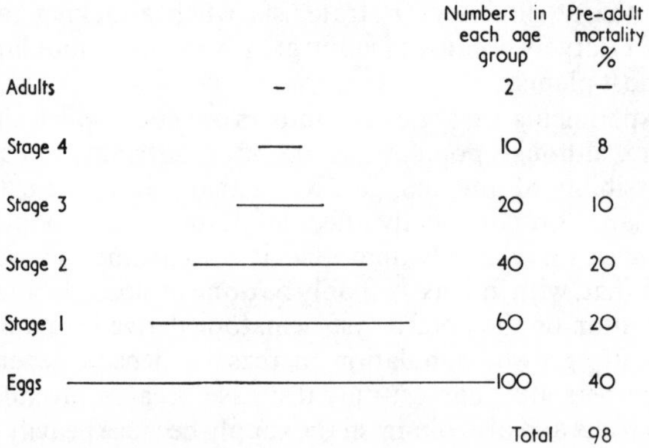

		Numbers in each age group	Pre-adult mortality %
Adults	–	2	–
Stage 4	——	10	8
Stage 3	———	20	10
Stage 2	————	40	20
Stage 1	————	60	20
Eggs	—————	100	40
		Total	98

Fig. 1.4 The age structure of a hypothetical population in real equilibrium. Note the excess of juvenile stages and the small proportion of adults.

mortality rates for each immature stage would total 98%, leaving two adults, one male, one female, to continue the species. Although this is a hypothetical picture it shows the great excess of juvenile forms that is a common feature of many animal populations. The pyramidal age structure shown in Fig. 1.4 is not always immediately apparent because populations also have a time structure as in Fig. 1.5. Measurement of populations is often difficult because eggs and immature stages may require different methods of collection or extraction and are usually less easily identified than adults, and population estimates in fields are subject to large statistical errors.

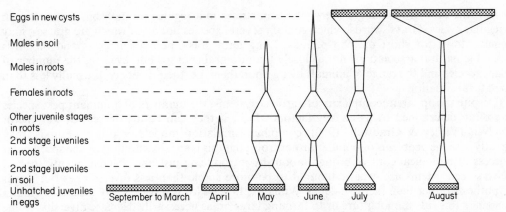

	September to March	April	May	June	July	August
Eggs in new cysts						
Males in soil						
Males in roots						
Females in roots						
Other juvenile stages in roots						
2nd stage juveniles in roots						
2nd stage juveniles in soil						
Unhatched juveniles in eggs						

Fig. 1.5 The time structure of a population of *Globodera rostochiensis*, diagrammatic. All stages except the second stage juvenile and the mature male are immobile. As there is one generation a year and no recruitment of stages, this picture is relatively simple.

Animals depend ultimately on green plants inhabiting the same area. At the base of the animal food chain are herbivorous species and dependent on them is a hierarchy of predatory and parasitic animals. The relationship between plant host and herbivore, prey and predator, host and parasite, parasite and hyperparasite is basically the same; one is the food for the other. When members of these pairs of organisms occur without other species to complicate the issue, their numbers tend to oscillate. If circumstances allow the food organism to increase, that consuming the food follows suit shortly afterwards and its increase usually exceeds those that the food supply can support, so a population crash follows. Ideally, the numbers of the organism providing food and that consuming it follow separate sine curves which are out of phase with each other. The amplitude of the oscillations depends on the extent to which increase exceeds the food supply and is modified by the extent to which growth of a plant host is stunted or the replacement of victims curtailed, i.e. the food supply is both consumed and diminished.

This relatively simple relationship becomes more complicated when the herbivorous animal has a hierarchy of other animals dependent upon it. When the animals are all specific feeders on those next below them in the food chain, the conceptual model is two dimensional – a triangle with the population of host plants at the base and decreasing numbers of each successive hierarchy of animal dependents reaching to the apex in successive layers. Because different species of animals differ in size from each other and from the plants on which all are ultimately dependent, numbers of individuals are not a true guide to their interrelationships which are better expressed in terms of biomass or energy flow. Nor are two-dimensional models adequate to express the complexities of an ecosystem. In natural situations, the base is occupied by a plant association which itself may be layered. Moreover some of the hierarchies of animals feed on more than one plant or animal species and some are omnivorous. The model becomes multidimensional, difficult to grasp, and is further complicated because animals and plants are used as food sources by endo- and ectoparasitic micro-organisms such as fungi, bacteria, protozoa and viruses, and some organisms consume and degrade the bodies of dead plants and animals.

In agriculture the ecosystem is usually simplified and less diverse, i.e. contains fewer species, in that the complex plant association at the base of the system is replaced by a crop which is mainly one species and its associated weeds which nowadays are usually few and unimportant. In arable farming systems, the crop is often short lived and the land bare for part of the annual cycle during which the food supply is absent. Except when a crop is 'ranched' as is continuous barley, the food crop changes every year, sometimes in a regular rotation but more often in irregular cycles. Grass swards, which contain several species not all of which are grasses, may remain down for short or long periods. Long term grass, permanent grass and orchards resemble natural associations more closely than do arable crops but, even so, the number of plant species and the range of animals living upon them, i.e. their diversity, is usually less than in natural situations.

Despite the apparent complexity of agro-ecosystems, the numbers of dominant pest species are often determined by a few key factors which, if they can be discovered, by key factor analysis (Varley & Gradwell, 1960), computer simulation models or other methods, may greatly ease the problems of population control. That this is so is exemplified by the spectacular success of biological control brought about by introducing just one effective enemy species. Diverse ecosystems are usually thought to be more stable than less diverse ones but this is hypothesis rather than fact (Emden & Williams, 1974). Nor is it necessarily true to say that a tendency towards monoculture or to farming large areas necessarily decrease diversity. Some robust monocultures exist, e.g. cereals grown continuously in the U.K. and elsewhere.

All the major pests and many other non-pest species are able to exploit rapidly any situation that presents them with more food, diminishes those factors that depress their reproduction or provides a suitable unoccupied ecological niche. Birds, being most mobile, are able to exploit situations over a wider area and to move with seasonal climatic changes, but other less mobile species or even immobile species invade and exploit distant favourable sites in time.

The occupation or re-occupation of territory by an immigrant species is slow at first, speeds up as sites become colonized and numbers rise and slows down again as sites remaining to be colonized become fewer and fewer. The process is essentially the same whether a country is being colonized for the first time, an area of annual or perennial crop is being recolonized or a few square yards are being re-populated. The speed of the process of colonization depends on many factors of which the mobility of the species concerned is one of the most important. The potato cyst-nematode (*Globodera rostochiensis* (Woll.)), a relatively immobile soil pest dependent very largely on human agency for transmission, took between sixty and a hundred years to colonize the potato-growing areas of England, Scotland and Wales. The immature stages of insect pests are also relatively immobile; they are not primarily concerned in dispersal. Adult insects differ greatly in mobility. The Colorado beetle (*Leptinotarsa decemlineata* (Say.)), with its winged adult stage, spread from the Bordeaux region of France, partly helped by human agency, over areas of Europe greater than the British Isles in about 30 years. Considerable areas of eastern England are recolonized by aphids each year in the space of a few weeks. Eggs on winter hosts or females that have passed the winter in shelter provide numerous and scattered sources from which re-infestation occurs, but the rapidity with which great areas of crop are colonized is remarkable and explained by the winged females becoming airborne in thermal upcurrents. Birds are also remarkably mobile and noxious species may operate from places far removed from the crops on which they feed. The tendency to concentrate on areas of crop attractive to them at a particular time produces dense and injurious populations locally, although the average population may be small. Similar aggregations occur but on a smaller scale in insects and in nematodes; for example, the flocking of stubby root nematodes (p. 202) around root tips.

Populations in fields are never evenly or randomly distributed. If they were randomly distributed, the numbers found in a series of samples of equal size would fit the Poisson distribution, assuming there were no serious errors in extracting and counting. Usually population counts approximate to some other distribution (e.g. the negative bionomial; Southwood, 1978) because individuals breed and cause local aggregations (e.g. the colonies that develop around migrant aphids, concentrations in soil from recent infestations with cyst-nematodes). Other causes of aggregation are food, soil type, soil structure, shelter and crop edge effects. The most noticeable effects of food are those which spring from planting crops in rows and from preference for particular parts of plants. The overwintered adults of pigmy mangold beetle are almost totally distributed around the sprouts of old beet crowns whereas the incidence of root ectoparasitic nematodes depends greatly on soil structure (see p. 186).

Populations of nematodes, slugs, isopods, myriapods and flightless insects are all relatively immobile and therefore tend to be inbred. Although agriculture aids dispersal, the annual rate of immigration is small. Consequently new pockets of infestation tend to preserve the idiosyncracies of their founders more than do populations of highly mobile animals. The dispersal of some strongly flying insects and probably also that of some birds and mammals is also circumscribed. Even when related animals occupy the same territory differences in behaviour, host preferences or the timing of mating may preclude interbreeding and ensure isolation which encourages race formation and speciation. Many pest species already consist of a series of races which become apparent only when genes for resistance are incorporated into new cultivars (e.g. cyst nematodes, Hessian fly). Regrettably these are not effective universally. Even where effective, their cultivation encourages the multiplication of pre-existing races and may select new ones. Races of parthenogenetic species in which males are rare or apparently ineffective, also exist. The existence of races and the possibility that new ones may arise are important considerations when breeding for resistance (Jones, 1979).

Animal phyla containing pests

Animals ranking as pests belong to four phyla: Nematoda, Mollusca, Arthropoda and Chordata. The Arthropoda contains the Insecta (insects), Symphyla (symphylids), Arachnida (mites), the Diplopoda (millepedes) and the Crustacea (woodlice). Insects exceed all other pests in numbers and species. They compete with man by menacing his food supply and by transmitting disease to him, to his domestic animals and to crop plants. Their outstanding feature is mobility coupled with a high degree of adaptation to land life, which means that they can exist and be active in relatively dry habitats. Nematodes come second in importance. They are typically inhabitants of soil, have only limited mobility and, being microscopic and aquatic, are active only in moist habitats. The Mollusca contains slugs and snails which are incompletely adapted to land life and therefore, like nematodes, can live and be active only in moist habitats. Some birds and mammals belonging to the Chordata, sub-phylum Vertebrata are also important pests. They are active, mobile creatures sharply distinguished from invertebrate pests by their regulated body temperature, which enables them to be active when low temperatures inhibit cold-blooded animals, and by the greater complexity of their behaviour.

Nomenclature

The Latin names of animals change frequently as progress is made in the study of their taxonomy and systematics. Common names change less frequently but vary from place to place. The Latin and common names used in this edition are for the most part those to be found in Seymour (1979).

2

An introduction to insect structure and classification

Some of the characters responsible for the success of insects may be summarized as follows:

1. *Small size*. Their small size means they need little food and can readily find shelter in nooks and crannies and so avoid enemies and extremes of weather.

2. *Rapid reproduction*. Production of many young and short life cycles enable insects to take advantage of brief opportunities for increase such as are provided by annual crops or brief periods of weather favouring the growth of host plants.

3. *Mobility*. Insects are extremely active. Most can crawl or run and some can leap, but of outstanding importance is the flight of adult insects. Mobility is of great advantage in seeking food, shelter and breeding places and in escaping from enemies. The greater mobility of adult insects has made it possible for them to live in habitats and exploit food supplies denied to their less mobile, juvenile stages, and from this has arisen the great divergence of form between juveniles and adults in the more advanced orders and families.

4. *The cuticle*. The insect ectoskeleton has contributed much to their success. It is characterized by the presence of chitin which, together with tanned proteins, lipids and water makes a horny and protective integument combining strength with lightness. Most of the diverse characters by which insects are recognized are those of the exoskeleton.

5. *Efficient water conservation*. This is achieved by waterproofing the cuticle and by having an excretory system which minimizes water loss. Waste nitrogenous material is voided as crystalline uric acid.

6. *Adaptability*. The adaptability of insects in response to their external environment has helped them to invade successfully almost every possible land habitat, including fresh water. They do not occur in the sea, but they are found in natural pools of crude oil, and in warm springs. Different insects use a wide range of materials for food including crops, domestic animals, stored food, beeswax and wool.

With no mechanism to regulate body temperature, the activity of insects depends greatly on temperature. They become less active as the temperature falls and the winters of high latitudes enforce a quiescent period. In the tropics, temperature is less important in determining seasonal activity, but drought, excessive heat and lack of food may enforce a seasonal dormancy. To pass successfully through adverse periods some insects enter diapause, when growth is arrested and metabolism depressed. In any one species diapause is limited to a single stage of development: egg, larva, pupa or adult. When it occurs in the adult, reproduction is arrested. In some species diapause occurs automatically in each generation, in others it may be induced by some environmental stimulus that precedes unfavourable conditions, such as fewer hours of daylight, falling temperatures or scarcity of food. When in diapause an insect resists cold or drought and is not immediately activated by the return of favourable external conditions but remains dormant until a temperature-dependent process of physiological development has been completed. For example, beet fly pupae enter diapause in autumn and even when temperatures are favourable in late autumn, no development occurs. In temperate zones, diapause ends after exposure to moderately low temperatures, 5–10°C, which induce the secretion of a growth hormone and only after that does the interrupted life cycle continue when the temperature becomes favourable. Most pest species that undergo diapause have two or more generations a year, and, under favourable circumstances, the non-diapausing generations increase rapidly. The release from diapause occurs when there are the plants needed for food and oviposition, and after the period of short days in spring. There is a general correlation between the temperature range most effective for the ending of diapause (usually a few degrees below the developmental threshold) and the normal winter temperature. Differing responses to different temperatures in different places help to secure the most favourable length of diapause under the average climatic conditions in each place. Impending drought induces diapause in many insects in the tropics.

Diapause therefore enables insects to survive when environments do not favour them all the year round and is an important factor determining insect distribution. The species is protected during adverse weather and is synchronized with its food supply (Lees, 1955, 1962; Way, 1962; de Wilde, 1962).

Insect structure

For practical reasons every agriculturist should know something of the structure of an insect, for this helps in understanding pest problems, in the recognition of specific pests and in deciding on suitable control measures. As there are several detailed textbooks on insect structure (Imms, 1977; Imms, 1978) and physiology (Wigglesworth, 1972; Chapman, 1982), no attempt will be made to deal with these subjects in detail: only an outline is given.

External structure

The insect exoskeleton is secreted by an underlying layer of cells, the hypodermis (Fig. 2.1). In section, it is made up of three layers:

1) An outer waterproofing layer, the epicuticle which contains a lipoid layer almost impervious to water.
2) A middle layer of hardened protein, and chitin, the exocuticle.

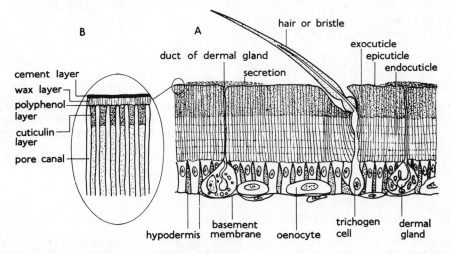

Fig. 2.1 A, section of a typical insect cuticle; B, details of the epicuticle. (Redrawn from Wigglesworth (1972), *Principles of Insect Physiology*, 6th edn., Methuen.)

3) An inner endocuticle, composed of chitin and protein but remaining flexible and forming the hinges between the hardened portion of the skeleton (Beament, 1961; Wigglesworth, 1972).

To be effective, certain types of spray used against insect pests must wet and not run off the outermost layer, and some component of the spray must penetrate it. Abrasive action may also scratch through the lipoid layer and increase water loss which may be serious for a small insect in a drying environment.

The hard portions of the epicuticle and exocuticle are known as sclerites or plates. One sclerite can move on another at the joints by muscles attached internally. Inpushings of the exoskeleton known as apodemes provide leverage and points of attachment.

The insect body is divided into three regions: head, thorax and abdomen, each made up of ring-like segments. The head consists primitively of six segments, the thorax of three and the abdomen of eleven (Fig. 2.2). There are three pairs of walking legs, and in adults, two pairs of wings are usually present. Insects respire by means of segmentally arranged spiracles or pores which lead into tubes called tracheae that ramify in all organs of the body.

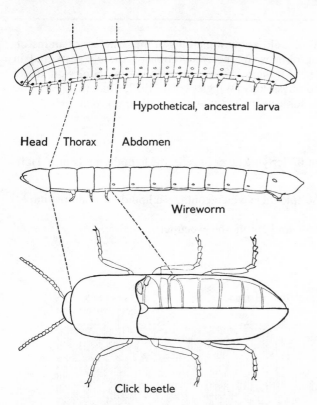

Hypothetical, ancestral larva

Head / Thorax / Abdomen

Wireworm

Click beetle

Fig. 2.2 Diagram of insect segmentation to show the relationship between a hypothetical ancestral larva, a wireworm, and a click beetle.

Head

The head is formed from chitinous plates fused together to form a capsule (Fig. 2.3A). The dorsal epicranium is separated from the median frons by the frontal suture. The frons articulates with the clypeus which joins the labrum or upper lip, and conceals the mouth parts. The

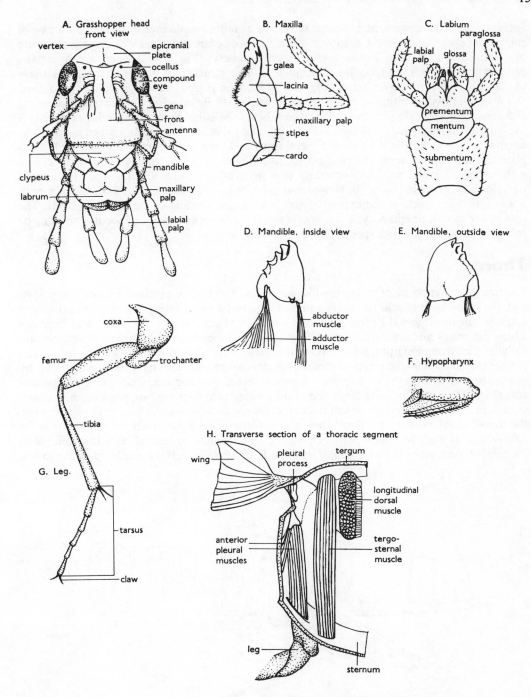

Fig. 2.3 A, front view of head of a grasshopper; B–F, unspecialized mouth parts of cockroach (*Blatta*); G, generalized insect leg to show structure; H, diagrammatic cross section of a winged segment to show flight muscles. (A, after Simpson, Pittendrigh & Tiffany (1957), *Life*, Routledge and Kegan Paul, London, Harcourt, Brace & World, New York; H after Wigglesworth (1972), *Principles of Insect Physiology*, Methuen.)

antennae are sense organs, and there may be one or more simple eyes or ocelli, and a pair of compound eyes with from six to thousands of facets or ommatidia; each having the structure of a simple eye with a lens, pigment, sensory cell and nerve (Fig. 2.14C). There are three pairs of mouthparts which are modified appendages. Their structure is best seen in the relatively unspecialized cockroach or locust (Fig. 2.3 B–F). The paired mandibles consist of a hard chitinous plate with a biting or cutting edge moved by strong muscles. The maxillae are paired and consist of a basal joint, the cardo, which bears the stipes from which arise externally a five-jointed maxillary palp and internally a hoodlike galea which protects a lobed lacinia. The labium or second maxilla is a pair of appendages fused together to form the lower lip. From a plate-like submentum arises the mentum, its distal region or prementum bearing on either side a three-jointed labial palp and centrally two pairs of plate-like sclerites, the glossae and paraglossae corresponding to the lacinia and galea of the maxillae. On the floor of the mouth is a tongue-like structure, the hypopharynx, bearing the opening of the salivary glands. The roof of the mouth or epipharynx is a simple structure in cockroaches and locusts but may be enlarged into an organ with special function in some insects.

Thorax

The thorax is composed of three ring-like segments. The first or prothorax bears one pair of legs, the second or mesothorax bears a pair of legs and a pair of wings, as also does the third segment, the metathorax. Figure 2.4 shows a simplified side view of the thorax and indicates where the wings and legs are inserted; Fig. 2.3H is a diagrammatic transverse section and includes the strong longitudinal and vertical muscles which act as indirect wing muscles. The greater elaboration of the meso- and metathorax is associated with the presence of wings; by comparison the prothorax is simple. In cursorial insects (Orthoptera, Coleoptera) the prono-tum (Fig. 2.4) is usually quite large and is subdivided into four regions, prescutum, scutum, scutellum and post scutellum. When the vertical muscles contract, the wings are raised; when the longitudinal muscles contract, they are lowered. The dorsal plate or tergum may be subdivided, as may the lateral plate or pleuron. The sternum or ventral plate is smaller than the tergum, because of the insertion of the legs. Each leg (Fig. 2.3G) is made up of five joints,

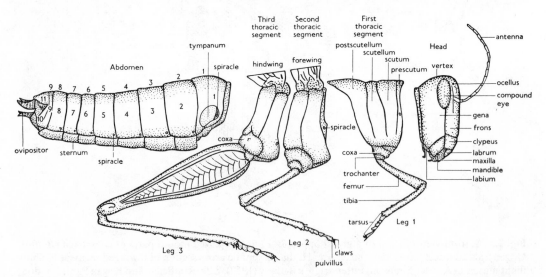

Fig. 2.4 Diagram of a grasshopper to show insect structure. (After Metcalf, Flint and Metcalf (1965), *Destructive and Useful Insects*, McGraw-Hill Inc., used by permission.)

coxa, trochanter, femur, tibia and tarsus. The number of tarsal joints ranges from one to five and the tarsus may end in claws, between which there is usually a soft pad or pulvillus, or occasionally in vesicles, as in the Thysanoptera. The wings are developed from sac-like, out-foldings of the integument. During development the two surfaces of the wing come together leaving only channels or veins, which strengthen the wing and usually contain blood spaces, nerves and a trachea. The pattern of venation is important in identifying insect species. Many insects have devices to ensure that both wings act together, e.g. hooks in aphids and bees, and an overlap in butterflies. Spiracles occur on one or more thoracic segments (Fig. 2.4).

Abdomen

Primitively the abdomen consists of eleven segments with an additional end portion or telson (Figs 2.2, 2.4), but the last two or three segments may be telescoped together, so that fewer are visible when the insect is viewed either dorsally or ventrally. The last segment usually bears a pair of many-jointed cerci. The aperture of the female sexual organs is on segment eight and that of the male is on segment nine. In transverse section (Fig. 2.5) the tergum and sternum are

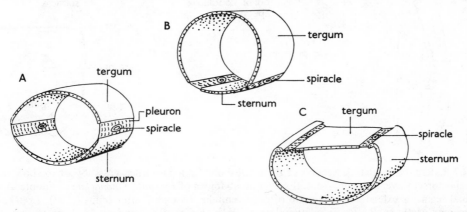

Fig. 2.5 Diagrammatic representation of the relative sizes of the terga and sterna in insect abdomens. A, equal; B, large tergum, small sternum; C, small tergum, large sternum.

quite distinct, but the pleural region is somewhat smaller and its position varies. A pair of spiracles usually occur on the pleural regions of the first eight abdominal segments (Fig. 2.4).

In insect embryos, each segment bears a small appendage, which is retained as a vestige in the older stages of some primitive insects. The older stages of most insects only have them on the segments around the genital apertures. In the female the appendages from segments eight and nine are often retained to form an ovipositor, as in the long-horned grasshopper (Fig. 2.6). Not all insects have a well developed ovipositor. Sometimes the tip of the abdomen serves as an ovipositor and the terminal segments are tubular and telescopic, e.g. Lepidoptera, Diptera (Fig. 2.7B).

The male genitalia occur on segment nine (Fig. 2.7C) and consist of an aedeagus with a central penis and two lateral parameres with a pair of external claspers. The arrangement of the parts forming the tip of the abdomen in different species, however, varies greatly. The elaborate male sexual organs of some insects may prevent intermating of related species which might upset the genetics of their populations.

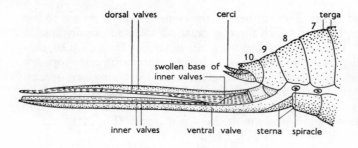

Fig. 2.6 Ovipositor and abdomen of long-horned grasshopper (Orthoptera). (After Snodgrass in Imms (1957), *Textbook of Entomology*, Methuen.)

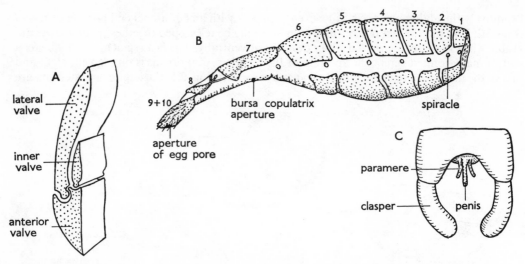

Fig. 2.7 External genitalia. A, section of the ovipositor of a long-horned-grasshopper to show valves (lateral = dorsal valve, anterior = ventral valve); B, abdomen of *Lymantria monacha* (Lepidoptera) with terminal segments extended; C, diagram of male genitalia. (A and C after Imms (1957), *Textbook of Entomology*, Methuen; B after Eidmann in Imms (1961), *Outline of Entomology*, Methuen.)

Internal structure

To understand the internal structure of an insect it is best to dissect one of the larger kinds such as the cockroach or locust. Figure 2.8A shows the digestive system of a cockroach. After being chewed by the mouth parts, digestive juices are added from the salivary glands and the food passes into a narrow oesophagus, which widens into a large crop where food may be stored and partly digested. The entrance from the crop to the muscular gizzard is small, and the gizzard contains six chitinous teeth which grind the food finely. The food is filtered by fine hairs before passing into the enteric caeca and midgut. The foregut, which extends to the gizzard, is really an inpushing of the integument and is therefore lined with cuticle. In the midgut and caeca, digestion is completed and absorption takes place through the endodermal lining. Many insects produce a cuticular membrane, the peritrophic membrane, that extends backwards from the gizzard. In the hindgut, absorption of food is completed and waste material is stored in the rectum before it is expelled. Here, too, the last traces of water are removed to conserve the water balance. The Malpighian tubules are the excretory organs. Waste nitrogenous material from the blood passes into the tubules which excrete it from the anus as uric acid crystals. Associated with the digestive system are a set of salivary glands, the

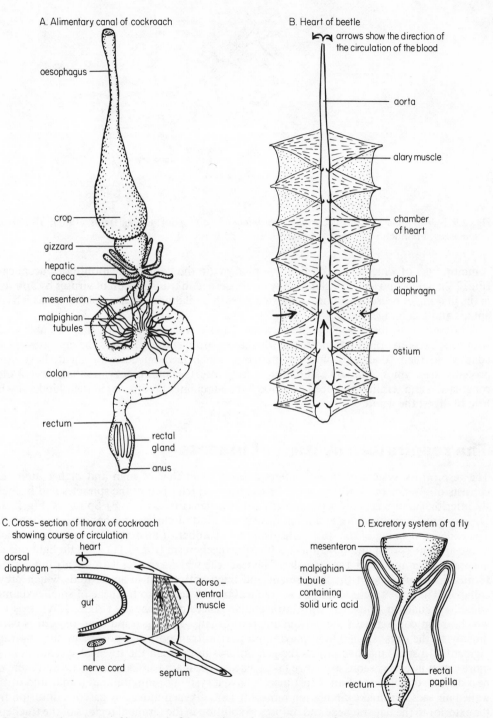

A. Alimentary canal of cockroach

oesophagus

crop

gizzard

hepatic caeca

mesenteron

malpighian tubules

colon

rectum

rectal gland

anus

B. Heart of beetle

arrows show the direction of the circulation of the blood

aorta

alary muscle

chamber of heart

dorsal diaphragm

ostium

C. Cross-section of thorax of cockroach showing course of circulation

dorsal diaphragm

heart

gut

nerve cord

dorso-ventral muscle

septum

D. Excretory system of a fly

mesenteron

malpighian tubule containing solid uric acid

rectum

rectal papilla

Fig. 2.8 Internal structure of insects. A, from cockroach (*Periplaneta*); B, dorsal blood vessel of a beetle; C, from cockroach (*Blatta*); D, diagram showing Malpighian tubules containing uric acid as in a fly or mosquito. (B, C, D after Imms (1961), *Outlines of Entomology*, Methuen.)

18

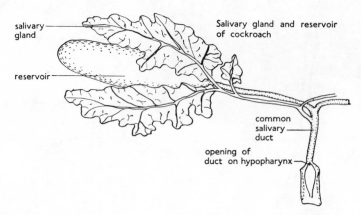

salivary gland

Salivary gland and reservoir of cockroach

reservoir

common salivary duct

opening of duct on hypopharynx

Fig. 2.9 Salivary gland from cockroach (*Periplaneta*) (one side). (After Imms (1957), *Textbook of Entomology*, Methuen.)

common duct of which opens on the hypopharynx in the floor of the mouth or buccal cavity (Fig. 2.9). Saliva and salivary glands may be concerned in transmission of viruses by aphids, or in the making of necrotic wounds, distorted growth or elaborate galls in plants on which some aphids and gall midges feed.

An omnivorous insect such as the cockroach has a wide spectrum of enzymes able to deal with fats, carbohydrates and proteins, whereas in more specific feeders the enzymes are best adapted to their food, e.g. blow fly maggots, which feed on decaying meat, have strong proteolytic enzymes. The saliva of blood-sucking insects contains enzymes that prevent blood coagulating and termites that feed on wood harbour ciliates (Protozoa) in their hind gut which help to digest the wood.

The respiratory system of insects

The respiratory system of insects differs entirely from that of man and higher animals. It consists of a series of ramifying tubes or tracheae, which open at the spiracles and branch in the interior, supplying every organ directly with oxygen (Fig. 2.10 A, B). Spiracles (Fig. 2.10 C, D) vary greatly in structure and often have some kind of closing mechanism, e.g. honey bee. The external opening of the spiracle leads into a chamber or atrium which may contain hairs. These are waterproof and function as a filtering mechanism (Fig. 2.11). Certain light oils used in sprays enter and block the spiracles. The tracheae which connect with the atrium are tubes formed as inpushings of the integument and lined with chitin in spiral folds which prevent collapse. The large tracheae known as tracheal trunks, lead into tracheae of smaller diameter which terminate in 'end cells' with intracellular spaces or tracheoles (Fig. 2.12A). It is these which are in contact with the various internal organs. During respiration oxygen is carried directly to the organs, the blood playing no part in transport and possessing no respiratory pigments. In small inactive insects oxygen diffuses in through the tracheal system; in larger, more active insects gaseous exchange takes place in the main trunks and air sacs and there may be a definite circulation of air. Most insects have a system of longitudinal trunks and of thin-walled air sacs that make circulation more efficient. Oxygen moves by gaseous diffusion from the exterior to the fine tracheae and diffuses in solution in the terminal regions of the tracheoles, a much slower process. When at rest the tracheoles are full of fluid, but after exertion the fluid is withdrawn, probably because of the osmotic pressure produced by the by-products of muscular contraction, so that oxygen can move by gaseous diffusion to the site where it is

19

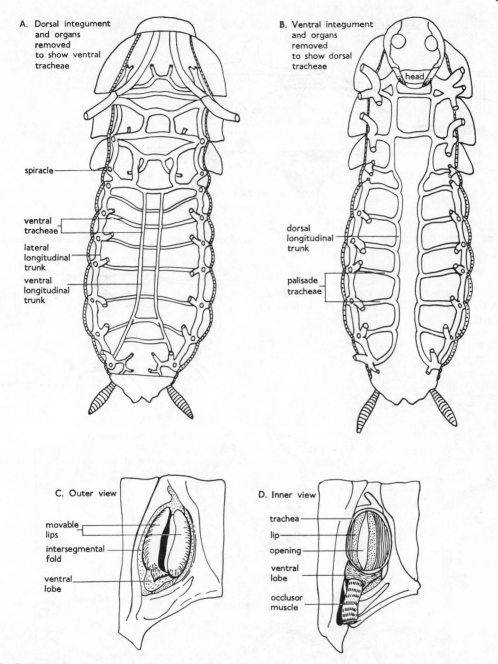

Fig. 2.10 Respiratory system A and B, tracheal systems of cockroach (*Periplaneta*) to show longitudinal trunks, spiracles and transverse tracheae; C and D, outer and inner view of a metathoracic spiracle of a grasshopper to show movable lips. (A and B after Miall & Denny in Imms (1957), *Textbook of Entomology*, Methuen; C and D after Snodgrass (1935), *Principles of Insect Morphology*, McGraw-Hill Inc., used by permission.)

20

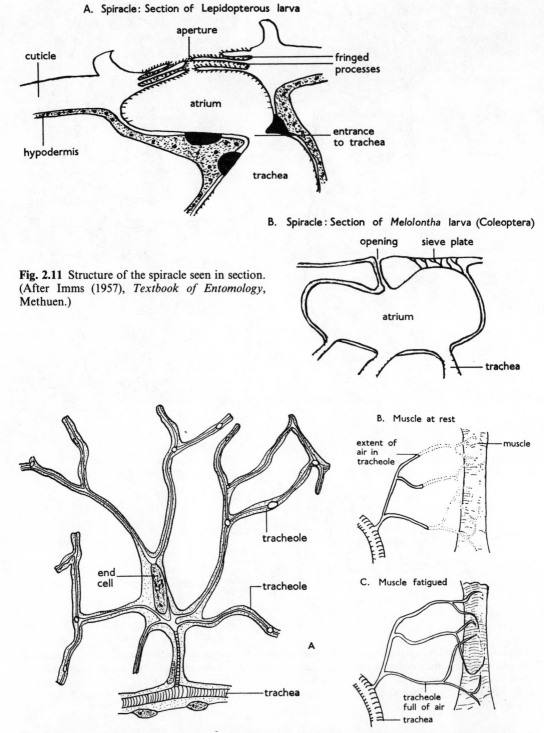

A. Spiracle: Section of Lepidopterous larva

aperture

cuticle

fringed
processes

atrium

hypodermis

entrance
to trachea

trachea

B. Spiracle: Section of *Melolontha* larva (Coleoptera)

opening   sieve plate

Fig. 2.11 Structure of the spiracle seen in section.
(After Imms (1957), *Textbook of Entomology*,
Methuen.)

atrium

trachea

B. Muscle at rest

extent of
air in
tracheole

muscle

tracheole

end
cell

tracheole

A

C. Muscle fatigued

trachea

tracheole
full of air

trachea

Fig. 2.12 A, end cell and tracheoles from the silk gland of a lepidopterous larva; B and C, movement of air in tracheoles during muscle fatigue. (A after Imms (1957), *Textbook of Entomology*, Methuen; B and C after Wigglesworth (1972), *Principles of Insect Physiology* 6th edn., Methuen.)

needed (Fig. 2.12 B, C). Carbon dioxide appears to diffuse more readily than oxygen through the cuticle and much waste carbon dioxide probably goes out of the body by that route. The distribution of spiracles varies; most adults and larvae possess ten pairs but larvae sometimes have fewer (Fig. 2.13). In the *holopneustic* condition, there are ten pairs of spiracles, two on the thorax and eight on the abdomen. This is usual for adult insects and many immature forms. The *peripneustic* condition has all but a few spiracles open. There is a tendency for the anterior thoracic spiracle to move forwards from the metathorax and for the eighth abdominal spiracle to move backwards. The *amphipneustic* condition is found in the larvae of the higher Diptera, where there is an anterior and a posterior pair of spiracles only. The front pair is lost in the *metapneustic* condition found in leatherjackets and mosquito larvae. Three less important types of distribution are: (1) the *propneustic* type where the anterior pair of spiracles only is open as in the pupae of mosquitoes; (2) the *apneustic* type where no spiracles are open as in some aquatic insects which breathe by means of 'gills' (flat, lobe-like structures richly supplied with tracheae); (3) the *atracheate* type where there are neither spiracles nor tracheae, e.g. springtails and some parasitic insect larvae.

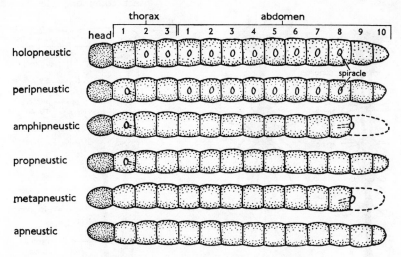

Fig. 2.13 Diagrammatic representation of the spiracular systems of insects.

The circulatory system

Blood is not concerned in the transport of oxygen and is free from respiratory pigments except in some larvae, such as *Chironomus*, which live in mud where the partial pressure of oxygen is small. Blood serves, therefore, to transport food, excretory products and hormones. Alterations in blood pressure caused by muscular contraction play a part in moulting and are often used to evert or extend various organs, e.g. the ptilinum, a small sac-like organ on the housefly head at the time of emergence from the puparium. There is only one blood vessel, the heart, which is situated dorsally above the gut. In the abdomen it consists of a series of chambers with valves, and in the thorax it is prolonged forward as a plain tube, the aorta (Fig. 2.8B). The heart of most insects is short and the chambers are fewer than the abdominal segments. In the cockroach the heart is primitive and is composed of thirteen chambers, whereas in the housefly there are only three. Blood enters the heart through lateral inlets or ostia and moves forward into the aorta which leads into the head. The blood then escapes into the spaces between the various muscles and other organs, so that they lie bathed in blood which percolates slowly back and eventually reaches the heart from the dorsal or pericardial sinus. Although there are no arteries the blood follows a fairly definite track (Fig. 2.8C).

Execretory system

The principal excretory organs are the Malpighian tubules which range in different species from two to twenty. They are long, slender, blind tubes opening into the front end of the hind gut (Fig. 2.8 A, D). Nitrogenous waste products are taken from the blood by the walls of the tubes which lie in the haemocoel and are passed into the lumen as sodium or potassium urate. At the lower ends of the tubes, uric acid crystals are precipitated and the bases and water are reabsorbed. The uric acid crystals pass out of the anus with the faeces. Other excretory products are sometimes deposited in the cells of the fat body.

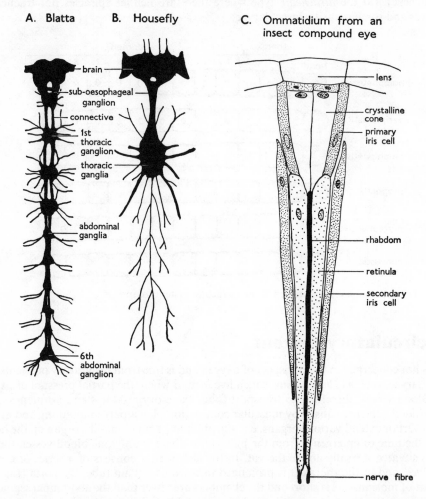

Fig. 2.14 Two types of insect nervous systems. A, cockroach (*Blatta*) with paired ganglia in every segment; B, housefly (*Musca*) showing fusion of connectives and of abdominal ganglia; C, diagram of an ommatidium (eucone type) from an insect compound eye. (A–C after Imms (1961), *Outlines of Entomology*, Methuen.)

Fat body

The fat body consists of irregular masses or lobes of cells surrounding all the main organs of the body. Its chief function is to store glycogen, protein and fat, so it corresponds roughly to the liver in higher animals.

The nervous system

The nervous system is comparatively simple (Fig. 2.14). There is a dorsal supra-oesophageal ganglion or 'brain' in the head which is formed from the nervous elements of three segments. It is linked with the sub-oesophageal ganglion by the para-oesophageal connectives which run one on either side of the oesophagus. From this a double ventral nerve chord runs backwards connecting up a pair of ganglia in each thoracic segment and six (or sometimes eight) abdominal ganglia. The nervous systems of different species vary greatly because of differences in the fusion of ganglia. The supra-oesophageal ganglion receives impulses from the outside world via the eyes and antennae and overrides and coordinates the thoracic and abdominal ganglia. The large suboesophageal ganglion is chiefly concerned with feeding. Connected with the nervous system are ductless glands, the corpora allata, which regulate moulting.

Reproductive organs

The reproductive organs of both sexes occur in the abdomen. The paired testes of the male are small, ovoid bodies situated either lateral or ventral to the alimentary canal and kept in position by the surrounding fat body (Fig. 2.15B). Each testis consists of a number of follicles or testicular tubules. The paired genital ducts or vasa deferentia lead from the testes and each becomes enlarged in its course to form a sac, the vesicula seminalis, where the spermatozoa congregate. The two vasa deferentia unite posteriorly to form a median ejaculatory duct. The wall of the ejaculatory duct contains powerful muscles and encloses distally the male intro-

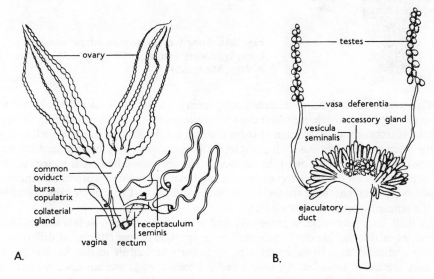

Fig. 2.15 Reproductive organs of insects. A, female of a lepidopteran; B, male of *Blatta orientalis* (L.). (A, after Imms (1957), *Textbook of Entomology*, Methuen; B, after Imms (1961), *Outlines of Entomology*, Methuen.)

mittent organ or aedeagus. This is formed from the membrane between the ninth and tenth sterna and strengthened with chitinous sclerites.

The ovaries of the females (Fig. 2.15A) are paired, compact bodies lying in the body cavity of the abdomen on either side of the alimentary canal. Each ovary is made up of several ovarian tubules or ovarioles in which the eggs develop. The ovarioles on either side open into a short oviduct which unites with its partner to form a median duct, the common oviduct which is continuous with the vagina. Opening into the dorsal wall of the vagina is the spermatheca, a pouch which receives and stores spermatozoa. One or more pairs of accessory or collateral glands occur in most insects and these open into the distal part of the vagina. These glands produce a secretion able to stick the eggs to the substrate or together where necessary, forms the egg stalks of lacewings' eggs and forms the non-chitinous ootheca or egg case of the Orthoptera. The bursa copulatrix is that part of the vagina adapted to receive the aedeagus and ejaculatory duct of the male during copulation.

The sex cells

Each spermatozoan consists of a large head which is made up chiefly of chromatin, a middle piece and a long vibratile tail. Each egg contains an abundant store of yolk and is invested by a delicate vitelline membrane (Fig. 2.16). Outside this there is a hardened shell or chorion

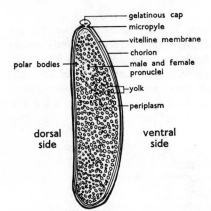

polar bodies

gelatinous cap
micropyle
vitelline membrane
chorion
male and female pronuclei
yolk
periplasm

dorsal side   ventral side

Fig. 2.16 Longitudinal section of mature egg of housefly. (After Imms (1957), *Textbook of Entomology*, Methuen.)

which may exhibit some kind of external sculpturing. In the chorion is found one or more pores, the micropyle(s) which enable(s) the spermatozoa to enter the egg during fertilization.

When the insects are ready to emerge from the egg, all the fluid within the shell is absorbed so that the young insect fills the shell completely. The chorion may be ruptured by muscular force or, owing to the contraction of the abdomen, the corresponding enlargement of the head causes a cap to burst off from the top of the egg shell. An egg burster is present in aphids, some young insects possess 'hatching spines' on the skin, and in the Lepidoptera the mandibles are used by the young insect to bite its way out.

The egg shell, like the cuticle, is protective and serves for a time to isolate the embryo from the adversities of the outside environment. The properties of egg cuticles of different species vary widely and differ also from the cuticles of juvenile and adult stages. Substances used to destroy eggs are called ovicides. Although well protected, eggs are not always proof against drought, minute egg parasites (e.g. *Trichogramma*) or predatory arthropods.

Reproduction and growth

Reproduction

In most insects, reproduction depends on the fertilization of eggs by spermatozoa, each gamete having half the adult number of chromosomes, but there are some important exceptions. In *parthenogenesis*, eggs develop without having been fertilized, as in the honey-bee where the unfertilized egg develops into a male. Eggs are formed without a reduction division to halve the number of chromosomes in many aphids, some weevils and some moths, and these develop into females. Some aphids and cynipid gall wasps have one or more parthenogenetic generations alternating with a bisexual one, thus ensuring rapid reproduction during the favourable season. *Paedogenesis* occurs also in a few species, e.g. *Dasineura brassicae* (p. 148), when eggs develop parthenogenetically from unfertilized eggs in the larva. In both parthenogenesis and paedogenesis the young are usually born alive, i.e. they are viviparous. *Polyembryony* occurs when more than one embryo develops from a single egg, an extreme example being the chalcid, *Litomastix truncatellus* (Dal.), a parasite of the silver Y moth, in which one egg may produce up to 1000 larvae.

Growth and metamorphosis

The horny integument of insects is capable of limited extension only, and has to be shed or moulted from time to time during growth. Moulting is termed *ecdysis*, and the stage between the moults is called the *instar*. The final instar is the *imago* or adult. In many insects a definite number of moults or instars precedes the adult condition, in others, e.g. the small wax moth, *Achroia grisella* (Fab.), the number is variable. The process is controlled by hormones produced by the corpora allata (p. 23) and the formation of the new exoskeleton begins before the old one is shed. The lower layers of the old cuticle are digested and partly absorbed under the influence of the exuvial fluid poured out by special glands in the hypodermis, while the new cuticle is protected by its epicuticle. The new cuticle is soft and white at first but it soon stretches, hardens and develops colour. In the primitive wingless insects, sometimes called the Apterygota, the change at each moult is slight and moulting continues after the adult condition is reached. In the more advanced winged insects, the Pterygota (p. 28), the change at each of the earlier moults is slight, then comes the final moult which gives rise to the perfect, sexually mature, flying insect. The final moult or moults bring about a more profound change of form than any of the earlier ones and is called *metamorphosis*.

When some insects emerge from the egg they resemble the imago in form and do not pass through a metamorphosis. These insects are termed Ametabola, e.g. Collembola and other primitive wingless insects, but most insects undergo metamorphosis and belong to the Metabola. The immature forms of the Metabola are termed *larvae*, and are divided into two types, *nymphs* and true *larvae*.

Nymphs are immature forms that are in a relatively advanced state morphologically when they leave the egg, and generally resemble the adult in appearance and mode of life, but they cannot fly and are sexually immature. They have compound eyes and their wing buds develop externally. Nymphs are usually active creatures, and the growth from nymph to imago is usually simple and continuous with no resting or pupal stage. Nowadays the term 'nymph' is less used but it is retained here for convenience.

Larvae on the other hand are in a relatively early stage morphologically when they emerge from the egg and often differ greatly in appearance and mode of life from adults, and may have different mouth parts. Their wing buds develop internally and they rarely have compound eyes, but may have a few simple eyes or ocelli. Larvae are often relatively sluggish and inactive and may live in concealed positions.

The degree of change at metamorphosis is greatest in those insects where the larvae have a different mode of life from the adults and, in consequence, have diverged considerably from the adult in structure and appearance, e.g. house fly, *Musca domestica* (L.). Very often the difference between young and adult is so great that an intermediate stage intervenes, a resting stage during which the transformation from larva to imago is brought about. This stage is known as the pupa, or pupal instar, and is characteristic of the life history of many insects of economic importance.

Insect orders with metamorphosis are divided into two groups:

1) EXOPTERYGOTA (Hemimetabola), where the young are nymphs which usually resemble the adults.
2) ENDOPTERYGOTA (Holometabola), where the young are larvae and a pupal instar precedes the emergence of the imago.

The nymphs and larvae of insects must be recognized, for they are often the stages that cause damage to crops. It is invariably the case in the Lepidoptera and Diptera, where the adults do not possess mouthparts able to injure plants.

Types of larvae

The three main types of larvae usually found are as follows (Fig. 2.17):

1) *Oligopod larvae* are active and peripneustic, with well-developed antennae, mouth-parts, legs and cerci. They have simple eyes and nine or ten abdominal segments are visible. Often the tenth segment bears an anal projection or pseudopod. This type of larva is characteristic of beetles (Coleoptera).
2) *Polypod larvae* are less active and live on or near their food. They possess abdominal 'legs' and a peripneustic tracheal system. The antennae and thoracic legs are small and the mouth-parts simple. A cluster of simple eyes occur on the head. This type of larva is characteristic of butterflies and moths (Lepidoptera) and sawflies (Hymenoptera, Symphyta).
3) *Apodous larvae* have no legs and are least active. They usually live in or entirely surrounded by their food and they have small or no sense organs, such as antennae and eyes. This type of larva is characteristic of weevils (Coleoptera), bees and wasps (Hymenoptera) and flies (Diptera) and a few Lepidoptera.

Consideration of the developmental stages passed through in the insect egg throws light on the types of larvae mentioned above. Eggs richly supplied with yolk develop to an advanced stage and give rise to a well-formed embryo, which hatches as an oligopod larva immediately able to move actively and feed, e.g. ground beetle larvae. Eggs less liberally endowed with yolk produce embryos that hatch in the polypod condition, an earlier developmental stage than the oligopod. Such eggs are usually laid on or near their food supply so that active food finding is unnecessary, e.g. most caterpillars. Eggs with very little yolk, as for example in hymenopterous parasites, are usually laid on or in their food supply. The embryo hatches at an early stage in development, the protopod stage, and is apodous. The protopod larvae of some parasites hatch at a very early stage indeed, and are aberrant, but later, acquire a more normal apodous form, e.g. *Apanteles* spp., the common parasite of the large white butterfly caterpillar. There is a tendency amongst burrowing larvae to lose their legs, e.g. weevils in plant tissue or soil, leaf and stem miners in plant tissues.

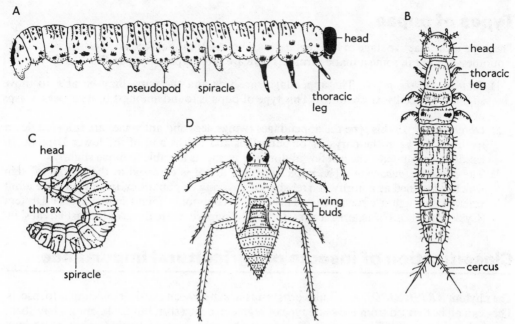

Fig. 2.17 Types of larvae to show the number of appendages. A, polypod larva of gooseberry sawfly, *Nematus ribesii*; B, oligopod larva of rove beetle, Staphylinidae; C, apodous larva of weevil; D, fourth instar or nymph of *Lygocoris pabulinus* (Exopterygota) to show developing wing buds. (A after Imms (1957), *Textbook of Entomology*, Methuen; B after Imms (1961), *Outlines of Entomology*, Methuen; D after Petherbridge and Thorpe in Imms (1961), *Outlines of Entomology*, Methuen.)

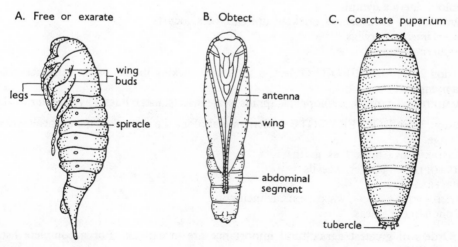

Fig. 2.18 Types of pupae. A, free or exarate, *Melolontha* sp. (male); B, obtect, noctuid moth pupa; C, puparium, Anthomyiidae.

Types of pupae

The pupa is the inactive stage of holometabolous insects. It cannot feed and is normally almost motionless (except in some aquatic forms). There are three types of pupae (Fig. 2.18).

1) *Exarate or free pupa*   The appendages are free and the pupa may be able to move slightly, especially its abdomen. This type of pupa is found amongst beetles, bees, wasps and sawflies.
2) *Obtect pupa*   In this type the appendages (wings, legs and antennae) are soldered down to the body, as in the chrysalis of butterflies and moths and of the lower Diptera. In lepidopterous pupae, the last few abdominal segments are able to move slightly.
3) *The coarctate puparium*   The pupa itself is free but is enclosed in the last larval skin which is retained as a horny, barrel-shaped covering or puparium, from which the adult emerges through a circular cap. This type is commonly found in the higher Diptera (Cyclorrhapha). The external characters of the puparium are those of the last larval skin.

Classification of insects of agricultural importance

The phylum *ARTHROPODA* includes organisms which have an exoskeleton similar to insects. They can all be derived from the same type of segmented ancestor, but the segments have been grouped together in different ways. Some breathe by means of tracheae, but most have breathing systems of a different type.

SUPERCLASS HEXAPODA, Insects

Class COLLEMBOLA: Springtails, wingless, ametabolous, i.e. no metamorphosis, moulting continues in adults, antennae with few segments, four-jointed legs.

Class PTERYGOTA: Winged insects, wings not always fully developed. Metamorphosis. No moulting after the adult stage is reached. Legs with more than four joints.

Division 1. BLATTOID-ORTHOPTEROID Orders, insects with biting mouth parts, hemi-metabolous, larva a nymph.
Orthoptera – cockroaches, crickets, grasshoppers, locusts
Dermaptera – earwigs
Isoptera – termites

Division 2. HEMIPTEROID Orders, insects with sucking mouth parts, hemimetabolous, larva a nymph.
HEMIPTERA – capsids, leafhoppers, aphids, scale insects, mealybugs Thysanoptera – thrips

Division 3. ENDOPTERYGOTE Orders, metabolous, i.e. complete metamorphosis with pupal stage.
LEPIDOPTERA – butterflies, moths
COLEOPTERA – beetles, weevils
DIPTERA – flies
HYMENOPTERA – bees, wasps, ants, ichneumons
Siphonaptera – fleas

Note: Orders of greatest agricultural importance are in capitals. For a complete list see C.S.I.R.O. (1970).

Characteristics of the more important orders of insects

PLANT BUGS (*Hemiptera*)
1) Two pairs of wings (wings are not always developed).
2) Mouth parts drawn out into a piercing organ (proboscis).
3) Young similar to the adults, with externally developing wing buds.
4) No pupal stage.

Chiefly plant feeders although a few are blood suckers (e.g. the bed bug). The piercing and sucking mouthparts make them ideal agents for the transmission of viruses.

BUTTERFLIES AND MOTHS (*Lepidoptera*)
1) Two pairs of broad wings clothed with scales.
2) The adults have coiled, sucking mouthparts. They feed on nectar from flowers.
3) The young stage is a caterpillar with biting mouthparts and is usually a plant feeder.
4) Pupa (chrysalis) with its appendages stuck down, sometimes enclosed in cocoon.

Apart from white butterflies, few are agricultural pests. There are many pests amongst the moths.

BEETLES (*Coleoptera*)
1) Two pairs of wings, the fore wings modified into horny wing-cases.
2) Biting mouth-parts in both young and adult stages.
3) Young stage a larva which varies greatly in form in the different families of beetles.
4) Pupa with its appendages free.

There are many different kinds of beetles. Damage to plants may be caused by young, adults or both.

TWO-WINGED FLIES (*Diptera*)
1) A single pair of wings. Hind wings modified into halteres.
2) In the adult the mouth-parts are elongated into a proboscis used for piercing and sucking, or for sucking only.
3) Young stage a legless larva.
4) Pupa of higher Diptera enclosed in an oval, horny case which is the remains of the last skin cast by the larva.

The adults are never plant feeders. Some are blood suckers (e.g. horse flies, mosquitoes, tsetse flies), others suck nectar from flowers or sweep sap exuding from damaged fruits. Those which damage crops do so only in the larval stage. Blood suckers transmit diseases of man and animals.

SAWFLIES, BEES, WASPS, ANTS, ICHNEUMONS (*Hymenoptera*)
1) Two pairs of membranous wings.
2) Biting mouth-parts, except in bees where they are elongated for sucking.
3) Young stage a caterpillar (sawflies) or a legless larva.
4) Pupa with its appendages free as in beetles. Often enclosed within a cocoon.

Sawfly larvae are plant feeders and many species are injurious. Gall wasps and seed chalcids injure plants. Bees are important pollinators; wasps and various parasitic Hymenoptera are beneficial; and ants are mostly beneficial but occasionally cause harm by protecting harmful species of insects (mostly Hemiptera).

3

Insects of minor economic importance – Collembola, Orthoptera, Dermaptera, Thysanoptera

Class Collembola

Springtails are small, wingless insects, rarely exceeding 5 mm long with six abdominal segments and biting mouth-parts which can be withdrawn into the head. The outstanding feature of most of the Collembola is the forked springing organ or furcula on the fourth abdominal segment which, when not in use, is kept beneath the abdomen by a hook or hamula on the third abdominal segment. The head bears a pair of short, jointed antennae (usually four joints) and up to eight simple eyes or ocelli on each side. A small sucker-like tube arises from the ventral surface of the first abdominal segment. Many springtails are atracheate and there are two main body shapes, elongate and globular (Figs. 3.1, 3.2, Table 3.1). Springtails are often extremely numerous and normally occur in moist situations in soil, under bark, in garden refuse, or under pots and boxes in glasshouses. Between 5×10^7 and 8×10^8 springtails ha^{-1} are commonly found in arable soils.

Table 3.1 A key to the onychiurid Collembola commonly found in arable land (after Brown, 1982). See Figs 3.1 and 3.2.

1) Body globular, thorax and first four abdominal segments fused	Sminthuridae
Body elongated, segments separate	2
2) First thoracic segment without dorsal setae, often reduced in size, cuticle smooth	Entombryidae and Isotomidae
First thoracic segment with some dorsal setae, cuticle granulated	3
3) Third antennal joint with small sensorial structures. Pseudocelli absent. Body usually pigmented, rarely white or yellow. Eyes sometimes present	Poduridae
Third antennal joint with two or three large sensorial structures, often partly hidden by papillae. Pseudocelli present. Often white or yellow. Eyes absent	Onychiuridae

Onychiuridae

4) Sensorial structures on third antennal joint directed towards each other. Claws without unguiculus. Adults often shorter than 1 mm	*Tullbergia*
Sensorial structures on third antennal joint directed dorsally. Claws with unguiculus. Adults often longer than 1 mm	*Onychiurus*

Onychiurus

5) Post antennal organ complex	6
Post antennal organ simple	7
6) Anal horns present	*Onychiurus ambulans* gp.
Anal horns absent	*Onychiurus fimetarius* gp.
7) Furca developed	*Onychiurus furcifer* (Borner)
Furca simple	*Onychiurus armatus* gp.

To distinguish the types within O. *armatus*, see Pitkin (1980).

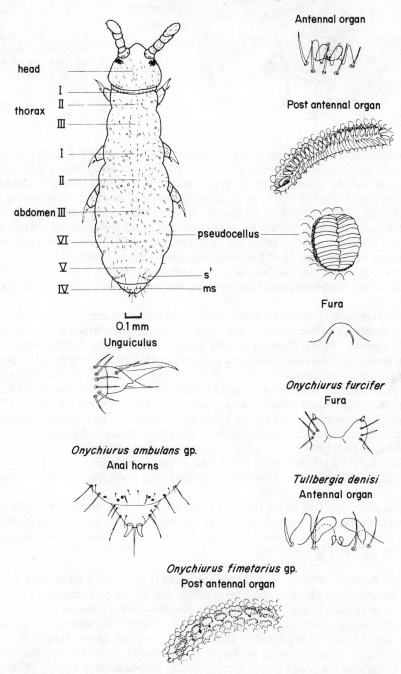

Onychiurus armatus gp.

Antennal organ

Post antennal organ

head

thorax

I
II
III

abdomen

I
II
III

pseudocellus

VI
V
IV

s'
ms

Fura

0.1 mm

Unguiculus

Onychiurus furcifer
Fura

Onychiurus ambulans gp.
Anal horns

Tullbergia denisi
Antennal organ

Onychiurus fimetarius gp.
Post antennal organ

Fig. 3.1 Some morphological details of a species in the *Onychiurus armatus* group and of some related species. Courtesy R.A. Brown.

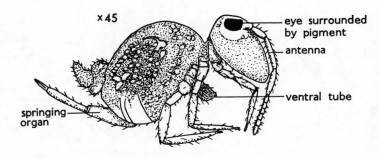

×45

eye surrounded
by pigment

antenna

ventral tube

springing
organ

Fig. 3.2 Side view of *Bourletiella hortensis* to show the globular shape and forked springing organ. After Folsom in Smith (1948).

Onychiurus spp. are slow moving, blind, white springtails, between 0.8 and 2.0 mm long (Fig. 3.1). They are rarely seen above ground and if disturbed do not jump as they lack the springing organ. Members of the species group damage sugar-beet seedlings (Jones & Dunning, 1972; Baker & Dunning, 1975; Ulber, 1978). *O. armatus* spp. are general feeders, eating fungi and yeasts (McMillan, 1975) as well as tissues of higher plants. In the laboratory, Ulber (1980) found that *O. armatus* spp. would eat weed seedlings and fungi but preferred sugar-beet seedlings; this has been confirmed in the field (Brown, 1982). In British sugar-beet fields, *O. armatus* spp. have two peaks of breeding each year. In East Anglia these are in early May and September; in Yorkshire late May and November–December. Juvenile mortality is high, due to predation by mesostigmatid mites but adult mortality is low, some individuals surviving up to 12 months.

Damage to sugar-beet is typically many small, rounded, blackened pits in the seedling roots. If damage is extensive before the four leaf stage, plants may be stunted or die and this results in yield loss (Brown, 1981). Damage can be severe in cool, damp springs, when sugar-beet crops grow slowly and are vulnerable for a longer time than in warm springs. *O. armatus* spp. damage sugar-beet most frequently in low-lying, alluvial soils and, on the Yorkshire wolds, they occur in association with symphylans (*Scutigerella immaculata* (Newport)) and the millepedes *Blaniulus guttulatus* (Boscowan) and *Brachydesmus superus* (Latzel) (see p. 178).

Sugar-beet crops can be protected by applying, during drilling, granular preparations of carbofuran, carbosulfan, aldicarb, bendiocarb, oxamyl or sprays of gamma-HCH incorporated into the seed bed. Only fields with a history of poor establishment through the depredations of soil-inhabiting seedling pests are at risk. With the exception of an isolated attack on beans (Edwards, 1962) reports of *Onychiurus armatus* spp. damage to other crops are unconfirmed.

Onychiurus and other species are predatory on female cyst-nematodes (Murphy & Doncaster, 1957) and in captivity some species can live and reproduce on a diet of nematodes but they may prefer other foods. In some soils they may be too large to follow nematodes through soil spaces (Jones, 1973a) (Plate XVIII).

Bourletiella hortensis (Fitch)   The garden springtail (Fig. 3.2) is a small black to dark green insect that attacks many plants making tiny holes in the leaves, cotyledons and stems, and has been accused of causing 'strangles' in beet, but plants suffering from strangles may show no signs of insect feeding whatsoever, the epidermis in the constricted regions being intact. In fact strangles is now known to be caused by the drying of seedling stems exposed during singling and by the flexing action of wind on larger plants (Boyd, 1966). Some soil incorporated herbicides cause a similar condition, e.g. quintozene, propham (Boyd *et al.*, 1970; see Leaflet no. 547).

Sminthurus viridis (L.)   The lucerne flea occurs commonly on grass, clover, lucerne and other herbaceous vegetation where it makes tiny holes in the leaves (Plate I, E).

Leaf-feeding springtails on pot plants, lettuce and tomatoes in glasshouses can be controlled by derris, HCH, malathion or nicotine. Outdoors, although controlled to some extent by a range of insecticides applied for other purposes, no products are specifically approved as control is unnecessary.

Xenylla welchi **(Folsom)** The mushroom springtail occurs in cucumber beds where it damages young leaves. It is readily controlled by malathion sprays.

Order Orthoptera

This order contains large insects with biting mouth-parts (Fig. 2.3) and two pairs of wings. The anterior pair, known as the tegmina, is horny and shiny and covers the hind wings or true flight wings, which are characterized by well-developed anal lobes. Apterous (wingless) and brachypterous (short-winged) forms are common. A pair of well-developed terminal cerci occurs on the abdomen, and the female has a well-developed ovipositor.

Cockroaches (Blattidae) are essentially insects from warmer lands. *Blatta orientalis* (L.) and *Blattella germanica* (L.) occur in bakehouses and kitchens but can be controlled by modern insecticides, although *B. germanica* especially develops resistance to organochlorine compounds (Green, 1962). Crickets (Gryllidae) are found out of doors, in refuse and on rubbish heaps. They migrate into houses, bakehouses and other warm spots where they chirp (stridulate) by rubbing the tegmina together. The left tegmen bears small teeth on a thickened vein which rasps a smooth area on the right tegmen. Apart from being a nuisance they are unimportant economically (see Leaflet no. 383). The mole cricket *Gryllotalpa gryllotalpa* (L.) possesses enlarged, digging forelegs and injures crops on light sandy soils in France and elsewhere on the Continent; it also occurs in many places in Britain, mostly south of the Thames but is local and rare.

The family ACRIDIDAE contains short-horned grasshoppers and locusts. Fortunately, Great Britain is not troubled by locusts although just inside the range of the migratory locust. In many countries they are of great importance.

Order Dermaptera

This order contains the earwigs, which are not very important economically. Their outstanding features are: (*a*) biting mouth-parts; (*b*) very short horny forewings covering the hindwings, which are kept folded underneath when not in use; (*c*) the abdomen terminates in a pair of forceps which are modified cerci.

Forficula auricularia **(L.)** (Fig. 3.3) Earwigs are usually most abundant on rough ground and near shelter. They are general feeders and are more troublesome in gardens, eating the leaves of flowers, dahlias and other plants, than they are on a farm scale. However, they do eat holes in the heart leaves of sugar-beet, and other crops. The female earwig lays eggs in a cavity below ground and remains with the nymphs when they hatch, exhibiting a measure of 'parental care'. Eventually the 'family' disperses, and it is at this stage, about July, that injury becomes most noticeable in crops such as sugar-beet, vegetables and flowers.

Control measures are not easy, as the nymphal stages are passed below ground and the sporadic nature of attacks makes the timing of control measures difficult. Gardeners commonly employ traps such as upturned flower pots to catch earwigs, taking advantage of their tendency to hide in crevices. Injury to mature field crops such as sugar-beet is usually slight and control measures unnecessary. HCH and carbaryl are used in orchards and HCH is used to protect chrysanthemums, dahlias and pot plants. Some earwig populations are resistant to HCH. Earwigs feed on aphids and other small insects and therefore are to some extent beneficial.

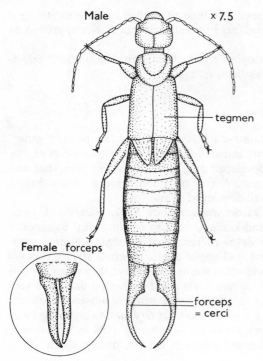

Male   × 7.5

— tegmen

Female forceps

forceps
= cerci

Fig. 3.3 Common earwig, *Forficula auricularia*, male. Also forceps of female.

Order Thysanoptera

This order contains thrips, minute, elongate insects abundant in vegetation everywhere and common in all kinds of flowers (Lewis, 1973). They are characterised by two pairs of narrow wings fringed with setae, antennae with from six to nine joints, complicated asymmetrical mouth-parts modified for superficial piercing and sucking (Plate IB). The modified mouth-parts or stylets are enclosed in a short cone or rostrum projecting downwards from the ventral surface of the head which is applied to the surface of the plant during feeding (Fig. 3.4 B, C). There are no cerci on the abdomen. The tarsi are two-jointed and terminate in a vesicle (Fig. 3.4A).

The life cycle is simple. There are five instars, in the first and second the nymphs, usually yellow or orange, feed actively but, during the third and fourth instars, they remain quiescent and overwinter in soil, litter or in crevices in bark. The wing buds appear in the third instar, sometimes called the 'pre-pupa' to distinguish it from the fourth or 'pupa' which, although usually motionless, can move if disturbed. The fifth instar or adult appears the following spring or early summer (May, June). Some species are parthenogenetic.

Kakothrips robustus **(Westwood)** The pea thrips, is troublesome in dry seasons and in the drier counties such as Essex and Suffolk, where there are important pea-growing areas. Garden peas suffer especially. Damage by the yellow nymphs is partly to the flowers, which fail to develop, but more commonly to the pods. Such pods as are produced are misshapen and have markings on the surface with a silvery sheen resulting from the empty epidermal cells filling with air. The life cycle follows the typical thrips pattern; the nymphs feed during the first and second instars, then after 3 or 4 weeks descend into the soil, and overwinter. The adults appear in May and June and lay their minute, kidney-shaped eggs, usually in the stamen sheath of the pea flowers, from which the nymphs soon hatch.

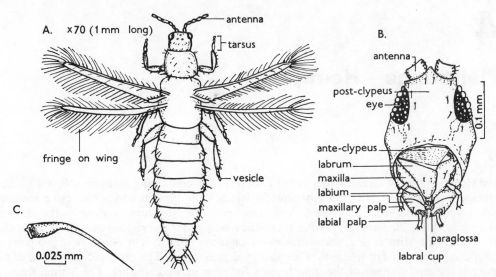

Fig. 3.4 Structure of thrips. A, dorsal view of *Thrips tabaci* showing fringed wings and vesicles at the ends of the tarsi; B, front view of head of *Chirothrips hamatus* (Trybom) to show the assymetry of the mouthparts; and C, structure of the only functioning mandible, the left. B and C after Jones (1954).

Pea thrips can be controlled by spraying with azinphos-methyl mixtures, dimethoate feni-trothion or HCH. Spraying should be done under dry conditions. Rain is unfavourable to the thrips and may terminate an attack (see Leaflet no. 170).

Thrips tabaci **(Lind.)**   Onion thrips (Fig. 3.4) is very similar to the pea thrips and is injurious in dry years. It is commonest on flowers and in glasshouses where it transmits the virus that causes 'spotted wilt' disease of tomatoes. In dry weather, the young leaves of onions may be attacked by the nymphs, the tiny punctures causing the leaves to assume a silvery white appearance so that they shine in the sun, and also causing twisting and curling. A good shower of rain soon stops the attack and the later leaves grow normally. Control can be obtained by spraying with the insecticides recommended for pea thrips.

Limothrips cerealium **(Hal.)** and *L. denticornis* **(Hal.)**   Grain thrips are commonly found feeding as young and adults inside the glumes of developing cereal florets. The nymphs are yellow, the adults are dark and males apterous but the females are able to fly. In most years they swarm in great numbers during hot sultry weather in July. Young and adults overwinter in coniferous bark and elsewhere. The economic status of grain thrips is uncertain but they are known to damage wheat and barley grains and affect germination. Feeding tends to be concentrated on the radicle on either side of the grain, probably because the thrips enter between the lemma and the palea near the rachis (Oakley, 1980). Spraying from the air with fenitrothion to control thrips and wheat blossom midges is effective (see Leaflet no. 788).

Aptinothrips rufus **(Halid.)** and *Anaphothrips obscurus* **(Muell.)**   Grass thrips, are found on grasses and oats, sometimes causing sterility in the latter. They overwinter in grass litter.

Thrips angusticeps **(Uzel)**   The field or cabbage thrips, attacks beet, brassicae and peas in some localities, mostly in eastern England. In Denmark it is a rather important pest of beet seedlings. Short-winged adults overwinter in the soil and emerge to feed on seedlings in April and May. When the small heart leaves of beet seedlings expand they are strap-shaped, roughened with irregular reddened or blackened margins and tips. There are also small silvery lesions on the leaf surface. Attacks are rarely discovered early enough for preventive measures to be taken.

4

Plant bugs – Hemiptera

The Hemiptera is the largest Order in the Exopterygota, containing more than 40 000 species. Members of the Order are primarily plant feeders both as nymphs and adults, but a few suck blood, such as the bed bugs (*Cimex* spp.), and some are predatory, feeding on the juices of other insects, e.g. *Reduvius personatus* (L.) (Reduviidae), *Anthocoris nemorum* (L.) (Anthocoridae) and some Miridae (e.g. *Blepharidopterus angulatus* (Fall.)). The adults bear two pairs of functional wings, except where they are lost by degeneration. The outstanding feature of the order is the modification of the mouth parts for piercing and sucking. The labium forms a grooved sheath in which lie two pairs of needle-like stylets, the modified maxillae and mandibles articulated together by a series of grooves to form a composite organ (Fig. 4.1). The inner pair of stylets are the maxillae and, where their innermost surfaces come together, two continuous channels are formed, the anterior one is the food channel and the posterior the salivary canal.

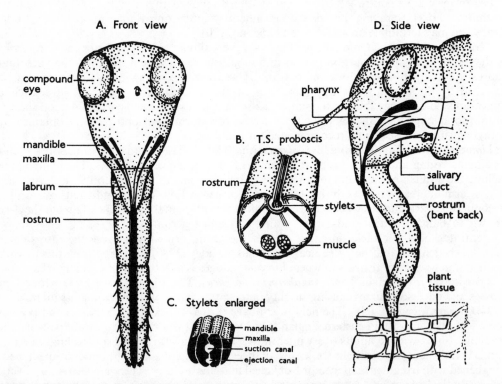

Fig. 4.1 Diagrammatic mouthparts on head of a hemipteran. A, front view of head; B, transverse section across the proboscis showing the stylets in the groove of the rostrum; C, stylets enlarged to show the suction canal of the pharynx and ejection canal with salivary duct; D, heteropteran feeding.

The tip of the labium contains muscles and acts as a clamp during the act of piercing. Saliva is forced down the posterior canal into the wound and contains ferments, which may attack the contents of the cells. Sometimes other substances are present in the saliva which are toxic or which affect plant growth. In sedentary feeders such as aphids, sap is forced up the stylets by the turgor pressure of the phloem sieve tubes which they must penetrate to feed satisfactorily. When the embedded stylets of large aphids are cut from the heads, drops of sap continue to exude from the cut ends of the stylets for many hours. The rate at which the plant forces the sap up the stylet food canal determines the rate of sap uptake and excretion of honey dew.

Aphids are able to tap the sap as required, for the controlling mechanism is in the insect's head and thorax. When the muscles of the anterior alimentary canal wall contract, sap under pressure is admitted from the stylet food canal into the oesophagus and stomach and, when the muscles relax, the anterior wall returns by its elasticity to its normal position, and so shuts off the flow of sap. Nitrogen in the form of free amino-acids and amides is removed from the sap, so the honey dew contains less than the ingested sap. The honey dew also contains sucrose, fructose, glucose and melezitose, while the only sugar normally ingested is sucrose. For reviews of aphid feeding, nutrition and virus transmission see Carter (1961), Auclair (1963) and Harris & Maramorosh (1977).

The great economic importance of the Hemiptera is related to their peculiar mouth parts. Damage to plants is of two kinds:

1) Tissues are injured directly during piercing and sucking and the injected saliva also gives rise to necrotic spots (mirids), bleached spots (jassids), leaf curl (some aphids) or galls (*Pemphigus bursarius* (L.)).
2) Indirect damage by transmission of viruses (e.g. potato leaf roll, and potato Y, beet yellows and beet mosaic viruses).

Direct injury of economic importance is usually caused by small Hemiptera only when they are very numerous. Indirect injury by virus transmission, however, may take place at small population densities and, as viruses are normally systemic, the whole plant may become infected from the feeding of one viruliferous individual. Because of the form of the mouth parts, stomach poisons are ineffective against Hemiptera, but three other types of insecticide can be used: (1) contact insecticides; (2) systemic insecticides acting through the host plant; and (3) fumigants.

Classification

The order is divided into two suborders, based on the condition of the forewing.

Suborder a. Heteroptera

Forewing partly horny, partly membranous. Wings usually overlapping over the abdomen. Pronotum large. Tarsi usually 3-jointed.

Series 1 Antennae concealed. Aquatic insects (e.g. water boatmen).
Series 2 Antennae conspicuous and freely movable in front of the head.
Contains plant feeding and blood sucking forms. Pests of economic importance occur in several families.

Family MIRIDAE (Capsidae). Capsid bugs (Fig. 4.3). Forewing with a cuneus but no distinct embolium (Fig. 4.2). Ocelli absent. Toxic saliva causes necrotic spots. Rarely carry viruses.

Family PIESMIDAE. Lace bugs (Fig. 4.4). Flattened insects with prothorax and upper surface of forewing adorned with cell-like reticulation. One or two species transmit viruses.

Suborder b. Homoptera

Forewings entirely of one consistency, either membranous or horny throughout. Wings usually sloping over the sides of the abdomen. Pronotum small.

Series 1   STERNORRHYNCHA

Proboscis appears to issue from between the forelegs. Antennae well developed without conspicuous arista. Tarsi 1- or 2-jointed.

Family APHIDIDAE. Plant lice (Figs 4.7–4.10, 4.12, 4.13). Wings transparent, often wanting in the female. Siphunculi (cornicles) and cauda usually present. Antennae usually 6-jointed. A family of great economic importance. Almost every known plant is a host for one or more species of aphid. Parthenogenesis and viviparity common. Enormous powers of multiplication.

Family PSYLLIDAE. Jumping plant lice. Forewings of firmer consistency than the hind pair. Antennae 10-jointed.

Family ALEYRODIDAE. White flies (Fig. 4.15). Body and wings covered with white powdery wax. Nymphs degenerate, scale-like. Incipient pupal stage.

Family COCCIDAE. Scale insects and mealy bugs (Fig. 4.16). Female wingless and degenerate, scale-like, gall-like or with a powdery or waxy coating. Males wingless or with only the anterior pair functional, mouth parts wanting. Possessing a pupal stage. Contains many important pests, few of which are British.

Series 2   AUCHENORRHYNCHA

Proboscis plainly arising from the head. Antennae very short with terminal arista. Tarsi 3-jointed.

Family CICADELLIDAE. Leafhoppers or jassids (Fig. 4.17). Cause 'hopper burn'. Important vectors of viruses.

The Cercopidae and Delphacidae also belong to this series.

Suborder Heteroptera

Family MIRIDAE (Capsidae). Capsid bugs have conspicuous 4-jointed antennae, the forewing has a cuneus and a peculiar arrangement of veins in the membrane (see Fig. 4.2 where a capsid forewing is shown with that of the Lygaeidae and the Anthocoridae). Among the species of capsids considered below, the potato capsid (*Calocoris norvegicus* (Gmel.)); common green capsid (*Lygocoris pabulinus* (L.)); and the tarnished plant bug *Lygus rugulipennis* (Poppius) are the most important (Fig. 4.3) (Plate IF).

The life histories of *L. pabulinus* and *C. norvegicus* are similar. They have woody winter hosts on which the eggs are laid, and herbaceous summer hosts, on which an additional summer generation develops. Damage to crops is often confined to fields surrounded by high hedgerows or near orchards. In feeding on plants such as sugar-beet or potato, the toxic saliva causes necrotic spots so that stems and petioles twist and leaves tear during growth. Fruit becomes 'scabby' when attacked. In sugar-beet and its allies, stab wounds in the principal veins lead to yellowing of the distal portions of the lamina, a condition simulating beet yellows. Attacks are only occasionally severe enough to warrant treatment and then usually to field margins only.

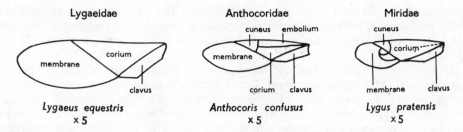

Fig. 4.2 Diagrams of hemielytra of Lygaeidae, Anthocoridae and Miridae.

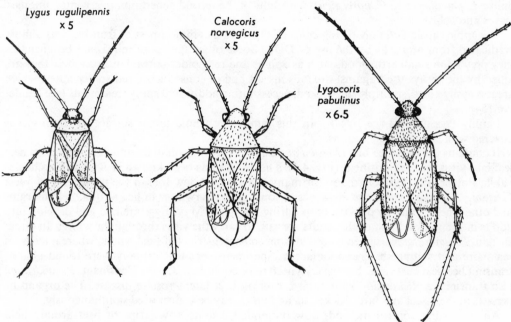

Fig. 4.3 Three common capsid bugs.

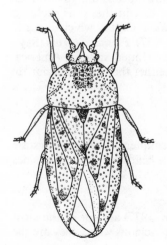

Fig. 4.4 Dorsal view of *Piesma quadratum* (× 14), a typical lace bug.

Phorate granules or dimethoate spray are effective, DDT and HCH should be avoided as they kill aphid predators. 'Blindness' in seedlings may also be caused by *L. rugulipennis*, but attacks cannot be forecast. Aldicarb granules applied in the seed furrow decrease attacks (Dunning, 1957).

L. pabulinus seems to have changed its habits, for early works give no reference to it as a pest of fruit trees on which it is now commonly found. Control is rarely necessary on field crops but is essential on fruit, for blemished fruit is less easily marketed. Eggs on the fruit trees may be attacked with winter washes such as DNOC in petroleum oil while the buds are still dormant. If used at any other time, this treatment would scorch the leaves. At a later stage, small capsids can be killed with carbaryl, chlorpyrifos, fenitrothion, HCH, nicotine or triazophos.

The tarnished plant bug is cosmopolitan. It passes through two generations a year and, unlike *L. pabulinus* and *C. norvegicus*, the adults of the second generation overwinter amongst leaves and debris.

The apple capsid (*Plesiocoris rugicollis* (Fall.)) was a common pest of fruit but was almost eradicated from orchards by the use of DDT. Some other species of mirids are beneficial as they prey upon small arthropods such as aphids and red spider mites. Unfortunately they are killed by the sprays used against noxious insects and in consequence red spider mites, which are less easily killed, multiply unchecked, necessitating additional spray treatments (see Leaflet no. 154).

Family PIESMIDAE (Lace bugs).   In this family the whole upper surface of the insect is covered with lace-like punctations.

It contains the beet lace bug (*Piesma quadratum* (Fieb.)) a small insect which transmits beet leaf curl virus on the Continent (Fig. 4.4). The disease has been known in Russia, Czechoslovakia, Poland and East Germany for many years with a few foci in the north east of West Germany. In areas where the disease is serious, bugs overwinter in hedgerows, field margins and other shelter, and fly into the crop during April and May. Eggs are laid and develop into adults in 3-8 weeks. Most of the adults are said to carry the virus through the winter. In Great Britain, *P. quadratum* has been recorded from coastal districts of East Anglia where it feeds on sea-shore and saltmarsh Chenopodiaceae. Specimens sent to Germany were found able to transmit beet leaf curl virus, but the tests need to be confirmed. On the Continent, *P. quadratum* also transmits a *Rickettsia*-like organism causing beet latent rosette disease. The organisms persists in the insect and both Rickettsia and virus may be transmitted simultaneously.

An interesting control method, now outmoded, was to sow strips of beet around field margins. These germinated and attracted the lace bugs. The strips were then ploughed in deeply and the main crop sown. Nowadays, control is achieved by spraying the field margins with an organophosphate insecticide before the bugs disperse. In Great Britain the virus vector is absent from beet fields, but a second species of *Piesma*, *P. maculatum* (Costa), which does not transmit the virus, occurs around the coast and inland (Kershaw, 1957). *Piesma cinerum* (Say.) transmits the virus which causes 'savoy disease' of sugar-beet in the United States of America. Thanks to the Importation of Plants Order (1971) (see p. 318) neither the Rickettsia nor the viruses yet occur in Britain (Heathcote, 1972; Proesler, 1982).

Suborder Homoptera

The Homoptera is by far the most important suborder of the Hemiptera. The forewings are either completely horny or completely membranous and at rest are held sloping over the sides of the body.

Series 1. STERNORRHYNCHA

Family APHIDIDAE. Aphids (Lowe, 1973; Blackman, 1974; Dixon, 1977) are one of the most important families of insects economically, especially in temperate regions where they are the chief vectors of plant viruses. Aphids are also among the most prolific of insects. Many have alternate woody winter hosts for egg laying and herbaceous summer hosts on which reproduction is asexual, parthenogenetic and viviparous. They are efficient virus vectors partly because of (1) their fecundity and rapid maturation, (2) the mobility of the winged forms and their ready transport in air currents, and (3) the form of the mouth parts, which act as a hypodermic syringe reaching into the phloem (Pollard, 1971).

Aphids (Figs 4.7–4.10, 4.12, 4.13) are characterized by (1) a pair of dorsal organs known as cornicles or siphunculi in a dorso-lateral position on the fifth abdominal segment, (2) a well-defined cauda, or tail, (3) a pair of 4–6 jointed antennae, (4) a proboscis which appears to issue from between the coxae of the front legs, (5) a number of secondary sensoria (plate-like sense organs) found on the antennae, especially of the winged forms, and (6) 2-jointed tarsi.

The life cycle of the black bean aphid (Fig. 4.5) is typical of those that have an alteration of woody winter host and herbaceous summer hosts. This is set out below and followed by a list of technical terms used extensively for aphids.

Reproduction on the summer hosts is parthenogenetic and viviparous. Winged forms arise that are unable to accept the host plant on which they were born and so migrate and found fresh colonies on new herbaceous host plants during summer. By migrating in this way from the crowded, parental host plant, aphids escape their enemies that multiply in the old colonies and are able to make use of available host plants in the most favourable condition for rapid multiplication.

TECHNICAL TERMS USED EXTENSIVELY FOR APHIDS

Form	*Description*	*Progeny*
Fundatrix -ices	The first form on the winter host. Hatches from the egg	Produces the spring colonies
Fundatrigenia -ae	Progeny of the fundatrix Winged or wingless	Winged or wingless females
Alata -ae	Winged female. Summer migrant	Usually wingless females
Aptera -ae	Wingless female	Wingless and winged females
Sexupara -ae	Forms giving rise to the true males and females	True males and/or females
These are:-		
Gynopara -ae	Winged sexupara which migrates to the winter host and there produces the egg-laying female. The autumn migrant	Egg-laying females
Andropara -ae	Winged or wingless sexupara giving rise to the true males	Males
Sexuales	The true sexual forms	Eggs
Ovipara -ae	The true sexual female, capable of mating and producing eggs	Eggs

Parthenogenesis Reproduction from unfertilised eggs: without the intervention of the male.
Viviparity The production of living young as opposed to the laying of eggs.
Virginoparae Alatae and apterae which reproduce parthenogenetically and viviparously.

In *Aphis fabae*, and other aphid species too, the distinction between apterae and alatae is not as clear cut as might be supposed. In fact, the two states obscure a range of forms and of behaviour patterns. Extreme apterae lack the apparatus for flight which appears to have been sacrificed in favour of the early development of ovarioles and embryos. At the other extreme are the alatae which do not reproduce on the host plant on which they developed. These are the migrants proper which are best equipped to exploit distant habitats. In between are apterae with wing pads, alatae which lack the ability to fly, and fliers which reproduce on the host

42

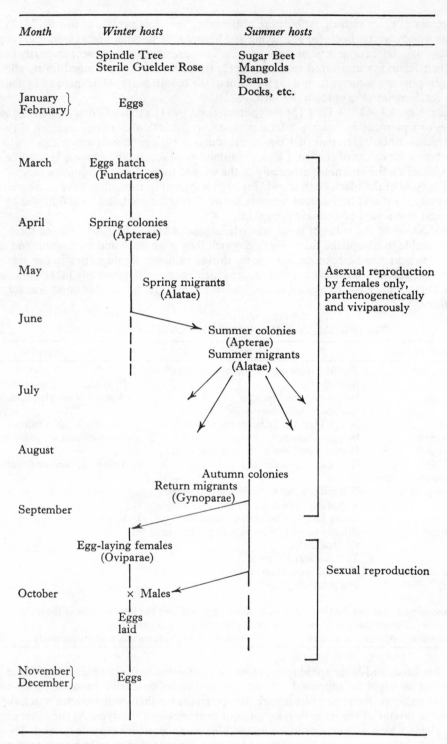

Month	Winter hosts	Summer hosts
	Spindle Tree Sterile Guelder Rose	Sugar Beet Mangolds Beans Docks, etc.

January } February }   Eggs

March   Eggs hatch (Fundatrices)

April   Spring colonies (Apterae)

May

      Spring migrants (Alatae)

June

      Summer colonies (Apterae)

      Summer migrants (Alatae)

July

Asexual reproduction by females only, parthenogenetically and viviparously

August

      Autumn colonies

      Return migrants (Gynoparae)

September

      Egg-laying females (Oviparae)

Sexual reproduction

October   × Males

      Eggs laid

November } December }   Eggs

Fig. 4.5 Life-cycle of the black bean aphid.

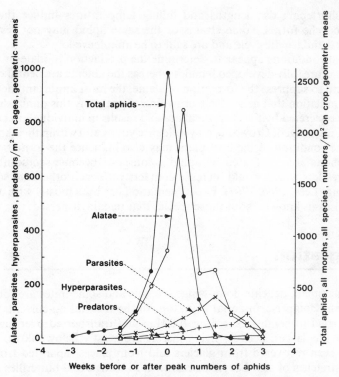

Fig. 4.6 Numbers of aphids, their parasites, hyperparasites and predators caught in emergence cages over winter wheat at Rothamsted. Geometric means of numbers caught. Time scale synchronized with peak number of aphids for the years 1973 to 1977 inclusive. (Redrawn from M. G. Jones, 1979a.)

before flying. The last are more concerned with local spread than distant migration (Shaw, 1970a, b, c). Evidently *A. fabae* and other species of aphid are well adapted to produce, at relatively short notice, forms best suited to the local situation. This plasticity ensures that the food supply is fully exploited and that dispersal to new hosts occurs before the food supply has deteriorated. Doubtless this ability is a feature which has helped to make aphids so successful.

There are other types of life cycle as in cabbage aphid where one type of herbaceous host is used throughout the year, and there is no alternation of woody and herbaceous host plants. Others, such as *A. viburni* (Scop.) on Guelder rose and *A. euonymi* (F.) on spindle tree are examples of aphids confined to woody hosts throughout the year. Some aphids have no sexual phase, the aphid continuing throughout the winter as the adult or nymphal viviparous female, e.g. glasshouse potato aphid (*Aulacorthum solani* (Kaltb.)), the Japanese spindle aphid (a variety of *A. fabae*). Many aphids, e.g. *Myzus persicae* (Sulz.) which have a sexual stage are able to overwinter as apterae when the weather is mild or shelter available.

Polymorphism in aphids

The number of chromosomes in aphids ranges from $2n = 4$ to $2n = 22$ but most have multiples of 4, e.g. the genus *Aphis* with $2n = 8$. The longest chromosomes are the sex chromosomes and males usually result when one of these is lost at meiosis, but chromosome cycles and chromosome variations in different groups make generalisation difficult. Parthenogenetic reproduction is based on endomeiosis which permits crossing over and explains variability within aphid clones (Clark, 1973).

In autumn, shortening day lengths and falling temperatures induce the production of sexuales in holocyclic forms. Some strains of the same aphid may not respond and these continue the parthenogenetic cycle and are said to be anholocyclic.

Environmental conditions appear to determine the production of alate or apterous virgin-oparae. The alatae are fully developed females whereas the apterae are neotinic. Although high temperatures tend to suppress the formation of alatae, the most important factor is thought to be the tactile stimulation that arises from crowding. Apparently this diminishes the output of juvenile hormone secreted by the corpora allata and results in individuals in crowded colonies becoming adult, i.e. winged. Crowding may influence young still within their apterous mothers. Variations in the condition of the host plant may also influence the formation of apterae or alatae. When plants are debilitated by aphid colonies or become senescent the aphids are possibly more restless which would increase contacts with each other and so have the same effect as crowding (Lees, 1966, 1967). Experimentation has been mostly with aphids that form dense colonies. Less is known about those species that remain dispersed.

Aphid migration

Aphids and other small, delicate, long-winged insects and some mites are found in the air in large numbers and comprise an aerial plankton. In the thermal up-currents of turbulent air they may be carried to heights exceeding 1000 km and transported over many hundreds of kilometres while the larger, actively flying insects are mostly below 100 m. Aphids in good condition have been recovered from glaciers and from places separated from the points of origin by wide stretches of sea. Larger insects such as locusts and butterflies are also assisted by air currents but must supply power for sustained flight and are not carried to great heights above the ground. The aphid component of the aerial plankton is greatest in July over the southern half of Britain with subsidiary peaks sometimes occurring in spring and autumn, which may represent migration to and from winter hosts. Near the ground diurnal peaks tend to occur in late morning and afternoon. Higher up the peaks are delayed because of the time taken in rising. Aphid flight and the thermal up-current assisting it tend to die out at night and the numbers caught in traps fall rapidly.

In spring and early summer the alatae migrate to new summer hosts; in autumn the alatae of those species that possess them migrate to their primary or winter host which are often woody. Johnson (1969) reviewed the factors that determine aphid flight. Alate aphids are weak fliers. Unaided they move at speeds between 1.6 and 3.2 km per hour and, once air-borne, are dependent on the speed and movement of air masses. During daytime the layer of air immediately above the ground is warmed and rises, carrying alatae upwards in thermal upcurrents to high altitudes as if they were inert particles and few remain near the ground. At sunset temperatures near the earth's surface fall and up-currents cease, aphids cease flying and descend: few remain air-borne overnight.

The aerial population builds up anew each day, the numbers leaving the crop being determined partly by population density and partly by weather. In many species the numbers airborne reach peaks in mid-morning and early afternoon. The peaks are determined partly by weather, especially temperature, and partly by the rate at which individuals complete the last moult and metamorphose into alatae. After this moult they are not immediately able to fly but require a further short period of development, the teneral period. Many *A. fabae* complete the last moult in the early morning, others move slowly during the rest of the day and again move rapidly in the evening. In the morning when temperature and light intensity have risen above the threshold for flight, aphids that matured overnight fly away to cause the mid-morning peak. The second peak includes aphids that moulted in the morning and completed their general development by afternoon. The number airborne decrease as flight ceases when temperatures fall and daylight fades. The time spent in the air is determined by the altitude reached. Few aphids fly for longer than two hours and those at middle heights fly for an hour or less. Distance travelled is a function of time and windspeed and the intensity of migration is determined by response to light, the speed of flight and the velocity and direction of wind. The aphids have little control over the direction of migration, the distance travelled or where they settle.

Not all aphids behave as outlined above. For example, the sycamore aphid (*Drepanosiphum platanoidis* (Schr.)), which flies at low temperatures, has peaks of migration in early morning and after sunset, and flies at wind velocities smaller than those suitable for *A. fabae*. It therefore is better able to control its migrations and remain near its host plants, behaviour described as boundary layer migration (Taylor, 1965). The lime aphid (*Eucallipterus tiliae* (L.)), which flies at relatively high temperatures (18–26 °C), has only one migration peak shortly after mid-day (Dixon, 1971).

After flying, alate aphids tend to land on leaves, reflecting long wavelength light, being especially attracted to yellow light emitted from young actively growing crops or to senescing plants. On alighting, aphids probe the plant and settle only on suitable ones. The substance that arrests the cabbage aphid is the glucoside, sinigrin, which probably has no nutritional value for the aphid.

During the first hour or so of flight glycogen reserves provide the muscular energy and thereafter fat is metabolized. After settling and starting to reproduce the indirect flight muscles of some aphids degenerate, so no further flights are possible.

Aphids and their host plants

Despite the observed abundance of summer hosts colonized by polyphagous aphids such as *A. fabae* and *M. persicae*, aphids, in common with other Homoptera and other plant feeding groups (e.g. Thysanoptera, p. 34; Agromyzidae, p. 161), are more than 99% host specific (Table 4.1). About 90% of aphids do not have alternate woody winter hosts and their summer migrants colonize relatively few species of summer hosts. The summer migrants of heteroecious species, i.e. those with host alternation, are a good deal more polyphagous. Even so, the host range of polyphagous species is restricted: *A. fabae* and *M. persicae* rarely occur on grasses and *A. fabae* is absent from oats. So many plants are not colonized by aphids that it seems probable that colonization of new hosts is by capture, the result of positive adaptation of the aphid to the plant rather than vice versa, for it is difficult to conceive that the many unattacked

Table 4.1 Host specificity in European aphids (data from Eastop, 1971).

Type	No. of species of aphids with hosts in:			More than one plant family
	One plant family			
	1 genus	2–5 genera	>5 genera	
Without host alternation	407	23	10	15
*With host alternation	40	17	9	11
% with alternation	9	43	47	42

* Primary host not included.

plants are unsuitable nutritionally although some may contain repellent or antifeedant principles. For example, the Solanaceae contains many medicinal plants and, considering its size,
harbours relatively few aphid species. Possibly the medicinal principles and sticky hairs of
some species are inimical to many aphids.

It is thought that aphids first lived on trees and that they evolved on an extinct group of
angiosperms and transferred to the sub-class of dicotyledons known as the Hamamelidae
which contains elms, walnut, beech, oak, birch and other tree families. Early aphids may have
transferred to a few conifers. The aphid–plant relationship seems to have evolved further by
the acquisition of new hosts. Often new hosts are specifically or generically related, for there
may be chemical and physical barriers which preclude adaptation to more distantly related
hosts.

Coevolution on early hosts seems not to have made aphids 'parasitic prisoners', a fate that
tends to befall highly adapted parasites, for despite their specializations, aphids have remained
free-living and have retained the ability to colonize new environments (Eastop, 1971).

If aphid taxonomy were more advanced, it would probably be found that some species exist
as a series of races with well defined host specificities. This situation seems likely to be
uncovered as breeding for host-crop resistance intensifies. Indeed, sub-races of some species
are already known, e.g. the Japanese spindle aphid (p. 50), biotypes of *Acyrthosiphon pisum*
(Ham.), *Aulacorthum solani* (Kltb), and *Macrosiphum euphorbiae* (Thos.), and the races of the
rubus aphid *Amphorophora rubi* (Kltb.) distinguished by resistance genes used in raspberry
breeding (Briggs, 1965). A gene-for-gene relationship may possibly exist between aphids and
their hosts. Such a system is well-known in other obligate parasite–plant host systems, e.g. rust
fungi, the Hessian fly (p. 147), cyst-nematodes (p. 301) and possibly in viruses (R. A. C. Jones,
1981).

Aphid enemies

Aphids have many enemies. Table 4.2 lists the enemies of cereal aphids but those of other
common aphids are much the same. Just as the numbers of cereal aphids vary greatly from
year to year (M. G. Jones, 1979a) so do those of their enemies. Moreover, the relative
importance of the groups is not constant nor is that of species within groups. Stary (1976)
thought that there are key parasites for each aphid species and concluded that *Aphidius
uzbekistanicus* was the key parasite of *Sitobion avenae*. However, M. G. Jones (1979b) found
that *A. picipes* was more common in that species which was itself the main parasite emerging
from *Metopolophium dirhodum*. Possibly habitat, season and numbers at the close of the
previous season are as important as species/host relationships in determining which parasite
or predator is dominant at a particular place and time. (See Powell, 1982.)

Because aphid numbers peak at different times every year, averages over several years are
more meaningful when synchronized about peak numbers. The resultant curve (Fig. 4.6) shows
how successfully aphids exploit the host crop during a period of approximately three weeks of
warm dry weather. Then, their capacity for increase outstrips their enemies and is almost
exponential until competition, ageing and deterioration of the host, and the switch to the alate
condition brings about a population fall as rapid as was its rise. Figure 4.6 also shows that
numbers of enemies follow rather than match or precede those of the aphid prey, i.e. they are
not well enough synchronized to exert maximum effect. Parasites are more numerous than
predators but it must be remembered that each kills only one aphid whereas individual
predators may kill many. Hyperparasites are abundant in some years. These decrease the
numbers of parasites at the beginning of the next year and there is a suspicion that, in
consequence, aphid numbers are sometimes greater (M. G. Jones, 1972, 1979b; Chua, 1977).

More needs to be known of the events that lead to outbreaks. Apart from non-resident

Table 4.2 Some enemies of aphids.

General predators

Forficula auricularis (L.)	(Dermaptera: Forficularidae)	
Anthocoris nemorum (L.)	(Hemiptera: Anthocoridae)	
Tachinus rufipes (Deg.)	(Coleoptera: Staphylinidae)	
Tachyporous hypnorum (F.)	(„ : „)	I
Ground beetles	(„ : Carabidae) See Table 6.1, p. 95	
Insect-feeding birds		

Dasysyrphus lunulatus (Meig.)	(Diptera: Syrphidae)	
Epistrophe grossulariae (Meig.)	„ : „	
Episyrphus balteatus (Deg.)	„ : „	
Mellanostoma mellinum (L.)	„ : „	
Metasyrphus corollae (F.)	„ : „	IIb
M. luniger (Meig.)	„ : „	
Platycheirus albimanus (F.)	„ : „	
Scaeva mellina (Harris)	„ : „	
Sphaerophoria scripta (L.)	„ : „	
Syrphus vitripennis (Meig.)	„ : „	

Aphid-specific predators

Chrysopa carnea (Steph.)	(Neuroptera: Chrysopidae)	
Adalia bipunctata (L.)	(Coleoptera: Coccinellidae)	
A. 10-punctata	„ : „	
Coccinella 7-punctata (L.)	„ : „	
C. 11-punctata (L.)	„ : „	IIa
Halyzia 16-guttata (L.)	„ : „	
Myrrha 18-guttata (L.)	„ : „	
Propylae 14-punctata (L.)	„ : „	
Thea 22-punctata (L.)	„ : „	

Aphid-specific parasites

Primary		
Aphidius spp.	(Hymenoptera: Braconidae)	
Praon spp.	„ : „	
Ephedrus spp.	„ : „	
Secondary (Hyperparasites)		
Asaphes spp.	: Ceratophronidae	III
Dendrocerus spp.	„ : „	
Phaenoglyphis spp.	„ : „	
Alloyxsta spp.	„ : „	
Fungal parasites		
Entomophthora spp.	„ : „	

I **A.** Pea seedlings stunted by *Thrips angusticeps*, middle plant unattacked (Crown copyright). **B.** Microphotograph of a thrips (Crown copyright). **C.** Female delphacid, *Javesella pellucida* (courtesy Ryoti Kissimoto). **D.** Delphacid eggs (courtesy Ryoti Kissimoto). *cont.*

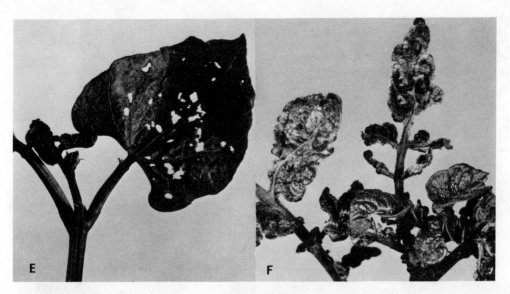

I *cont*. **E.** Bean leaf holed by spring-tails (Crown copyright).   **F.** Mirid (Capsid) damage to potato foliage (Crown copyright).

II **A.** Sugar-beet with leaves severely curled by colonies of black bean aphid on the underside (courtesy W. E. Dant). **B.** Eggs of black bean aphid about the buds of spindle tree (courtesy F. D. Cowland, Rothamsted). **C.** Immature winged black bean aphid feeding on bean stem (courtesy F. D. Cowland, Rothamsted). **D.** Leaf petiole galls caused by *Pemphigus bursarius* on poplar. The aphids are concealed within the galls (courtesy J. A. Dunn, N.V.R.S.).

III **A.** A colony of *Aphis schneideri* on *Ribes* attended by the ant *Lasius niger* and with a male syrphid, *Syrphus balteatus*, feeding on honey dew (courtesy C. J. Banks, Rothamsted).   **B.** A colony of black bean aphids, wingless viviparous females and their offspring, on a bean leaf (courtesy F. D. Cowland, Rothamsted).   **C.** Spindle tree in spring with leaves curled by colonies of the black bean aphid (courtesy W. E. Dant).

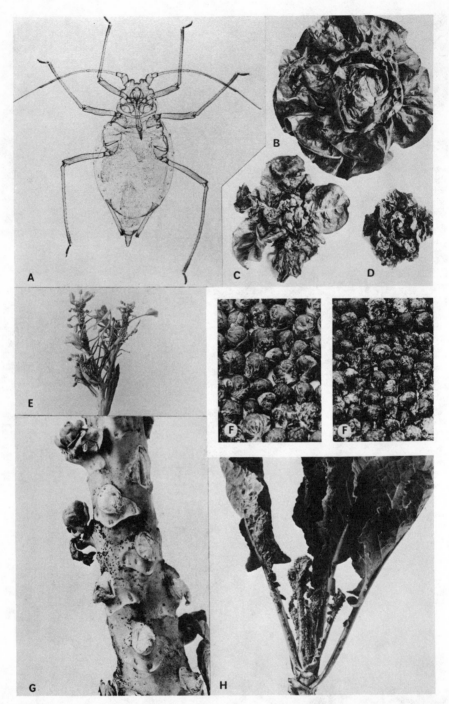

IV **A.** Photomicrograph of the peach-potato aphid, *Myzus persicae* (copyright V. Stansfield, Rothamsted). **B.** Healthy lettuce (courtesy, D. W. Wright, N.V.R.S.). **C.** Lettuce attacked by aphids (courtesy D. W. Wright, N.V.R.S.). **D.** Lettuce attacked by aphids and virus (courtesy D. W. Wright, N.V.R.S.). **E.** Spring colony of cabbage aphid in brassica flower head (courtesy W. E. Dant). **F.** Brussels sprout sprayed and unsprayed against cabbage aphid (courtesy W. E. Dant). **G.** Cabbage aphid eggs on Brusselss sprout stem (courtesy W. E. Dant). **H.** Cabbage aphid colony on cauliflower plant (courtesy W. E. Dant).

enemies that follow aphids into newly planted crops, there is a large and rather diverse resident fauna of predatory ground beetles (p. 95, Table 6.1). Many of these feed on aphids that alight on the soil or are dislodged from plants by rainfall, and it is presumed that ground beetles climb plants in search of prey, especially at night. Resident predators and early immigrants probably destroy most of the alate aphids (and their progeny) that first arrive in the crop. Probably the arrival of relatively few migrants over a protracted period during somewhat adverse weather enables resident predators to prevent the establishment of a significant number of aphid colonies. Alternatively the arrival of many aphids over a short period is more likely to lead to establishment of aphids in the crop and to set the scene for a potential epidemic.

When applying control measures it is obviously desirable not to use pesticides likely to discourage aphid enemies too drastically. Insecticides applied to the soil before or at sowing, e.g. soil incorporated HCH sprays, overall treatments with granular insecticides or nematicides may remove many ground predators. Some organophosphates applied later to the foliage, e.g. chlorpyrifos to control caterpillars are harmful to coccinellids, syrphids, lacewings and parasites. These materials improve the prospects for aphid survival and thus assist the spread of aphid-borne viruses and increase the risk of outbreaks.

Viruses transmitted by aphids

Viruses transmitted by aphids can be classified as (*a*) non-persistent, (*b*) semi-persistent, (*c*) persistent (Watson & Plumb, 1972), or as (*a*) stylet-borne and (*b*) circulative (Kennedy, Day & Eastop, 1962; Sylvester, 1980; Harris & Maramorosh, 1977). Non-persistent viruses are transmitted immediately after the aphids have fed on infected plants for a short time only and ability to infect is short lived. The particles of the virus were thought to be carried on the stylets, hence Kennedy *et al.* called them 'stylet-borne' (but there is now some evidence that particles are inside the head). Such viruses spread locally around a source of infection. Semi-persistent viruses are acquired after about ten minutes of feeding but they are transmitted more successfully after feeding longer. Their vectors remain infective for several days but not after moulting and the virus cannot be recovered from the vector's haemocoel. These characteristics, which are intermediate between those of non-persistent and persistent viruses, were not separated from them by Kennedy *et al.* Persistent viruses can be transmitted only after a latent period which lasts from several hours to a few days. They cannot be acquired from an infected plant by brief probes but only after periods of feeding, usually from the phloem, lasting 24–48 hours. Once acquired the aphids remain infective for a long time, even after moulting. Persistent viruses can be carried considerable distances. As they pass into the insect's body, Kennedy *et al.* called them circulative. Sometimes a virus which is not normally transmitted by a vector becomes transmissible when another virus is present in the same or in another plant on which the aphid has fed. This applies to persistent and non-persistent viruses. Many aphids are virus vectors, the most important species in the temperate zone being *Myzus persicae* (Sulz.) For an account of vector behaviour and virus spread see Swenson (1973) and Thresh (1982).

Aphid pests

In this section, accounts of aphids of agricultural importance and of a few other species are given. Some of the diagnostic features of each species are mentioned along with details of life history and bionomics. Because of its importance, *Myzus persicae* comes first and is followed by groups of species occurring on potatoes, on beans and peas, the cabbage aphid, carrot aphids, lettuce aphids, cereal aphids and strawberry aphid. A logical grouping based on crops is impossible because aphid host ranges sometimes overlap. Control measures on crops are at the end of the section on aphid pests.

The peach-potato aphid

Myzus persicae (Sulz.)   The peach-potato aphid (Fig. 4.7) is cosmopolitan in temperate regions and is found widely in Great Britain, Europe, the U.S.A. and elsewhere. It rarely occurs in populations sufficiently dense to cause direct injury by its feeding but is able to transmit at least a hundred viruses including potato leaf roll, virus Y of potato, mosaic, yellow-net and the yellows viruses of sugar-beet, cauliflower mosaic plum pox, cucumber mosaic, lettuce mosaic and turnip mosaic of brassicas, and is therefore one of the most serious of our crop pests.

M. persicae is green or sometimes slightly reddish, with a pair of well-marked frontal tubercles (Fig. 4.7B) which project inwards. The siphunculi are long, cylindrical, may be slightly swollen and emit an alarm pheromone (Nault *et al.*, 1973). The cauda is prominent and half or less than half the length of the siphunculi. The total length of the insect is 2–2.5 mm (Plate IV).

M. persicae is found on a wide range of summer hosts, such as potatoes, sugar-beet, turnips, swedes, spinach and mangolds. It is also found on stored potatoes and onions, in mangold clamps and in glasshouses. Weeds such as shepherd's purse, black nightshade, charlock, and fat hen are also hosts. The woody winter host is peach. In normal winters in Great Britain, the egg phase is relatively unimportant, for *M. persicae* is able to overwinter as the adult or nymphal apterous viviparous female, especially in the centres of plants (e.g. spinach, lettuce, sugar-beet, seed turnips, savoys), in mangold clamps, gardens, glasshouses and other sheltered places. The efficiency of *M. persicae* in spreading viruses is related to its habits. Its ability to overwinter in the adult and nymphal stages enables it to start rapid reproduction as soon as weather conditions are favourable, without waiting for development from an egg, and to migrate between summer host plants at an early date. Because of its wide host range it is never at a loss for suitable hosts at any time of the year. Although its powers of reproduction are very great, it rarely forms large compact colonies but shows a marked tendency to scatter and

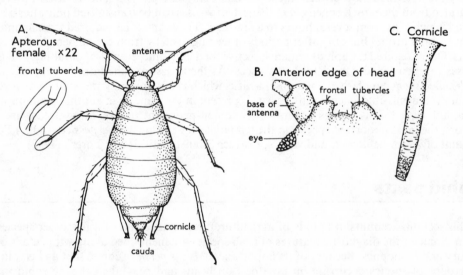

Fig. 4.7 A, apterous female of *Myzus persicae*; B, anterior edge of head; and C cornicle (siphunculus).

so to infect new plants. A winged form may alight on a plant, feed, produce two or three young and then fly off to infect more plants.

Other aphids found on potatoes

Aulacorthum solani (Kltb.) the glasshouse and potato aphid (Fig. 4.12) overwinters on many plants, is found on potatoes and is abundant on the sprouts of stored seed tubers in early spring in company with *M. persicae* and *Macrosiphum euphorbiae* (Thomas). It is green, with a dark head and thorax. The siphunculi are pale green and cylindrical and the banding on the abdomen of the alate female is very marked. The alata is about 23–25 mm in length.

Macrosiphum euphorbiae (Thomas) The potato aphid (Fig. 4.12) occurs on many host plants including *Spiraea* sp., potato, roses, tomato, chrysanthemum and lettuce and is generally distributed over Great Britain. It is especially abundant on potatoes. The asexual stage may persist throughout the winter under glass, on lettuces and on potatoes in store.

The alata is green or pink, with brownish lobes on the thorax. The siphunculi are long, green, with dark apices. The cauda is knob-like and about half the length of the siphunculi. The pink alata has a yellowish head and a yellow-pink thorax. The aptera is similarly green or pink, and is longer (4.0–4.1 mm) than the alata. The ovipara is apterous but the male is alate.

M. euphorbiae is even more restless than *M. persicae* and moves more readily from plant to plant. The nymphs are usually found on the basal leaves of potato and on the tips of the shoots, but the adults are more randomly distributed about the plant. The apterae are found more frequently on the lower surface of young, growing leaves and on the tips of the shoots than are those of *M. persicae* particularly in the cool of the morning, but less so on hot, dry afternoons, suggesting daily movement in hot weather. On days when the temperatures are less extreme, there are fewer apterae on the middle and bottom leaves than on the upper leaves, a distribution differing from that of *M. persicae*. Apterae are easily washed off by rain and are occasionally blown off by wind. Fresh colonies are formed on new host plants by both alatae and apterae, and the rate of reproduction is very great. Prolonged infestations of potato leaves induce top curl (sometimes confused with symptoms of protection with potato leaf roll virus) and decrease the yield of tubers.

Aphis nasturtii (Kltb.) (Fig. 4.12) The buckthorn-potato aphids overwinter as eggs on buckthorn (*Rhamnus catharticus* (L.) and *Frangula alnus* (Mill.)). These are usually laid towards the bases of the bush. It migrates to potatoes in June–July and in some years builds up very large populations. The progeny of migrants tends to remain in colonies on the leaf on which they were produced, moving to new foliage only when overcrowding occurs. *A. nasturtii* is a vector of viruses, and appears a more efficient vector of virus Y than of leaf roll. However, except in epidemic years *A. nasturtii* is not an important carrier of viruses; districts where it occurs abundantly grow seed potatoes successfully. During epidemic years, much of the injury to the potato is due to large smothering colonies.

The head and thoracic lobes of alatae are black, although the pronotum is green. There are black lateral abdominal spots, and the siphunculi and cauda are black. The cauda is acuminate and not as long as the siphunculi. The aptera is bright green or yellowish-green and 1.4–1.7 mm long. The siphunculi and cauda are pale green, although the siphunculi and femora may have dusky apices. The cauda is rather thick and blunt. The males are alate but the oviparae are apterous, with swollen hind tibiae covered with secondary sensoria. On buckthorn this species curls the leaves and the tips of the shoots.

Rhopalosiphoninus latysiphon (Davids.) The bulb and potato aphid is a dark green aphid occurring in potato stores but has been found on potato stolons in one or two areas. It does not normally infest the foliage because it does not tolerate high air temperatures (Gibson, 1971). Its position as a virus vector is obscure but some virus transmission might occur below ground. If so, the rate of transmission would be slow compared with overground transmission. Both apterae and alatae have characteristically swollen siphunculi (Fig. 4.12).

Myzus ornatus **(Laing)**   The violet aphid is a small yellowish or whitish green aphid with black spots. It overwinters as adults or nymphs in sheltered places. Numbers on potatoes are few from May to July but large numbers may build up on senescing or damaged haulms in August and September. It is found mostly on the upper surfaces of leaves where it produces few alatae.

Sugar-beet aphids

The most important aphids found on sugar beet are *Myzus persicae* (p. 61) and *Aphis fabae* (see below).

Pea and bean aphids

Aphis fabae **(Scop.)**   The black bean aphid (Figs 4.8, 4.9, 4.10) was confused for many years with a whole complex of closely related black species under the name of *Aphis rumicis* (L.). The true or permanent dock aphid, now called *A. rumicis*, a one-host aphid occurring only on docks, is one of the easiest to separate from the complex (Jones, 1942) (Plates II and III).

A. fabae has a heterocyclic life history typical of many aphids. Eggs are found on a woody winter host plant, the spindle tree, *Euonymus europaeus*, and also the sterile Guelder rose, *Viburnum opulus* var. *roseum*. Colonies of viviparous females are found on many summer host plants such as beans (*Vicia faba*), beet, docks, thistles and poppies, but the black aphids on these plants are not necessarily *A. fabae*. Another related black aphid, the Japanese spindle aphid, which lives wholly on evergreen *Euonymus japonicus* and lacks a sexual phase, does not transfer to beans or sugar-beet.

The life cycle of *A. fabae* is shown in Fig. 4.5 (p. 42). The shining black eggs are found in the axils of buds and in the crevices in the twigs of the spindle tree (Plate II). They hatch in April, or earlier in mild springs, into apterous females, known as fundatrices. These fundatrices differ slightly from the later apterae, being somewhat smaller with 5-jointed antennae. They mature in about three weeks, depending on the temperature, and then produce young parthenogenet-ically and viviparously. The first generation produced by the fundatrices are apterous, but the second generation contains some winged or alate forms, the spring migrants which fly away from the spindle tree to found summer colonies on one of the summer host plants. Here apterae are first produced, but in the succeeding generations alatae appear, and in turn migrate to new host plants. The original colony on the spindle tree drags on during May and June, becoming smaller, and increasingly parasitized and preyed upon by ladybirds and spiders, and eventually disappears. The colonies then continue on the herbaceous plants and the woody plants cease to be attractive to the migrating aphids (Kennedy & Booth, 1951). Generally the young growing points or the senescent leaves of the summer host plants attract more aphids than the mature leaves.

If the primary infestations on summer hosts are numerous and the weather is favourable, the aphid population begins to take on epidemic proportions. Summer host plants are increas-ingly colonized, beans and sugar-beet (seed and root crops) suffer badly and the epidemic spills over on to the less favoured 'reserve hosts' like potatoes, peas and even apple twigs. Sometimes epidemics are local but sometimes widespread and may involve all the countries bordering on the North Sea. By mid-July or later in the north and west, the epidemic reaches its peak. During early August, the aphid population suffers a catastrophic decline and is reduced almost to extinction by late August, when colonies may become very difficult to find.

If, on the other hand, the primary infestation is light and the weather adverse, that is, wet and rather cold, there is no epidemic. Odd plants become infested throughout the summer,

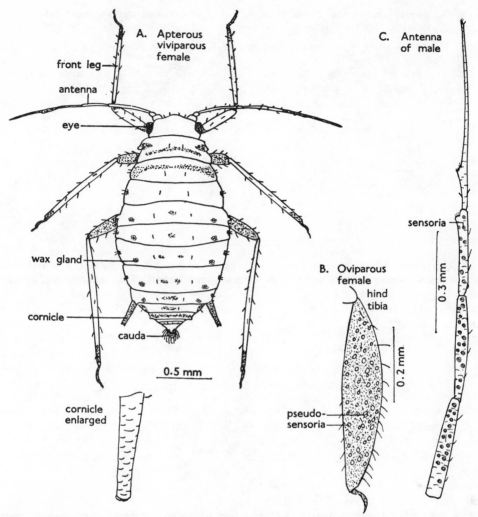

Fig. 4.8 *Aphis fabae*. A, apterous viviparous female (dorsal view) with cornicle (siphunculus) enlarged; B, flattened hind tibia of oviparous female to show the plate-like sensoria; C, antenna of alate male to show large numbers of secondary sensoria.

and by late August colonies the size of a 10-pence piece can be found on the lower surfaces of beet leaves. These colonies, although not large, e.g. 10 per plant in a field of 65 000 plants per ha equals 650 000 aphids per ha, are vastly bigger than the population of aphids at the same level at the end of an epidemic year.

In the autumn, a change occurs in the colonies on the summer host plants. Return migrants or gynoparae are produced. These are alatae, slightly bigger than the summer alatae with a large number of secondary, plate-like sensoria on the antennae, which return to the winter host plant, chiefly spindle tree, the senescent leaves of which are now attractive, and there give rise to apterous oviparae. These are distinguishable from other apterous forms by the thickened hind tibiae, bearing plate-like sensoria (Fig. 4.8B). At the same time as the production of gynoparae, winged males are found in the colonies on the summer host plants. These males also fly to the winter host and mate with the oviparous females. As a result, fertile winter eggs

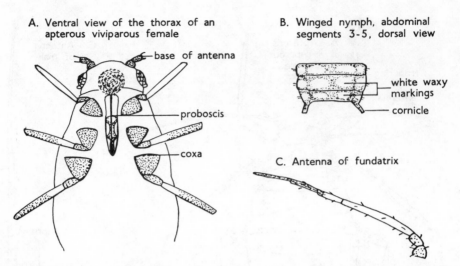

A. Ventral view of the thorax of an apterous viviparous female

base of antenna

proboscis

coxa

B. Winged nymph, abdominal segments 3-5, dorsal view

white waxy markings

cornicle

C. Antenna of fundatrix

Fig. 4.9 *Aphis fabae*. A, ventral surface to show the proboscis and bases of the legs; B, dorsal view of 3–5 abdominal segments of alate nymph to show white waxy markings; C, 5-jointed antenna of fundatrix.

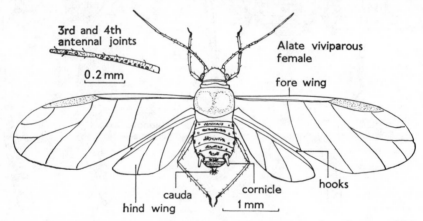

3rd and 4th antennal joints

0.2 mm

Alate viviparous female

fore wing

hooks

cornicle

cauda

hind wing

1 mm

Fig. 4.10 *Aphis fabae*. Dorsal surface of alate viviparous female with enlarged antennal joints to show the sensoria.

are laid by the oviparae in sheltered positions on the twig. At first they are dark green but soon become black and shiny. They remain dormant until the spring.

Using the knowledge that the spindle tree is the chief winter host, attempts were made in the 1940s and again more recently by Cammell *et al.* (1977, 1978) to forecast outbreaks by estimating the number of winter eggs. Epidemics tend to occur in alternate years, because following an epidemic year there is a slight deposit of eggs on spindle trees (Fig. 4.11) (Way, 1967, Jones & Dunning, 1972). During the following year, e.g. 1943 after 1942, the aphids slowly recover and give rise to a heavy deposit of eggs in the following autumn. Epidemics depend very much on the weather. Figure 4.10 shows the relationship between the abundance of eggs and the primary infestation from 1941 to 1948. There was an epidemic threat in 1946 and 1947 but it was suppressed in 1946 by bad weather, and in 1947 by 7–8 days of bad weather in early July. The great epidemic years are usually drought years, or years in which the weather has prevented the early sowing of crops. Fine weather in spring and autumn favours the migration away from and back to the winter host plants.

There is a close interrelationship between spindle tree and aphid life history. Bud burst and egg hatching are synchronized and leaf fall does not occur until return migration has taken place and eggs laid. The more exotic species of spindle trees introduced into our gardens usually do not fit in with the aphid life cycle; bud burst is late and leaf fall early.

Control of *A. fabae* by eradication of the winter host can be considered possible, for here is an obligate, well-defined host, the spindle tree. Eradication might not be effective, as the aphid might evolve other methods of overwintering or become adapted to new winter hosts, although considering the heavy mortality that normally occurs each autumn during the return to spindle trees, this seems unlikely. There is, however, the practical difficulty of locating and removing all the spindle trees in the country. Suckering from roots left behind in the hedgerows would occur. There is also the possibility of migration from the Continent, the alatae being able to replenish the depleted native stocks, particularly in the south-eastern areas of England. (See leaflet No. 54 and Cammel, 1981.)

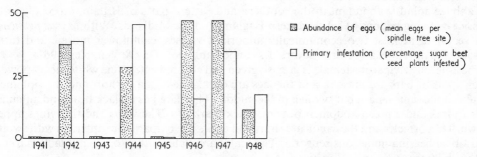

Fig. 4.11 Diagram to show the relationship between the abundance of eggs and the primary infestation on sugar-beet seed plants between 1941 and 1948 before the widespread use of aphicides.

Acyrthosiphon pisum (**Harris**) (Fig. 4.12)   The pea aphid is large and green with long antennae, legs, cauda and siphunculi. The alate female is about 2.5–3 mm long with deep red or black eyes, and yellowish green antennae. The apterous females are similar with a smooth, shiny or even glaucous cuticle. The oviparous female is apterous and has swollen hind tibiae, while the alate male has brownish markings on the dorsal surface of the head and thorax, and dusky markings on the abdomen. The markings on the apterous male are less well defined but the antennae of both types of males bear many secondary sensoria.

Pea aphid is found on many leguminous plants, especially cultivated and ornamental species of *Pisum*, clovers, vetches and lucerne. It forms large smothering colonies which cause direct injury and, in bad years, may overwhelm the growing points of pea plants and stunt growth. Later infestation affects flowering and podding. The aphid is also a vector of lucerne mosaic, pea leaf roll, pea enation mosaic and pea mosaic virus in Great Britain, and of pea enation mosaic virus in the U.S.A. (McEwen, Schroeder & Davis, 1957; Heathcote & Gibbs, 1962; Cockbain, 1971; Cockbain & Gibbs, 1973).

A. pisum has no woody winter host. Eggs are laid on standing forage crops such as sainfoin, trefoil and lucerne. They hatch in early spring, give rise to colonies from which alatae migrate in late spring and summer, and in favourable weather large colonies are formed. By the end of July, there are few aphids left on peas or beans because the crops have matured or been harvested. Sexual forms appear later in the season from aphids living on forage crops, especially lucerne. Eggs are laid but the aphid may pass through very mild winters asexually.

In North America the life cycle varies with latitude. In Canada, winter is passed in the egg, in Central North America, eggs and asexual forms occur, but in the south only asexual forms are found.

In England attempts have been made to correlate epidemics with the preceding winter egg-population without success. There is probably much predation of eggs, especially in mild winters.

Cabbage aphid

Brevicoryne brassicae **(L.)** (Fig. 4.12)   The cabbage aphid is a serious pest of cabbage, cauliflower and Brussels sprout. It feeds on the leaves, stems and flowers, and forms large, compact colonies. It also colonizes radish, swedes and mustard. Slight infestations occur on kale and rape, but turnips appear immune. Colonies are also found on cruciferous weeds such as charlock and shepherd's purse. A heavy attack on young plants may check growth beyond recovery. Attacks on larger plants spoil the crop by contaminating them with wax, cast skins and with honey dew which they excrete, and which provides a good medium for growth of fungi. Aphids sometimes penetrate into the hearts of cabbage, Brussels sprout and cauliflower, so lowering their market value (Plate IV).

Cabbage aphid is abundant in the southern and eastern parts of Britain, especially in the brassica-growing regions of south-eastern England and the Midlands. With *Myzus persicae* it is a vector of at least two economically important viruses, cauliflower mosaic and turnip mosaic, and probably a vector of several less important ones (Day & Venables, 1961).

The apterous viviparous female is greyish green and covered with a fine white mealy powder. Black spots or bars are present, and the legs are dark. The cauda is short and the siphunculi are about the same length but swollen in the middle. The alata has a black head and antennae, dark thorax and a green abdomen bearing black markings. The veins and the wings appear brown. The cornicles and the cauda are short and dark. The oviparae are apterous with swollen hind tibiae bearing numerous sensoria. The males are alate, with prominent black markings on the abdomen.

B. brassicae is a 'one host aphid' spending all its life on various species of brassicas and other Cruciferae and has no woody winter host. The egg phase is passed on the stalks near leaf scars and on lower leaves of such brassicas as kale and Brussels sprout which stand throughout the winter, the black eggs often occurring in masses. The eggs hatch in late February or March and the colonies build up, appearing in the flower heads of seed crops or neglected crops that have been left until they flower. In mid-May alatae appear and fly to new hosts. They are produced periodically throughout the summer and, in years of great abundance, can be so numerous that when favourable flight conditions occur, the air is thick with them.

In the autumn the sexuales appear and the eggs are laid over a long period with a peak in October. During mild winters asexual forms survive especially in the south-west.

Many crops such as Brussels sprout and kale are cleared during the winter, and the soil ploughed before the eggs hatch. Plants saved for seed, or neglected farm and garden crops, therefore, are mainly responsible for bringing *B. brassicae* safely through the winter. For this reason it is inadvisable to grow brassicas close to overwintering seed crops unless the seed crop is kept clean by spraying.

Climatic conditions affect the build up of *B. brassicae* and, during dry seasons, cabbage aphid is abundant, whereas dull, cool years do not favour outbreaks. Parasites and predators help to keep the colonies in check, the chief predators being hover-fly larvae. (See Leaflet no. 269.)

Cereal aphids

Sitobion avenae **(F.)** (Fig. 4.13)   The grain aphid is a red-eyed, green to reddish-brown species with black siphunculi and a long pale green or yellow cauda. The aptera is 2.4–2.8 mm long and has long antennae. The eggs are laid in the autumn on grasses or self-sown wheat, and the alatae fly to the crops in June. Colonies build up and, in August, alate males appear on the cereals and migrate to wild grasses, where they fertilize the oviparous females.

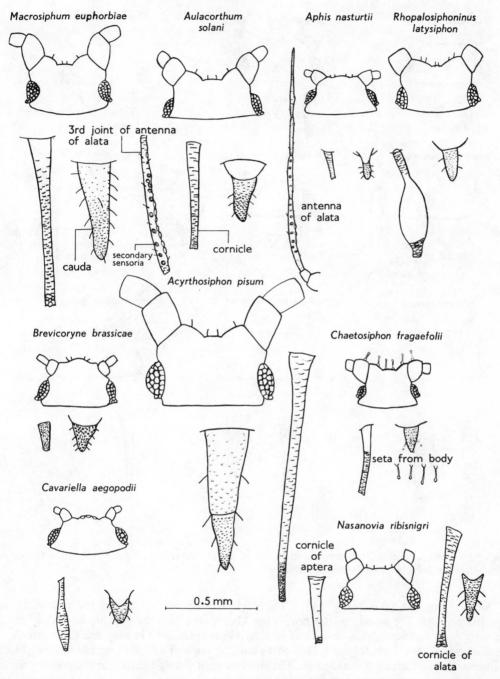

Fig. 4.12 Head, cornicles (siphunculi) and cauda of aphids. The third antennal joint of the alata of *Macrosiphum euphorbiae* and whole antenna of alata of *Aphis nasturtii* to show the difference in size between these aphids.

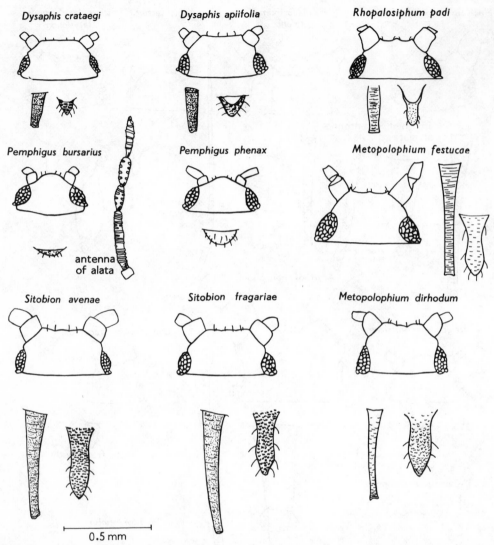

Dysaphis crataegi

Dysaphis apiifolia

Rhopalosiphum padi

Pemphigus bursarius

antenna
of alata

Pemphigus phenax

Metopolophium festucae

Sitobion avenae

Sitobion fragariae

Metopolophium dirhodum

0.5 mm

Fig. 4.13 Head, cornicles (siphunculi) and cauda of aphids. Antenna of alate viviparous female of *Pemphigus bursarius* to show the annular sensoria.

Sitobion fragariae **(Wlk.)** (Fig. 4.13) The blackberry-cereal aphid is widely distributed over the British Isles. The woody winter host is blackberry on which the eggs are laid. They hatch in early April and fundatrices cause leaf curl. Alatae are produced in May and fly to cereals or grasses, where they form colonies. The aptera is dull green and 2.5 mm long and has long black siphunculi. The alatae are dark green. The siphunculi of young nymphs are unpigmented. In some years, large damaging colonies infect wheat ears.

Rhopalosiphum padi **(L.)** (Fig. 4.13) The bird cherry aphid is found on cereals and grasses in the summer and on *Prunus cerasus* and *P. padus* in the winter. It is greenish brown and 1–2 mm long, with dark markings on the abdomen with short stubby siphunculi. The summer form has longer antennae and more vasiform siphunculi than *S. fragariae*. The alata has a dark brown thorax, a green abdomen with black lateral and two posteriorly placed black bars.

There are dark rusty patches at the bases of the siphunculi. The short cauda is about half the length of the siphunculi. Aphid colonies curl the leaves of the winter host. *R. padi* is the main vector of barley yellow dwarf virus which has increased greatly in importance now that winter cereals are sown early (e.g. mid-September).

Metopolophium dirhodum (Walk.)   The alata of the rose-grain aphid (Fig. 4.13) is about 2 mm long, light green with a darker, green stripe down the centre of the back. The siphunculi are long, slender and pale. The aptera is a little smaller. This species is found on cereals, sheep's fescue grass (*Festuca ovina*), and other grasses but does not infest the ears. Serious injury may occur to stands of re-seeded grass in the years when the apterae overwinter in abundance.

Metopolophium festucae (Theob.) (Fig. 4.13)   Although usually found on grasses on which it overwinters, there being no other winter host, the grass aphid can damage cereals if sufficiently numerous. The body is yellow or pinkish green with green, cylindrical siphunculi each with a slight depression underneath the expanded tip.

For a review of cereal aphids in Europe see Vickerman & Wratten (1979).

Carrot aphids

Cavariella aegopodii (Scop.) (Fig. 4.12)   The willow-carrot aphid is small, green and inconspicuous. The alate viviparous female is 2 mm long, has relatively short antennae, a dusky cauda and a supracaudal process. The clavate siphunculi are dark in colour and about twice as long as the cauda. The apterous viviparous female is similar in structure. The oviparae are apterous and have swollen posterior tibiae bearing secondary sensoria. The males are alate.

The eggs are laid on willow, chiefly *Salix fragilis* and *S. alba*. Occasionally apterae overwinter on carrot seed crops, which suffer severely. They also overwinter on carrot root crops left in the ground: such crops provide an abundant source of viruliferous migrants after mild winters. The eggs hatch in the spring, in phase with bud burst of the winter host, and the winged alatae appear early and migrate to carrots. Alatae from aphids overwintering on carrots appear about three weeks before those from willows and many are viruliferous. Peak migration occurs in late May and early June, and, by the first week in July, migration usually has stopped. The colonies build up on the leaves, and many aphids can be found in the third and fourth weeks of June (Wright & Kirkley, 1963). Affected carrot foliage becomes shiny and sticky with honey dew, and the soil underneath the infested plants is covered with cast skins. Besides carrots, summer hosts include parsnips, parsley, celery and a range of wild Umbelliferae. Ladybird larvae and adults, hover fly larvae and parasites help to clear infested plants by mid-August.

The willow-carrot aphid is by far the most important aphid on carrots. Large aphid colonies cause twisting and malformation of the foliage, reduction in the size of the root and can stunt and kill the plants. The aphid is also the vector of the two component viruses that cause carrot motley dwarf disease, symptoms of which are yellowing and reddening of the foliage accompanied by severe stunting (Plate V). The viruses concerned are carrot mottle and carrot red leaf, but the mottle virus is not transmitted by the carrot-willow aphid except when the red leaf virus is also present. Aphids appear to pick up both viruses from wild umbelliferous host plant *en route* from willow, the woody winter host, to carrot crops mainly during July.

Dysaphis tulipae (B.d.F.)   The tulip bulb aphid is a pest of stored bulbs such as tulips and gladioli. Apterous viviparous aphids are fawn while the alatae have darker brown markings on the head, thorax and abdomen. The siphunculi are short and cylindrical and the cauda is also short. Males are unknown. *Dysaphis cragaegi* (Kltb.), the hawthorn-carrot aphid (Fig. 4.13), is very similar to *D. tulipae* but attacks carrot leaf bases and tap roots, causing them to crack. *D. apiifolia* (Theob.), the hawthorn-parsley aphid, attacks celery. *Pemphigus phenax* (Börner and Blunck), the carrot root aphid, attacks carrots. The winter host is poplar.

Lettuce aphids

Nasonovia ribisnigri **(Kltb.)** (Fig. 4.12)   The currant-lettuce aphid occurs in company with *Myzus persicae, Hyperomyzus lactucae* (L.) the currant-sowthistle aphid and *H. pallidus* the gooseberry-sow thistle aphid on the leaves and in the hearts. In a dry spring, large numbers of aphids on lettuces lessen their market value. Later on in the year, attacks are not so important. Both species occur on lettuces under glass.

The eggs are laid on *Ribes* spp., chiefly currants and gooseberries. The alatae migrate in May and June to Compositae such as lettuce, endives, sow thistle and species of *Crepis*. The apterae produced on lettuce are yellow to light green with dark lateral spots on the abdomen and are 2–2.5 mm long. The head and thorax are dark. The long siphunculi are yellow with dusky apices and the cauda is yellow. There are dark markings on the legs. The third antennal joint bears a few secondary sensoria. The body of the alate female is much darker and on the third–fifth antennal joints are about sixty secondary sensoria. Alate males and apterous oviparae are found from August to November and eggs are laid anywhere on the shoots of currants or gooseberries. In the spring, the first generations cause curling of currant leaves, but the damage to lettuces is more important economically. Apterae can be found overwintering on lettuces in company with *M. persicae* and *Macrosiphum euphorbiae*, and so infestations can build up early without the necessity of the winter hosts.

Pemphigus bursarius **(L.)** (Fig. 4.13)   The lettuce root or poplar-lettuce aphid attacks the roots of lettuces and causes plants to wilt and dry up. The woody winter host is *Populus nigra*, and its varieties, preferably Lombardy poplar (Plate II & Plate V).

The eggs, which are laid in fissures in poplar bark, hatch when the buds break. The emerging fundatrices form closed galls on the leaf petioles, mature within and produce 100–250 young. These are all potential alatae, and when adult escape through one or more small holes along the original line of closure, migrate to lettuce or to sow thistles and there produce apterae which migrate to the roots. Wax is produced and gives the aphids a mealy appearance and covers the roots and surrounding soil. The apterae are yellowish white or pale yellow in colour, with short dark antennae and legs. The siphunculi are absent, and the cauda short. Peak numbers occur about mid-August. Alatae produced in August have dark brown heads and thoraxes, and abdomens greener than those of the apterae. There are elongate, secondary sensoria on the short antennae, and the short proboscis reaches only to the base of the second coxae. These alatae are all sexuparae which fly back to poplar. There they produce the sexuales, which do not feed and so remain very small. Each ovipara produces only one egg, which is deposited in a well-protected crevice. When conditions are favourable apterae overwinter on the roots of lettuce and allied plants. They may also remain in the soil for several months during the winter, so quickly infesting new lettuce crops if rotations are not practised. The butterhead lettuce, Avondefiance, the crisphead lettuce Avoncrisp and the leaf lettuce Salad Bowl are highly resistant and remain uncolonized where root aphid is a severe pest (Plate VA).

Strawberry aphids

Chaetosiphon fragaefolii **(Cock.)** (Fig. 4.12)   Strawberry aphid is the most important species found on strawberries. Feeding by nymphs, apterae and alatae cause the leaves to twist, but more important is its part in the transmission of 'yellow edge' and crinkle virus complexes of strawberry, found everywhere in strawberry-growing districts. Because strawberry is perennial and propagated from runners, it is essential to use virus-free clones, grown in isolation, for raising new plants. (See Leaflet no. 530.)

The aphid is primrose yellow and has reddish-brown eyes. On the body and appendages are characteristic clubbed hairs (Fig. 4.12). The apterae can be found on strawberry plants all the year, but peak numbers occur in the summer. Apterae are found on the lower surface of the young leaves and move to new leaves as those they are feeding on mature. Numbers drop in the autumn. Alatae can be found in June and also in late summer. Autumn populations are highest on strawberries planted during the previous autumn or spring, and aphids overwinter on the lower surfaces of the leaves. No egg stage has been found.

Control measures against aphids

Natural control

Although the wrong use of modern insecticides has an adverse effect on predators and parasites of aphids and especially on ground predators (pp. 95, 167), it cannot be said that natural enemies were conspicuously successful before they were used. Indeed, before the 1950s the farmer was almost helpless once aphid outbreaks got underway whereas now he has at his disposal and at a price an array of aphicides to prevent or curtail outbreaks and to lessen the spread of viruliferous aphids within crops. Within limits this is possible without discouraging natural enemies and pollinating insects too much. In any case, not all crops are sprayed and there is a great reserve of aphid enemies in the countryside generally, although crops in which they are depleted for the time being may suffer from the effects of aphids unnecessarily.

Potato aphids

Myzus persicae is found widely on potatoes, but other species such as *Aulacorthum solani*, *Macrosiphum euphorbiae*, *Aphis nasturtii*, *Rhopalosiphoninus latysiphon* and the violet aphid (*Brachycaudus helichrysi* (Kltl.)) may be present also. *M. persicae* is the most important vector of potato viruses.

Before anything was known about aphid transmitted viruses, the value of obtaining fresh seed potatoes, preferably from hill land in the north and west of Great Britain, was fully appreciated. Seed potatoes from the south and east soon lost their vigour and were useless. This loss of vigour is due to virus infection. About twenty-five distinct potato viruses have been described, but not all occur in the U.K. (See Leaflets nos 139, 278 and 575.) Important aphid-borne viruses are leaf roll and virus Y. Virus X, mop top and tobacco rattle are transmitted by other means. The symptoms caused by any one virus vary in different cultivars and with the age of the infection (R. A. Jones, Fribourg & Slack, 1982). Whereas virus Y can be transmitted after an aphid has fed for a short time on an infected plant and retains infectivity for about 12 hours, a feed of about 6 hours is necessary for leaf roll virus and to transmit this virus the aphid must feed on a healthy plant for at least 2 hours and remains infective for the rest of its life. So plants can be protected by insecticides from aphids that spread leaf roll but not so easily against those that transmit virus Y. Leaf curling which may be confused with the symptoms of leaf roll virus, is caused by the feeding of the aphid *Macrosiphum euphorbiae* (p. 49), by the diseases caused by the bacterium *Erwinia carotovorum* and the fungus *Rhizoctonia*, and by the potato shoot borer *Hydraecia micacea* (p. 90). Even moderate infestations of aphids cause loss of crop especially when peak numbers are reached while tubers are forming in late July and August. In eastern England trials over several years showed that more than $2\frac{1}{2}$ tonnes a hectare of ware were lost and the yield lost was greater in better crops. King Edward suffered more than Majestic. It was also observed that spraying with copper and tin-based fungicides decreased injury possibly by deterring migrants.

To protect ware crops in eastern England where aphids are numerous, from direct damage by aphids, a single routine spray in June or early July is justified in most years. In the south and west routine spraying rarely pays. There it is cheaper to spray when aphids arrive and, when possible, to combine an aphicide with a fungicide to control potato blight. The threshold for spraying is three to five aphids per true leaf based on a sample taken from equal numbers of upper, middle and lower leaves. When granules (aldicarb, oxamyl, carbofuran) are applied at planting to control potato cyst-nematodes, aphids that migrate into the crop are killed up to the end of June. Unfortunately, granules are less effective in drought years when aphids are most likely to be numerous. The most commonly used aphicide is demeton-S-methyl but *Myzus persicae* is resistant to it in some areas. There pirimicarb can be substituted but in a few localities aphids are not killed by it either. Whether pyrethroid insecticides can be substituted and will retain their potency has still to be discovered.

Seed potato production

The lowlands of Scotland, and the east and south of England, with their relatively low rainfall and high summer temperatures, are ideal for aphid multiplication. Here fresh seed rapidly degenerates and is usually worthless if grown twice, as virus infection of the seed potato stocks causes serious losses of yield. Deterioration of potatoes has forced seed production into those districts, i.e. the north and west of Britain, where climatic conditions do not favour aphid migration, multiplication or overwintering. Although these areas are not aphid free, heavy rainfall and low temperatures enable sound seed to be raised. Even so, crops intended to be used as seed must be carefully rogued to lessen the small percentage of infected plants which are always present. Good seed production is to a considerable degree dependent on the skill of the seed grower. Seed grown in Britain consists 50% of Scotch, Irish or Dutch seed and 50% of once-grown seed mostly raised in England.

The characteristics of the good seed-producing areas are low summer temperatures, high relative humidity and exposure to wind. Aphids are affected by variations in climate, but they can be found in flight over a considerable climatic range. The climatic conditions around the aphid colonies on the plant, i.e. the microclimate and the physiological state of the aphids, probably decide whether flight shall begin, whereas macroclimate decides whether it shall be prolonged and extensive. Probably the density of the winged aphid population in any particular area is the most important factor of all, causing some migration to take place even when the macroclimate is unfavourable.

Apart from the purely climatic factors which affect aphid reproduction and migration, the best seed growing areas are: (1) away from large towns which provide abundant shelter and host plants for aphids and are, therefore, a source of virus and of vectors early in the season; (2) in districts where there are few crops harbouring overwintering aphids, e.g. brassicas, seed crops of sugar beet or mangolds; and (3) where aphids infest crops late in the year (mid-July to mid-August).

In trying to find seed-producing areas in Wales in the 1930s the '100 leaf index' was devised to distinguish between good and bad seed-growing areas. In successful areas there were less than 20 aphids per 100 leaves in July, whereas in unsuccessful areas there were more than 100 aphids per 100 leaves in July. An estimate of the number of aphids per plant is a better index of infestation. The aphids present on a unit, consisting of one upper, one middle and one lower leaf, taken at random from 33 plants is counted and the average number of leaves per plant calculated. Upper, middle and lower leaves occurr in the ratios 2:1:1 so the total number of leaves divided by four gives the number of complete units per plant. Thus the average number of aphids per plant is

$$\frac{\text{Total number of aphids}}{33} \times \frac{\text{Average number of leaves per plant}}{4}$$

This method was modified by Hollings (1960). However good the area chosen and trouble taken, seed crops always contain a small percentage of virus-infected tubers which even the most vigorous roguing fails to remove.

The types and spread of virus diseases may be classed as either extrinsic, due to infective aphids migrating into the crop, or intrinsic, caused by movements of aphids within the crop which occurs chiefly in June or July. The most important source of virus infection is usually that within the crop itself. The seed producing areas in Scotland are ones where aphids arrive late in the season, just before the haulm is burnt off, so any viruses transmitted have little time to pass down to the tubers.

Obviously seed potatoes should be true to type with as small a percentage of virus-infected tubers as possible. For seed production a better grade is required than for ware production (for grades see Leaflet no. 139). Seed in store or being chitted must also be protected by keeping the eyes and sprouts free of aphids. This is done by fumigating with nicotine smoke or (if the temperature is below 16°C) with HCH smoke.

To produce 'once-grown' seed in a non-seed producing area, the field chosen should be isolated from ware potato crops and from sources of overwintering aphids (gardens, market gardens, overwintered brassica crops, winter rape). Proximity to the previous year's potato fields should be avoided because of the numerous virus-infected groundkeepers that survive. A granular aphicide should be applied in the furrow at planting (e.g. disulphoton, phorate) or aldicarb granules should be broadcast and mixed into the top soil before planting. One or more sprays of demeton-S-methyl or preferably pirimicarb should be applied as necessary to keep the haulm as free as possible from aphids until it is burnt off (see Leaflet no. 575).

Sugar-beet aphids

The two species most commonly found on sugar-beet are *Myzus persicae* and *Aphis fabae*. Both are vectors of beet viruses but *M. persicae* is by far the most important. Yellows viruses occur widely in Europe and the U.S.A. In Britain years of abundance occur irregularly followed by periods of relative scarcity. On the Continent, severe infections occur somewhat more often. Beet yellows symptoms are caused by several viruses. The beet yellows virus (BYV) causes most yield loss but now rarely occurs in Britain. Beet mild yellows virus (BMYV) is common in some years but much less harmful. BYV is semi-persistent and BMYV fully persistent (p. 47). The former causes vein etching whereas the latter does not. Field symptoms are similar; in June it takes 10 to 14 days at least for symptoms of BYV to appear after infection and more than 30 days for BMYV. In older plants, later in the year or in dull weather it takes longer. Yellows symptoms are most conspicuous on the tips of middle-aged leaves but may affect all the outer leaves, only the heart leaves remaining green. Infected leaves become thickened and brittle and crackle when crushed. The yellowed areas of older leaves turn brown and necrotic due to attacks by weakly parasitic fungi and finally collapse. Symptoms vary somewhat from plant to plant and may be confused with yellowing caused by capsids, downy mildew and mineral deficiency diseases. (See Leaflet no. 323.)

Beet western yellows virus is prevalent in sugar-beet in the U.S.A. but in Europe it is found only in lettuce and a few other crops including some cultivars of rape.

Mosaic virus is non-persistent and gives the leaves a mottle of dark and light green. When the leaves are held up to the light, small light green rings on a darker background are visible, or dark green bands along the veins with lighter areas between. The symptoms are most pronounced in the heart leaves and usually fade as the leaves grow older. Infected plants are not noticeably stunted or deformed. The virus occurs mostly in the vicinity of seed crops.

Viruses lower the vitality and the yield of sugar-beet and decrease the sugar content of the root. The earlier the infection the greater the reduction of the potential sugar yield. A 100% infection in July leads to a 50% reduction in sugar, while 100% infection in mid-September leads to a 15–20% reduction in sugar. In addition to the loss of yield for the farmer, virus diseases interfere with the extraction processes in the factory by decreasing the percentage of sugar and increasing the 'noxious nitrogen' present. The common beet yellows viruses are not seed borne, so that the sugar-beet crop starts free from infection each year and the viruses must be introduced into the crop from outside sources. As yet, there are no known resistant varieties of beet, but tolerance and resistance are being sought by plant breeders. For many reasons, the most promising methods of controlling the disease are those directed towards the elimination of overwintering sources of virus, and especially those which are infested with aphids in early spring.

The seed stecklings of sugar-beet, mangold, fodder beet and red beet sown in July are often heavily colonized by aphids and thoroughly infected with virus before the end of the season when aphid flight ceases. Thus, even if all the aphids die during the winter, the stecklings are a reservoir of viruses when re-colonized in the spring.

Weeds, groundkeepers, fragments of roots at cleaner-loader sites and plants in gardens such as spinach, spinach beet and derelict red beet may also provide sources of aphids and virus early in the season. Weather in late winter and early spring has an important effect on overwintering aphids. The number of asexual forms available in the important period April–May may be related to the number of days with frost between January and March, except when December is unusually cold as in 1981.

Measures to lessen losses in sugar-beet

Root crops

Crops sown early in well cultivated and correctly fertilized soils grow vigorously. By the time aphids begin to arrive in late May and early June, these crops are becoming increasingly resistant to aphid colonization and virus infection. Factory fieldmen inspect beet crops each day for aphids from May to July and, when the infestation reaches a critical level, fewer than one aphid to every four plants in the south and one per plant in the north, a warning is sent to growers who should then spray the crop with one of the recommended insecticides. As there is considerable local variation in infection growers should also inspect their own crops and act, if necessary, before they get their warning card.

Seed crops

All sugar-beet seed crops should be in isolated areas away from root crops which are mainly east of the Great North Road. They should be sown in April under barley cover crops or not before late July in open ground. All beds should be sprayed in autumn with pirimicarb. 'Stecklings', i.e. transplants, from open ground should be planted into furrows to which aldicarb or thiofanox granules have been applied. This is because they are liable to contain a higher proportion of virus-infected plants than stecklings raised under cover. A voluntary scheme exists for the certification of sugar-beet steckling; the beds are inspected to assess the proportion of plants showing yellows. Those beds with less than 1% infected plants are certified as satisfactory for planting; those with more than 10% are condemned, and those between 1% and 10% are used only if necessary to complete the acreage.

Chemical control of aphids in sugar-beet

The earliest successful chemical control of *A. fabae* on sugar-beet was with nicotine, used as dust (4%) or vapour, and liberated under a drag sheet drawn behind a vehicle with low (beet crop) or high (beet seed) clearance above the crop. Three pounds per acre was used to safeguard beet seed crops in the early part of the 1939–1945 war. The rate of movement and the dosage of nicotine was adjusted to give satisfactory kill and action was selective, killing the aphids and leaving the parasites and predators. The main drawbacks of this treatment were the cost and the cumbersome, specialized equipment necessary for application. These methods have been entirely replaced. A number of organophosphate and carbamate insecticides can be used for spraying aphids on the root crops, steckling beds and seed crops. One or two applications of demeton-S-methyl, or preferably pirimicarb spray, if properly timed, give effective control of aphids and lessens yield losses from yellow viruses. The more mature the crop, the greater the persistent effect of the insecticides. With small, young plants, the effect disappears in a few days, probably because of the rapid growth of young tissue. Spraying against *M. persicae* follows closely the advice given by factory fieldmen and can vary greatly from year to year and place to place. The percentage of fields sprayed with various aphicides in two contrasting years is in Table 4.3 and the effects of spraying in Fig. 4.14.

Spraying against *A. fabae* only should be undertaken before aphids average two per leaf. After that yield losses develop rapidly as the aphids increase, curl the leaves and blacken the crop.

Table 4.3 Usage of insecticides/nematicides on sugar-beet in 1981 and 1982. Percentage of the crop treated (R. A. Dunning *in litt.*).

Seed furrow granules	1981	1982	*Foliage sprays (aphicides)*	1981	1982
Aldicarb	25%	25%	Demephion	2%	0%
Bendiocarb	6%	5%	Demeton-S-methyl	40%	6%
Carbofuran	8%	10%	Dimethoate	5%	1%
Oxamyl	1%	2%	Ethiofencarb	2%	0%
Thiofanox	3%	1%	Oxydemeton methyl	3%	0%
			*Pirimicarb	43%	6%
	43%	43%		95%	13%
Soil incorporated spray			*No pesticide applied*	5%	39%
Gamma-HCH	18%	18%			
Gamma-HCH + granules	1–2%	1–2%			
	20%	20%			

* Mostly in areas where *M. persicae* is resistant to demeton-S-methyl.

The average increase in the sugar yield as a result of treatment against *M. persicae* and yellows viruses is 15–20%, more in bad years, when the infection in September and October reaches 100%. Most of the benefit is from the first spray application if well timed. Occasionally, however, the second application is the important one. When the beet is sown early, spraying gives less improvement in yield, but for late sown crops the improvement is considerable.

General considerations

The seed crops of sugar-beet, fodder beet, mangold and spinach beet which once were a potent source of viruliferous aphids in spring, are much less important nowadays since they are grown under barley cover crops or late in open ground protected by granular aphicides and all are carefully sprayed in autumn. However, beet harvesters leave 5 to 10% of plants in the field and unwanted tops are rarely fed off by stock and on the field or at loading sites no one bothers to remove the remaining roots once the lorries are loaded: these crop residues are a potential source of viruliferous aphids. Weeds also may be important but their effects have not been quantified. Whether the large expansion of the winter rape acreage has greatly increased the numbers of viruliferous aphid migrants in spring is also unknown. No amount of spraying prevents the ingress of viruliferous aphids whatever their source, hence their importance. The

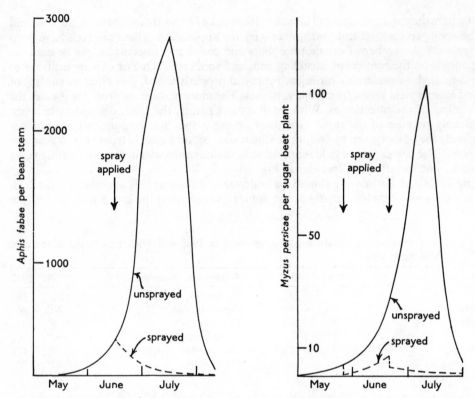

Fig. 4.14 The effect of spraying with a systemic organophosphate insecticide. Left, *Aphis fabae* on beans: one spray timed at the end of migration from spindle tree gives effective control. Right, *Myzus persicae* on sugar beet. Spraying is to control the spread of beet yellows viruses by relatively small populations. Two sprays correctly timed give effective control in early sown crops and delay infection enough to increase yield appreciably.

appearance of resistant aphids and the disappearance of useful insecticides (for economic reasons often connected with patent rights), the dearth of new classes of insecticides, the inability of plant breeders to make progress with resistant or tolerant varieties of beet, the increased susceptibility to aphids and virus, and of crops 'sown to stand' all add to the current difficulties of excluding virus from sugar-beet crops. Plant breeding should perhaps provide the solution but yellows viruses are a problem only one year in four or five. Other attributes of the crop are equally or perhaps more important and, initially, every resistance factor introduced in breeding carries a penalty usually expressed as a somewhat decreased sugar yield. Farmers prefer to plant the best yielding varieties. That way they avoid the penalty inflicted by resistant varieties in the years when resistance is inoperative.

Pea and bean aphids

Aphis fabae forms conspicuous black colonies on field beans and broad beans. The 'green' aphids, *Myzus persicae*, *Acyrthosiphum pisum* and the vetch aphid, *Megoura viciae* (Buckt.), are less conspicuous and easily overlooked. All four species are concerned in the transmission of the viruses of bean leaf roll, bean yellow mosaic and a variant of pea mosaic (Kennedy, Day & Eastop, 1962).

    Aphis fabae is the most important and most damaging species. Preventive control may be attempted with a persistent aphicide. Disulphoton or phorate granules may be applied to the foliage from the beginning of June and these give protection during the flowering period. In the Midlands, where aphids tend to arrive later in the season, it may be sufficient to treat the headlands only (9–10 m). A careful watch on the centre of the field is essential in case this becomes substantially infested in epidemic years. When preventive treatments have been omitted granules or sprays must be applied if aphid colonies develop. Late treatments coincide with flowering and, as pollinators are essential especially for areas greater than 4 ha, steps must be taken to avoid destroying them. Granules are safe. Of the spray materials, pirimicarb is least harmful to bees and coccinellid beetles. Demephion, demeton-S-methyl and thiometon are usually safe but are best applied late in the day preferably in damp dull weather when bees are less active. Granular insecticides are expensive and give a slow kill in dry years; for these reasons they are not favoured. To facilitate the application of sprays or granules and to minimize damage to the crop a 'tram-line' system and high clearance machines are desirable. If local beekeepers are given advance warning that spraying is intended, bees can be confined to their hives and the risks of poisoning minimized.

Economies can be made and unnecessary treatments avoided by paying attention to area forecasts made on the basis of infestations on spindle trees (Cammell & Way, 1977). As winter beans are more mature and less attractive when aphids migrate from spindle trees in May–June, it is rarely necessary to apply sprays or granules to them.

Treatments used against *A. fabae* effectively control 'green' aphids on the bean crop except in localities where there are insecticide-resistant populations of *M. persicae*. There pirimicarb is currently the best aphicide to employ.

Heavy rain readily removes pea aphids from their host plants. Pea crops should be examined about the first week in June, and if substantially infested and not due to be picked for 3–4 weeks, should be sprayed with a suitable organophosphate or carbamate insecticide.

Cabbage aphids

Two virus diseases affect brassica crops grown on farms and market gardens: cauliflower mosaic and turnip mosaic. Cauliflower mosaic virus is persistent and is found mostly in cauliflower where it first causes vein clearing, and later vein banding and stunting. The older leaves turn yellow and fall, the curds are small and infected plants may die in the winter (see Leaflet no. 3701). The symptoms of turnip mosaic, a non-persistent virus, are more severe in cabbage and Brussels sprout than in cauliflower and take the form of necrotic lesions. There are several strains of both viruses which may occur together. Many species of aphid transmit one or both viruses, but the important vectors are *Myzus persicae* and *Brevicoryne brassicae*. As *M. persicae* overwinters viviparously on many herbaceous plants, it is probably the main agent transferring virus from overwintering crops to seed crops. Cucumber mosaic virus, which occasionally infects brassica crops, is also aphid transmitted. It causes slight distortion of turnip leaves but is carried by other brassicas without causing obvious symptoms. Turnip yellows is a virus that is highly concentrated in the sap of infested plants. It is readily transmitted by many phytophagous insects, chiefly flea beetles (p. 109).

In the counties of Cambridgeshire, Bedfordshire and North Hertfordshire, a local Pest Order, the Cabbage Aphid Order, attempted to enforce the clearance of unwanted brassicas by 15th May and the disinfestation of plants saved for seed, but was rarely invoked and fell into abeyance. Nevertheless, the rotation, discing and ploughing under of unwanted brassicas is thoroughly desirable as a measure of hygiene mainly against the cabbage aphid, as it removes potent sources of viruliferous aphids which would transfer to new crops. Seed beds should be sited well away from seed crops including oil seed crops, and should be protected with aphicides when necessary. Carbofuran applied for cabbage root fly control gives protection against aphids arriving early. Crops that are directly drilled or transplanted after mid-May should be protected by a soil application of phorate, disulphoton or disulphoton plus fonofos granules. If not so protected they should be watched and treated as necessary with aphicidal sprays or pumice-based granules of disulphoton as the last treatment. The aim of all these treatments is to terminate cabbage aphid attacks during the early stages of colony formation before plant damage and leaf distortion occur and to minimize the spread of aphid transmitted viruses.

Brussel sprouts are subject to a longer period of attack by the cabbage aphid than most other brassicas. The main danger period is in June or early July. Best results have been obtained from granules applied to the soil before transplanting or direct drilling, followed by sprays or pumice-based disulphoton granules applied as late as possible before buttons (developing sprouts) become infested and not later than six weeks before harvest.

Brassica seed crops need to be examined from mid-April to May and sprayed promptly if large infestation of cabbage aphid begins to appear. It is usually unnecessary to treat winter-sown oil seed rape. Spring rape matures later and is more vulnerable. Action is necessary when infestations develop before the pods have started to fill. Brassica seed crops are attractive to bees and precautions to avoid bee poisoning are necessary as for field beans (p. 65). Turnips are not attacked by cabbage aphid but swedes can be adversely affected. Control is sometimes worthwhile on crops grown for fodder.

Cereal aphids

There are two main periods when aphids invade cereals: in Autumn and in spring and summer. In the autumn *Rhopalosiphum padi* is the important species because it is the principal vector of the severe strain of the persistent virus barley yellow dwarf (BYDV). The virus has increased greatly in importance now that winter cereals are sown much earlier than formerly, e.g. mid-September. At this time, numerous aphids migrate from grasses and a proportion carry the virus. In coastal and estuarine areas extending southwards from Suffolk to Kent and Sussex, then westwards to Devon and Cornwall and also including the Vale of Glamorgan, the Gower, Pembroke and Herefordshire, BYDV is endemic, possibly because winters are milder, grasses are abundant and *R. padi* readily overwinters (as do the grain aphid and the grass aphid). Hence, in these areas routine spraying is advisable. Pyrethroids (e.g. permethrin, deltamethrin, cypermethrin) are currently better than other aphicides and the first week in November is the best time for spraying. In other areas, BYDV is important only in some years. For use here, an index of infectivity has been developed which is obtained by multiplying the numbers of *R. padi* caught in suction traps by the proportion able to transmit BYDV to test plants. Surprisingly only a very small fraction of aphids caught are viruliferous. In Hertfordshire, spraying is advisable when the index is 50 but whether this applies elsewhere has yet to be determined (Plumb, Lennon & Gutteridge, 1982). More rapid methods of testing aphids than applying them to bait plants have been developed (Torrance & Jones, 1982) which should improve the speed with which the index can be made available and increase the numbers of aphids which can be tested. Means of predicting outbreaks of BYDV inland are desirable because examining crops is not effective: many aphids land on the soil or insinuate themselves into leaf sheaths and are difficult to detect.

When winter cereals follow directly after leys or stubbles containing dense patches of volunteer cereals, direct transfer of aphids to the succeeding crop is possible. In Wales and the southwest it is always desirable to spray in such circumstances as severe BYDV may ensue. Often there is insufficient time to clean the stubbles and the ploughing of leys does not always ensure that all grass is buried; i.e. proper hygiene is not practised.

After the turn of the year, cereals are seldom infested before tillering and *R. padi* is less important than the other species attacking cereals, although it is often the first to arrive in late May when it colonizes the lower leaves, especially of oats and barley. When numerous it may spread over the whole plant. The rose grain aphid colonizes lower leaves just as the ears are emerging. The grain aphid arrives late in June and is found on the upper leaves and ears. The blackberry-cereal aphid also attacks the ears but is less important than the grain aphid. The grass aphid is usually less important than the preceding species. The cereal leaf aphid, *Rhopalosiphum maidis* (Fitch) has been recorded occasionally on barley leaves.

Early and heavy infestations before tillering, although uncommon, can greatly reduce yields. Much damage is also done when ears become heavily infested by the grain and/or blackberry aphids. More than five aphids per wheat ear at the beginning of flowering warrant spraying immediately or yield losses of 10 to 20% may ensue (George & Gair, 1979) accompanied by losses in bread making quality (Wratten, 1978). Treatments applied after the grain is milky/cheesy are too late. Therefore cereal crops need close watching when catches from suction traps indicate that aphids are migrating and the weather is warm and settled. Unfortunately trap catches give no indication of the extent to which aphids will increase on cereal crops.

Sprays currently approved for use on cereals include demeton-S-methyl, dimethoate, formothion, heptenophos, oxydemeton methyl, phosalone, thiometon, pirimicarb and thiometon, with demeton-S-methyl and permethrin (not winter oats) to control BYDV in autumn. (See Leaflet no. 586.)

Carrot aphids

The chief carrot aphid is *Cavariella aegopodii*, a vector of several viruses including the two which cause carrot motley dwarf disease. The other aphids, *Dysaphis crataegi*, the hawthorn carrot aphid and *Pemphigus phenax* are much less important. Disulfoton, carbofuran, or phorate granules drilled in a narrow band in front of the seeder coulter (bow-wave technique) prevents severe infestations from developing quickly but one or more sprays with a systemic aphicide are usually necessary to give really effective control. The more persistent insecticides, e.g. demephion, demeton-S-methyl and thiometon, are the more effective. It is difficult to control immigration from heavily infested old crops: these sources should be found and sprayed.

Lettuce aphids

Myzus persicae and *Nasonovia ribisnigri* are the commonest aphids found on lettuces, and both are vectors of lettuce mosaic virus which stunts the affected parts and hardens the leaves which are easily torn. This virus is also seed-borne. Lettuce seed beds should be sown with seed containing less than 0.1% infection and should be sited well away from lettuce crops. Old, unwanted crops should be ploughed in. If aphids appear the seed bed should be sprayed with demeton-S-methyl, dimethoate or malathion. Prior to planting out, transplants may be dipped in malathion solution. Winter lettuce remain relatively free of aphids if free when transplanted but summer lettuce are subject to infestation by migrants from May to September, so they must be watched carefully. If harvesting is some way off, the more persistent aphicides may be used to spray them. If cutting is near less persistent chemicals may be used and within a week of cutting only shortlived chemicals, e.g. malathion, mevinphos are permitted.

Beet western yellows virus, transmitted by *Myzus persicae* is also a problem on outdoor lettuce. It causes yellowing and dwarfing; infection often occurs late in the season.

Pemphigus bursarius feeds on the roots. When lettuces wilt it is too late to attempt control, for treatment to be effective it must be applied in anticipation of an attack. As this aphid can overwinter in the soil, crop rotation is desirable and land should be cleared immediately after harvest. Resistant varieties (e.g. Avoncrisp and Avondefiance) are available but not favoured by growers. Their resistance needs to be incorporated into better varieties. (See Leaflets nos 392 and 603.)

Strawberry aphids

These are controlled on runners by dipping them in malathion or nicotine solution before planting, and on older plants by spraying with a systemic aphicides as required.

Whiteflies

Family ALEYRODIDAE, whiteflies. Whiteflies are small, moth-like insects covered with a white meal of powdery wax. The wings are about 3 mm across and the venation much reduced. They feed on the under surfaces of the leaves of their host plants and rise in a cloud when disturbed. In the tropics and subtropics, whiteflies tend to replace aphids as vectors of viruses on field crops.

Aleyrodes proletella (L.)   Cabbage whitefly is common on various brassicas, e.g. Brussels sprout, savoy, broccoli and cabbage in the Midlands, south-east and south-west of England, but is less common in the north. There are three or four generations a year and adults from the November generation overwinter on the under surface of brassica leaves. Feeding causes white or yellow patches on the leaves of the host plant. The honey dew excreted is less abundant than from aphids but may cover the leaves and impair photosynthesis (Fig. 4.15). When abundant, whiteflies replace aphids as the main vectors of brassica viruses.

The oval eggs possess pedicels and are found grouped together in a circle on the under surfaces of leaves from mid-May until autumn. The emerging nymphs have six legs and are mobile for a short time. Then they insert their stylets into the plant tissues and suck the juices,

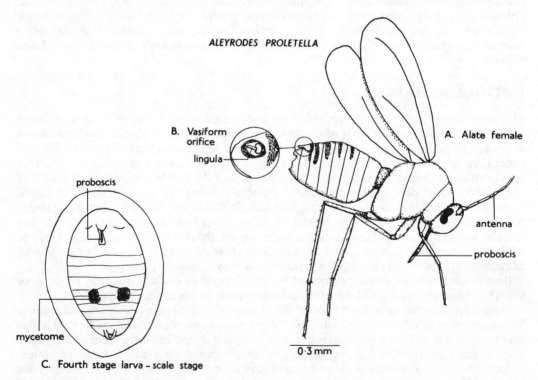

Fig. 4.15 Structure of whitefly (*Aleyrodes proletella*). A, adult, side view with inset; B, vasiform orifice showing lingula and operculum; C, fourth stage larva or 'pupa' with mycetomes.

thereafter becoming degenerate, scale-like and secreting a waxy covering. On the dorsal surface of the last abdominal segment is a vasiform orifice, an organ characteristic of the group (Fig. 4.15B). The fourth instar, sometimes called the pupa is thicker and more opaque than the earlier instars with pigmented areas or mycetomes (cells with symbiotic fungi) (Fig. 4.15C). The adult emerges from the nymphal cuticle through a T-shaped dorsal rupture. The life cycle from egg to imago lasts about a month in the summer. Honey dew is excreted by all stages but is flicked away by the lingula up to a distance of 2 cm and so does not always accumulate on the leaf. In the summer there is a definite light threshold for flight, which ceases in the evening, while in the autumn, flight occurs whenever the temperatures are favourable, and lasts longer (El Khidir, 1963, 1972).

Cabbage whitefly has little effect on yield and control measures for field crops are unnecessary. It is sometimes a nuisance to those harvesting brassica crops in mild winters and is perhaps best described as a 'cosmetic' pest.

Trialeurodes vaporariorum **(Westw.)** The glasshouse whitefly is similar to *A. proletella.* Although it thrives best in the shelter of glasshouses, where it is troublesome on tomato, cucumber and other plants, it also occurs in the field. Care should always be exercised when introducing new plants into a glasshouse that they are not infested with *T. vaporariorum* which transmits lettuce pseudoyellows virus found in France and the Netherlands.

T. vaporariorum can be controlled biologically by the introduction of the parasitic chalcid wasps, *Encarsia formosa* (Gahan) and *E. partenopea* (Masi) that lay eggs in the scale stage which is subsequently killed by the parasitic larva (see p.138). Affected scales turn black and an adult parasite emerges from each through a circular hole in the cuticle. This is one of the few examples of successful biological control in Great Britain and is being used increasingly. To ensure its success, stocks of whiteflies must be kept as hosts so that the parasite does not die out. In the rearing house or supplies obtained from commercial producers, the presence of an occasional tobacco plant, on which the parasite does not readily lay its eggs helps to preserve the stock of whitefly. A programme for the production and handling of glasshouse whitefly and its parasite has been worked out and both are mass-produced (Scopes & Biggerstaff, 1971). The temperature of the glasshouse must be kept up in the winter because neither host nor parasite are truly native and are unable to survive the winter outdoors.

Some growers still fumigate for whitefly by using smokes which consist of a simple pyrotechnic into which HCH has been incorporated but HCH is injurious to cucumber, melons and marrows (all Cucurbitaceae). After fumigation thorough ventilation is essential, and crops should not be picked for at least 24 hours. Control by spraying or by applying granules is complicated by the fact that whiteflies have developed resistance to several insecticides. (See Leaflet No. 86.)

Scale insects and mealy bugs

Superfamily COCCOIDEA.

This superfamily contains several families with a waxy or powdery covering. The proboscis is short but the stylets are very long and when at rest are looped to fit into a special pocket or crumena. The mature males have only two functional wings, fly actively, mate and die without feeding. Scale insects give little trouble to field crops in Great Britain, but are important in tropical and subtropical lands and in glasshouses. They occur more commonly on trees and shrubs than on herbaceous plants.

Family PSEUDOCOCCIDAE, mealy bugs, contains many species injurious to tropical and glasshouse crops. The adult female lays 300–600 eggs in a sack placed beneath the hind end of the body. The nymphs crawl over the plant after hatching. They are oval, light yellow, 6-legged

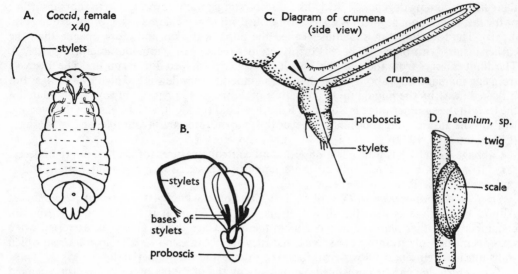

Fig. 4.16 A, female coccid (*Carulaspis juniperi* (Bouché)) to show reduced legs; and B, stylets and proboscis enlarged; C, diagram to show the structure of the crumena (side view) a pocket, into which fit the stylets; D, a scale *Lecanium* (= *Coccus*) sp. on a twig.

bugs with smooth bodies, but after feeding they secrete a white waxy material which covers them and forms elongate filaments which radiate from the sides of the body. They are able to move sluggishly over the plant. When fully grown, the females are about 7 mm long. In greenhouses one generation is completed in about a month. A mealy bug commonly found on citrus plants in the U.S.A. is *Pseudococcus citri* (Risso), the citrus mealy bug. *Pseudococcus njalensis* (Laing) is a vector of a serious virus disease of cocoa called swollen shoot. The females are elongated with distinct segmentation and are covered with a mealy or waxy filamentous secretion extended into lateral and terminal filaments. These bugs are attended by ants, *Crematogaster* sp., which imbibe the honey dew. When the ants are killed by chemicals the honey dew accumulates, and bacteria and fungi grow on it, adversely affecting the bugs.

Family COCCIDAE, scale insects (Fig. 4.16). The apterous female is obscurely segmented, immobile and degenerate. It is more common than the male in which the mouth parts are lacking, wing venation is degenerate and only the front pair is functional (also in mealy bugs). Like aphids, coccids are very prolific and are sometimes parthenogenetic. They often produce copious honey dew attractive to ants. Unlike aphids, they are immobile most of the time and are not viviparous. A single female may produce up to 2000 eggs and these are often protected by her body and by heavy secretions of wax. The first-stage nymphs have legs and are mobile but subsequent instars remain attached by their mouth parts to their host plant. Males undergo four moults and females only three. In fact the adult female stage seems to have been suppressed and the last nymphal stage to have taken on its reproductive function (neoteny). There are two types of scale insects, one type has a hard shell or scale over the body, and the other does not possess a hard separable scale but the body is smooth, rounded and very often convex.

Because scale insects are immobile, they rely for transport on animal and human agencies. Many species have been spread about the world on planting stock, and once in a new locality, free from enemies, they reproduce rapidly.

In Great Britain the mussel scale, *Lepidosaphes ulmi* (L.), is found on unsprayed fruit trees, and the soft scale, *Coccus hesperidum* (L.), occurs on apple, pear and plum (Fig. 4.16D). Effective control is obtained by tar oil or petroleum oil winter washes or lime sulphur spray. (See Leaflet no. 88.)

Many injurious species occur in North America, including the red scale of citrus, *Aonidiella aurantii* (Maskl.) and *Chrysomphalus aonidum* (L.), and the San Jose scale, *Comstockaspis perniciosus* (Com.), which attacks deciduous trees. *Kerria lacca* (Kerr), the Indian lac insect is commercially important and is cultivated because it provides a wax from which shellac is prepared. *Dactylopus coccus* Ferris syn. *Coccus cacti* (L.) lives on various species of cacti in Mexico, chiefly prickly pear, and provides a red dyestuff, cochineal. 140 000 insects are needed to make 1 kg of dye. Oriental species such as *Ericerus pela* (Char.) and *Ceroplastes ceriferus* (F.) yield wax commercially.

Control measures for scale insects and mealy bugs are similar. Deciduous woody plants are sometimes scraped during the dormant season or sprayed with tar oil washes. During the growing seasons, plants can be sprayed with malathion, deltamethrin or petroleum oil emulsion, for mealy bugs two or three times at 14-day intervals. Isolated groups may be painted with malathion or nicotine-white oil. Aldicarb granules may be applied to the soil of ornamentals.

Series 2. AUCHENORRHYNCHA

Leaf and plant hoppers are second in importance to aphids as virus vectors especially in warm climates. They also transmit mycoplasmas, spiroplasmas and rickettsias (Harris & Maramorosh, 1977; Conti, 1981).

Family CICADELLIDAE   Leafhoppers or jassids are very abundant in field crops and may be collected by sweeping foliage. They are small, slender insects that rest in a position ready for jumping. When disturbed they jump several feet and take to the wing. The heads are blunt, the proboscis obviously issues from the head and the antennae are short with terminal arista. The hind tibia are triangular and possess a row of spines (Fig. 4.17D) along the margins by means of which they can be distinguished from the more robust frog hoppers (cercopids).

The life cycle is simple. Elongate eggs are deposited in longitudinal rows on stems, under the leaf sheaths or on the leaves. There are six instars and the wing rudiments are evident in the

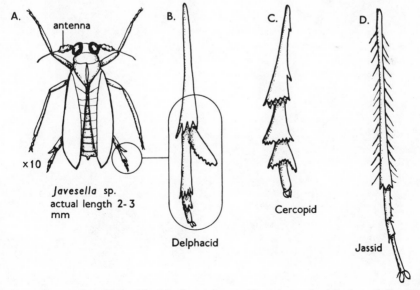

Fig. 4.17 A, structure of a delphacid with the armature B, compared with that of C, cercopid, and D, jassid.

third instar. Some species pass through more than one generation in a season. Although early feeding is limited to certain food plants, the adults are usually polyphagous.

Jassids are not important as pests in Britain although they are present in large numbers on many crops such as potatoes, sugar-beet and clover. Potato leaves show white bleached spots due to toxic action of the saliva of *Eupteryx aurata* (Z.). Some species transmit mycoplasmas which cause diseases such as strawberry green petal, clover phyllody and witches' broom. In the United States heavy infestations of potatoes leads to damage known as hopper burn.

Several species occur on apple and plum where the leaves are damaged, e.g. *Typhlocyba quercus* (F.) with white forewings bearing pillar-box red spots, and *Alnetoidia alneti* (Dahl.) with yellow forewings.

In the United States *Cerculifer tenellus* (Baker) spreads the virus causing curly top of sugar-beet, which is controlled by growing tolerant-varieties. *Endria inimica* (Say.) transmits wheat striate mosaic virus from infected wheat to wheat seedlings, barley and oats.

Family DELPHACIDAE   Plant hoppers or delphacids (Fig. 4.17A,B) are similar in structure and life history to jassids, but each hind tibia has a large, mobile apical spur (Fig. 4.17B). A common species, *Javesella pellucida* (F.) (Plate I), the cereal leafhopper transmits the organism, causing European wheat striate mosaic in England, a disease widespread in Europe and Eurasia. Infected wheat plants are stunted and produce no grain. Once it acquires the disease organism, *J. pellucida* remains infective for life and some individuals transmit the organism to their offspring via the egg. Epidemics of the disease are rare (Plumb, 1971).

Family CERCOPIDAE, frog hopper, are similar to jassids but more robust and lack the rows of spines on the hind tibiae (Fig. 4.17C). The nymphs of *Philaenus spumarius* (L.) produce froth or cuckoo spit, which probably serves to protect their soft bodies from the sun, and also from predaceous arthropods. Cercopids are unimportant economically in Great Britain, although they are sometimes abundant on crops.

5

Butterflies and moths – Lepidoptera

The Lepidoptera is a large Order containing many species of economic importance, especially in tropical and subtropical countries. In Britain alone, some 2000 species are known but only a few are serious pests. The structure and habits of adults and larvae are remarkably uniform although there is much superficial variation. The eggs are often sculptured with characteristic patterns. Typically the adults feed on nectar and are capable of prolonged flight. In contrast the larvae or caterpillars are fleshy and sluggish, rarely straying far from their food. Notable exceptions are the army worms or processionary caterpillars that form into bands and migrate. Because of their mouthparts and mode of feeding, adults do no direct economic damage although it is sometimes necessary to control them. Caterpillars, however, are voracious eaters. Usually they feed in an exposed position and therefore are relatively easy to kill. Some, however, feed in concealment as stem and leaf miners, others live in the soil surface or deep in soil and are more difficult to reach with insecticides. Although primarily plant feeders, some caterpillars feed on wool, beeswax, stored grain and other stored products. Fumigation and other special measures are necessary against them.

The adults possess two pairs of membraneous wings clothed in scales (Fig. 5.1). Mandibles are usually absent and the functional mouthparts form a long coiled tube or proboscis composed of the elongated galeae (maxillae) (Fig. 5.1B). The antennae are usually long and many jointed. The caterpillars are polypods with biting mouthparts. False legs or pseudopodia are present on abdominal segments 3–6 and 10 in *Pieris* spp., or on 5, 6 and 10 as in semilooper caterpillars (e.g. *Autographa gamma* (L.) (Fig. 5.8)), or on segment 6 and 10 only as in looper caterpillars (Geometer moths). The pseudopodia bear varying numbers of hooks or crotchets arranged in different patterns useful in identification (Fig. 5.1D, E). The pupae are obect (Fig. 2.18, p. 27), enclosed as a rule in a protective, silken cocoon or earthen cell. The silk is produced by glands in the head of the caterpillar and may be regarded as hardened saliva. It issues from a spinneret situated on the labium (Fig. 5.1A, C).

Classification

No insects of economic significance in Britain belong to the suborders Zeugloptera and Dachnonypha.

Suborder Monotrysia

Wings aculeate or pointed, venation of front and hind wings the same. Spiral proboscis never developed. Female with one, rarely two, apertures on the ninth abdominal sternite.

Family, HEPIALIDAE Swift moths (Fig. 5.2A, C). The antennae are short, the mouth-parts vestigial and the wings coupled by overlapping lobes. The caterpillars are subterranean, white with ochreous, chitinized heads and possess a full set of abdominal appendages armed with

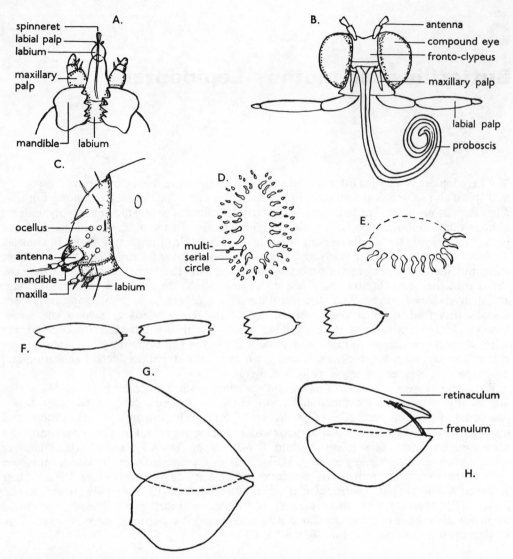

Fig. 5.1 Structure of Lepidoptera. A, caterpillar mouth parts, dorsal part of head removed; B, head of adult – front view; C, caterpillar head, *Agrotis segetum* – side view; D, crotchets from false leg of *Hepialus*; E, crotchets from Noctuid caterpillar; F, scales from butterfly wing; wing coupling G, of butterflies and H, of many moths.

crotchets arranged in several circles (Fig. 5.1D). They are unspecialized feeders and attack the roots and underground portions of many plants.

Suborder Ditrysia

In this suborder the venation of the front and hind wings is different, a spiral proboscis is usually present and wing coupling is normally by hook and frenulum (Fig. 5.1H). This suborder contains over 90% of all Lepidoptera; only important superfamilies can be mentioned.

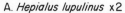

A. *Hepialus lupulinus* x2

B. *Acrolepiopsis assectella* x5

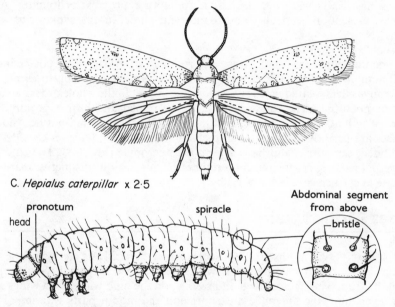

C. *Hepialus caterpillar* x 2·5

Fig. 5.2 A, Swift moth, *Hepialus lupulinus*, to show short antennae and wing markings; B, *Acrolepiopsis assectella*, leek moth; C, caterpillar of *Hepialus*. The head and pronotum are chitinized and appear light brown. The rest of the body is white. The bristles are black.

Superfamily TINEOIDEA

Tineid moths are small with narrow wings adorned with long hair fringes. The larvae have a full set of abdominal appendages (Fig. 5.2B). Those of many species are pests, e.g. diamond-back moth, Angoumois grain moth, leek moth, cocksfoot moth, clothes moths.

Superfamily TORTRICOIDEA

Tortrix moths are small with broad wings adorned with narrow hair fringes. The forewing often has a terminal lobe and, in the male, may possess a costal fold bearing expansible hairs functioning as scent organs. The larvae usually live concealed in leaves spun together, in seed pods and flower heads or tunnel into plant tissue. They have a full set of abdominal appendages with two or three circles of crotchets. The pupae have rows of spines on most abdominal segments and force their way out of the cocoon before emergence. Pests include the pea moth, codling moth and tortrix moths of fruit trees (Fig. 5.5A, B).

Superfamily PYRALOIDEA

Pyralid moths are small to medium sized with slender legs and bodies. Maxillary palps and abdominal tympanal organs are present, e.g. grass moths; flour moths.

Superfamily PAPILIONOIDEA. Butterflies

The antennae are clubbed and the wings often large and highly coloured (Fig. 5.5C–E). Wing coupling is by overlap, not hook and frenulum (Fig. 5.1G). Although the caterpillars are plant feeders, only the white butterflies are pests in Britain.

Superfamily GEOMETROIDEA

Geometer moths are slender with relatively large wings. There are tympanal organs on the abdomen and the maxillary palps are atrophied or absent. Females of some species have rudimentary wings. The larvae are elongated caterpillars that progress by 'looping'. Abdominal appendages are present on segments 6 and 10 only, e.g. winter moths, canker worms.

Superfamily NOCTUOIDEA

Owlet or noctuid moths are usually large and robust, with broad wings possessing a typical hook and frenulum coupling (Fig. 5.1H). A coiled proboscis is usually present. The most important family is the Noctuidae (Figs 5.8, 5.9), the largest in the whole order. The moths are crepuscular or nocturnal. The forewings are square tipped and bear a prominent kidney-shaped mark. The hind wings are paler and more uniformly coloured but occasionally their upper surfaces are brightly coloured as in the yellow underwing (*Noctua pronuba* (L.)). The eggs are rounded, sculptured and have a terminal micropyle (Fig. 5.8). The caterpillars are mostly nocturnal feeders, cryptically coloured but having various distinctive markings. They usually pupate in earthen cells in the soil.

Suborder Monotrysia

Family HEPIALIDAE   Swift moths are crepuscular and with rapid and characteristic flight. The British species are sombrely coloured. The wings have a special type of wing coupling. The jugal lobe of the front wing is elongated and rests upon the hind wing. Both wings are similar in shape and venation, the antennae are short and the mouth parts wanting (Fig. 5.2A). Females have two genital openings on the ninth abdominal segment (see Leaflet no. 160).

Hepialus lupulinus (L.)   Caterpillars of the garden swift moth (Fig. 5.2C) are active, deep soil insects, white and devoid of any protective colouration. The orange-red head is heavily chitinized, and there is a chitinized shield on the prothorax. The pupa is reddish-brown and cylindrical.

The nocturnal moths appear at the end of May and in early June and about 200 eggs per female are laid in fields, being dropped singly while in flight. Young caterpillars feed in summer and autumn, but damage to roots of garden plants and grasses by older caterpillars occurs only during the late autumn and spring of the following year especially in February and March. Pupation occurs in April, and the pupae rise actively to the surface of the soil where the adults emerge in May.

The caterpillars are common soil insects in south-west England. They are found especially in nurseries, garden land, waste land, strawberry beds, lettuce beds and hop gardens. Flower beds of anemones, peonies, and phlox also suffer, and the caterpillars can damage wheat and grass leys. On a field scale they are sometimes troublesome after the land has been left rough for a year, especially in hop gardens. The caterpillars cut off plants just below ground level or tunnel into fleshy roots or burrow in the basal stems of chrysanthemums. Treating the soil with gamma-HCH before planting gives effective control.

Hepialus humuli (L.)   The ghost swift moth, a larger species measuring 40–65 mm across the wings, in which the male is silvery white and sought by the buff-coloured female, is otherwise similar to *H. lupulinus*. The caterpillars feed on the roots of many plants including hops, strawberry and raspberry for two years before pupating.

Suborder Ditrysia

Superfamily TINEOIDEA. Tineid moths

Plutella xylostella (L.)   The diamond-back moth (Fig. 5.3, 5.4C) is world wide and a major pest of cruciferous crops in America, Australia, New Zealand and South Africa.

The moths are small with a wing span of about 16 mm. They have three light brown to white, triangular marks on the posterior margin of the brownish front wing which, when the two wings are at rest, form a diamond pattern. The hind wings are grey with long hair fringes. The female is lighter than the male.

The caterpillar is light green and sometimes has irregular dark markings on the head. It is about 12 mm long, widest in the middle and moults four times during its life of 16–21 days. If disturbed, it wriggles violently and may fall off the leaf on which it is feeding, remaining suspended to it by a silken thread. When fully grown, the caterpillar constructs an open-work cocoon on the host plant or in some protected place nearby. After one or two days quiescence

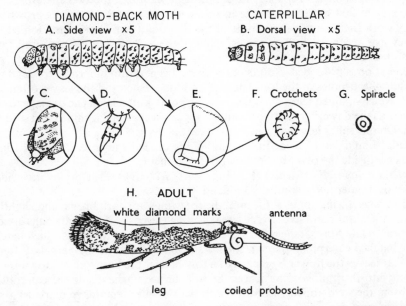

Fig. 5.3 Diamond-back moth, *Plutella xylostella*. A–G, larval structure. The older caterpillars are green while the younger ones are yellow; B, shows the characteristic spindle shape of the caterpillar from above; C, D, E, F and G show details of structure; H, side view of moth to show diamond markings.

in the prepupal stage, it forms a light yellow or green pupa. The prepupal and pupal stages last about 8–11 days.

Moths overwinter as adults and appear in the spring. They are weak flyers, active mainly in the evening. Minute eggs, about 160 per female, are laid in May and June after dusk, first on cruciferous weeds such as charlock or on turnips and later on brassicas such as cabbage, broccoli and Brussels sprout. The newly-hatched caterpillars crawl to the lower surface of the leaf and bore into the spongy mesophyll, where they become leaf miners. After the first moult, they emerge and feed on the lower surface eventually making holes or windows right through the leaf. The complete cycle takes 2–3 weeks, according to temperature, and there are two to three generations in a year in Britain.

The caterpillars destroy the foliage of cruciferous vegetables, particularly turnip, cabbage, Brussels sprout and broccoli, and the seed heads of turnips, swedes and cabbages. Warm, dry weather in July favours an attack, and transplants are often caught at a disadvantage and suffer severely before their root systems develop and growth can commence. Whereas other kinds of caterpillars (e.g. large white butterfly) do most damage to well-grown brassicas in the autumn, the diamond-back moth caterpillar is more active early in the year and especially in July. Occasionally there are serious epidemics. The year 1958 was noteworthy for an influx from Central U.S.S.R. when enormous numbers of moths were caught in Scotland. (See Leaflet no. 195.)

Acrolepiopsis assectella (Zell.) (Fig. 5.2B)   The leek moth is an introduction from the Continent, first detected along the Kent and Sussex coasts on leeks and onions. It has now been found along the east coast, north of the Thames estuary, and inland and causes slight damage to occasional crops (Becker, 1961). The moth is small with a wing span of 16–18 mm, dark brownish-grey with a triangular white spot on the hind edge of each forewing. When the wings are folded the white spots unite to form a large triangular mark. The adult hibernates through the winter and becomes active in April or May when the female lays eggs on onions or leeks at or near ground level. These hatch after 5–8 days and the small caterpillars burrow into the leaves making tunnels in the leaf tissue, but leaving intact the two epidermes. The caterpillars are grey at first but soon change to yellowish green, and when fully grown are 13 mm long. They work their way to the middle of the leek plant and feed on the folded leaves. On onions they live inside the hollow leaves. After 3 weeks they spin cocoons and form brown chrysalids on dead vegetation around the plants. The adults emerge in July, lay eggs and a further generation is produced. Adults of the second generation hibernate through the winter.

As a result of the mines in the leaf, bacteria and fungi enter and the leaves often rot around the lesions. Newly-planted leeks die if the growing point is attacked. The damage to seed plants can be severe as the flowering stalks of both onions and leeks may be mined and seed formation prevented. Onions suffer less than leeks.

Insecticide should be applied on the first signs of damage in early June, for once the caterpillars are inside the plant it is difficult to reach them.

Glyphipterix simpliciella (Steph.) (F.) (Fig. 5.4A, B)   Cocksfoot moths are widely distributed but most numerous in southern England.

The moths have a wing span of 7–9 mm, the forewings are dark brown and slightly metallic with a black apical spot and five silvery streaks along the costa. The hind wings are dark grey. Moths appear in late May, and eggs are laid inside the florets of the grass which hatch in June. The newly-hatched caterpillars feed in the florets and hollow out the kernels. After a month the caterpillar leaves the flower by means of a tiny hole at the base, crawls down the stem and bores a hole into the hollow stem nearer the soil surface. Once inside the caterpillar spins a cocoon closing the stem above and below with pads of silk, remains in diapause through the winter and emerges in the following year. Pupation occurs in March or April. Characteristic emergence holes can be found if cocksfoot seed or straw is examined (Fig. 5.4B).

As with grass seed gall midges (p. 143) the longer the stand of grass is kept the greater the

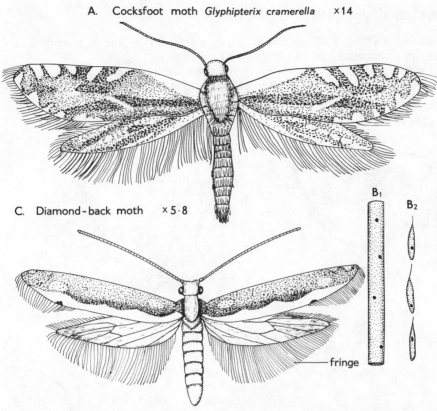

A. Cocksfoot moth *Glyphipterix cramerella* ×14

C. Diamond-back moth ×5·8

B₁

B₂

fringe

Fig. 5.4 A, cocksfoot moth, *Glyphipterix simpliciella* (Steph.) to show structure and wing markings; B₁, cocksfoot grass stem showing entrance holes of caterpillars; B₂, damage to grain of cocksfoot grass; C, diamond-back moth, *Plutella xylostella*, dorsal surface to show wing markings.

infestation tends to become. Burning the stubble and destroying all infested straw are practical control measures that prevent overwintering. New fields of cocksfoot for seed should be sited well away from old infested stands. Grazing in the autumn is ineffective.

In the family TINEIDAE, the proboscis is short or absent and the maxillary palps are long. Tinaeid moths occur all over the world. The caterpillars of most European species feed on animal or plant materials, and are primarily pests of stored products. The clothes moths, *Tinea pellionella* (L.), *Tineola bisselliella* (Humm.) and *Trichophaga tapetzella* (L.), are household pests attacking clothes, carpets, feathers and fur causing untold damage to articles of animal origin in store unless adequately treated with insecticides.

The family GELECHIDAE contains several species of economic importance:

Sitotroga cerealella (**Ol.**)   The Angoumois grain moth is found all over the world and its caterpillars destroy stored grain and even grain in the field. It occurs commonly in European granaries, but not outdoors.

Pectinophora gossypiella (**Saund.**)   The pink boll worm is a destructive pest of cotton. The small pink caterpillars soil the lint and eat the seeds. Development from egg to adult takes 25–30 days in summer, and there may be 4–6 generations during the growing season.

Phthorimaea operculella (**Zell.**)   The potato tuber moth, is a widespread pest of potatoes in warm countries.

Holcocera pulverea (**Meyrick**)   This is an enemy of the lac insect in India (p. 71); the caterpillars feed on the scales.

Superfamily TORTRICOIDEA

Tortrix moths (Plate VIC) are small, crepuscular moths occurring in temperate regions. They are more important to fruit growers than to farmers.

Cydia nigricana (F.) (Fig. 5.5, Plate VI)   The pea moth is one of the most serious pests of field and garden peas. It is prevalent in pea-growing areas of east England, in Essex, Kent, Suffolk, Lincolnshire, Cambridgeshire and Bedfordshire.

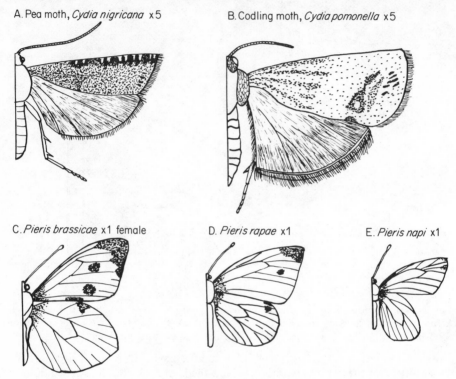

A. Pea moth, *Cydia nigricana* x5

B. Codling moth, *Cydia pomonella* x5

C. *Pieris brassicae* x1 female

D. *Pieris rapae* x1

E. *Pieris napi* x1

Fig. 5.5 Wing structure. A, pea moth, *Cydia nigricana*; B, codling moth, *Cydia pomonella*; C, large white butterfly, *Pieris brassicae*; D, small white butterfly, *P. rapae*; E, green-veined white butterfly, *P. napi*. Notice the clubbed antennae of the butterflies.

The adults emerge from old pea fields in June and are active until the end of July. The forewings, 12–16 mm across, are brown, satiny with black and white markings along the anterior margins. The peak of the egg laying is from mid-June to mid-July. Eggs are laid on exposed parts of the plants, either singly or in batches of two or three. They take 8 days to hatch, and the active caterpillars wander to the young pods and bore into them. There are rarely more than two in one pod. The caterpillars are pale yellow with a black head and a brown ring on the prothorax, with eight brown dots on the following segments. They are about 10 mm long when fully grown and have black legs. After feeding within the pod for 17–20 days they cut their way out, fall to the soil and burrow down to a depth of about 10 cm. There they form oval silk cocoons and hibernate through the winter. In April or May the caterpillar emerges from the cocoon and comes nearer to the surface where it spins a second cocoon, out of which it cuts its way before pupating. The caterpillars pass through a diapause (p. 10) which is broken by cold. This keeps the insect in step with its normal host plants. Because the moth population cannot be estimated by the usual sampling method nor by sweeping, the period of

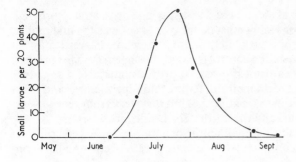

Fig. 5.6 Activity curve of pea moth (after Wright, Geering & Dunn, 1951).

summer activity was determined from the population of tiny caterpillars estimated by sampling peas at regular intervals (Fig. 5.6).

Pea moth is primarily a pest of the dry-harvested pea crop, rather than of peas picked and vined while green. Nevertheless even slight infestations are important on crops for quick freezing as the newly-hatched caterpillars soon enter the pods where they are difficult to control. The correct timing of spray applications is vital and this has become even more essential now that persistent organochlorine insecticides are no longer permitted. If the optimum time is passed by a few days, control falls off significantly. The problem of timing has been solved satisfactorily by the Rothamsted hormone trap (Fig. 5.7). These are set up by farmers to monitor the appearance of males: it is not enough to rely on records from neighbouring farmers or on an area basis (Perry, Macaulay & Emmett, 1981). The optimum time for spraying is based on the sustained capture of more than 10 moths per trap in traps set up in fields of 20–30 ha combined with simple calculations of egg development in day degrees from local maximum and minimum air temperatures (Biddle *et al.*, 1983). These calculations predict egg hatch. The Agricultural Development and Advisory Service also provides a phone in service for which the reply tape is brought up to date daily. Prognosis is based on trap catches and prevailing air temperatures. Apparently latitude is unimportant. Information from traps

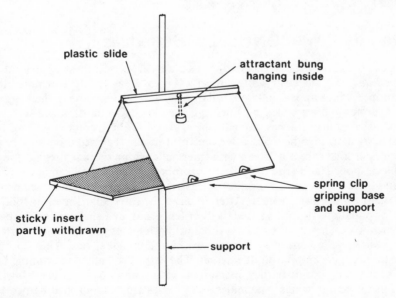

Fig. 5.7 Pheromone trap for males of pea moths as used in monitoring abundance and in timing spray applications.

and the Advisory Service enables farmers to avoid unnecessary spraying and to adjust spraying to the best time when it is necessary. Peas for vining must be sprayed as soon as the first moths are caught in pheromone traps. Here control is absolutely essential as whole harvests may be rejected if they contain even a few caterpillars: such control is strictly cosmetic. Peas for protein production and not for human consumption are not sprayed. Approved materials for spraying include azinphos-methyl mixtures, carbaryl, deltamethrin, fenitrothion, permethrin and triazophos. Spray dates are usually near to June 29th (see Fig. 5.6) but they vary from year to year and are earlier in the south than the midlands and the north. (See Leaflet no. 334.)

Aphelia paleana (**Hb.**)   The caterpillars of this small moth sometimes attack seed crops or timothy grass (*Phleum pratense*). They feed on the leaves which they web together and also on the seed and so lessen yield.

Cnephasia interjectana (**Haworth**)   Caterpillars of the flat tortrix moth bind leaves or leaf parts of sugar-beet together and feed on the inner surfaces. Control is rarely necessary. Aldicarb in the seed furrow and probably other granular insecticides are effective as is trichlorphon spray. A number of tortrix moths are pests of fruit in Britain, e.g. the codling moth, *Cydia pomonella* (L.) (Fig. 5.5). (See Leaflet no. 532.)

Superfamily PYRALIDOIDEA. Pyralid moths

Ephestia kuehniella (**Zell.**)   The Mediterranean flour moth is common in stored flour (Fig. 13.1, p. 271).

Chrysoteuchia culmella (**L.**)   The caterpillars feed on grass but sometimes where cereals follow grassland ploughed late they may attack young cereal plants. Damage is first noticed in April and May when round holes are found in the stem and leaves; later the shoots yellow and fall. The caterpillars feed inside the larger shoots and are 13 mm long when fully grown, brown or purplish brown with a dark brown head and brown spots on the back. Feeding tunnels are lined with fine silk. Caterpillars feed in the autumn and spring and hibernate during winter, emerging from cocoons inside the silken tunnels in June and July. Attacks can be prevented by early ploughing; burying the turf deeply is helpful.

Superfamily PAPILIONOIDEA. Butterflies

Only the species commonly found on cruciferous plants are serious pests in Britain.

Pieris brassicae (**L.**)   The large white butterfly (Fig. 5.5C) is widely distributed throughout Asia and Europe. Adults appear in late April or May and the females lay their eggs in groups on the under surface of brassica leaves. Each egg is enclosed in a hard, ridged shell. After 8–10 days, the young caterpillar hatches out, eats the egg case and feeds on the leaf. To begin with, the caterpillars remain together, but after the third moult they separate and crawl to new leaves. These caterpillars are blue green with three yellow longitudinal streaks along the body and many black spots. Hairs arise from fleshy protuberances. On the head is a well-developed spinneret, and there are six pairs of ocelli. Prolegs are present on abdominal segments 3–6 and a pair of claspers on the last segment. After 30 days, the caterpillar leaves its food plant and seeks a site for pupation. It climbs a tree, wall or fence and, in a sheltered spot, spins a silken carpet to which it attaches itself by silken threads and pupates. The pupa is shorter than the caterpillar, green or yellow green with well-marked abdominal spines. After 2 or 3 weeks in summer the thorax splits and the adult emerges. The wings are small and crumpled at first but they soon expand and become harder, reaching a wing span of 6–7 cm. The wings are cream with black tips to the fore wings. The males and females are distinguished by two black dots, on the upper surface of each forewing in the female but on the lower surface in the male. When the butterfly alights the wings are held at right angles to the body, but in repose they are closed vertically above the body. Eggs are laid again in July and August and the second generation passes through the winter as the pupa (Plate VII).

The caterpillars of the second generation cause most damage to cabbages, Brussels sprout and cauliflowers when the numbers are augmented by migrants from the Continent. Plants in gardens and small enclosed fields suffer most, and leaves may be skeletonized. In coastal regions damage may be particularly severe when migration from the Continent is heavy. On a farm scale inland, the caterpillars are not very troublesome.

In garden and allotments, hand picking of caterpillars from brassicas is still practised, but insecticides can be used very effectively. Sprays containing azinphos-methyl mixtures, chlorpyrifos, deltamethrin, derris, dichlorvos, diflubenzuron, etrimphos, iodofenphos, mevinphos, permethrin, pirimphos-methyl, quinalphos, triazophos, and trichlorphon all give good control.

Pieris rapae (L.) The small white butterfly (Fig. 5.5D) is similar to the large white but smaller. The life cycle differs only slightly. Adults appear earlier in March. Eggs are deposited singly and the caterpillars are green and feed singly; more in the centre of the plant and greatly decrease the saleable value of cabbages and other leaf brassicas.

Pieris napi (L.) The green-veined white, is similar to *P. rapae* but much less common (Fig. 5.5E).

Superfamily GEOMETROIDEA. Geometer moths

The looper caterpillars of the Geometridae are notorious defoliators of fruit and forest trees. They move by drawing the posterior segments close to those of the thorax, so the body forms a loop; the head and thorax are then extended in the desired direction and the looping action is repeated. *Operophtera brumata* (L.), the winter moth, *Alsophila aescularia* (Denis & Schiff.), the March moth and *Erannis defoliaria* (Clerck), the mottled umber moth, whose caterpillars feed on the foliage of trees and shrubs, is injurious to orchard trees. The males are alate but the females apterous (Fig. 5.10). (See Leaflet No. 11.)

Superfamily NOCTUOIDEA. Owlet or noctuid moths

Cutworms

Cutworms are caterpillars of several species of noctuid moths. They inhabit the surface layers of the soil and are voracious and polyphagous feeders on plants just above or below the soil surface. Lettuces seem especially vulnerable, probably being attacked at different times by more than one species. Field vegetables, especially carrots, also suffer a great deal. All other field crops are attacked either as seedlings or later when the damage is 'cosmetic'.

Populations are always present in arable land and local outbreaks sometimes occur when masses of caterpillars can be collected (Plate VIII). (See Leaflet No. 225.)

Agrotis segetum (Den. & Schiff.) (Figs 5.8 A–D & 5.9) The turnip moth is the commonest, most injurious and most widespread species of cutworm. The adult moths have a wingspan of 30 mm and fly in June and July. The colour of the forewings varies from grey to brown or reddish brown. The reniform (kidney-shaped) mark on the forewing is distinct, and there are other characteristic markings. The hind wings are white in the male and off-white in the female. Each female lays over 1000 eggs close to the ground on plants or among soil surface litter. These hatch in 7–14 days. When fully grown the caterpillars are approximately 36 mm long, pale grey or greyish brown with a dark, pale-edged line running down the back. The under surface is light grey and there is an X-shaped mark on the head (Fig. 5.8). On the sides are black glossy spots each bearing a small hair. The pupa is 10–12 mm long, smooth and brown and heavily chitinized with two spines at the hind end. Exceptionally there is a second generation but is doubtful whether its progeny pass the winter successfully.

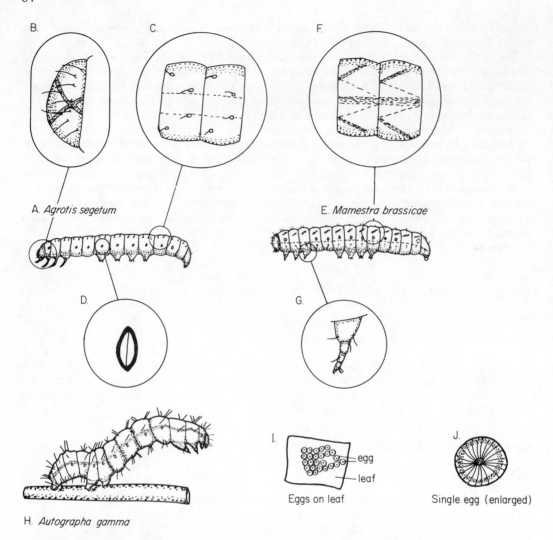

Fig. 5.8 Noctuid caterpillars. A, side view of a cutworm, *Agrotis segetum*, with B, dorsal surface of head to show the cross; C, dorsal view of the abdominal segments and D, spiracle; E, caterpillar of cabbage moth, *Mamestra brassicae*, with F, dorsal view of the abdominal segments, and G, thoracic leg; H, caterpillar of *Autographa gamma*; I and J, eggs of *Xestia c-nigrum*.

The first and second instar caterpillars feed above ground and later instars mostly below ground. Late sown or backward crops of beet or mangolds, late sown carrots or any crop with relatively thin stems and roots suffer severely if caterpillar populations are great. Plants are cut off or crippled and thin patches appear. Forward crops with thicker stems and roots (e.g. normal sugar-beet crop at this time) become more or less pitted and potato tubers may be hollowed. Feeding continues until autumn and caterpillars increase in size but decrease in numbers as parasites and other enemies take their toll. By October caterpillars are fully fed and overwinter in the top 5 cm of soil as fully grown caterpillars or pre-pupae. They do not feed again. Pupation occurs in early April to mid-May next year.

Euxoa nigricans **(L.)** (Fig. 5.9)   The garden dart moth was occasionally important in the Fens on sugar-beet, carrots and onions.

The adults fly in July and August when the eggs are laid on clover, weeds and other plants. The forewings of both sexes are ochreous brown although colour varies through red-brown to pale brown. The markings are usually obscure and sometimes there are additional spots and pale streaks. The eggs are globular, striated and shiny. The caterpillars appear in March or April and are greenish-brown with dark green dorsal lines edged with black and double white lines laterally.

The eggs are laid mainly in August but tiny caterpillars do not appear until the following spring. Whereas egg laying by *A. segetum* is earlier and the crop on which the eggs are laid suffers shortly afterwards, the conditions which lead to an attack by *E. nigricans* depend largely on the previous crop. Once important and sometimes extremely numerous in beet fields, this pest is now rarely found in them possibly because of the insecticides used to protect the seedling crop.

Other noctuid moths have caterpillars that behave as cutworms and are briefly mentioned below.

Agrotis exclamationis **(L.)** (Fig. 5.9)   The heart and dart moth has a varied colouration of the wings but the markings are distinct. The caterpillars can be distinguished from those of *A. segetum* by the large size of the black spiracles compared with the black spots surrounding them and by the pear-shaped markings on the back. This cutworm is uncommon in open fields.

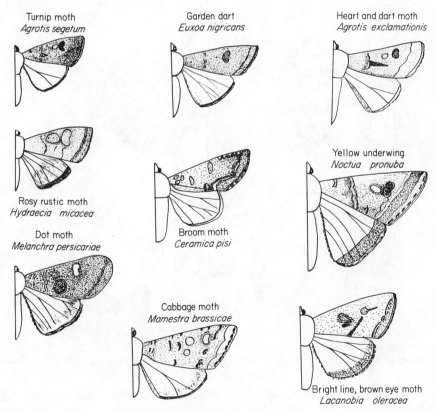

Fig. 5.9 Wing patterns of noctuid moths (× 1·5).

A. puta (Hubrei) the shuttle shape dart moth and *A. vestigialis* (Hufu.) have life cycles similar to *A. segetum* but their caterpillars are much less common in fields.

Euxoa tritici (L.) (Fig. 5.10)   The white-line dart moth is widely distributed throughout the British Isles. There is a well-marked white line along the forewing. The larvae are grey or brown and are most common on sandy soils. Its life cycle is similar to that of *E. nigricans*.

Noctua pronuba (L.) (Fig. 5.9)   The large yellow underwing moth is also widely distributed. The forewings are brown but the hind wings are yellow with a narrow dark brown border. The caterpillar is brown or greenish-brown with ochreous lines on the back. The main flight period is from July to September and there is a pre-oviposition period of at least 6 weeks. Caterpillars appear in September and continue to feed throughout the winter during mild spells or in protected situation. Much cutworm damage in winter months is from this species. That in April, May and June is mostly from *E. nigricans*.

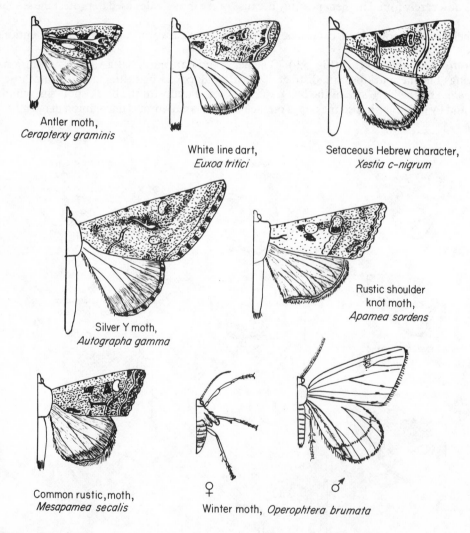

Antler moth,
Cerapterxy graminis

White line dart,
Euxoa tritici

Setaceous Hebrew character,
Xestia c-nigrum

Silver Y moth,
Autographa gamma

Rustic shoulder
knot moth,
Apamea sordens

Common rustic,moth,
Mesapamea secalis

♀   ♂

Winter moth, *Operophtera brumata*

Fig. 5.10 Wing patterns of noctuid moths (× 2) and of a winter moth, *Operophtera brumata* (Geometridae), with wingless female.

Control of cutworms

Poison baits were once used to kill cutworms because they have unspecialized feeding habits and wander over the soil surface. Paris green and bran or beet pulp were mixed and moistened with water. Molasses was sometimes added but was not essential. The bait was spread over affected parts of fields late in the day so that it would remain moist and attractive as long as possible. These baits worked well and gave an 80% kill. Later HCH wettable powder was substituted for Paris green, but baits are no longer used.

Until recently it was difficult to control cutworms with sprays other than rather heavy applications of those containing DDT or HCH. DDT is virtually banned and HCH may taint carrots or potatoes planted within 18 months of application. It is also harmful to ground predators and other beneficial insects.

In outbreak years, *A. segetum* especially, may cut off and kill seedlings leaving bare patches. More often however 'cosmetic' damage is done to carrots, potatoes and other vegetables which results in samples unacceptable to supermarkets and other important outlets for prepacked produce. The main problem in obtaining control with non-persistent insecticides was timing applications to catch the first and second instar caterpillars while still feeding on plants above ground. Once below ground they are difficult to reach and, in any case, later instars of these and some other caterpillars become increasingly resistant to insecticides.

The problem of timing has now been solved by a predictive model (Bowden *et al.*, 1983) based on catches of moths in light traps and on an index derived from local temperature and rainfall which enables ADAS to inform growers when is the correct time to spray. Fairly precise timing enables non-persistent insecticides to be used which formerly could not have been employed (when timing is imprecise, early spraying with persistent materials is essential). Currently the recommended insecticide is chlorpyrifos but this seems likely to be replaced by pyrethroid insecticides which are less harmful to beneficial insects.

Foliage-feeding noctuids

Although the caterpillars of noctuid moths have different host preferences, mainly determined by the plants on which the adult females lay their eggs, nevertheless these ranges are rather catholic and include many field and vegetable crops.

Autographa gamma **(L.)** (Fig. 5.10)   The silver Y moth is a well-known migrating species and is found all over the British Isles. Apparently it does not survive the winter and adults migrate into Britain every year, occasionally in great numbers. A few adult moths sometimes appear early in the year and can be seen taking nectar on bean flowers. These are all gravid female immigrants.

The forewings have a span of 3.6–4 cm and a marbled appearance, the colour ranging from a silvery grey to a velvety black. The silver Y mark is distinct. The hind wings are light brown with a dark border. The moth is diurnal unlike many noctuids. Oviposition occurs in late May or at the beginning of June, and the caterpillar, a semi-looper, feeds from June to October on sugar-beet and other low growing plants. The colour of the caterpillar varies from bright green to dark olive green, and a dark green dorsal line edged with white is present (Fig. 5.8H). There are several white transverse lines between the yellow spiracular line and the dorsal black line, and the head bears black markings. When full grown the caterpillar pupates on the leaf in a loose web-like cocoon. The pupa is black and shining and differs from that of the cutworms and their allies in not being formed in an earthen cell.

In cycles of about 10 years large numbers of moths fly in from the Continent chiefly in June and July when the leaves of sugar beet and other crops suffer severe defoliation in late July and August (Plate VII). The epidermics decline at the end of August and sugar beet recovers rapidly, for the pest is attacking a mature crop with plenty of reserve. Although thousands of tonnes of leaves may be consumed, the effect on the crop is small. Control is rarely necessary and not attempted.

Lacanobia oleracea **(L.)** (Fig. 5.9)   The bright-line, brown-eye or tomato moth is a variable moth. The brown-eye and white submarginal line with the W mark may be indistinct. The caterpillar varies in colour, has minute black and white spots, and a yellow lateral stripe edged with black running along the body. In most years this species is commoner than the silver Y moth on sugar-beet.

Melanchra persicariae **(L.)** (Fig. 5.9)   The dot moth is common is southern Britain. The forewings are blueish black with a well-defined white reniform mark. The caterpillar is pale green, with pink markings and a hump at the posterior end.

Xestia c-nigrum **(L.)** (Fig. 5.10)   The cetaceous Hebrew character moth. The life cycles of the two foregoing species are similar to that of *Agrotis segetum* as are those of the nutmeg moth, *Dicestia trifolii* (Hufuagel) and the angle shades moth, *Phlogophora meticulosa* (L.). The caterpillars of all these species occur commonly on sugar-beet and other leafy crops.

Mamestra brassicae **(L.)** (Fig. 5.9)   The cabbage moth occurs all over the British Isles and on the Continent. The caterpillars feed on many kinds of plants, but greatest damage is to cabbages, where they feed on the leaves and bore into the hearts, fouling them with masses of frass. Attacks are most common on gardens and allotments but occur occasionally on a field scale.

The adults appear in May or June and oviposit on the leaves of cabbages or other herbage. The round eggs are deposited singly but in an orderly manner on the leaves. Usually there is only one generation a year but there may be a partial second, the caterpillars feeding until October. The forewings are grey, brown or black, the kidney-shaped mark is distinct and has a white outline. There is an irregular white transverse line near the margin. The wings have a span of 4.2 cm and the hind wings are brown. The species is very variable. The caterpillars are green when first hatched, turning to dark green or to grey and even to brown as they mature. There is a dusky central line speckled with white on the dorsal surface and a lateral yellow or light green stripe (Fig. 5.8E, F, G). After 4–5 weeks the caterpillar (Plate VIIF) is fully grown and pupates in the soil forming a glossy brown chrysalis.

Evergestis forficalis **(L.)**   The garden pebble moth is an important pest of brassicas. The forewings span 2.5 cm and are pale yellow-brown with two almost parallel transverse lines of darker brown thickening near the apex of the wings to form two brown marks. The hindwings are paler with a thin marginal line and an inner indistinct line. The grey-green caterpillar has a pale head and a body tapering posteriorly with backwardly directed anal claspers (prolegs). Hence the name *forficalis*. The body is clothed with long pale hairs, bears a black spot above each spiracle and confused areas of white pigmentation on each segment. The body turns bright green just before pupation.

Cabbage caterpillars

Leaves of brassica crops are regularly attacked by the caterpillars of three species of butterfly (*Pieris brassicae, P. rapae* and *P. napi*, see p. 82) and six species of moth (*Plutella xylostella*, p. 77, *Autographa gamma, Lacanobia oleracea, Melanchra persicariae, Mamestra brassicae* and *Evergestis forficalis*) (for key see Table 5.1). Of these possibly the most important on cabbages are *M. brassicae, E. forficalis* and *P. rapae*. In some years *P. brassicae* defoliates brassicas in gardens. Caterpillars that shot-hole the leaves and feed on the heart leaves of cabbages render them unsaleable. Usually early crops suffer less than mid-season and later crops but attacks may occur throughout the growing season (see Leaflets nos 69 and 70).

Preventive control is essential as once inside cabbage hearts caterpillars are difficult to kill and quality has already been sacrificed. The small white and the cabbage moth are the worst offenders. Pyrethroid insecticides (cypermethrin, deltamethrin and permethrin) are possibly

V A. Lettuce plants attacked by the lettuce root aphid *Pemphigus bursarius*. Left, resistant variety; right, susceptible variety (courtesy J. A. Dunn, N.V.R.S.). B, C. The effects of the willow-carrot aphid and carrot motley dwarf virus on carrot plants (courtesy D. W. Wright, N.V.R.S.).

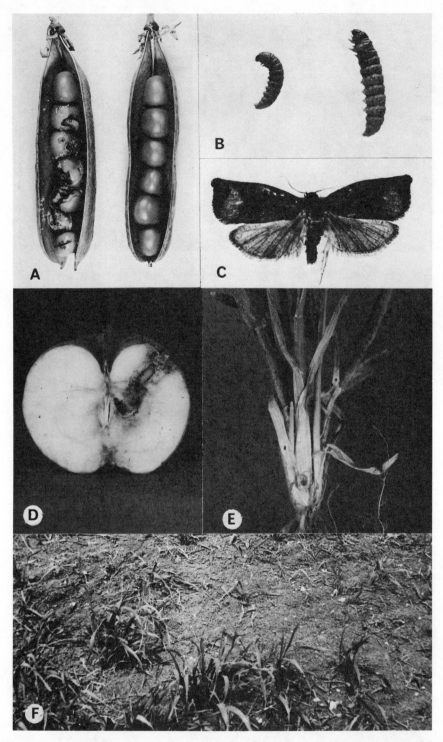

VI **A.** Pea moth damage to peas (courtesy J. A. Dunn, N.V.R.S.). **B.** Pea moth caterpillars (Crown copyright). **C.** A tortrix moth (courtesy R. A. French, Rothamsted). **D.** Apple damaged by codling moth (Crown copyright). **E.** Crambid caterpillar damage to wheat (Crown copyright). **F.** Winter wheat damaged by Crambid moth caterpillar.

VII **A.** The large white butterfly (Crown copyright). **B.** Eggs of the large white butterfly (Crown copyright). **C.** Caterpillars of the large white butterfly on a brassica leaf (Crown copyright). **D.** Brussels sprout skeletonized by caterpillars of the large white butterfly (Crown copyright). **E.** Cabbage moth (Crown copyright). **F.** Caterpillar of cabbage moth (Crown copyright). **G.** Eggs of cabbage moth (Crown copyright). **H.** Beet leaf skeletonized by caterpillar of silver Y moth (courtesy W. E. Dant).

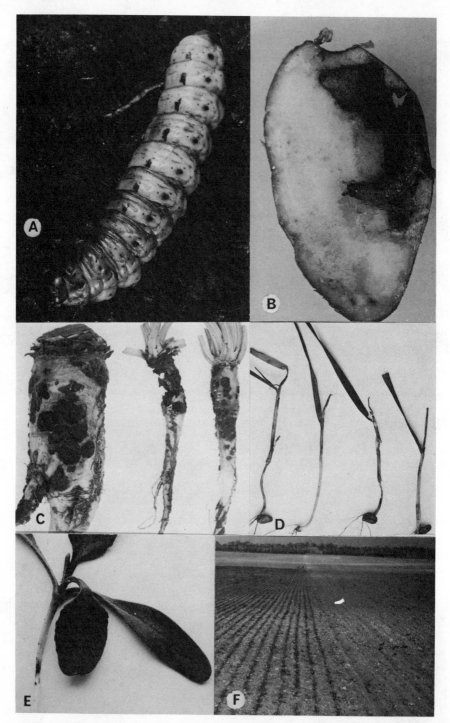

VIII **A.** Cutworm (*Agrotis* sp.) (courtesy H. C. Woodville).   **B.** Cutworm damage to potato tuber (Crown copyright).   **C.** Sugar beet damaged by cutworm (courtesy W. E. Dant).   **D.** *Helophorus nubilus* damage to wheat seedlings (Courtesy W. E. Dant).   **E.** Pigmy mangold beetle damage to beet seedling (courtesy W. E. Dant).   **F.** Wheat field attacked by *Helophorus nubilus* after sainfoin. Left, sown mid-October; right, sown early November (courtesy W. E. Dant).

Table 5.1 Key to caterpillars feeding on brassicas in the U.K. Mean maximum length in parentheses. For use with instars other than the first. (Adapted from Emmett, 1980.)

1. Three pairs of pseudopodia (30–40 mm)	*Autographa gamma*
Five pairs of pseudopodia	2
2. Head black, black and grey or green, with black and white hairs	3
Head glossy with hairs of one colour	5
3. Body fawn with black patches and yellow longitudinal stripes (40 mm)	*Pieris brassicae*
Body green with black and white spots. Yellow patches on or near spiracles	4
4. Yellow patches elongate, diffuse. Spiracles with pale centres. Yellow dorsal line (34 mm)	*Pieris rapae*
Yellow patches circular, discrete. Spiracles with dark centre. Indistinct green dorsal line (34 mm)	*Pieris napi*
5. Body tapering with backwardly directed terminal pseudopodia (less than 25 mm)	6
Body not tapering (more than 30 mm)	7
6. Body pale green. Black spots on prothorax and terminal segment. White spots on other segments. Spots on prothorax and all but terminal segment each with one black hair (14 mm)	*Plutella xylostella*
Body grey green. A black spot above each spiracle. Confused patches of white on each segment. Body hair long, pale, sparse. Body bright green when about to pupate (22 mm)	*Evergestis forficalis*
7. Body green or brown with distinctive dark markings	8
Body green without distinct dark markings	10
8. Body brown, mottled with black and white spots. Lateral line orange with dark upper edge (40 mm)	*Lacanobia oleracea*
Body brown, not mottled. Lateral line not orange	9
9. Prothorax with three longitudinal white stripes. Abdominal segments with chevron-like markings, 8th humped (45 mm)	*Melanchra persicae*
Prothorax without stripes, 8th abdominal segment not humped but with dark patch. Body pale fawn to almost black. Lateral stripe fawn, yellow or brown (45 mm) Instars 4–6	*Mamestra brassicae*
10. Body bright green, often with yellowish bands between segments. Dorsal and lateral lines white. Instars 1–3	*Mamestra brassicae*

the best to use as contact insecticides: both upper and lower surfaces of the leaves must be sprayed.

Grass-feeding noctuids

Apamea sordens **(Huf.)** The rustic shoulder knot moth (Fig. 5.10) is widely distributed in the British Isles. Caterpillars feed chiefly on grasses (*Agropyron* sp.) and cereals (wheat and barley). These are two types of injury: (1) hollowing of wheat grains in the early autumn; (2) destruction of stems in spring. Severe attacks sometimes occur. The adults fly in May and June. The forewings are pale brown and each has a single black dash running from the base called the 'shoulder knot'. The hind wings are brown with a slightly darker margin. The caterpillar is brown or olive brown with brown lines on the back and black marks around the dark spiracles. It feeds at night concealing itself by day. Young caterpillars feed in the autumn, go into diapause, feed again in the spring. When fully grown the caterpillar forms an earthen cell in the ground where it pupates as a shining brown chrysalis.

Mesapamea secalis **(L.)** (Fig. 5.10) The common rustic moth is essentially a grass-feeding species, the usual hosts being cocksfoot grass (*Dactylis glomerata*), annual meadow grass (*Poa annua*), meadow fescue grass (*Festuca elatior*) and woodrush (*Luzula maximus*). The caterpillars bore into the stems. Occasionally the central shoot of wheat and oats is destroyed by the caterpillars, causing injury like that produced by frit fly and wheat bulb fly (Gough, 1947). Attacks on cereals are sometimes locally important in Norfolk coastal areas, where up to 90% of the plants may be destroyed when wheat follows a rye grass–clover ley. Wheat and oats may also be attacked following the ploughing up of old grassland. The moth is widely distributed throughout the British Isles.

The adults fly in July and August. The colour of the forewings varies from light brown to black but has a white reniform mark. The slender caterpillars are yellow green with three faint red lines running along the back. They are active from April to May, and when they mature enter the ground to pupate. In areas liable to an attack, rye grass leys should be ploughed early in August before eggs can be laid upon them.

Cerapteryx graminis (**L.**) (Fig. 5.10)   The antler moth, caterpillars of which occur on hill pastures in the Pennines, Scotland and Wales, feed on all kinds of grasses, chiefly mat grass (*Nardus stricta*). The adults are found in August and September. They are dull grey to reddish brown with the white stigma enlarged and bearing three branches rather like the antlers of a stag. The hind wings are brown. The eggs are laid in August by the female when in flight. The eggs may pass through the winter unhatched or hatch in the autumn and the caterpillar pass the winter in diapause and start feeding again in the spring. The caterpillar is bronzy brown, glossy and much wrinkled with longitudinal yellow stripes. Below the black spiracles, the ventral parts are ochreous. In June, when the caterpillar is fully fed, it forms a cocoon below the soil surface and pupates. On rare occasions the caterpillars appear in great numbers behaving as army worms. As they move forwards they are much preyed upon by birds. Outbreaks can be controlled by cutting trenches with vertical sides in the path of the advancing 'army', by burning the grass when dry, or by using poison baits and insecticides as for cutworms. Hill grass has little commercial value, however, and is often left to its fate.

A stem boring noctuid

Hydraecia micacea (**Esp.**) (Fig. 5.9)   The rosy rustic moth or potato shoot borer is found most frequently in the coastal regions, but is widely spread over the British Isles. It occurs sporadically as a pest in potato, rhubarb, iris, sugar-beet, gladioli and onions. The young larvae also mine raspberry canes and cause wilting of the apical leaves. Occasional caterpillars are found most years in potatoes, beet and hop, but are rarely of economic importance.

The adults appear in the autumn and may be seen sipping nectar from ragworts and other flowers. The ground colour of the forewings varies from pink to reddish brown. There is a terminal grey margin, bounded on the inside by a dark line which is continued through the grey hindwings and is visible when the wings are opened. The wing expanse is 2.4–3.6 cm. Eggs are laid on the lower surface of the plants in the early autumn. They hatch and the young caterpillars hibernate in the soil, becoming active again in the spring when they tunnel into the stems of the host plants. In potato haulms the caterpillar is concealed and gives rise to virus-like symptoms which may confuse the experts. It is flesh-coloured with a central red stripe. On the body are black or dark brown wart-like markings, each bearing a bristle. The head is reddish brown. Pupation takes about 5 weeks. No control methods have been worked out: none is usually required.

Enemies of caterpillars

Caterpillars have many predators and parasites. Those of the large white butterfly have been much studied and recently reviewed by Feltwell (1982). Predators range from birds (sparrow, thrush, seagull), through small mammals (field mice), many insects (ants, wasps, pentatomid and reduviid bugs, coccinellid beetles, the earwig, lacewing larvae, predatory Diptera) to spiders and possibly slugs. Because of difficulties of nomenclature and identification there is much uncertainty about hymenopterous parasites. Feltwell (1982) considered there are only five regular parasites in the U.K. that have a significant impact on numbers and concluded that this is what would be expected of a migratory host producing transient populations. Moreover, these parasites, which include the egg parasite *Trichogramma evanescens* (Westwood), the larval parasites *Apanteles glomeratus* (L.) and *Hyposoter ebeninus* (Graven.), and the pupal parasites *Pteromalus puparium* (L.) and *Pimpla instigator* (Fab.) have other abundant hosts. Common dipterous parasites are *Compsilura concinnata* (Meig.) and *Phryxe vulgaris* (Fall.) to which similar remarks apply. Feltwell (1982) makes no reference to predation of early instars by ground beetles (carabids). This is a possibility, especially when caterpillars are dislodged by wind, rainfall or crowding.

Caterpillars also become infected with strains of *Bacillus thuringiensis* (Berl.), with granulosis virus and, more rarely, with protozoa and nematodes.

Moss (1933) listed the mortality of 10 000 large white butterfly caterpillars (Table 5.2). Unspecified disease killed more than 60% (59% caterpillars, 2% pupae), *Apanteles* killed 34%, *Pteromalus* 1% and birds 4%. Probably about twice as many eggs (20 000) were laid to produce 10 000 caterpillars and of the ensuing 32 adults possibly as few as 10 (5 males, 5 females) survived to reproduce. Total mortality was of the order of 99.9% and this represents an 'overkill' of the kind that usually terminates large insect outbreaks. To replace such a population each female would have to lay about 4000 eggs. However, in the U.K. large white butterfly populations are usually augmented by immigration and must rarely depend on local balance.

Table 5.2 Mortality in *Pieris* caterpillars (Moss, 1933).

Cause	Mortality		Number remaining out of 10 000
	actual	*cumulative* %	
Disease of caterpillars	5917	59.2	4083
Apanteles	3438	93.6	645
Disease of pupae	174	95.3	471
Pteromalus	14	95.4	457
Birds	425	99.7	32

Until recently little was known about cutworm populations. It now appears that populations of *Agrotis segetum* are not subjected to heavy, predation, parasitism or disease but that weather (temperature and rainfall) are the primary factors determining the survival of early instars while feeding on plants above ground (Bowden *et al.*, 1983; Sherlock, 1983). Just how weather 'kills' small caterpillars is uncertain. As the eggs are laid on the soil rather than on host plants, caterpillars may starve if weather prevents them from finding hosts or may make them more prone to the attentions of ground predators. There is also a large mortality of overwintering caterpillars and prepupae especially in clay soils: the causes are unknown. Many fewer survive harsh winters than mild ones.

Outbreaks of leaf-feeding caterpillars such as those of the silver Y moth suffer parasitism and predation as do caterpillars of the large white butterfly. In some years *Phryxe vulgaris* (Meig.) is an important parasite of silver Y moth caterpillars. Possibly one reason that cutworms suffer less from parasitism is that the later instars operate mainly below ground.

6

Beetles – Coleoptera

In this large group of insects, adult structure is fairly uniform but species vary greatly in size and habits. The prothorax is large, the forewings of adults are hardened into wing cases or elytra which meet in a straight mid-dorsal line at rest and are held outstretched in flight. The large membraneous hindwings which produce the motive power for flight are kept folded underneath the elytra when not in use (Fig. 6.1). Beetles are not usually sustained flyers but are mainly cursorial; flight is used chiefly for dispersal. The mouth parts are adapted for biting and the ligula is lobed. The Coleoptera is one of the largest Orders in the animal kingdom and contains many injurious pests. Crops and stored products may be damaged by the adult only, by the larva only, or by both. In temperate regions beetles tend to overwinter as adults, and after spring feeding pair and lay their eggs. The larval stage occurs during late spring and early summer and is followed by pupation, usually in the soil. Adults may emerge for a period of feeding before hibernation or they may not appear until the following year. The larvae are variable in habit and structure but, like the adult, always possess biting mouth parts. Most have three pairs of thoracic legs.

The basic larva is campodeiform (Fig. 2.17B), a predaceous type found in the Carabidae

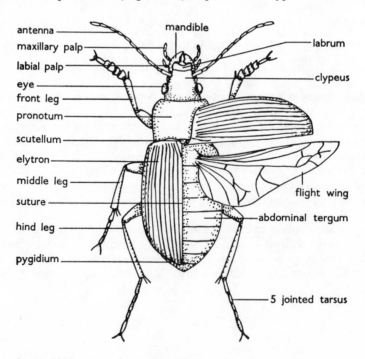

Fig. 6.1 Diagram of a carabid beetle to show general structure.

and the Staphylinidae. It is active, with chitinized dorsal plates, well-developed anal cerci, sickle-shaped mandibles and a somewhat flattened body. There is every gradation from this type to the more sluggish plant feeding eruciform larvae of the Chrysomelidae in which the mandibles are well developed but the abdomen is large and fleshy (Fig. 6.8D). The burrowing larvae of the Elateridae (wireworms) have no dorsal plates, no anal cerci and are less active. The legs are short and the larvae are cylindrical (Fig. 6.4). The fleshy, curved larvae (white grubs) of chafers are found in soil where there is an abundance of food (Fig. 6.20). Lastly there is the legless type where the fleshy, somewhat curved larva has lost its thoracic legs as in the weevils (Fig. 2.17). There is therefore a complete series of larval types from oligopod to apodous. In all, the head is well developed, thoroughly chitinized and bears mandibles. Spiracles are present on the prothorax or between the pro- and mesothorax but are absent from the metathorax, and are also present on abdominal segments 1-8, in the peripneustic condition (Fig. 2.13). The pupa is free (exarate, Fig. 2.18).

Classification

Classification of such a large number of insects, over 250 000 species, is difficult. Some of the characters used are: (1) the general shape of the body; (2) the length of the elytra; (3) the type of antennae, filiform (threadlike), clavate (clubbed), geniculate (elbowed) or lamellate (plate-like); (4) the form and the number of joints of the tarsi; and (5) whether the head is elongated anteriorly to form a snout or rostrum.

Only an abbreviated classification to include the main pests is given below.

Suborder Adephaga

Superfamily CARABOIDEA

Family CARABIDAE. Ground beetles (Figs 6.1, 6.2). Adults and larvae inhabit the soil surface. Predaceous, mostly beneficial. Larva an active oligopod, often called campodeiform because of the resemblance to the primitive insect *Campodea*. Cerci present, tenth abdominal segment modified into an anal projection or pseudopod. Tarsal claws double.

Suborder Polyphaga

Superfamily STAPHYLINOIDEA

Family STAPHYLINIDAE. Rove beetles (Fig. 6.3). Adults with short elytra. Larva campodeiform, predaceous, beneficial, tarsal claws single.

Superfamily HYDROPHILOIDEA

Short clubbed antennae, long tactile maxillary palps (Fig. 6.6). Many are aquatic.

Superfamily ELATEROIDEA

Family ELATERIDAE. Click beetles and wireworms (Fig. 6.4). Hind angles of the pronotum project acutely, fore coxae are small.

Superfamily CUCUJOIDEA

Section (*a*) Clavicornia. Tarsal joints numbering 5-5-5 or 3-3-3.
Family CRYPTOPHAGIDAE. These are very small beetles with clubbed antennae (Fig. 6.7A).
Family NITIDULIDAE. Blossom or pollen beetles (Fig. 6.7B). Feed on flowers.
Family COCCINELLIDAE. Ladybirds (Fig. 6.8A, B). Feed as larvae and as adults on adults of aphids and scale insects, beneficial. Occasionally larvae feed on plants.

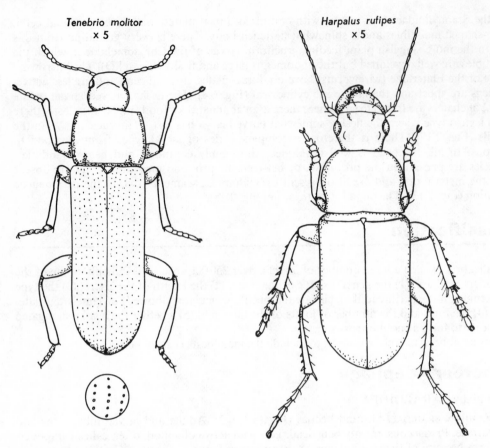

Fig. 6.2 *Tenebrio molitor* (Cucujoidea) – mealworm, tarsal joints 5, 5, 4, and *Harpalus rufipes* (Caraboidea) – tarsal joints 5, 5, 5.

Section (*b*) Heteromera. Tarsal joints 5–5–4 in both sexes.

Family TENEBRIONIDAE. Mealworms (Fig. 6.2). Adults and larvae feed on a variety of stored products. Larvae resemble wireworms, injurious.

Superfamily CHRYSOMELOIDEA

Nearly all members of this group are plant feeders. The tarsi are apparently 4-jointed, the third joint being deeply lobed and carrying the fourth which is vestigial (phytophagous tarsi, Fig. 6.14D). The fifth joint bears two claws.

Family BRUCHIDAE. Pulse or seed beetles (Fig. 6.14E–G).

Family CHRYSOMELIDAE. Leaf eating beetles, flea beetles (Figs 6.8C, D, 6.10–6.14A–C). This is a very large family and the second most important in the Coleoptera from an agricultural standpoint. Larvae fleshy or sluggish or cylindrical for burrowing.

Family CERAMBYCIDAE. Longhorn beetles. Mostly wood feeders.

Superfamily CURCULIONOIDEA

Body is pear shaped and often covered with brightly coloured scales. Head produced into a rostrum. Tarsi phytophagous. Antennae clavate and often geniculate (elbowed). Larvae fleshy with chitinized heads and the labium bears a transverse bar on the inner side.

Family CURCULIONIDAE. Weevils (Figs 6.15, 6.16, 6.17, 6.18). A very important family agriculturally. Larvae apodous with well-developed heads and mandibles. Usually live in concealment surrounded by their food.

Superfamily SCARABAEOIDEA

Antennae lamellate, forelegs fossorial.

Family SCARABAEIDAE. Chafers and dung beetles (Fig. 6.19). Larvae live in soil or dung and are white, curved and fleshy.

Suborder Adephaga

Family CARABIDAE Ground beetles are active, predaceous and inhabit the soil surface. Several species attack strawberries in Great Britain, e.g. *Harpalus rufipes* (Degeer) (Fig. 6.2), *Pterostichus cupreus* (L.), *P. madidus* (Fab.) and *P. melanarius* (Illiger). *Zabrus tenebrioides* (Goeze), a pest of cereals in parts of Russia, Italy and France, occasionally damages wheat and barley seedlings in the U.K. Leaves are shredded to a fibrous mat on the soil surface and dragged to 5 cm below ground by third instar larvae during early May. Common grasses are also attacked and adults, which emerge in July from pupae formed at the end of May, may damage developing grain in the ear: so far this has not been observed in the U.K. A DDT spray applied in May was effective against larvae, prepupae and pupae (Bassett, 1978) but as DDT is no longer approved; HCH could probably be substituted.

Apart from these examples, the usual food of larval and adult stages is insects and other small animals, in fact the complex of ground beetles that occurs in field soils is an important factor on the natural control of many insect pests, e.g. aphids (p. 40), caterpillars (p. 90) and some Diptera (p. 167).

Table 6.1 gives the relative abundance from 1971 to 1975 of eleven species of carabids and

Table 6.1 Relative abundance of twelve common predatory ground beetles in a Rothamsted wheat field from 1971 to 1975. (Data from M.G. Jones, 1976.)

	Agonum dorsale (Pont.)	*Bembidion* spp.	*Clivina fossor* (L.)	*Pterostichus madidus* (Fab.)	*P. melanarius* (Ill.)	*Harpalus rufipes* (Deg.)	*Loricera pilicornis* (Fab.)	*Nebria brevicollis* (Fab.)	*Notiophilus biguttatus* (Fab.)	*Philonthus* spp.	**Tachinus rufipes* (Deg.)	*Trechus quadri-striatus* (Schr.)	Totals
1971	150	63	17	25	216	490	0	2	11	2	7	0	983
1972	293	28	12	22	72	110	12	34	44	13	3	1	644
1973	65	11	16	60	105	417	18	12	16	1	2	15	738
1974	40	13	11	16	98	127	12	67	11	81	34	2	512
1975	737	192	45	1	35	67	41	159	183	38	0	2	1500
Totals	1285	307	101	124	526	1211	83	274	265	135	46	20	4377

Totals of the twelve species

	1971	1972	1973	1974	1975	Totals
April	39	26	20	6	81	172
May	173	129	239	105	918	1564
June	288	271	212	193	361	1325
July	483	218	267	208	140	1316
Totals	983	644	738	512	1500	4377

* Staphylinidae, all others Carabidae.

one staphylinid beetle in a wheat field at Rothamsted. The beetles were captured in pitfall traps and so the numbers cannot be expressed as population densities. Rather they are densities multiplied by a factor related to their activity. As temperature seems to be the main factor influencing activity, the large numbers of some species, e.g. *Agonum dorsale*, caught in 1975 which was hot and dry, may exaggerate the abundance of these beetles in that year. Nevertheless, the more active the beetles the greater their effectiveness as predators. In a general way, Table 6.1 shows how variable species composition is from year to year and that activity is much less in April than in the three following months. The peak activity as indicated by pitfall traps is in June–July for most species but some species continue to be active in August and September, e.g. *Trechus quadristratus* (M.G. Jones, 1976).

Although it is presumed that ground beetles climb plants in search of prey, they are rarely seen on plants during the day when the Hemipteran *Anthocoris nemorum* and some staphylinid and coccinellid beetles are found feeding on aphids.

Suborder Polyphaga

Superfamily STAPHYLINOIDEA

This large group is characterized by the entire absence of cross veins in the hindwing and by having at least some of the abdominal tergites exposed beyond the elytra. The beetles are chiefly mould eaters or predatory and rarely phytophagous. The larvae are oligopod and usually campodeiform.

Family STAPHYLINIDAE There are many species of small predaceous rove beetles (Fig. 6.3A); dung contains enormous numbers of them. The elytra are short and the abdomen is dorso-ventrally flexible. The larvae resemble those of the ground beetles but the tarsi bear only

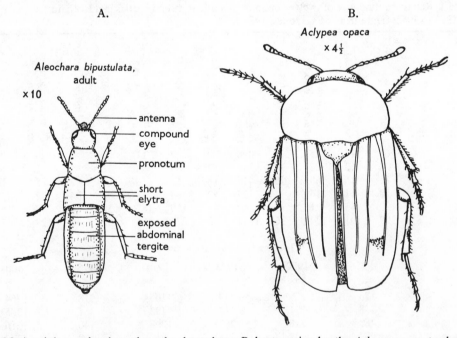

A.

Aleochara bipustulata,
adult

× 10

— antenna
— compound eye
— pronotum
— short elytra
— exposed abdominal tergite

B.

Aclypea opaca
× 4½

Fig. 6.3 A, adult rove beetle to show the short elytra; B, beet carrion beetle, *Aclypea opaca*, to show the flattened form.

one claw. Like the adults most larvae are also predaceous. *Quedius* spp., *Philonthus* spp. and other small forms inhabit the soil surface where they feed on small insects, arthropods, worms, etc. In captivity some feed on nematode-cysts (*Philonthus* spp., *Omalium* spp.), and the larvae of *Aleochara bilineata* (Gyll). *A. bipustulata* (L.) and *A. algarum* (Fauv.) are parasites of cyclorrhaphous flies. *A. bilineata* parasitizes cabbage root fly pupae and has two well-marked larval forms. Larvae of *Tachinus* spp. and *Tachyporus* spp. climb plants and feed on aphids. The devil's coach horse beetle, *Staphylinus olens* (Muell.), is a large, black aggressive predator reaching 3 cm long.

Family SILPHIDAE   Carrion beetles are mostly scavengers, feeding on decaying flesh. The genus *Nicrophorus* includes the well-known burying beetles, which bury the bodies of small mammals and other vertebrates by excavating beneath them. The eggs are laid in galleries leading from the corpse and the first instar larvae are fed by the female. Some species, however, have become plant feeders. The beet carrion beetle, *Aclypea opaca* (L.), attacks cotyledons and foliage leaves of sugar-beet, turnip and other chenopodiaceous plants.

Aclypea opaca (L.) (Fig. 6.3B)   The adult beet carrion beetle is about 12 mm long, dull black, with a reddish pubescence. The elytra are smooth and slightly ribbed. Adults overwinter in woods and other deep shelter, dispersing early in spring when they attack the cotyledons of the beet and mangold seedlings. The eggs are laid in the soil from the end of April onwards and the larvae appear in May. They look rather like black woodlice. Larvae and adults feed together on the foliage leaves and, if sufficiently numerous, may cause serious defoliation especially in May. This is another pest which has virtually disappeared since the introduction of modern insecticides.

Superfamily ELATEROIDEA

Most of the noxious species belong to the Elateridae

Family ELATERIDAE, click beetles or skipjacks (Fig. 6.4)   When the beetles are laid on their backs they struggle for a while and suddenly jump in the air with an audible click. Leaping is connected with the pronounced joint between pro- and mesothorax (Fig. 6.4). A spine from the sternum of the prothorax fits into a cavity in the sternum of the mesothorax. This spine is held with some force but is suddenly released when the body is bent. This causes both the click and the rise into the air. Other characteristics of the family are: (1) insertion of the antennae near the eyes; (2) the serrations of the antennae; (3) the concealment of the head from above by the pronotum; (4) the dull brown to nearly black colouration; (5) the elongate, parallel-sided body with elytra tapering to a point; (6) the elongate, cylindrical larvae with very tough yellow skins.

Larvae of certain species are exceedingly injurious to crops and are known as wireworms (Fig. 6.4). They live in the soil and are most numerous in grass. In the 1914–18 and 1939–45 wars, when much grass was broken for arable crops, many crop failures resulted from wireworms and they were studied extensively.

Wireworms

The natural home of wireworms is permanent grassland, where the abundance of food and shelter provides them with ideal living conditions and leads to the development of large soil populations. When grass is ploughed, wireworms, being polyphagous, turn to arable crops for food, and because of their long life cycle injury persists for about three years, the worst damage often being in the second and third years, declining thereafter. Wireworms attack cereals and all types of root crops. Arable land of long standing does not, as a rule, contain sufficient wireworms to cause trouble, but there is nevertheless an 'arable wireworm' problem in sugar-beet, potatoes and some other crops (Plates IX, XC).

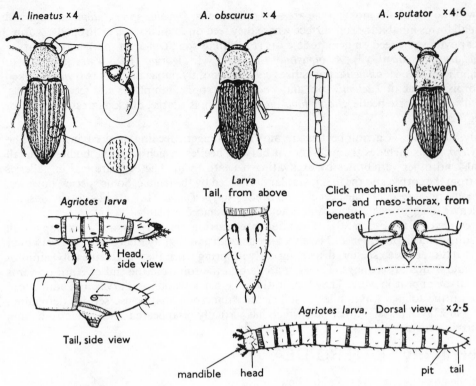

A. lineatus ×4   A. obscurus ×4   A. sputator ×4·6

Agriotes larva

Head, side view

Tail, side view

Larva
Tail, from above

Click mechanism, between
pro- and meso-thorax, from
beneath

Agriotes larva, Dorsal view ×2·5

mandible   head   pit   tail

Fig. 6.4 The three common species of click beetles (*Agriotes*), wireworm and details of the wireworm structure.

There are three common species: *Agriotes lineatus* (L.), *A. obscurus* (L.) and *A. sputator* (L.), all very similar in appearance, life cycle and habits, and a fourth species *Athous haemorrhoidalis* (F.) is also often considered a pest but is less common. *Ctenicera* spp. occur in upland fields. The genus *Agriotes* can be distinguished from other wireworms by the form of the tail (Fig. 6.4). The relative abundance of the three species varies in different parts of a field and in different parts of the country (Brian, 1947). In the north, *A. obscurus* is dominant, but the others become more common in the Midlands and the east. In the south, *A. lineatus* is the commonest species.

The adult click beetles are dull brown insects, 10–16 mm in length. They are found running actively over the surface of the soil and through the grass from April to July. In the daytime they spend much of the time sheltering under small objects and are more active by night, particularly in damp weather. They feed on the leaf bases of grasses and cereals but do no harm. After mating and oviposition, the adults die, so that few can be found after July. The eggs are small, white and spherical and hatch in 3–4 weeks, generally in July. The young wireworms are barely 1 mm long and move little at first. They moult once, a month after hatching, and again later in the year. The life cycle lasts 4–5 years. Moulting occurs after feeding in the early summer and late autumn. The larger wireworms usually observed are 2–5 years old and from 13–37 mm long. They have tough yellow skins, powerful jaws, three pairs of thoracic legs and a bluntly pointed tail. Figure 6.5 gives a pictorial representation of the growth of the larva. Larvae move downwards in summer and winter following temperature and moisture changes in the soil surface. When fully grown the larvae burrow more deeply into the soil where pupation takes place in the late summer of the fourth or fifth year, and the

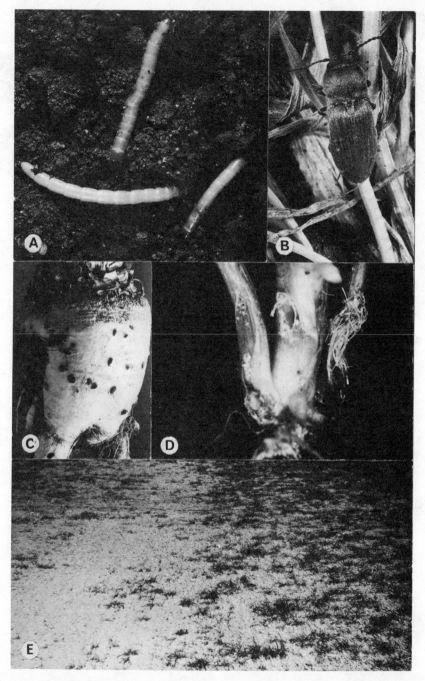

IX **A.** Wireworms in soil (courtesy H. C. Woodville).   **B.** Click beetle (courtesy H. C. Woodville).   **C.** Sugar-beet with wireworm holes (courtesy W. E. Dant).   **D.** Wheat plant attacked by wireworm (Crown copyright).   **E.** Wireworm damage to wheat. Left, after sainfoin; right, after trefoil (less damage) (courtesy W. E. Dant).

X **A.** Tortoise beetle and pupa on beet leaf. Note the shot-hole type of feeding (courtesy W. E. Dant). **B.** Mangold flea beetle injury on sugar-beet seedling (courtesy W. E. Dant). **C.** Wireworm injury on hypocotyl of a beet seedling (courtesy W. E. Dant). **D.** Cabbage stem weevil larva mining the stem of a brassica plant (courtesy J. A. Dunn, N.V.R.S.).

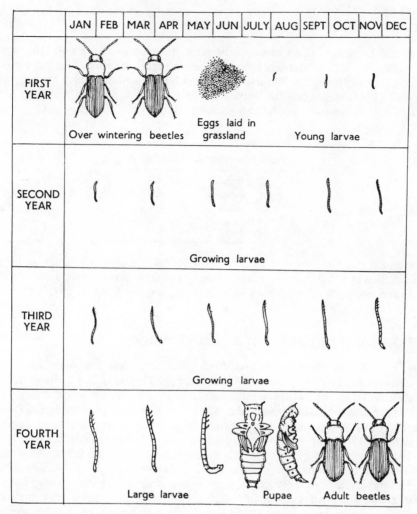

Fig. 6.5 The life history of a click beetle.

adult beetles emerge from the pupa the same autumn but hibernate through the winter and disperse the following spring.

There are two periods of active feeding by wireworms during the year, one in the spring, which coincides with the critical seedling stage of many crops, and another in the autumn, when most crops are mature. The type of injury varies with the crop attacked, but wireworms feed only on juices from the wounds they cause. In cereals the wireworms rasp ragged holes in the ensheathing leaves at the base of the stems, but they burrow right into potato tubers and into carrots. They also burrow into the pith of the haulms of tomato and potato. In sugar-beet seedlings, small black holes are produced on the hypocotyl or on the young root or stem just below ground level and, although the wound is small, it is sufficient to cause wilting and death (Fig. 14.1). Sugar-beet plants are not able to withstand attack until well after singling. Injury to the fully grown roots takes the form of small black pits eaten into the surface.

Wireworm populations and cropping

In wireworms there is a fairly clear relationship between numbers and crop injury. The economic threshold (p. 2) for various crops has been roughly assessed. This assessment varies not only for the individual crop but also with the conditions under which the crop is grown. Salt & Hollick (1944) estimating the population of a meadow on the Cambridge University Farm at 20 000 000 wireworms per ha and grouped them in the following sizes.

Table 6.1 The percentage of wireworms of different sizes.

Length of wireworm	% of total
2–6 mm	59
6–10 mm	26
10–14 mm	9
14–18 mm	5
18–24 mm	1

The farmer and the casual observer usually find only those exceeding 10 mm in length. The number of adults was 400 000 per ha, that is about 2 per cent of the total population, the level expected from a population roughly in balance assuming the females lay 100 eggs (see Fig. 1.4).

Estimations of population in advisory work

During and immediately after the Second World War, the wireworm population of the soil was estimated in many fields as grassland was broken up. This was done by Regional Advisory Entomologists; although wireworm sampling is rarely done nowadays it is interesting to review the methods used and the results obtained because they illustrate important principles.

The method of sampling was to take 20 cores of soil from a field with a 10.2 cm diameter corer to a depth of 15.2 cm from unploughed grassland and to plough depth, about 20 cm from ploughed land. From fields of 10 ha or more, additional cores were taken. An average of 1 wireworm per core represents a population of 1 250 000 wireworms per ha. The method was crude, as the samples ignored the seasonal migration and the increase from eggs hatched from July onwards was not always taken into account. Results also varied with the method of extraction. When the direct, dry method of hand sorting was used the samples accounted for about 80 per cent of wireworms over 10 mm in length and very few of those under 5 mm. The efficiency also depended on the type of soil and the position of the core in the field and on the operators. The wet method of extraction by flotation avoided variables associated with the dry method and almost doubled the rate of extraction (Salt & Hollick, 1944; Cockbull, Henderson, Ross & Stapley, 1945) and gave 95 per cent of the wireworms over 6 mm, that is nearly all the wireworms responsible for injury.

The steps in the process were as follows:

1) Soaking the sample or even freezing it for several days to shatter clays.
2) Washing the soils on three sieves, coarse, medium and fine (about 60 meshes to the inch) from which the wireworms and residue were removed.
3) Flotation of the residue from the fine sieve by brine or concentrated magnesium sulphate.
4) Collection of wireworms and debris from flotation, boiling in water to cause plant debris to sink.
5) The addition of paraffin to the water and the final removal of the wireworms from the water-paraffin interface. (Benzene was also used but the fumes are inflammable and poisonous.)

After sampling and extracting the wireworms by the wet method the field was placed in one of four categories and the farmer advised which crops could be grown safely:

I. 0–1 500 000 wireworms per ha — Crop injury slight.

II. 1 500 000–2 500 000 wireworms per ha — Most crops can be grown with safety provided growing conditions are good. Grow only early varieties of potatoes to avoid autumn injury to tubers.

III. 2 500 000–4 500 000 wireworms per ha — Considerable risk of crop failures.

Suitable crops — Winter: wheat sown early under good conditions, rye and beans.
Summer: peas, linseed, flax, mustard, silage mixture and barley.

Unsuitable crops — Sugar-beet, mangolds, swedes, carrots, spring wheat, spring oats and potatoes.

IV. More than 4 500 000 wireworms per ha — Grow only resistant crops, e.g. beans, peas, clover, lucerne, linseed, flax, mustard and silage mixture.

The resistant crops are less attractive to the wireworms than cereals and grasses. They fall into two main groups: peas and beans which have large seeds and thick stems; and mustard, flax and linseed which have small seed and large plant populations (flax 10 000 000–25 000 000 plants per ha). Potatoes succeed at almost any population density but tubers may be riddled with holes. Early potatoes, lifted early, escape the second, autumn period of feeding, and tubers form after the peak of the first feeding period.

No one now believes that a million wireworms per ha do no damage to crops, or that most crops can be grown with safety when two and a half million are present. Standards have changed and risks which had to be taken before modern insecticides were available are no longer tolerated. 'Cosmetic' injury to root crops, e.g. to potatoes and carrots, makes produce unacceptable. Drilling to a stand has made sugar-beet unsafe without protection if there are as few as 250 000 ha^{-1} (about four per plant).

Chemical control

The chemical control of soil organisms is more difficult than control of above-ground pests because they are protected by a great mass of soil, because incorporation of materials is difficult and because the soil may destroy, detoxicate or adsorb them before they take effect.

Many substances have been tried as soil insecticides but the first real success was with gamma-HCH which has one drawback: it sometimes taints potatoes, carrots and other vegetable crops. For this reason aldrin, which is almost equally as effective, though somewhat slower acting, was often substituted. However, its long persistence is objectionable because of the effect on soil fauna and of residues in crops. At one time aldrin was included in fertilizers (about 0.2% to give 1–2 kg ha^1) but this led to more aldrin being applied than was justifiable. At present aldrin is used only to control wireworms in potatoes after ploughed up grass. It is applied as a spray and worked well into the soil before planting.

When sugar-beet or cereals are to be grown and wireworms are few, seed treatments give adequate protection but do little to diminish wireworm numbers. Powder seed treatments for cereals contain about 20% gamma-HCH and a fungicide. Those for beet contain methiocarb and a fungicide; treatments containing 40% gamma-HCH were discontinued when found phytotoxic to seedlings. To protect sugar-beet now that so little seed is sown, all seed is pelleted and the methiocarb is incorporated in the outer layers of the pellet. When wireworms are numerous, soil treatment is essential; for both cereals and sugar-beet gamma-HCH is the approved insecticide applied as a spray and worked into the seedbed by a single harrowing but if potatoes or carrots follow, 18–24 months should elapse before they are sown. Alternatively, apply bendiocarb, or carbosulphan granules in the seed furrow when sowing beet; such a treatment is less harmful to beneficial insects and also protects against other pests attacking the cotyledons and early true leaves. When, unexpectedly, wireworms are found attacking beet seedlings in April or May, damage can be alleviated by spraying with a gamma-HCH close to the plants and hoeing to mix the insecticide into the soil. As little insecticide is applied the risk of tainting a subsequent crop of potatoes or carrots is lessened. (See Leaflet No. 199.)

Superfamily HYDROPHILOIDEA

Many are aquatic but three species of *Helophorus*, the mud beetles, damage crops. These are drab brown with the ridges on the prothorax and elytra usually covered by mud stuck on their surface.

Family HYDROPHILIDAE

Helophorus nubilus **(F.)** (Fig. 6.6A)   The adult wheat shoot beetle is 4–6 mm long, and active from May to September. Eggs are laid in the autumn, for larvae can be found in November. The larvae resemble small carabid larvae but they are dirty white with characteristic dark, chitinous plates on the dorsal surface, two large ones on each thoracic segment and four on each abdominal segment (Fig. 6.6C). Posteriorly is a pair of cerci. The larvae feed throughout the winter, often when bitterly cold, and become fully grown in March or April when they pupate. The pupae are soft and white and bear bristles.

Helophorus damage is rare nowadays. It used to occur in districts where rabbits were abundant but any connection between rabbit grazing and *Helophorus* damage would be difficult to prove. The larvae attack winter wheat after ryegrass and clover leys but also after sainfoin, lucerne and trefoil leys, especially if broken late (Plate VIIID, F). The larvae are active during the low temperatures of January and February when most other pests are inactivated by cold. They are commonest in light soil overlying chalk in the eastern counties

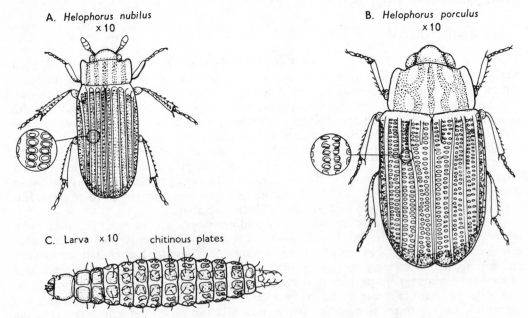

A. *Helophorus nubilus*
× 10

B. *Helophorus porculus*
× 10

C. Larva × 10   chitinous plates

Fig. 6.6 A, wheat shoot beetle, *Helophorus nubilus*, and C, larva; B, turnip mud beetle, *H. porculus*.

where short ley, rotational farming is practised but there are also records of damage from other wheat-growing areas. Injury caused is similar to wireworm feeding but the plant tissues are macerated or ragged. The larvae bite little slits or holes below the first node or even higher up, causing the central shoot to yellow, while the side leaves remain green; sometimes the whole plant turns yellow. Usually the damage is patchy but, in badly attacked fields, the wheat may fail entirely and have to be resown. The larvae are difficult to find but can be extracted from soil samples by the wet method used for wireworms (p. 100). Although *Helophorus* injury usually follows a ley, occasional damage has occurred when wheat has been grown after wheat. There are no reports of damage after sugar-beet, beans or potatoes. Larvae occur in old grassland but, as populations rarely exceed 60 000 per ha, there is no significant damage to wheat sown after ploughing.

To prevent damage by *Helophorus* it was customary to break the leys in July or August, for fields ploughed in September were liable to attack. It was essential also to sow wheat early, in mid-October: this tillered well and, even if attacked by the larvae, suffered little harm (Petherbridge & Stapley, 1944).

Helophorus rufipes **(Bosc d'Antic)** and *H. porculus* **(Bedel)** (Fig. 6.6B)   Turnip mud beetles are very similar, *H. porculus* being smaller than *H. rufipes* and having rounded humeral angles to the elytra. The larvae and pupae are similar to *H. nubilus*. The winter is passed as the larva, and pupation occurs in March and April in an earthen cell 5 cm below ground. The adults emerge from June onwards, eggs are laid in July and August, and the resulting larvae feed on turnips until the following March.

Adults and larvae cause injury chiefly to white turnip (*Brassica rapa*) in which the growing point may be destroyed. They also attack swedes, beans, lettuce, kale and cabbage. They bore tunnels into the roots and stems and both the adults and the larvae eat the edges of the leaves. Damaged swedes and turnips tend to rot in the ground.

Damage has been reported from various parts of the country but attacks are sporadic. Trouble occurs in fields cropped too often with brassicas and does not usually occur to turnips grown in a normal rotation unless susceptible crops, in which the beetles build up their numbers, have been grown frequently.

Cultural measures of control may be employed. Turnips should be sown early and should not be taken where a crop has been attacked. A late summer fallow is recommended to rid land of larvae. Control with gamma-HCH is possible.

Superfamily CUCUJOIDEA

The Cucujoidea includes many species with filiform or clubbed antennae.

Section Clavicornia

Family CRYPTOPHAGIDAE.

Atomaria linearis **(Steph.)** (Fig. 6.7A)   The pigmy mangold beetle is a tiny brown beetle about 2 mm long which before 1935 ranked as the most important single seedling pest of sugar-beet. Where beet was grown after beet, large populations occurred and hundreds of acres had to be resown. After 1935 the pest declined in importance following the introduction of a rotational clause into factory contracts forbidding the cultivation of beet after beet, mangolds or red beet, although this clause was aimed primarily against beet cyst-nematode (p. 223) (Plate VIII). This clause has now been rescinded and there is no longer any direct control over cropping. Farmers in beet growing areas should be aware of the peril of growing beet after beet.

Young adults are pale yellowish-brown while the older ones are dark brown (Bombosch, 1955a and b). The elongate body has parallel sides and the thorax is finely punctured. The elytra are covered with a pale short pubescence. The brown antennae have three dilated apical segments and the legs are reddish brown. The beetles appear in large numbers in beet fields in the autumn, and populations exceeding 2 500 000 per ha have been detected by soil sampling. The main dispersal flights are in spring and early summer on warm, calm days. Little flight occurs in autumn and the beetles spend the winter in the soil of the old beet fields. Many may be found under the clods of earth, in debris or feeding around old beet crowns, especially the last, where they aggregate in large numbers. Dispersal flights occur with the first fine spring weather from mid-April to May or June. Eggs are laid in the soil around plants in new beet fields where the larvae feed on the roots. When the larvae are fully grown they are 3 mm long, grey in colour and weakly sclerotised.

Beetles already present from autumn flight or when beet follows beet begin to feed on sugar beet and mangold in March, chiefly below ground but under moist conditions above ground. Characteristic pits are eaten into the hypocotyl or small circular areas are eaten from the cotyledon, the terminal bud and the foliage leaves. The pits in the hypocotyl turn black. As the foliage leaves are folded in the terminal bud, injury is often symmetrical. The most serious

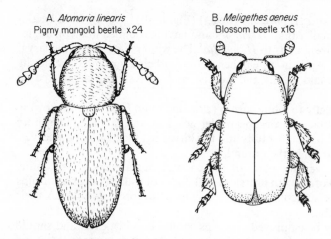

A. *Atomaria linearis*
Pigmy mangold beetle × 24

B. *Meligethes aeneus*
Blossom beetle ×16

Fig. 6.7 A, pigmy mangold beetle, much enlarged; B, blossom beetle, much enlarged.

injury arises from bites below ground from germination to the early cotyledon stage (Fig. 14.1). Beetles may chew the radicle before it has emerged by entering splits in the seed pellet. The effect on the crops depends on the beetle population and the growth stage of the plants. If large numbers of overwintering beetles are present, every seedling has many bites and many wither and die. Once past the cotyledon stage the plants usually survive. Later injury to the tap root and feeding on the terminal shoots of seed crops are of little importance. Red beet, fodder beet and spinach are also attacked, and although the beetles can be induced to feed on a variety of plants in the laboratory they pay little attention to them in the field. The large numbers of adults that accumulate in beet fields in July and August cause little damage as the plants are long past the susceptible stage.

A. linearis does not appear able to transmit viruses. This is fortunate because, before dispersing in the spring, many feed first on old beet crowns which have survived the winter. In part-cropped fields where the current crop overlaps beet grown in the previous year, the area where seedling are destroyed often coincides within a metre or so of the previous beet, showing that the beetles do not move far in winter. (See Leaflet No. A589.)

Not all crop failures occur after beet, mangolds or red beet, for severe thinning and loss of stand can arise in fields grown in rotation. These losses are probably due to beetles that have dispersed from last year's beet fields and have become more serious as seed rates have decreased and drilling to a stand becomes more prevalent; they occur most frequently in intensive beet growing areas. The population of affected fields may rise to 7 500 000 beetles per ha, and may even cause the failure of the second drillings of beet. As a rule, however, too few beetles arrive early enough to cause severe damage.

Rotation obviously provides the most economic method of control. The standard insecticidal seed treatment of methiocarb incorporated in seed pellets is fairly effective. Gamma-HCH applied as dust or spray and worked into the soil is advisable where injury is prevalent but routine treatment cannot be recommended since this type of attack is so sporadic. For crops under attack, band spraying the seedling rows with gamma-HCH in enough water to ensure a run off gives partial control. The methiocarb treatment now applied to pelleted beet seed gives a measure of protection.

Family NITIDULIDAE (Fig. 6.7B)   Blossom or pollen beetles are small, active, flat insects. The antennae have eleven joints, the last three forming a club. The coxae are cylindrical and the tarsi 5-jointed. They are unusually active flyers.

Meligethes aeneus (F.)   The bronzed blossom beetle occurs in large numbers in the flower heads of cruciferous crops grown for seed and in many other flowers. It is a pest of cabbage, swede and turnip seed, brown mustard, white mustard and oilseed rape.

The adult is a small, 1–1.5 mm long oval beetle, black with a greenish bronze tint, finely punctured and with a fine pubescence. It is usually well dusted with pollen. The outer margin of the anterior femur is finely toothed. The adults hibernate and emerge during the first warm days of spring. In April, May and June they feed on unopened flower buds and oviposit in them. An important wild host was charlock but this has almost disappeared from cereal fields as a result of hormone weedkillers. The eggs hatch in 7–10 days and the larvae live concealed in the buds and flowers, feeding mainly on pollen. If numerous, the pistil is destroyed and often the tips of the racemes as well. There are two instars and pupation occurs in the soil. There is one generation a year, the adults hibernating in the soil during the winter.

The chief damage is destruction of the flowers, mainly by adults, leading to loss of seed. The larvae are important only when present in abnormally large numbers. If the infestation is less than 25 per cent of the flowers they may even have a beneficial effect by acting as pollen carriers. In mustard, the tips of the racemes suffer most severely from larval feeding. If the beetle appears when the flowers are in bloom, it often eats the nectaries when feeding on the pollen and does little real harm to the gynoecium. If it appears before flowers open it makes its way into the flower buds in search of pollen and destroys them. Injured buds wither and fall off leaving blind stalks. The beetle and its close relative *M. viridescens* (Fab.) are abundant and maintain themselves on many flowers moving from one flowering crop to another when the weather is favourable for flight.

Winter oilseed rape and early sown brassica seed crops suffer less from feeding by pollen beetle adults and larvae than spring sown oilseed rape. The former are more than 1 m tall and have well-developed racemes and may be flowering when pollen beetles arrive in May. They have many potential flowers and are well able to compensate for any damage done by the feeding of adults and larvae. Spring rape and late sown brassica seed crops may be no more than 10 cm tall with only the beginnings of an inflorescence at this time. If there are 15–20 beetles per plant in the green bud stage, control is necessary and one of the approved sprays (azinphos-methyl mixture, endosulfan, HCH, malathion) should be applied. If flowering has begun, damage has occurred and the best time for spraying is past. Because of the possibility of killing bees, no spraying should be done while the crop is in flower nor if there is an 'undercroft' of flowering weeds. Early sprays of azinphos-methyl mixture or gamma-HCH also control stem weevil, seed weevil and cabbage aphid. Endosulfan, although not entirely harmless, is less toxic to bees than azinphos-methyl mixture, gamma-HCH or malathion. (See Leaflet No. 576.)

Family COCCINELLIDAE (Fig. 6.8A, B)   Ladybird beetles are moderately sized beetles with convex bodies and heads partly concealed by the pronota. The third tarsal segment is deeply

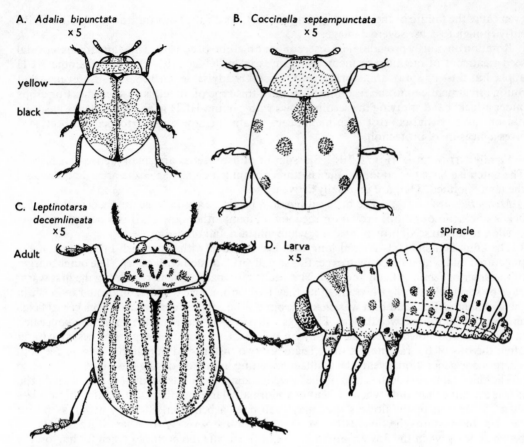

A. *Adalia bipunctata*
x 5

yellow

black

B. *Coccinella septempunctata*
x 5

C. *Leptinotarsa decemlineata*
x 5

Adult

D. Larva
x 5

spiracle

Fig. 6.8 Two common ladybird beetles, and the adult and larva of the Colorado beetle.

bilobed. Ladybird larvae and adults feed on aphids, coccids and occasionally on other soft-bodied insects. The adults are usually brightly coloured. The larva is elongate, slightly fleshy, black or grey with red or yellow markings. The beetles hibernate, often in large groups, although the meaning of this gregariousness is obscure, and become active in the spring and solitary in habit. After pairing, yellow eggs are usually laid in batches in or near aphid colonies, e.g. *Aphis fabae* on beans, so that the oligopod larva does not have to search far for its food. However, when the aphid infestation is light, the newly hatched larva has to search for its food. The larva of *Adalia bipunctata* (L.) consumes about 15–20 aphids daily and the number of aphids consumed during the entire larval period varies between 200 and 500. The adults are even more voracious, the number consumed varying with the temperature. When fully fed the larvae pupate on leaves or nearby structures forming a plump chrysalis, attached by the end of the abdomen, and often conspicuously coloured. The pupal skin eventually splits down the back and the adult emerges.

The commonest species is *Coccinella septempunctata* (L.) which is red with seven black spots on the elytra. The 'two spot' *Adalia bipunctata* (L.) is also common and varies greatly in colour from the red form with two black spots to a yellow form with two brown spots. Once the ladybirds arrive on beans with heavy aphid infestations, they destroy aphid colonies rapidly. The date of their appearance is determined by the distance of the bean plants from shelter or from the generation reared on nettles, so if their arrival is late they may fail to control the bean

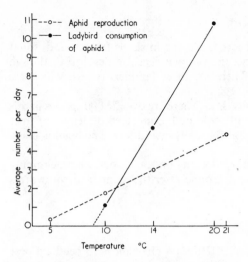

Fig. 6.9 Graph to show the relationship between aphid reproduction and ladybird feeding at different temperatures (Dunn, 1952).

aphids (Banks, 1955). Dunn (1952) has shown that the temperature affected the feeding rate of *C. septempunctata*, but the reproductive rate of the pea aphid *Acyrthosiphon pisum* (Harris) was not equally affected (Fig. 6.9). Above 10°C the consumption potential of the ladybird exceeded the reproduction potential of the aphid and therefore was capable of controlling the aphid infestation. Below 10°C the aphid populations built up more rapidly than could be checked by the ladybird. Warm summer weather, therefore, increases the feeding capacity of the ladybird and the control of the aphid populations, whereas low spring and autumn temperatures aids the build-up of aphid colonies. During favourable weather ladybirds and other enemies may become so numerous that aphids have difficulty in coming through the winter. A small black ladybird, *Stethorus punctillum* (Weise), feeds on fruit tree red spider mites.

Biological control of the scale insect, *Icerya purchasi* (Mask.) in California as a result of the introduction of *Rodolia cardinalis* (Muls.) has been so successful that the beetle has been imported into all countries where the coccid has become injurious.

Section Heteromera

The tarsi of the fore and middle legs have five joints but those of the hind leg only four joints.

Family TENEBRIONIDAE (Fig. 6.2) Tenebrionid beetles resemble ground beetles and their larvae resemble wireworms. The family contains over 10 000 species of which *Tenebrio molitor* (L.) and *T. obscurus* (F.) are almost cosmopolitan. All stages occur in meal, flour and stored goods hence the larvae are called 'mealworms'. *Tribolium castaneum* (Herbst) and *T. confusum* (J. du V.) are likewise widespread in granaries and stores.

Mealworms usually pass the winter as larvae; adults appear in the spring and die after mating and oviposition. There is usually one generation a year, but when temperature and relative humidity are favourable breeding goes on throughout the year. The larva of *T. obscurus* is dark brown and that of *T. molitor* is bright yellow. Larvae moult from 12–22 times with an average of 14 or 15 times and are very resistant to starvation. They are often found with mites, *Caloglyphus berlesei* (Mich.), attached.

Periodic thorough cleaning of granaries and depositories is desirable. Fumigation is effective and removes infestations that have insinuated themselves into unreachable crevices but must be done by approved contractors (p. 274).

Superfamily CHRYSOMELOIDEA

This is a large group of plant and wood feeding beetles, ranging in shape from round, squat chrysomelids to elongate cerambycids. In the larva the thoracic legs are developed, though often small and the antennae have three joints. There are three families, two of which are mentioned.

Family CHRYSOMELIDAE (Figs 6.8C, D, 6.10–6.14A–D)   There are over 26 000 species in this family. The adult beetles of many species are brightly coloured. Some feed on leaves of trees and others on herbage. The larvae are soft bodied and usually cylindrical. The head is black but the body is often red or yellow.

Injury to plants may be caused by beetles and larvae feeding on leaves, or by larvae mining within stems or leaves, or feeding on the roots below ground after the manner of wireworms (p. 97).

Colorado beetle

Leptinotarsa decemlineata (Say.) (Fig. 6.8C, D, Plate XIIE–H)   The Colorado beetle, a pest of potato, is well known to everyone because of the publicity given on posters, advertisements and radio. It was originally confined to the semi-desert areas in Colorado, U.S.A., where it fed on wild species of *Solanum*, especially *Solanum rostratum*. It spread rapidly when the country was opened up, helped by human transport and by the cultivation of potato which proved an excellent host. It travelled about 80 miles a year and reached the Atlantic Coast in 1874. From there it invaded Europe. In 1877, breeding colonies appeared in Germany and a single beetle was found on Liverpool Dock. An introduction at Tilbury in 1901 was successfully eradicated, but later the beetle gained a hold in the Bordeaux region of France, where the terrain is difficult for eradication. Since 1933 there has been an increasing number of outbreaks in Europe. The Second World War brought a relaxation of vigilance and the vast movements of men and materials helped it to spread across Europe. The spread eastwards has continued into the U.S.S.R., Turkey and Baltic Sea area. In 1970, in the Colentin peninsula of France, the 70% infestation of the potato crops led to areas of complete defoliation. Spasmodic outbreaks have been notified in Britain and have been dealt with promptly (Bartlett, 1979). There is always a danger, however, that isolated outbreaks may be missed and the beetle remains a potential pest. Most field outbreaks have been small, consisting of isolated beetles or of one beetle with 200–300 larvae. The whole colony site including potato foliage is covered with a plastic sheet and fumigated with methyl bromide. If the site is too uneven or too steep, DD or some other fumigant is injected. All potato crops in the area are inspected and sprayed. The farmer concerned must leave the crop so that any beetles missed will not disperse, and must plant potatoes as a catch crop on the same land the following year to hold any beetles that overwinter in the field. There is an embargo on the import of potatoes and other produce containing beetles enforced by quarantine service at the ports. So far establishment has been prevented. No breeding colony was found in Britain between 1952 and 1976. In that year a colony was found in Kent and many beetles were imported in a consignment of rye seed from the Netherlands. Another colony was found in Hastings in 1977. These outbreaks followed two years that had been favourable for the beetle climatically (Aitkenhead, 1981; Bartlett, 1979).

Dunn (1949) studied the spread of Colorado beetle in the Channel Islands, where he found that the normal flight range of the adult was 1.5 miles. Occasionally swarms occur in the spring and autumn, and air currents may carry them beyond their unaided flight range. Adult beetles are able to survive immersion in sea water. Swarms and sea water survival were probably responsible for the invasion of the Channel Islands by these beetles.

After the Second World War, during periods when mass movements of beetles into Britain from the Continent seemed likely, potatoes around the ports (London, Tilbury, Harwich) and throughout Kent were given protective sprays with DDT. This was discontinued about 1955.

The adult is about 1.2 cm long, hemispherical, bright yellow with black markings on the thorax and five longitudinal lines on each elytron. The adults hibernate in the soil and become active in May. The females lay several hundred eggs in batches on the undersides of potato leaves. The eggs hatch in 4–9 days and the larvae feed on the leaves of the host plant. They are orange red, with lateral black spots and fleshy with a markedly humped abdomen. The larvae feed voraciously, pass through four instars and become full grown 10–22 days after hatching. The larvae then burrow into the ground to pupate. After 10–19 days, the adult beetle emerges and crawls out of the soil. Eggs may be laid after a few days' feeding, commencing a second brood, for there are usually two overlapping generations a year. Fortunately, proceeding northwards climatic conditions in Europe gradually become more adverse. In northern Germany and Holland there is usually only one generation per year. The climate, especially in the north of Britain, and modern insecticides make establishment of this pest less serious than at one time thought.

Both larvae and adults feed on the leaves of the potato, leaving only gaunt stems and the larger leaf veins soiled with excreta.

In countries where the beetle is established, control is possible by using one of the many stomach or contact insecticides as both adults and larvae are exposed on the leaves. The old method was to spray with lead arsenate, when anything from 2 to 10 applications were given depending on the abundance of the beetles. Calcium arsenate was used as a dust. These have been replaced by organophosphorous and carbamate insecticides some of which are compatible with blight sprays. The number of applications necessary depends on the district and the strength of the local population. With care the number of applications may be reduced to a minimum. Only one or two are necessary in north Germany. (See Leaflets No. 71 and RB No. 53.)

Tribe HALTICINI   Flea beetles form a well-defined group of small chrysomelids with hind femora enlarged for leaping and elongate cylindrical larvae.

Cruciferous flea beetles

There are eight species of the genus *Phyllotreta* which feed on cruciferous plants such as turnips, swedes, radish and cabbage: *P. nemorum* (L.), *P. undulata* Kuts., *P. atra* (F.), *P. nigripes* (F.), *P. cruciferae* (Goez.), *P. consobrina* (Curt.), *P. aerea* (Allard) (=*punctulata* (Marsh.)), *P. diademata* (Foud.). Six of these species are common and all can be considered together (Fig. 6.11).

The adults hibernate from October to March in hedgerows, copses, stacks and other available shelter. When the temperature rises in the spring the beetles move out of hibernation. At first they hop about near the hedgerows, feeding on weeds, but gradually pass on to the adjoining fields. They tend to move with the prevailing wind, and general dispersal does not usually occur until the first spring heat waves raise the midday temperature to 20°C when they feed in large numbers on any nearby cruciferous crop. They are active on fine days, particularly towards the end of April, after which they disperse and become widespread, causing damage in open fields. They feed on the cotyledons and slender stems of the seedling cruciferous plants, beginning while the plants are still below ground, so that the plants wither in hot sunshine (Plate XIA, B). Characteristic pits are made in the upper epidermis. Towards the end of May the beetles become less active, they pair and eggs are laid, usually in the soil. The larvae hatch and feed on roots and resemble wireworms in form and habits (Fig. 6.10). The larvae of *P. nemorum*, however, are leaf miners, although the eggs are laid in the soil. Larvae feed during the summer and pupate after about four weeks. The pupal stage lasts another four weeks and the adults emerge at the end of July and early August, when they feed in large numbers on the mature crops. Plenty of food is available at this time but very large numbers may cause damage

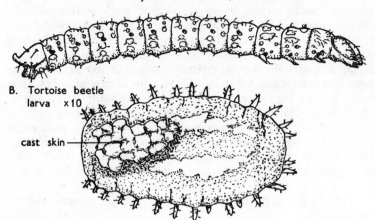

A. *Phyllotreta cruciferae*, Larva ×30

B. Tortoise beetle larva ×10

cast skin

Fig. 6.10 A, larva of a flea beetle (*Phyllotreta cruciferae*); B, larva of a tortoise beetle (*Cassida* sp.). (A, after Newton, 1928.)

to Brussels sprout, kale and broccoli. Heavy attacks sometimes occur when beetles migrate after brassica seed crops are cut.

In the autumn the beetles move into shelter to hibernate. On days suitable for flight they move, aided by the wind, into hedgerows and copses where they are trapped by the lower shade temperatures. They are most numerous in deep shelter and do not remain in the open on old cruciferous crops. The number of inactive beetles hibernating in a given shelter depends on its position relative to current cruciferous crops. In spring they emerge in the reverse order from that in which they went into hibernation. *P. nigripes* is usually the last into cover and the first out; *P. consobrina* is also an early beetle, *P. nemorum* and *P. undulata* come out next, while *P. atra* and *P. cruciferae* are the last to emerge.

The flea beetle population of any field is a drifting one and different species dominate in different parts of the country. *P. atra* and *P. cruciferae* are most damaging in the eastern counties, *P. nigripes* and *P. undulata* in the south and *P. undulata* in the north. On fine warm days in spring the upper layers of the soil are energized by the sun and the surface temperature rises especially, so providing a suitable microclimate for flea beetle activity while plant growth is slow in the relatively colder deeper soil. Injury suffered by the crop depends on the beetle population and its activity. Hot sunny weather with no wind is ideal for them but intermittent hot and cool weather with showers is adverse. In moist weather, plants grow better and escape serious injury. The period of maximum abundance and activity is usually between 10th April and 20th May when most damage is done.

As the period of greatest activity is between 10th April and 20th May, cruciferous crops are vulnerable between these two dates. At the first sign of activity the crop should be sprayed with gamma-HCH. However, preventive measures are best. Brassica seeds should be treated with HCH which gives control of the beetles feeding in the soil and also protects the small plants for some time after emergence above soil. When attacks are heavy further protection can be given by spraying with gamma-HCH.

Flea beetles are known to transmit turnip yellow virus (p. 65) as they regurgitate infective juices during feeding.

Descriptions of adults, eggs, larvae and pupae and injury to plants are summarized in Table 6.2. (See Leaflet No. 109.)

Table 6.2 Flea beetles.

FEEDING ON CRUCIFERAE

Name	Adult	Damage	Distribution	Egg	Larva	Larval habits	Pupa	Natural enemies
Phyllotreta nemorum (L.) Turnip flea beetle Length 2.5–3.5 mm	Oval, with two yellow, slightly wavy stripes on elytra, head closely punctured, thorax rounded, punctured. First three joints of antenna pale, tibiae reddish yellow	To leaves and hypocotyl below ground	Common	Yellow and rounded, surface finely pitted, laid on soil in June	Yellow with black head and dark brown chitinous plates, 6 mm long	Forms blister-like mines in mesophyll of leaf	In soil	Braconidae: *Diospilus morosus* (Reinh.) *Perilitus aethiops* (Nees) on larvae
Phyllotreta undulata (Kuts) Length 2–3 mm	Two broad yellow bands on elytra, antennae black with reddish basal joint, thorax densely punctured with coppery tinge, tibiae black, basal third yellow	To leaves and hypocotyl below ground	Common	Yellowish-white, laid singly on soil surface	White, black plate on last abdominal segment, roughened surface, pointed, 4–5 mm long	Feeds on roots underground	In soil	
Phyllotreta nigripes (Fab.) Length 2.5 mm	Long and flattened, greenish blue, black antennae, six or eight irregular rows of small punctures on head, fine punctures on elytra, legs black	To leaves and hypocotyl below ground	Common	Pale yellow laid singly on soil surface	White, last abdominal segment white with no dorsal black plate, well-formed hairs present, 6 mm long	Feeds underground	In soil	Braconidae: *Perilitus aeteolatus* (Thoms) *Mesochorus* spp.
Phyllotreta atra (Fab.) Length 1.75–2.25 mm	Black with black legs, head and elytra strongly punctured, second and third joints of antennae red or yellow	To leaves and hypocotyl below ground	Very common	Pale straw yellow laid singly on soil surface near wild mustard	White with shining black spots on thorax and abdomen, black head, prothorax, last abdominal segment has no dorsal plate and bears a small projection with spring, 5 mm long	Feeds underground on roots	In soil	Ichneumonidae: *Gelis carinatus* (Forst.) on larva
Phyllotreta cruciferae (Gz) Length 1.8–2.5 mm	Shining black with metallic lustre, punctures on head, thorax and elytra, punctures on elytra regularly arranged, second and third joints of antennae red	To leaves and hypocotyl below ground	Common	Pale yellow white, with polygonal pits and fine sculptures 0.3 mm long	White, elongate, first thoracic and ninth abdominal segments more highly chitinized than the rest, 5–6 mm long	Feeds underground on cruciferous roots	In soil	Ichneumonidae: *Gelis carinatus* (Forst.) on larva
Phyllotreta consobrina (Ct.) Length 2–2.5 mm	Dark, with bluish tinge, punctures on elytra not in rows, fourth and fifth joints of male antennae black and dilated	To leaves and hypocotyl below ground	Less common	Deep yellow and round	White with weak setae, pointed anal shield with posterior depression	Feeds underground on roots	In soil	

Table 6.2 Flea beetles (*continued*).

Name	Adult	Damage	Distribution	Egg	Larva	Larval habits	Pupa	Natural enemies
FEEDING ON CRUCIFERAE (*contd*).								
Psylliodes chrysocephala (L.) Cabbage stem flea beetle Length 4-4.5 mm	Oval, blue-black, with punctures in well-marked rows, antenna 10-jointed, first joint dark, legs brownish red, with dark posterior femora and tibiae	To stems	Not common	Ovoid yellow on soil	Creamy white head, pronotum and last abdominal tergite brown, three transverse rows of hairs on abdomen, two strong crotchets on last abdominal segment, 7-8 mm long	Bores in stems and midribs of cabbage, cauliflower and swede stecklings	In soil	
FEEDING ON CEREALS								
Phyllotreta vittula (Redt.) Barley flea beetle Length 1.25-2.5 mm	Black with greenish metallic tinge, yellow non-converging stripes on elytra, legs black, tibia slightly lighter at base, antennae of males not dilated	To leaves (barley)	Uncommon	On surface of soil	Yellow, eight small dark plates on dorsal surface of meso and metathorax, 4 mm long	Mines the stems and leaf petioles	In soil	
Crepidodera ferruginea (Scop.) Length 3-4 mm	Brown, 11-jointed antenna, punctures in regular rows, legs yellow, first tarsal joint enlarged	Polyphagous to stems of wheat, oats and barley	Common on weeds in Europe	Singly on soil	White, brown head and anal plate, pointed at each end, abdominal segments bear two transverse series of five plates	Bores into stems of wheat and grasses, attacked plants turn yellow	In soil	
FEEDING ON POTATOES								
Psylliodes affinis (Payk.) Length 2.25-3.75 mm	Yellow to reddish brown large punctures in rows on elytra, pale antennae and legs, posterior coxae dilated	To leaves	Common on wild solanaceous plants	Oval, yellow surface minutely sculptured	White with light brown head capsule, smooth plate on last abdominal segment, 5-6 mm long	Underground, feeds on roots forming mines	In soil	
FEEDING ON MANGOLDS AND SUGAR BEET								
Chaetocnema concinna (Marsh.) Length 2.5-3.5 mm	Bronze, punctures on thorax and in rows on elytra, head small, punctures near each eye, antennae and legs reddish brown, projection on hind tibia	To leaves, serious in seedling stage	Common	Elongate, laid singly on soil	Elongate, white, brown head, light brown anal plate, three strong setae on prothoracic sternal plate, 5-6 mm long	Underground, feeds on roots of Polygonaceae	In soil	

Other flea beetles

The genus *Psylliodes* can be distinguished by the insertion of the elongated first tarsal joint well before the end of the tibia.

Psylliodes chrysocephala **(L.)** (Fig. 6.12)   The cabbage stem flea beetle is 4–4.5 mm long. Adults appear in summer but rarely in sufficient numbers to injure crops. After a period of aestivation in July, egg laying begins in August and continues until winter. In autumn and winter, larvae mine the stems and petioles of spring cabbage, kale, winter oilseed rape and swede seed stecklings. When many larvae are present the plants become flabby, lose their leaves, collapse and die. The larvae leave the plants in February and burrow into the soil where they pupate. After laying eggs, adults die or hibernate and produce more eggs the following spring. Females may live for 18 months and lay over 800 eggs. Larvae hatching from spring eggs have a shorter feeding season than those from autumn eggs (Williams & Carden, 1961).

Crops sown early, before mid-July, suffer less than those sown later. New crops should be sown as far apart as possible from old ones. As swede seed stecklings in south Lincolnshire and west Norfolk may be riddled with tunnels, the seed is drilled in open fields and singled, rather than being raised in steckling beds and transplanted, so the plants are better able to withstand attack. Crowded beds of swede seed stecklings or brassica transplants seem to be more attractive to adults (Plate XIA). The pest has spread more widely and increased in importance with the increase in the area under winter oilseed rape.

If possible, adult beetles should be killed before they lay eggs, and fields on which the pest occurs can be sprayed in August with gamma-HCH. To kill larvae, infested plants can be sprayed with gamma-HCH between October and December. It is economic to spray winter rape when there is more than one larval mine per 7.5 cm of plant height (Williams & Carden, 1961).

Phyllotreta vittula **(Redt.)** (Fig. 6.11)   The barley flea beetle feeds and lays its eggs on barley and sometimes on wheat. Larval attacks are of short duration and treatment is not usually necessary.

Aphthona euphorbiae **(Schr.)** and *Longitarsus parvulus* **(Payk.)** (Fig. 6.12)   Flax and linseed flea beetles are abundant on flax and linseed and are also found on species of *Euphorbia* but they rarely cause crop failure. A good seed bed usually ensures a successful crop but gamma-HCH spray can be used.

Psylliodes affinis **(Payk.)** (Fig. 6.12)   The potato flea beetle is large and brown and is found on solanaceous plants such as *Solanum dulcamara* and *S. nigrum* where it breeds. It sometimes feeds in large numbers on potato leaves in August but is unimportant.

Chaetocnema concinna **(Marsh.)** (Fig. 6.12)   The adult mangold flea beetle feeds on sugar-beet, mangolds and polygonaceous weeds. Cotyledons and early leaves may be severely pitted and susceptibility to herbicide damage increased. Young seedlings may be badly set back by feeding damage, but rapidly become resistant if growing conditions are good. The incidence of severe injury varies greatly from year to year and certain districts, such as the Vale of York and central East Anglia, around Bury St Edmunds, suffer more than elsewhere (Plate X).

Outbreaks can be controlled by spraying promptly with gamma-HCH as soon as the first signs of injury appear. Seed dressings containing a high percentage of gamma-HCH, as used for flea beetle control on cruciferous crops, cause severe phytotoxicity and cannot be used. All beet seed is now pelleted and treated with methiocarb.

Crepidodera ferruginea **(Scop.)** (Fig. 6.12)   The normal host plants of the wheat flea beetle are thistles and related plants. The larva enters wheat plants by accident while moving through soil in search of suitable hosts. It causes injury to the shoot resembling that of wheat bulb fly. The centre shoot turns yellow while the leaves remain green. The attack is noticed in March and is usually unimportant.

114

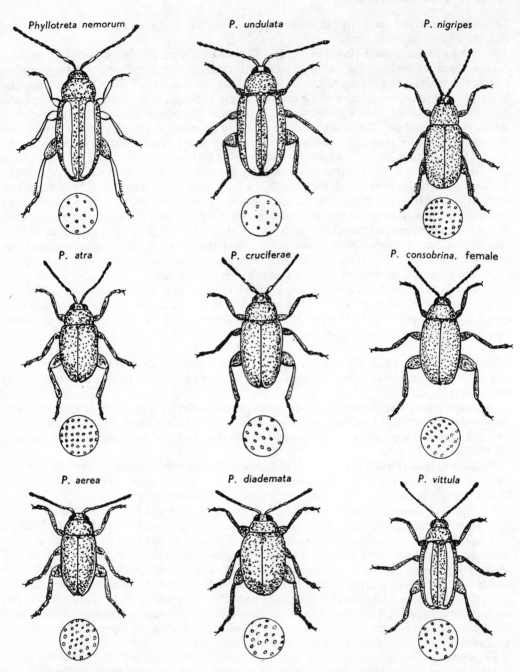

Fig. 6.11 Flea beetles with insets showing punctations on the elytra (all × 10).

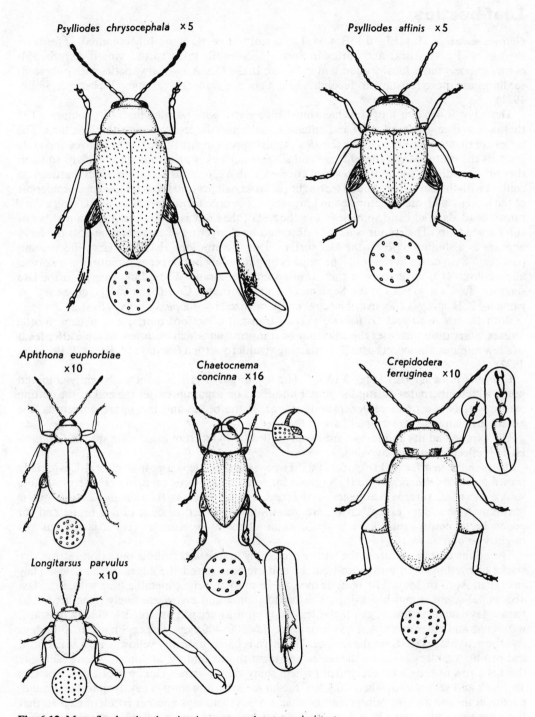

Fig. 6.12 More flea beetles showing important characters in inset.

Leaf beetles

Oulema melanopa **(L.)** (Fig. 6.13B, C)   The cereal leaf beetle is brightly coloured. It feeds on the leaves of oats, wheat and barley in Spring and readily takes to the wing if disturbed. It occurs sparsely but is locally abundant in Wales. In the U.S.A., wheat varieties with pubescent seedlings are more resistant to feeding adults than are glabrous varieties (Webster & Smith, 1971).

The adult is 4–5 mm long and has shiny, blue elytra with parallel lines of punctures. The thorax is orange, the head, eyes and antennae black and the legs yellow with black tarsi. The males are slightly smaller than the females. Adults appear in April, feed on oat leaves and mate towards the end of May. The eggs are laid singly in or near to the mid-rib a week or so later; they are cylindrical, shining yellow and covered with a glutinous secretion which hardens on contact with the air. After about a fortnight the larva hatches and feeds on the upper epidermis of the leaves which are skeletonized in long strips. The eruciform larva (Fig. 6.13C) has a dark brown head, dark legs and spiracles while the rest of the body is yellow and bears a number of stiff curved hairs. The colour is usually obscured by a covering of excrement and so the larva appears as a shining blob on the leaf surface. It is 4–6 mm long, has four larval instars and pupates 7–8 cm down in the soil. The pupa is bright yellow but darkens rapidly to the normal adult colour. It is covered by a thin, transparent membrane. The adults emerge in the late summer, feed on grasses until September and hibernate. Control is rarely necessary. A gamma-HCH spray is effective. The eggs are parasitized by *Anaphes flavipes* (Foster).

Both *O. melanopa* and *O. lichenis* (Voet.) transmit cocksfoot mottle and phleum mottle viruses. Starvation increases the efficiency of transmission which becomes possible after feeds of a few minutes, the insects usually remaining viruliferous for a few days (A'Brook & Benigno, 1972).

Crioceris asparagi **(L.)** (Fig. 6.13A)   The asparagus beetle is another brightly-coloured beetle which hibernates during the winter and feeds on asparagus after the end of the cutting season. It has two or three generations a year. If the beetle and larvae are numerous, the asparagus plant may be stripped bare of foliage.

The exposed adults and larvae are readily killed by pyrethrum insecticide sprays. Gamma-HCH is effective but best avoided.

Phaedon cochleariae **(F.)** (Fig. 6.13D)   The mustard beetle is a pest of mustard, especially brown mustard (*Brassica nigra* (L.)) grown for seed. It also feeds on many cruciferous plants such as swedes, turnips, cabbages, watercress and bitter-cress (*Cardamine amora*) but is uncommon nowadays except on watercress. Insecticides used to control flea beetle and for pollen beetle control may have lessened its numbers. *P. armoraciae* (L.) is larger but less brightly coloured.

The adults hibernate during the winter preferably in mustard stubble but also in straw, dry stacks of mustard seed and in soil cracks and crevices round the edges of the fields. They emerge in April or May. The body is oval, 3–4 mm long, bright metallic blue with a red last abdominal segment and black legs. The broad thorax and elytra are finely punctured. The female lays small yellow eggs on the leaves of various cruciferous plants, chiefly mustard, watercress and bitter-cress. A single female may lay 300–400 eggs. The eggs hatch in 8–12 days and the active larva feeds on the foliage. The larva is hairy, basically yellow, with a black head and prothorax, black spots on the meso- and meta-thorax and the abdomen. Along each side there is a row of brown tubercles from which shiny yellow glands can be protruded. The legs are black and yellow and, when fully fed, the larvae are 6 mm long. They drop to the ground, pupate in the soil and the adult emerges after 4–5 days and lays another batch of eggs so that there are two generations in a year. Feeding on watercress and mustard can continue from April to September when the beetles return to hibernation.

If watercress is badly attacked, raising the water level and flooding the cress beds drives the

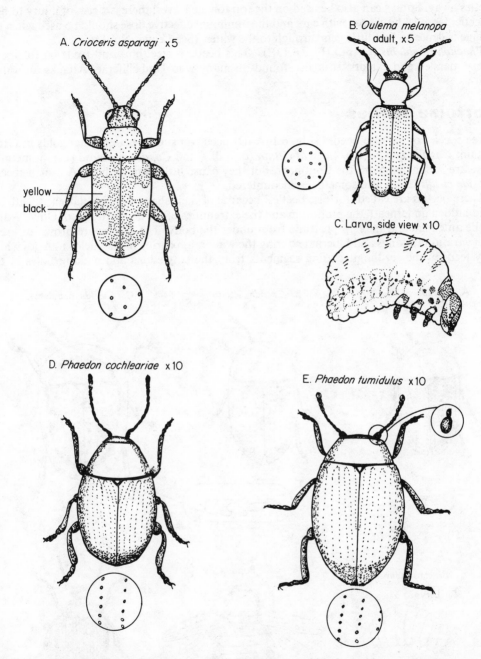

Fig. 6.13 A, asparagus beetle, *Crioceris asparagi*. The yellow and black markings on the elytra are very characteristic; B and C, adult and larva of cereal leaf beetle, *Oulema melanopa*. The adult is brightly coloured with orange thorax and blue elytra; D and E, Mustard beetles, *Phaedon cochleariae* and *P. tumidulus*. Head of *P. tumidulus* is bent forwards and is not visible from the dorsal surface. Note enlarged basal joints of antenna. Detail of punctations on elytra given in insets.

beetles away. Sprays can also be used on the top foliage leaves; there is a risk of injury to fish. Insecticides must be chosen with care and the minimum effective dose should be used when the foliage is dense so that little gets through to the water. (See Leaflet No. 144.)

Phaedon tumidulus (Germ.) (Fig. 6.13E)   This beetle is steel blue and feeds on foliage of celery, parsley and carrots. It is also found on many wild Umbelliferae. Attacks are rarely serious.

Tortoise beetles

Three species of tortoise beetles (Fig. 6.14A–D) occur on sugar-beet and mangolds in Great Britain: *Cassida nobilis* (L.), *Cassida vittata* (de Vill.), and *Cassida nebulosa* (L.). Sometimes these are serious pests on the Continent, but they cause little damage here. A fourth species *Cassida viridis* (L.) is also sometimes encountered.

These beetles are called 'tortoise beetles' because of their shape which is flattened, with the head hidden underneath the pronotum and the extremities of the elytra rounded. Only the tips of the antennae and the legs protrude from under the body. The beetles are brown or green with various markings and hibernate during the winter, appearing in beet fields in April where they feed on the seedlings, cutting segments from the cotyledons and making holes in the

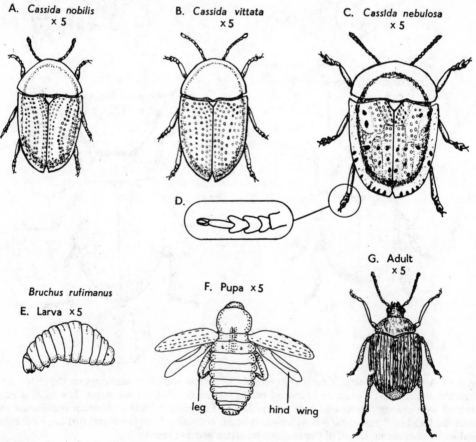

Fig. 6.14 A–C, three common species of tortoise beetle, and D, phytophagous tarsus (enlarged); E–G, larva, pupa and adult of seed beetle *Bruchus rufimanus*.

young foliage leaves. Eggs are laid on beet leaves from May onwards and the larvae which hatch are pale green, possess forked tails and bear lateral spines on each segment. After moulting the cast skin adheres to the tail and is carried over the back (Fig. 6.10B). Larval feeding causes injury similar to flea beetle damage, circular pits being eaten in the lower surface of the leaf, often leaving the upper epidermis intact. When the beetles and their larvae are numerous only the veins of the leaves remain (Plate XA). *C. nebulosa*, common on the Continent, is longer (3–7 mm long) than the others and has dark brown marks on the light green-brown elytra. *C. vittata* is 3–6.5 mm long, has yellow coxae, whereas *C. nobilis* has black coxae. *C. nebulosa* and *C. vittata* may have two generations a year but *C. nobilis* only one. Tortoise beetles are much less common since the introduction of modern pesticides.

Seed beetles

Family BRUCHIDAE   Pulse or seed beetles are pests of pods and seeds, particularly those of the Leguminosae. *Bruchus rufimanus* (Boh.), the bean seed beetle, lives in the seeds of broad or field beans and is common in Great Britain. Other foreign species such as *B. pisorum* (L.) the pea beetle and *Acanthoscelides obtectus* (Say.) the dried bean beetle are constantly being introduced in culinary peas and in dwarf beans but they seem to be unable to establish themselves and are renewed each year.

The elytra of seed beetles are shortened, leaving the tip of the abdomen exposed (Fig. 6.14G).

Bruchus rufimanus **(Boh.)** (Fig. 6.14E–G)   The bean seed beetle is a small, black oval beetle, 3–5 mm long with black markings and white hairs on the elytra. The first four joints of the antennae and the anterior femora are red and the pygidium is covered with a silvery pubescence. The adult flies readily and is common on beans in the spring, especially at flowering time. The eggs are laid on the young bean pods and the newly emerged larvae bore their way into the developing bean and reach the seeds. More than one larva may enter the same bean seed. The rest of the life cycle is completed in the seeds until the adults bite their way out after pupation. The adults may hibernate in the seed and therefore may be sown with the seed of the early winter beans.

The bean seed beetle occurs wherever beans are grown. Beans which are threshed and used for grinding suffer little loss from larval feeding, but in some seasons over 80% of the bean seeds may be attacked. If the beans are to be used for seed, sacks containing infested samples can be fumigated with 3 parts of ethylene dichloride to 1 part of carbon tetrachloride, at 3.5 l/ tonne. The sacks of seed should be treated either in an air-tight chamber or under a gas-proof sheet. Large tonnages can be fumigated with methyl bromide, aluminium phosphide or the mixture of ethylene dichloride and carbon tetrachloride. Fumigation is hazardous and best left to specialists.

Larval feeding does not appear to lower the germinating ability of the seeds as the radicle and plumule are rarely attacked, but damaged beans are more susceptible to moulds.

The life cycles and habits of *B. pisorum*, the Spanish bean beetle, and *B. brachialis* (Fahr.), the vetch bruchid, are very similar. They are important on the Continent and the dried bean beetle, *Acanthoscelides obtectus* (Say.), is important in the United States, where it breeds in stored seeds in addition to attacking beans in the field. Many larvae may occur in a single seed. (See Leaflet No. 126.)

Superfamily CURCULIONOIDEA

Family CURCULIONIDAE weevils   This group contains a large number of species and is closely allied to the Chrysomeloidea. Many species are pests.

Larvae are mostly concealed feeders, either in soil feeding on roots and nodules (pea and bean weevils), in plant tissue (turnip gall weevil), in unopened flower buds (apple blossom

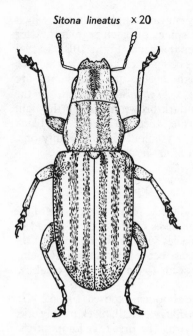

Sitona lineatus × 20

Fig. 6.15 Dorsal view of pea and bean weevil, *Sitona lineatus*, to show structure.

weevil) or in seeds (clover seed weevil). The concealed feeders are white but the few exposed feeders (lucerne weevil) are coloured.

The life cycle is typical of beetles. The adult hibernates in diapause and the larval stages occur in spring and summer.

Sitona spp. (Figs 6.15, 6.16) Pea and bean weevils are exceedingly common and feed on various leguminous plants such as pea, bean, clover, lucerne and sainfoin. They are among the earliest insects to become active as the soil warms up in the spring. The type of damage is highly characteristic, for it consists of U-shaped notches cut from the edges of the leaves. The notches may occur all the way round the leaf. If the plants are well grown, they can withstand damage, but when they are small and growth is slow, peas in particular can be severely checked. Weevils feed chiefly at night, although they can be found at any time. When disturbed, they drop to the ground and sham death, or hide under stones. Injury to sugar-beet by weevils resembling *Sitona* is generally due to the beet leaf weevil *Tanymecus palliatus* (F.).

The following species attack leguminous crops: *Sitona lineatus* (L.), *S. hispidulus* (F.), *S. macularius* (Marsh.), *S. puncticollis* (Steph.), *S. humeralis* (Steph.), *S. lepidus* (Gyll.), *S. sulcifrons* (Thunb.). Of these *S. lineatus* is the commonest. All species are short snouted and are 3–6 mm long.

Sitona lineatus (Fig. 6.15) The adults of this weevil overwinter in the hedgerows and fields and appear in the spring, feeding on beans and peas. They fly readily on bright days. The weevil is grey-black and clothed with fuscous scales. The short rostrum has a central furrow which is continued between the prominent eyes. The antennae are light brown and there are parallel lines, striae and very short hairs on the elytra.

Eggs are laid in the soil from April to July. They are oval and white when first laid but darken later. After 21 days the larvae appear and make their way to the root nodules of leguminous plants on which they feed, usually without affecting the plants greatly. The larvae are cream and on the tenth abdominal segment is a fleshy area which can be extruded and helps in movement. A fully grown larva is 4–5 mm long, after 6–7 weeks pupates in a cell about 5 cm below the surface of the soil and the adult emerges 17 days later. There is usually one generation a year.

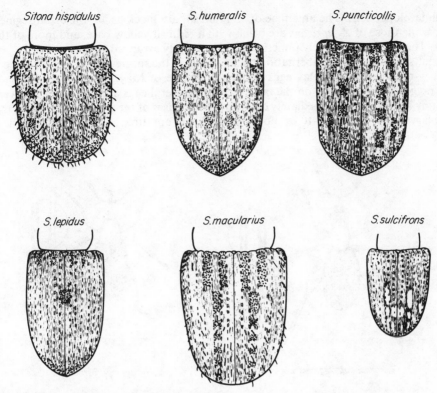

Fig. 6.16 The elytra of six different species of *Sitona*.

The adults do most damage to young peas and the larvae destroy the root nodules, but unless the weevils are very abundant, it is unnecessary to treat the crops (George, Light & Gair, 1962). Gamma-HCH, azinphos-methyl with demeton-S-methyl sulphone, fenitrothion, permethrin or triazophos spray can be applied to the young pea crop to kill the adults. Field beans are also attacked and there is evidence that larval feeding on nodules can diminish yields (Bardner & Fletcher, 1979). Aldicarb granules effectively control the feeding of adults and that of larvae on nodules. Sprays that kill the adults also result in fewer larvae feeding (McEwen *et al.*, 1981).

S. hispidulus, a very dark, hairy species; *S. macularius*, a reddish brown species with grey or yellow scales; and *S. sulcifrons*, a black species with copper-coloured scales, are pests of lucerne and clover. Much damage may be done to the roots of clover by the larvae of *S. hispidulus* (Fig. 6.16). *Sitona* spp. transmit broad bean stain virus and broad bean true mosaic virus but are much less efficient vectors than *Apion vorax* (Herbst.), the bean flower weevil. These viruses are common in field and broad bean crops in some years and diminish yields. (See Leaflet No. 61.)

Apion **spp.** (Fig. 6.17A, B)   Clover seed weevils. The two common species, *A. trifolii* and *A. apricans* (Herbst), are widely distributed. The chief damage is caused by the larvae eating the florets and seeds of all varieties of clover so that considerable loss of seed occurs. Clover seed is a highly priced product and well worth saving. One bad year a seed merchant in East Anglia removed 25% of seed husks from a consignment of 2500 kg of white clover seed and the total loss was probably greater than this.

The adults are tiny, blue pear-shaped weevils with a long snout, almost straight antennae and well-developed legs. The antennae of *A. aestivum* are black as are the distal segments of the legs, while those of *A. apricans* are black with a reddish yellow base, and more of the legs are dark. The elytra are coarsely punctured with regularly arranged striae.

The adults come out from hibernation rather slowly in the spring and move into clover fields where they mature, pair and lay eggs about the middle of May. Adults feed on the leaves making ragged holes and also on the inflorescences. Several eggs are laid in each flower head and the tiny legless larvae immediately eat through the bases of the florets and into the ovaries. A single larva may damage 6–10 or more ovules. There are three larval instars and the larval

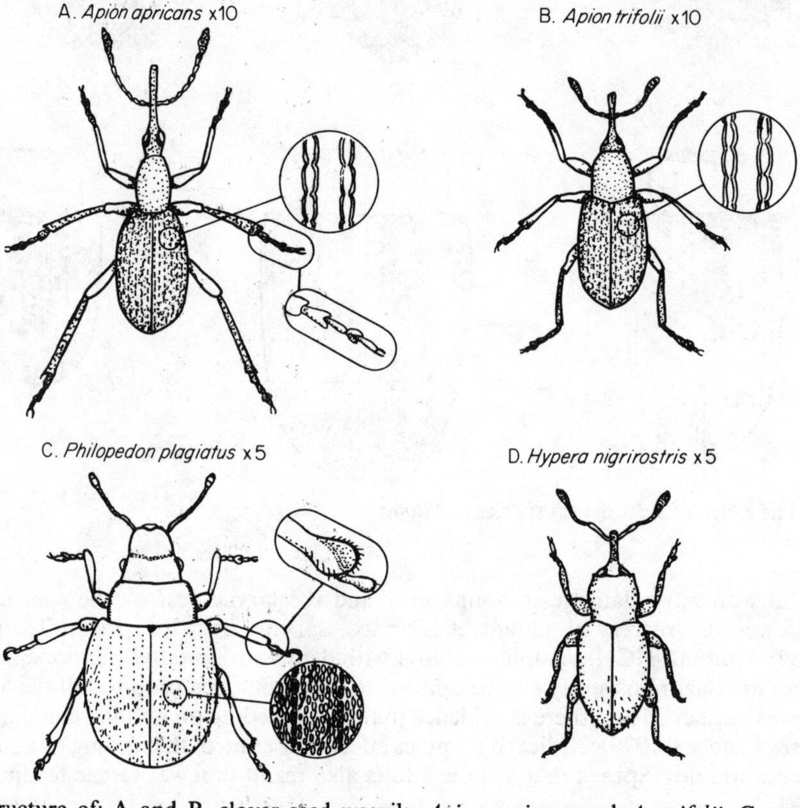

A. *Apion apricans* x10

B. *Apion trifolii* x10

C. *Philopedon plagiatus* x 5

D. *Hypera nigrirostris* x 5

Fig. 6.17 Structure of: A and B, clover seed weevils, *Apion apricans* and *A. trifolii*; C, sand weevil, *Philopedon plagiatus*; D, lucerne weevil, *Hypera nigrirostris*.

period averages 18 days. The larvae are easily killed by desiccation. The fully grown larva pupates, and after 6 days the adult emerges and starts laying eggs giving rise to another generation. Towards the end of the summer the adults accumulate around the edges of the fields and hibernate in suitable shelter.

Most injury occurs in the south-eastern parts of England where seed is taken. Cultural control is possible with early flowering clovers such as the broad red clover. Early cutting of the crop for hay when about 25% of the flowers are in bloom leads to the death by desiccation of many larvae. This lessens the incidence of attack on the clover left for seed, but this is not possible with the late-flowering red clovers as they are cut for hay one year and for seed the next (Plate XIC).

Chemical control is more effective. The chief difficulties are the penetration by chemicals of the great mass of foliage and the protected habitat of the larvae. A substantial number of adults can be killed by spraying the margins of clover fields with gamma-HCH towards the end of April before weevils have dispersed: this would be unpopular with conservationists. Treatment is more effective against the adults if applied immediately after the first cut when there is least foliage on the field and the larvae have been removed with the hay. On no account should insecticide be applied when the crop is in flower because of possible harm to bees.

Philopedon plagiatus (**Schall.**) (Fig. 6.17C)   The sand weevil is common in sandy soils. In the breckland district of East Anglia and the coastal areas of Suffolk it is sometimes a pest of sugar-beet and carrots.

The weevil is brown, 7 mm long and covered with fuscous scales that are alternately lighter and darker on the wing cases giving an appearance of longitudinal stripes. It has a short snout and clubbed, elbowed antennae. The adults, which cannot fly, appear in the spring and feed on a variety of plants and are injurious to carrot, sugar-beet and other crops with small seedlings. The eggs are laid on the soil in April and May, and the adults gradually die off so that few can be found by the end of June. The larvae develop slowly, feeding on the roots of various plants, such as *Lychnis* spp. and *Silene* spp. They live for about 18 months and finally pupate in the soil. The life cycle, therefore, takes 2 years (Plate XID, E, F).

Weevils cut segments from cotyledons and true leaves and, when numerous total failures of carrots and beet can occur with surprising speed. Gamma-HCH sprays are effective but timing is critical and a delay of 2 days may be disastrous. Attacks occur after crop sequences that give good ground cover and plenty of roots, e.g. cereals, sainfoin, lucerne, clover, ryegrass. Outbreaks were always uncommon but have been less frequent since organo-chlorine and other insecticides came into use.

Ceutorhynchus pleurostigma (**Marsh.**) (Fig. 6.18C)   Turnip gall weevil larvae cause galls on the roots of all cultivated Cruciferae: swedes, turnips, cabbages, cauliflowers, Brussels sprout, kohl rabi. The galls are marble-like swellings formed by the proliferation of the plant tissues, and if cut open are found to contain a weevil larva or a cavity and exit hole made by the larva. The gall weevil is not an important pest. The destruction of charlock by hormone weedkillers, the use of insecticidal seed treatments containing HCH and measures against brassica seed pests have all tended to decrease the weevil population in brassica fields (Plate XIIIB). If necessary, transplants in seedbeds can be protected with gamma-HCH spray. (See Leaflet No. 196.)

Ceutorhynchus assimilis (**Payk.**) (Fig. 6.18A, B)   The cabbage seed weevil, is 2–4 mm long and very similar to *C. pleurostigma*, but may be distinguished by greyish white scales which give it a lighter appearance. The elytra are heavily punctured and there are no teeth on the femora.

Adults come out of hibernation in the spring and feed on cruciferous seedlings but do little damage. In May the female lays eggs in the young pods of turnip and swede flowers. She bores a characteristic hole in the pod with her rostrum and inserts an egg. When the weevil larvae are fully grown, they drop down and pupate in an earthen cell in the soil. The adults emerge and feed a little before hibernating.

Much loss of seed is caused by the larvae feeding in the pods. The punctures in the pods made by the adults are also used by the brassica pod midge, *Dasineura brassicae* (Winn.), for oviposition (p. 148). The midge is more commonly found in seed pods of overwintering brassicae unless the spring crops are sown very early in the year. The larvae produced are paedogenetic, that is, become sexually mature and reproduce parthenogenetically. The mass of larvae in the pod cause distortion and bloating, the visible symptom being known as 'bladder pod'. Control of *C. assimilis* is essential, for effective control of *D. brassicae* and 'bladder pod'. Treatment against weevil damage is similar to that for blossom beetles (p. 105). Azinphos-methyl with or without demeton-S-methyl sulphone, carbofuran, endosulfan, HCH or malathion are recommended for seed crops other than winter rape to be applied before flowering. As the weevils fall down easily it is necessary to get the spray well down into the plants and onto the soil to destroy them. The onset of emergence of seed weevil larvae from eggs within pods

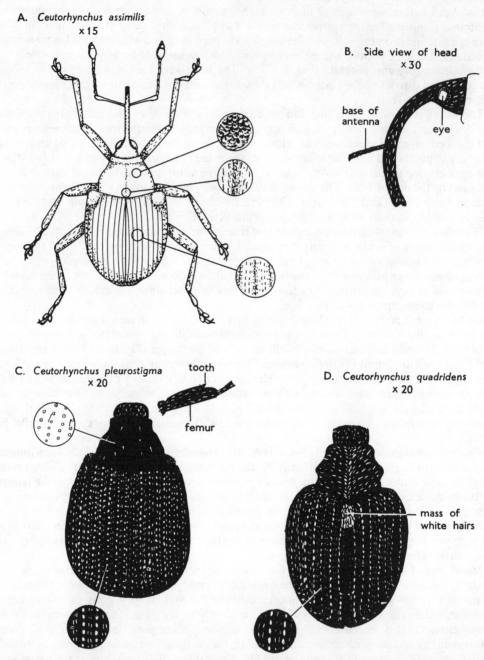

A. *Ceutorhynchus assimilis*
×15

B. Side view of head
×30

base of
antenna

eye

C. *Ceutorhynchus pleurostigma*
×20

tooth

femur

D. *Ceutorhynchus quadridens*
×20

mass of
white hairs

Fig. 6.18 *Ceutorhynchus* sp. A and B, *C. assimilis*, dorsal view and side view of the head to show the long rostrum. The weevil is black with short white hairs; C, *C. pleurostigma*, thorax and abdomen and anterior femur to show tooth; D, *C. quadridens*, thorax and abdomen, showing the mass of white hairs on the elytra.

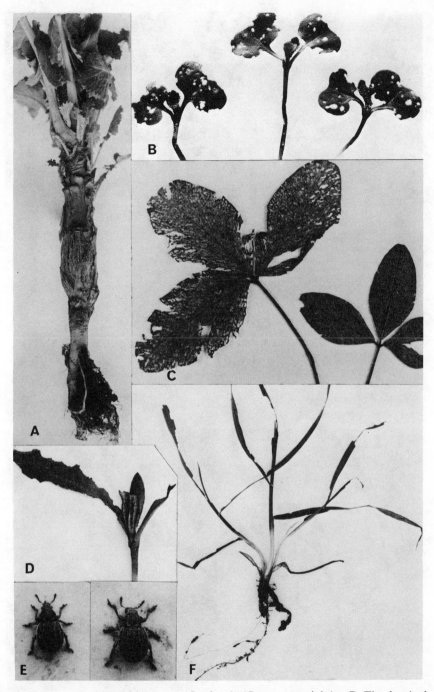

XI **A.** Rape plant attacked by cabbage stem flea beetle (Crown copyright). **B.** Flea beetle damage to brassica seedlings (Crown copyright). **C.** *Apion* damage to clover (Crown copyright). **D.** Beet seedlings attacked by sand weevils (courtesy W. E. Dant). **E.** Sand weevils (courtesy W. E. Dant). **F.** Wheat plant attacked by sand weevils (courtesy W. E. Dant).

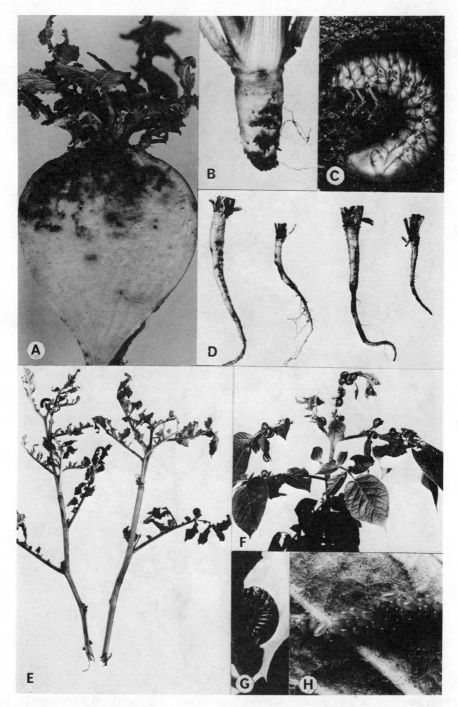

XII **A.** Swede attacked by stem weevil (Crown copyright). **B.** Tap root of sugar-beet severed by chafer grub (courtesy W. E. Dant). **C.** Chafer grub (courtesy H. C. Woodville). **D.** Sugar-beet damage by chafer grub (courtesy W. E. Dant). **E.** Potato leaves eaten by Colorado beetle (Crown copyright). **F.** Colorado beetles feeding on potato leaves (Crown copyright). **G.** Larva of Colorado beetle (Crown copyright). **H.** Eggs of Colorado beetle on potato leaf (Crown copyright).

coincides with the end of flowering of winter rape, i.e. when the field is predominantly green all over. So, a spray of phosalone at this time penetrates the pods and kills larvae of the weevil and the pod midge.

Brassica seed crops are pollinated by bumble bees, hive bees, flies, pollen beetles and other insects attracted to the flowers. Insecticides applied at flowering time destroy the pollinators. Treatment when the crop is in full flower can usually be avoided but local bee-keepers should be warned when operations are imminent for if many bees are killed ill-feeling is created locally. (See Leaflet No. 780.)

Ceutorhynchus quadridens (Panz.) (Fig. 6.18D)   The cabbage stem weevil and *C. picitarsus*, the brassica winter stem weevil, are common insects closely related to *C. pleurostigma*. The larvae burrow freely in the stems of brassicas without forming galls (Plate XIIA). *C. picitarsus* larvae are the more injurious as they kill growing points. Damage by stem weevil larvae may be confused with the cabbage stem flea beetle (p. 113).

Adults emerge from hibernation in the spring. The female bores a hole in the base of a petiole of a brassica leaf and lays several eggs. The larvae tunnel upwards and downwards into the midrib of the leaf and, as they get bigger, move to the main stem and growing point. When fully grown, the larvae bore their way out of the stems and drop down onto the soil for pupation (Winfield, 1961b).

All cultivated Cruciferae seem to be attacked. The presence of larvae in the stem is noted from May onwards when transplants of Brussels sprout, broccoli and cabbage are lifted from the seed bed. Swede seed stecklings may also be mined. Attacked plants are flabby, spongy to the touch and soon wilt and die after being transplanted. *C. picitarsus* attacks winter rape.

Gamma-HCH seed treatments control stem weevil on cabbage seedlings to some extent. The adult weevils can also be killed if the crops are treated before transplanting in the spring. Treatment with azinphos-methyl with or without demeton-S-methyl sulphone, endosulfan or malathion are effective. As attacks are local and sporadic, control is rarely undertaken but once the larvae are inside the plants the damage is done and control is more difficult. Phosalone and triazophos sprays are approved treatments for winter rape.

Hypera sp.   Clover leaf or lucerne weevils are very common and cause occasional damage to lucerne, trefoil, clover and tares when raised for seed. There are two main species, *Hypera postica* (Gyll.) on lucerne and trefoil, and *H. nigrirostris* (F.) on clover (Fig. 6.17D). The larvae are exceptional amongst weevils as they are exposed feeders with protective colouration and adhere to the host plant by means of a sticky secretion.

Hypera postica (Gyll.) is 7–10 mm long, brown with darker stripes on the elytra and thorax. During the winter it hibernates in hedges and other sheltered places becoming active in March and April. The adults feed on the leaves and stems and insert their yellow eggs in batches of ten. After hatching, the larvae move within the stems but eventually leave and feed externally on the buds which they destroy and so ruin both flowers and seeds. The larvae are green with a white line along the back and have a black head. In June and July the larva spins a wide-meshed cocoon in the foliage and pupates within it. When the beetles emerge, they feed on the foliage before hibernating.

Damage is not usually noticeable until the crops flower and by then it is too late to take action. Crops ruined for seed production can be cut for hay, which removes and kills the larvae and pupae. If the weevil is troublesome in the district, the seed crop can be sprayed or dusted as a precaution with HCH when the beetles become active in the spring.

Hypera nigrirostris (F.)   Adults and larvae feed on the leaves, stems and buds of clover but rarely cause appreciable damage. The larva is yellowish green and has a white line down its back. Pupation occurs in a cocoon which is eaten by the newly emerged weevil.

Both species of *Hypera* have been introduced into the U.S.A. from Europe where they are known as 'alfalfa weevils' and have become widespread and troublesome. Lucerne leys are left down for a considerable time and the leaves are eaten by the weevils. Most larvae and pupae are unable to survive the winter in Maryland.

Sitophilus granarius **(L.)**   The grain weevil (Fig. 13.1A), is a pest of stored grain, barns, silos, warehouses and granaries (see p. 270).

The adults are 4–6 mm long, elongated, cylindrical and dark brown. The wing cases are rigid and the beetles are unable to fly. The female is able to bore a hole in the kernel of grain with her rostrum which is a quarter the length of the body. She inserts an egg and the larva feeds inside and eventually pupates inside the grain. The adults live for 4–5 months and each female may lay up to 200 eggs. Generations follow each other while grain remains, their duration depending upon temperature. If the grain is cooled below 17°C by a forced draught of cold air, the life cycle of the weevil is interrupted. Weevils are unable to exist in Britain without shelter but have been here a long time. (See Leaflet No. 219).

Superfamily SCARABAEOIDEA

These are large stout beetles characterized by the lamellate club formed by the apical segments of the antenna. The fore tibiae are dentate with one apical spur and the eighth abdominal tergite forms an exposed pygidium. Many species are fossorial at some time during their lives and sexual dimorphism is common. The larvae are broad and fleshy, and occur in the ground, in dung or in roots. The body is curved like the letter C, and the last abdominal segment is enlarged. Legs are developed but rarely used as the larvae live surrounded by or near their food. The head is large and strongly sclerotized with powerful mandibles. The stag beetles, Lucanidae, have enormously developed mandibles in the male, and the larvae live in dead tree trunks.

The most important family economically is the SCARABAEIDAE, to which belong the chafers or white grubs (subfamily Melolonthinae). Adult chafers feed on foliage while the larvae feed on roots. There are three common species in Britain. Two others are less commonly found. (See Leaflet No. 52.)

Melolontha melolontha **(L.)** (Fig. 6.19A)   Cockchafer adults emerge from the soil in May or June and are called 'May bugs'. They are reddish brown, 22–26 mm long, with black heads bearing transversely directed bristles. The lamellate antennae are red and the thorax is black with white bristles. On the elytra are five raised lines with short white hairs. The tip of the abdomen is produced into a spade-like point. In the male the antennal club consists of seven plates, but in the female it is shorter, consisting of six plates only.

Chafers mate soon after emergence and the round white or yellow eggs are laid about 3 weeks later in batches of 12–30 in the soil around grass and cereals. The eggs hatch and the white fleshy larvae or 'white grubs' take 3 years to become full grown, when they reach 60 mm. The head is brown and shiny, with a 5-jointed antenna and well-developed mandibles. The pale brown legs are relatively short and the last enlarged abdominal segment appears dark as the intestinal contents are visible through the skin. The anal orifice and its spines are used in identification (Fig. 6.20). Pupation is in an earthen cell, 60 mm deep in the soil, and the adult emerges in late autumn but remains in the pupal cell in diapause until the spring, appearing 4 years after the egg was laid.

In Britain the adults, which feed on the fruit, leaves and flowers of various trees, do little damage, but on the Continent they often damage fruit. The polyphagous larvae damage roots of cereals, beet and also strawberries, especially when planted near woodlands. Injury to grass and sports turf is sometimes spectacular. The roots are completely severed so that the sward may be pulled off like fabric. Cattle and racehorses may kick up the turf in affected areas leaving large, bare patches. The powerful mandibles gnaw right through small roots and make large raw wounds on big ones, sometimes severing the tap roots a foot or more below the soil surface (Plate XIIB). Potato tubers may also be severely damaged.

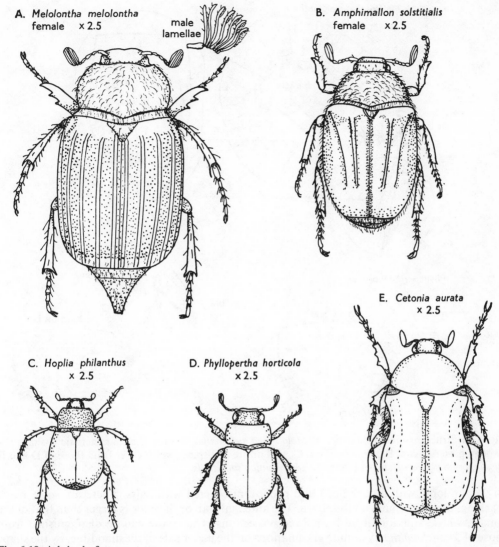

Fig. 6.19 Adult chafers.

As the life cycle lasts 3 years, there are three independent groups or 'régimes' existing side by side in the soil. One group may be more numerous than the others producing swarm years. The main damage to root crops is liable to occur in the 2 years following a 'swarm' year, especially in the neighbourhood of copses and woodlands.

Whenever an insect has an extended life cycle, the preponderance of certain 'régimes' may occur. *Magicicada septendecim* (L.), the periodical cicada of the United States, has a nymphal stage of 17 years in the north and 13 years in the south. The largest régime, called X, last occurred in 1953 and was due again in 1970. The régime, requiring 13 years for its development, last appeared in 1959. The adult swarms damage orchard trees by inserting eggs below the bark, so transplanting of young trees and shrubs, as well as pruning, is postponed until expected swarms have disappeared.

Amphimallon solstitialis **(L.)** (Fig. 6.19B)   The summer chafer or 'June bug' is light brown,

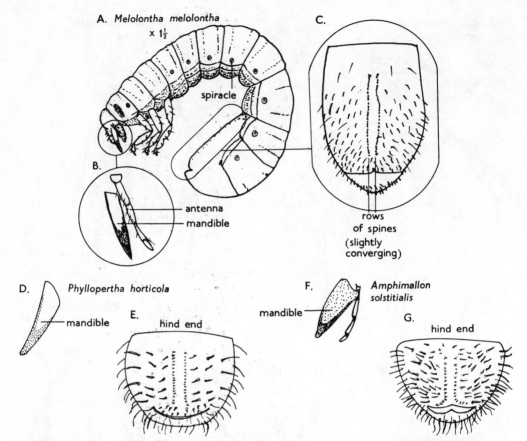

Fig. 6.20 Structure of chafer larvae. A, *Melolontha melolontha*, to show the characteristic U-shape and swollen posterior end; B, the mandibles, C, anal orifice and spines which differ from those of D and E, *Phyllopertha horticola*, and of F and G, *Amphimallon solstitialis*.

14–18 mm long, with a dark head and pronotum covered with hairs. There are some black markings and a few hairs on the elytra. The antennal club or the male is larger than that of the female, whose anterior tibiae bear three projections. The larvae can be distinguished from those of *M. melolontha* by minute granulations on the under side of the mandible, by the shape of the anus, by a triangular area of bristles and by two rows of small divergent spines (Fig. 6.20G). The life cycle takes 1 or 2 years in Britain. The adults appear at dusk and make a whirring noise with their wings as they fly. Larvae of this species are less troublesome than those of the cockchafer in grassland but sometimes damage arable crops.

Phyllopertha horticola (L.) (Fig. 6.19D)   The garden chafer is 7–11 mm long, and the head and thorax have a greenish metallic sheen. The red antennae have black clubs and the green thorax is narrower than the red-brown, shining elytrae; the legs are black. The larvae are 18 mm long. On the mandible is a pale oval area which has file-like ridges crossing it. The anus has a transverse opening, with two parallel lines of spines (Fig. 6.20E). The chafer is widely distributed and is common on poor quality grassland in hilly country where the soil is light (Raw, 1951).

The life cycle takes 1 year, pupation occurring in April and May. Adults appear in May or June and are active for 3–4 weeks. They feed on many kinds of plants, particularly on young bracken fronds. Larval feeding is injurious to pasture and farm crops.

Hoplia philanthus **(Fuess.)** (Fig. 6.19C)   The Welsh chafer is small and brown and has a life cycle similar to *P. horticola*. Its larva occurs in grassland.

Serica brunnea **(L.)**   The brown chafer is a local pest common in a few well-wooded places. The larva lives for two years.

Cetonia aurata **(L.)** (Fig. 6.19E)   The rose chafer is a shining, bright golden-green beetle, 14–20 mm long, and appears in June and feeds on flowers, particularly roses. The antennae are dark and each club has three lamellae. The larva has red legs and red hairs arranged in transverse rows along the body. The larvae feed for 2 or 3 years before pupating and occur in grassland.

Control of chafers

When a larval attack on grassland, on an arable crop, or in a tree nursery is under way, the larvae are often too deep in the soil to be reached by insecticides or mechanical means. Attacks can sometimes be prevented or minimized by careful planning. Fields near woods which are subject to chafer attack should not be sown with sugar-beet or potatoes. As the larvae are easily damaged, ploughing, discing and hoeing probably kill many. Gamma-HCH is known to kill larvae and, doubtless is helpful when it has been applied to and worked into the soil to control other pests. Adult beetles can be killed while feeding on fruit trees by insecticidal sprays. Flying cockchafer beetles converge on the highest points on the skyline and some economy in control has been made on the continent by spraying only the highest areas of woodland.

Grassland problems are different as grass is usually permanent and hill pastures in particular are difficult to treat. The encouragement of vigorous growth by good management gives the best results. Lowland pastures, if very badly attacked, can be ploughed and reseeded as the ploughing and repeated discing lessens the number of larvae in the top 7.5 cm of soil. The seed mixtures used for reseeding must contain deep-rooted grasses or forage plants; mixtures containing high proportions of lucerne and cocksfoot are the most resistant.

Birds such as rooks, magpies, thrushes, green plover and black-headed gulls prey upon the larvae after ploughing, but no one knows what proportion they kill. Tachinid flies have been bred from larvae on the Continent and they are also attacked by the fungus *Beauveria bassiana* (Bals.). *Bacillus popilliae*, *B. lentimorbus* and the nematode *Neoaplectana hoptha* (Turco) have been successfully introduced into the U.S.A.

Popillia japonica **(Newman)**   The Japanese beetle has not yet established itself in Britain, but is a potential pest. It has become widely distributed in the U.S.A. since its introduction in 1916.

The adults are shiny metallic green with copper brown elytra. They feed on foliage, flowers and fruits and damage ornamental and garden shrubs. The larvae are typical chafer grubs that feed on plant roots, injuring especially turf in lawns, parks, golf courses and garden, and also attacking many crop plants. The life cycle takes a year, the adults fly in June and July and eggs are laid in the soil in July; the larvae pupate the following May or June.

Biological control measures have been tried in the U.S.A. with some success. A bacterial disease, *Bacillus popilliae* (Dutky) has decreased the number of larvae in old infested areas. Two larval parasites, *Tiphia vernalis* (Rohw.) and *T. popilliavora* (Rohw.), have been introduced from the Orient, and also a fly, *Hypercteina aldrichi* Mesnil which attacks the adults. Moles, skunks and birds also consume many larvae.

So far Japanese beetles have been found only in American aircraft in Europe during June and July, the flight period of adult beetles in the U.S.A. (See Leaflet No. 449.)

7

Sawflies, wasps, bees and ants – Hymenoptera

Few members of the Hymenoptera are pests and many are beneficial in various ways. The adults are usually strong flyers and the behaviour patterns of many are highly developed, especially among species that are social. Antennal sense organs are often large, complicated and numerous. Bees are known to possess colour vision similar in range to that of man but extending further into the violet end of the spectrum and less far into the red end. They are able to communicate the position of nectar and pollen to other bees in the hive by means of directional dances.

The main characteristics of the Order are:
(1) Two pairs of membraneous wings, the smaller hind pair interlocking by hooks with the larger fore pair (Fig. 7.2F);
(2) Mouthparts primarily adapted for biting, but often modified for lapping and sucking nectar;
(3) The first abdominal segment is fused with the thorax;
(4) Ovipositor, often conspicuous;
(5) Metamorphosis complete, the larva being a polypod caterpillar or entirely apodous;
(6) The pupa is free;
(7) A cocoon is generally formed.

The Order is divided into two suborders: (1) **Symphyta,** sawflies; and (2) **Apocrita,** bees, wasps, ants, ichneumon flies, gall wasps.

Suborder Symphyta

In sawflies the abdomen is broadly attached to the thorax and the second abdominal segment does not form a petiole or waist. The ovipositor of the female is adapted for sawing or boring. The larvae are sometimes apodous but more usually polypod caterpillars with six or more pairs of abdominal legs. The most common arrangement is with additional appendages on abdominal segments 2–7 and 10 (Fig. 2.17A). They differ from lepidopterous caterpillars in having more abdominal appendages, in lacking the characteristic crotchets, and in the conspicuous head with only a single pair of ocelli. Spiracles are usually present on the prothorax and the first eight abdominal segments (peripneustic, see Fig. 2.13). When they are exposed feeders, sawfly larvae are coloured or spotted, but when internal they are white. Most spin cocoons in which they pupate.

Family Cephidae

Cephus pygmaeus (L.)   Wheat stem sawfly (Fig. 7.1) adults appear in June; they are 8–10 mm long and shiny black with two yellow rings on the abdomen. The antennae are 21-jointed and slightly clubbed. The males have yellow mandibles, a yellow spot on the clypeus and on each

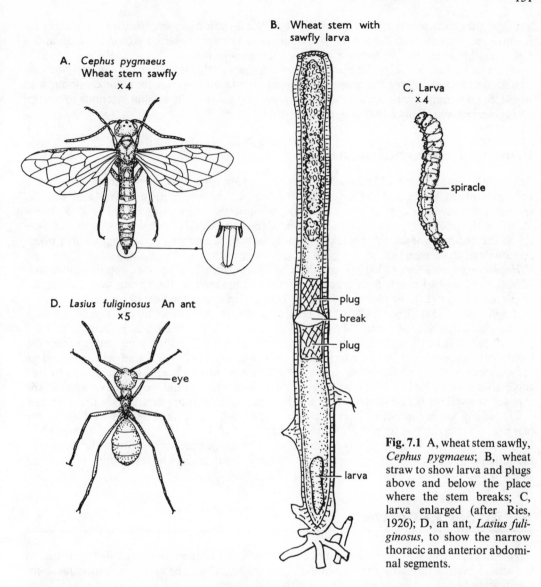

B. Wheat stem with sawfly larva

A. *Cephus pygmaeus*
Wheat stem sawfly
×4

C. Larva
×4

spiracle

D. *Lasius fuliginosus* An ant
×5

eye

plug
break
plug

larva

Fig. 7.1 A, wheat stem sawfly, *Cephus pygmaeus*; B, wheat straw to show larva and plugs above and below the place where the stem breaks; C, larva enlarged (after Ries, 1926); D, an ant, *Lasius fuliginosus*, to show the narrow thoracic and anterior abdominal segments.

interior margin of the eye, and yellow legs. The female is more sombrely coloured, the yellow patches being more ochreous than bright yellow.

The female inserts an egg into a wheat stem just below the ear, and the dull, whitish larva bores down the straw. The larva is legless except for tubercles representing the thoracic appendages (Fig. 7.1C). Metathoracic spiracles are present making ten pairs in all, and the last abdominal segment terminates in a tubular appendage which is extrusible and may help the larva in its progress down the wheat stem. When fully grown, the larva is 10–14 mm long and has reached the bottom of the straw. The larva cuts around the stem just above ground level so that it is easily broken off by the wind (Fig. 7.1B). The end of the straw is blocked with debris and excrement and in it the larva overwinters in diapause. It pupates early the following summer and the adult emerges in June.

The ears of attacked straws ripen prematurely and carry little grain, so that they stand out amongst the others which are still green. Up to 10% infestation causes little loss of yield. Barley is sometimes attacked. In Britain the pest appears to be kept in check by crop rotation and is rarely troublesome. Stubble burning kills overwintering larvae.

In the wheat belt of North America *C. pygmaeus* and a related species, *C. cinctus* (Norton), sometimes destroy 50% of the crop, losses being due chiefly to over-frequent cropping with wheat. Solid stemmed varieties of wheat have some resistance to attack, but attempts to control *C. pygmaeus* biologically have been unsuccessful (Beirne, 1972).

Family Tenthredinidae

This family contains most of the sawflies of economic importance. The eggs are usually laid in young shoots or leaves and the ovipositor is toothed. The serrations are large and stout in species which lay their eggs in wood or stems, fine in those ovipositing in leaves, and almost absent in the gooseberry sawfly, *Nematus ribesii* (Scop.), which attaches its eggs to a minute slit on the underside of gooseberry leaves. Sawfly larvae are caterpillars, and pupate in a silken cocoon or in an earthen cell.

***Hoplocampa testudinea* (Klug.)** (Fig. 7.2D)   The apple sawfly is a pest of apples and responsible for the loss of much fruit when trees are not sprayed in the spring (see Leaflet B). *Hoplocampa flava* (L.), the plum sawfly (Fig. 7.2C), is an important pest of plums.

***Athalia rosae* (L.)** (Fig. 7.2B)   The turnip sawfly is mentioned in old textbooks as a spectacular defoliator of turnips, but it died out in Britain and there was no record of it for over 40 years. It was found again in 1947 when the summer favoured immigration from the Continent, had widened its distribution in 1948, and by 1949 was found as far inland as Stowmarket, Suffolk, where several acres of turnips were infested. This outbreak was treated successfully with DDT but almost any contact insecticide would be effective against the exposed caterpillars (Plate XIIIA). This sawfly seems to have disappeared again from Britain.

The adults appear in May. They are orange and 6–8 mm long. Eggs are laid along the margins of turnip leaves, several hundred by one female. After 5–12 days the larvae hatch and feed on the leaf which is eventually skeletonized. The larvae moult three times and pupate in the soil in silken cocoons. In summer, pupation lasts about 21 days but the pupae of the third generation pass through the winter.

Suborder Apocrita

All members of this suborder, sometimes called the higher Hymenoptera, are distinguished by a waist or petiole formed by the second abdominal segment. The larvae are apodous, peripneustic and have narrow heads. Their habits are diverse.

The Apocrita was previously divided into (1) ACULEATA, in which the ovipositor is normally concealed, used as a sting (except in ants) and issues from the apex of the abdomen. The last abdominal sternum is not subdivided and the trochanters are single. (2) PARASITICA, in which the ovipositor, often elongated, issues ventrally from the abdomen, and is used for inserting eggs into or placing them upon another animal. The last sternum is subdivided and the trochanters are two jointed.

A later classification rejects subdivision into Aculeata and Parasitica and divides the Apocrita into eleven superfamilies which includes the Vespoidea (wasps), the Apoidea (bees), the Sphecoidea (solitary wasps), the Formicoidea (ants), the Ichneumonoidea (ichneumon flies), the Chalcidoidea (chalcid wasps) and the Cynipoidea (gall wasps). As the higher Hymenoptera are, with few exceptions (gall wasps, seed chalcids, some ants) beneficial, they are considered briefly using the older division into Aculeata and Parasitica.

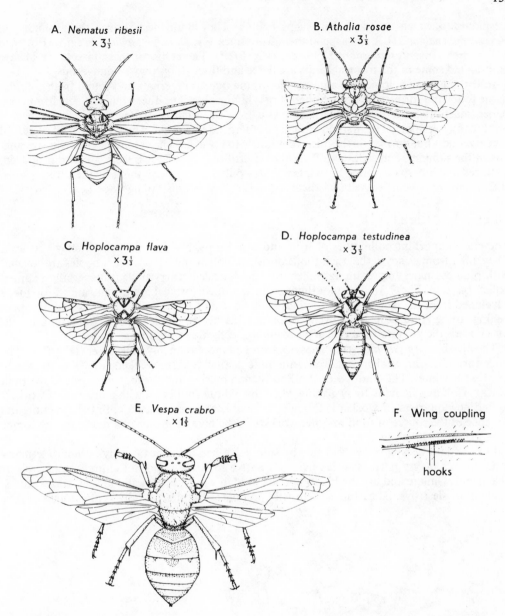

A. *Nematus ribesii*
× 3⅓

B. *Athalia rosae*
× 3⅓

C. *Hoplocampa flava*
× 3⅓

D. *Hoplocampa testudinea*
× 3⅓

E. *Vespa crabro*
× 1⅔

F. Wing coupling

hooks

Fig. 7.2 A–D, sawflies; E, hornet, *Vespa crabro*, with F, enlargement of wing coupling to show hooks.

ACULEATA, stinging forms

This group comprises the bees, wasps, fossorial wasps and ants.

Wasps (Vespoidea)

These are beneficial and feed themselves and their young on other insects. Sometimes they act as pollinators. They are mostly solitary, but social life has developed in one family, the

Vespidae, to which our common wasps belong. They build nests under the ground, e.g. *Paravespula vulgaris* (L.), or suspend them from trees, e.g. *Dolichovespula norvegica* (F.), or in hollow trees, e.g. *Vespa crabro* L. hornet (Fig. 7.2E). The stings of these insects are always painful and some of the tropical forms are fierce and their stings may be dangerous.

In Britain, wasp colonies are annual. The young queens hibernate and found fresh colonies in the following spring when the temperature is favourable. The queen builds a nest from wasp paper (masticated wood and saliva) and lays eggs in the cells. The larvae are fed by the parent until ready to pupate. The first individuals to emerge are workers, who care for the larvae and take over nest building while the queen continues to lay eggs. The colony increases in size and, late in the summer, males develop from the unfertilized eggs. True queens are also produced, fertilized and hibernate during the winter. Wasps only feed on sweet things in the autumn when their numbers have increased and their normal food supply is failing. (See Leaflet No. 451.)

Bees (Apoidea)

These are important pollinators of fruit and seed crops, besides producing honey and wax. They differ from wasps in that their food consists of pollen and nectar. Their bodies are covered with plumose hairs for retaining pollen and the hind legs carry pollen in specially adapted pollen baskets (Fig. 7.3). The mouthparts are modified for sucking nectar, and a tongue is developed from the labium in the families Halictidae and Apidae. Bumble bees and hive bees belong to the Apidae. The life cycle of the bumble bee resembles that of the wasp, but the queen feeds the young with a pollen paste and the cells are built of wax.

The hive bee, *Apis mellifera* (L.), has been studied more than any other insect (Fig. 7.3). The three forms, queen, worker and drone, are quite distinct and the workers as well as the queen survive the winter. The combs serve both for brood rearing and for food storage and are built of wax. Colonies increase by swarming when the old queen flies off with a number of workers to find new quarters. A worker is thought to be able to communicate both the direction and distance of the source of food to other workers in the hive by means of a dance on the combs (von Frisch, 1950).

Flowers and bees have evolved side by side. Hive bees are not the only pollinators: many other insects play a part. Hive bees are not native to Britain and not entirely suited to our changeable climate and are unable to operate in high winds, in rain and at low temperatures. Cultivation destroys the nesting sites of wild pollinators, and artificially kept colonies of hive

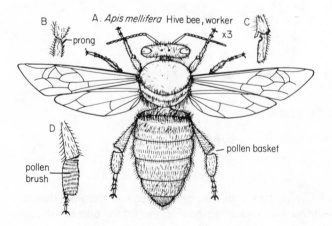

Fig. 7.3 Hive bee. A, worker, with B, second leg to show prong; C, first leg to show antennal cleaning apparatus; D, hind leg to show pollen brush.

bees are able to replace them in part only. The beekeeper benefits from the honey produced while the farmer benefits from a better set of fruit and seed. Spraying of orchards and soft fruit is so timed that bees are rarely killed and colonies do not suffer. The chief loss of honeybees results when pesticides are applied to field crops in flower, particularly field beans, oilseed rape and brassica seed crops, or when spray reaches an 'undercroft' of flowering weeds.

The danger of an insecticidal spray applied in the field to foraging bees is related to its intrinsic toxicity, the weight of active ingredient applied per ha, the proportion of dose picked up by the bees and the behaviour of the bees. This is influenced by the weather during the three previous days and during spraying, the attractiveness of the flowering crop and that of alternative sources of nectar available from other flowering crops and weeds in the vicinity (Smart & Stevenson, 1982). An indication of the likely toxicity in the field can be obtained by calculating the number of LD_{50} (p. 321) doses per ha from the LD_{50} and the amount of material applied at the recommended spray rate (Table 7.1). From the table it is clear why materials like azinphos-methyl, triazophos, malathion and gamma HCH may lead to honeybee poisoning if applied while crops are in flower. Low risk materials may safeguard bees but may not always kill the target pest effectively. Pyrethroid insecticides are very toxic to bees but recommended rates of application of the active ingredient are very small with the result that the risks to bees are much lessened.

Between 1974 and 1979 there were 611 confirmed incidents of honeybee poisoning in England and Wales: 14% were associated with cereals, 13% with field beans and 36% with oilseed rape. Half of all the incidents associated with rape occurred in 1978 and most were traced to spraying with triazophos and the factors thought to have contributed were: (1) recommendations that it should be used at the 'end of flowering' instead of after the end of flowering, (2) patchy flowering of uneven crops, (3) aerial spraying contractors trying to complete spraying in an area regardless of the state of flowering of crops, (4) unnecessary routine spraying to control seed weevils and (5) difficulties with the warning system that alerts beekeepers of the need to close hives. Most incidents occur in East Anglia because of the

Table 7.1 Toxicity to bees, field dosage rate and relative toxicity in the field (no. of LD_{50} doses ha^{-1}). Data from Smart & Stevenson, 1982.

Insecticide	Inherent toxicity to bees $LD_{50} \mu g$ bee^{-1}	Field rate recommended g ha^{-1}	No. of LD_{50} doses ha^{-1}, millions
High risk materials			
Azinphos-methyl	0.063	460	7300
Triazophos	0.055	400	7300
Malathion	0.270	1300	4800
Dimethoate	0.120	350	2900
HCH	0.200	280	1400
Demeton-S-methyl	0.260	240	920
Low risk materials			
Endosulfan	7.1	470	66
Phosalone	8.9	460	52
Pirimicarb	>50	140	<3
Pyrethroid insecticides			
Cyalothrin	0.027	13	460
Cypermethrin	0.056	25	450
Permethrin	0.110	50	450
Flucythrinate	0.270	75	280
Fenvalerate	0.230	50	220
Deltamethrin	0.051	10	200

concentration of bean fields and oils seed rape in the area. Incidents with cereals arise mainly because bees feed on aphid honeydew and sprays are applied to control aphids. Aerial spraying is more hazardous for bees than ground spraying (Stevenson, Needham & Walker, 1978).

The opportunities of honeybee poisoning have greatly increased with the rapid expansion of the area under oilseed rape. It seems clear that sprays should not be applied when the crop is noticeably yellow (i.e. not after the yellow bud stage and not before the petals have fallen and the crop is predominantly green). Spraying should be done in early morning or late evening and, if possible, in cool dull weather, the object being to avoid times when bees are most active. Beekeepers, especially those with hives near rape fields, should receive at least 24 hours warning so that they may close, water and ventilate their hives or remove them from the area. (See Leaflet Nos 328, 377, 485, 486.)

Further information on bees can be obtained from Butler (1962), Ribbands (1953) and Free & Butler (1959).

Ants (Formicoidea)

The role of ants is difficult to assess. They are partly beneficial but are also partly harmful as they protect aphids and coccids which they farm for their honeydew (Way, 1963).

The colonies are perennial and complex with an advanced caste system of queens, workers, males and soldiers. The nests or formicaries are of many different types, most being in the soil, and are perforated by galleries and chambers. A typical colony contains one or more egg-laying queens and thousands of workers. The workers are sterile females that look after the queen, larvae and pupae (Fig. 7.1D). In some species the worker ants have two or three forms. The so-called 'honey ants' have distendable abdomens which serve as a honey store for the rest of the colony (repletes). Certain workers among the seed gathering ants have large heads and jaws to crack seed. The soldiers have large heads. The males and mature females are winged and mate during a nuptial flight. At certain times of the year when weather conditions are favourable, large numbers of winged males and females leave the nest and, as this exodus happens simultaneously over a wide area, cross fertilization by members of different colonies is made possible. After flight, the queen finds a suitable place, tears off her wings, excavates a hollow in the ground and starts to lay eggs. She cares for the first larvae until they are fully grown. The larvae are white and legless with small heads. They develop into workers and from then on take care of the queen and larvae. The queen continues to lay eggs and may live up to 15 years.

Several species are pests. The Argentine ant, *Iridomyrmex humilis* (Mayr.), was introduced into the U.S.A. probably from Brazil and is a household pest wherever established. It swarms into houses and stores and injures fruit and citrus trees by attacking the blossoms and by distributing and protecting aphids, mealy bugs and scale insects. It even prevents people from sleeping and makes some residential areas undesirable.

Ants can be treated with insecticidal dusts and sprays containing HCH, or chlordane (DDT is less effective), but it is possible to destroy the nest with calcium cyanide or carbon disulphide. Slow-acting poison bait such as thallium sulphate and sugar solution, attractive to the workers so that it will be carried back to the nest and fed to the queen, larvae and other workers, is successful in houses and larders where other poisonous substances would be out of place and direct extermination of the nest is impossible.

Parasitica

This group contains parasitic insects many of which are very small and obscure. Because the larvae ultimately kill their hosts they differ from most parasites and might be better described as delayed predators; for this reason they are sometimes called parasitoids.

Primary parasites attack and ultimately kill their hosts and hence exert a measure of population control over them. Secondary or hyperparasites are parasites of primary parasites and have the opposite effect, although in complex ecological systems the interplay of host, parasite and hyperparasite may operate a series of mutual checks tending to an overall balance. Some parasites sting their hosts and deposit an egg or eggs against them, the larvae feeding ectoparasitically. Others oviposit within their hosts which continue feeding until the parasite is mature. True hyperparasites attack their primary host while it is feeding in its host and are usually endoparasites of endoparasites. False hyperparasites include a range of less specialized species that are often ectoparasitic. They attack any small cocoon they encounter and so may operate as primary or as hyperparasites. When primary parasite cocoons are numerous the hyperparasitic habit is predominant.

Some parasites are solitary some gregarious. Some attack eggs, others larvae, others pupae. Some enter the egg and complete their development in the larva, others enter the larva and complete in the pupa. Most parasites are not absolutely specific to one host but are found in a group of hosts bearing some relation to each other. Host range can be defined as a group of hosts in which the parasite is usually able to attack and develop successfully.

Superfamily Ichneumonoidea

All members of this superfamily are parasites on other insects or arthropods and therefore largely beneficial to man.

Most Ichneumonidae are parasites or less often hyperparasites (e.g. *Pimpla* sp.) of Lepidoptera: some parasitize other Hymenoptera and Coleoptera. The females of many species hibernate through the winter. Their behaviour patterns are the most highly evolved of all solitary insects, for they are able to find their hosts with remarkable precision.

The Braconidae are closely related in habit and structure to the Ichneumonidae, but are readily separated by wing venation (Fig. 7.4). They are found in many insects, but the Lepidoptera is most commonly parasitized. Pupation may occur within the host's body, but more usually the cocoon is constructed outside it. Polyembryony (see p. 25), the production of two or more embryos from one egg occurs in some braconids, e.g. *Ageniaspis* sp. As a result, one host may contain up to 3000 parasites.

Apanteles glomeratus (L.) attacks the caterpillars of many Lepidoptera including white butterflies (*Pieris* spp.) as many as 150 larvae occurring in one caterpillar. When mature the parasitic larvae gnaw their way out and construct sulphur yellow cocoons around the body of the host (Plate XIIID).

The Aphidiinae are parasites of aphids, usually of the apterous females in whose bodies a single parasite (*Aphidius* spp., Fig. 7.4) develops. If the aphid is parasitized before its second ecdysis, it fails to mature, but if parasitized later it matures and may produce young. When about to pupate, the larval *Aphidius* cements its host on to the leaf or stem where it is feeding, and the adult parasite later emerges through a circular hole in the host's body wall. *Praon* spp. usually leave the body of the host and pupate in a separate shelter, generally underneath the aphid body. For a theoretical study of the influence of aphid parasites see Knipling & Gilmore (1971).

Superfamily Chalcidoidea

This superfamily contains very many species including some of the very smallest ones in the Insecta. They are mainly parasites or hyperparasites of other insects, and are of economic importance as natural or biological agents of control. Some, however, destroy beneficial insects and are therefore indirectly injurious. There are also some plant feeding forms which infest seeds (e.g. clover seed chalcid) or produce galls (gall wasps). Lepidoptera and Coccoidea are attacked most frequently.

138

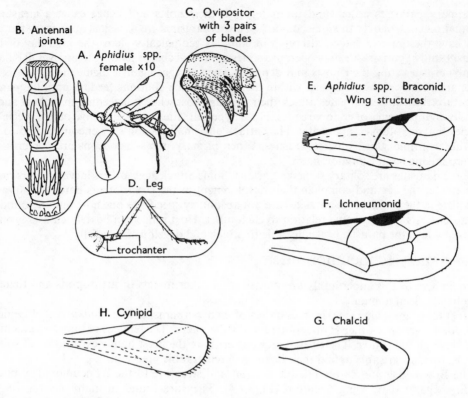

Fig. 7.4 Parasitic Hymenoptera A, *Aphidius* spp., with B, antennal segments enlarged to show elongated sensoria; C, ovipositor enlarged; D, leg enlarged to show double trochanter; E and F, forewing structure of a braconid, ichneumonid, G and H, chalcid and cynipid.

In temperate zones, chalcids have one to three generations a year. Pupation usually occurs within or near the remains of the host. The venation of the wings is reduced (Fig. 7.4G), and the antennae are elbowed. Some species such as *Perilampus hyalinus* (Say.) undergo hyper-metamorphosis, that is, the insect during its development passes through two or more different larval forms with differing habits. Some species, e.g. *Litomastix truncatellus*, are poly-embryonic.

Encarsia formosa (Gahan) and *E. partenopea* (Masi) are parasitic on the glasshouse white fly (p. 69) and their introduction into glasshouses gives efficient control of the pest in temperate regions.

Aphelinus mali (Hald.) is parasitic on the woolly apple aphid, *Eriosoma lanigerum* (Haus.) and belongs to the eastern states of North America. It was introduced into Britain to control *Eriosoma lanigerum* (Haus.) (p. 296) as it had done in New Zealand and Italy, but was only partly successful as it did not always survive the winter. If shoots bearing colonies of parasitized aphids are collected and placed in apple boxes at 4°–8°C, they overwinter safely and can be released in orchards in early summer, a little later than natural emergence (Massee, 1954).

Pteromalus puparum (L.) is common and widely distributed. It parasitizes the pupa of *Pieris rapae* and *P. brassicae*.

Trichogramma sp. is chiefly a parasite of lepidopterous eggs.

Members of the family Agaontidae live within the receptacles of figs and pollinate the flowers.

Superfamily Cynipoidea

The cynipoids are small, dark or black insects (Fig. 7.4H), many of which produce galls on plants which provide food and shelter for the larvae, or they are inquilines (guests) in such galls or parasites. The galls are produced in response to stimuli from the feeding of larvae. Over 80% of gall wasps gall oaks, 7% are restricted to *Rosa* spp. and the rest gall other plants, particularly Compositae.

Neuroterus quercusbaccarum **(L.)** is a common gall wasp in Britain, and produces flat galls on oak leaves. Females emerge in the spring and lay parthenogenetic eggs amongst the catkins and young leaves. The resulting galls are round and from these emerge both males and females. After copulation, the eggs are laid at the sides of the veins of the young leaves and the larvae produce the flat galls. There are therefore two generations which in the past were described as separate species, for there is considerable difference in the size of the ovipositor in the agamic (spring) females and the gamic (summer) ones.

Proctotrupoid Group

Members of this group of superfamilies are all small, slender parasites. The Proctotrupidae mostly parasitize beetles and the Platygasteridae mostly cecidomyids. The Ceratophronidae contains many secondary (hyper) parasites affecting aphids and coccids through their braconid and chalcid primary parasites.

8

Flies – Diptera

The Diptera is a large order of highly organized insects. The adults are usually strong fliers, mostly diurnal, distinguished by a pair of membraneous wings on the mesothorax and a pair of drumstick-like organs called halteres on the metathorax which are vestiges of the metathoracic wings and serve as balancing organs in flight. The head of the adult is mobile and bears large compound eyes. The antennae are least modified in the Nematocera where they consist of a many-jointed flagellum (Figs 8.2, 8.4), and most modified in the Cyclorrhapha where the swollen, 3-jointed base bears a bristle-like or feathery arista. The ptilinum, a structure characteristic of the Cyclorrhapha, is indicated externally by the arched ptilinal or frontal suture. When the adult is ready to emerge from the puparium, the bladder-like ptilinum is everted by blood pressure, presses against the wall and ruptures it. After emergence, it is withdrawn into the head and is no longer functional.

The mouthparts are modified for piercing and sucking or for sucking only. Mandibles are rarely present. The elongated, pointed mouthparts used for sucking blood in mosquitoes and gnats are analogous to those found in the Hemiptera, for the labium is elongated and grooved and encloses the stylets. They differ in having labella (labial palps) on the tip of the labium and in the possession of additional stylets, the labrum-epipharynx and the hypopharynx (Fig. 8.1). The mandibles and maxillae may disappear as in the tsetse fly, *Glossina palpalis* (Fig. 8.1F), but the labrum-epipharynx and the hypopharynx are always present. In tabanids (horseflies), the labium is also an organ for sucking up liquids from moist surfaces by means of pseudotracheae present on the labella.

In the predaceous Asilidae and Empidae, the labium is hardened and horny. The labella are small, the labium and hypopharynx large and strong and the maxillae rigid and blade-like for penetrating the prey. In the sucking type of mouthpart found in the higher Diptera (Cyclorrhapha), the labium is short and stout and the labella are enlarged bearing many canals or pseudotracheae strengthened by incomplete cuticular rings. The labrum-epipharynx and hypopharynx are less pointed and fit into the labial groove (Fig. 8.1D, E). Liquids and minute food particles are sucked up into the pseudotracheae and the prestomal teeth can be protruded and used as rasping organs. In the blood-sucking Muscidae and Pupipara the proboscis itself has become modified to form the principal organ for penetration.

The Diptera with piercing and sucking mouthparts transmit disease organisms to man and other vertebrates, such as malaria, sleeping sickness, yellow fever, and elephantiasis but not plant viruses. The partial eradication of the mosquito has helped to control the spread of malaria in many areas. Methods aimed against the larvae include drainage, introduction of fish into ponds and lakes, and the forming of an oil layer to the surface of the water to block their respiration. Adults are killed with contact insecticides (chiefly pyrethrins).

The feeding habits of the Cyclorrhapha enable them to disseminate bacteria and fungal spores and to contaminate human food. Contact insecticides are used extensively against the larvae and adults, and poisoned sweet sprays (sodium fluoride and molasses) have sometimes been used against the adults.

In adult Diptera, the pro-and metathorax are small and fused with the large mesothorax.

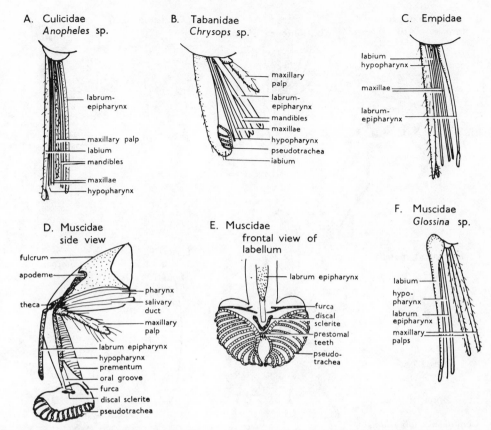

Fig. 8.1 Mouth parts of Diptera, stylet-like in piercing and blood-sucking Culicids and Tabanids, strong with poor pseudotracheae in predaceous empids, a proboscis adapted for sucking with well-developed pseudotracheae in Cyclorrhapha, stylets and reduced proboscis in blood-sucking muscid, *Glossina palpalis*. (Mouth parts of *G. palpalis* after Imms (1957), *Textbook of Entomology*, Methuen.)

The venation of the wings is reduced. The first segment of the abdomen is much reduced and, in many forms, four or five segments only are discernible externally. Metamorphosis is complete, the larvae are legless with small heads and usually a reduced number of spiracles. Most frequently the larvae are amphipneustic. The pupae are free (Fig. 8.7E) or enclosed in a hardened larval cuticle, the puparium (Fig. 2.18C).

The basic type of larva, the eucephalous type, is found in the Bibionidae (Fig. 8.7A–D). It has a normal head and a complete set of spiracles (holopneustic, Fig. 2.13), the first pair having migrated forwards to the prothorax and the last pair backwards to the ninth abdominal segment. There are three pairs of mouthparts, and the mandibles work horizontally. The hemicephalous larva, e.g. leatherjacket, has a reduced head partly withdrawn into the prothorax and is metapneustic (Fig. 8.5A). The acephalous larva of the Cyclorrhapha has a vestigial head. Mouthparts are replaced by sclerotized plates in the pharynx, the anterior pair being mouth hooks working vertically (Fig. 8.13). There are only two pairs of spiracles, one anterior and one posterior (amphipneustic).

Dipterous larvae thus show a gradual reduction in the complexity of structure of the head and spiracular system. The gall midges, Cecidomyiidae, have an anomalous type of larva which does not fit into the above scheme. The larvae are about 5 mm long and often brightly coloured (Fig. 8.2). The head is reduced but there are nine or even ten pairs of spiracles (peripneustic).

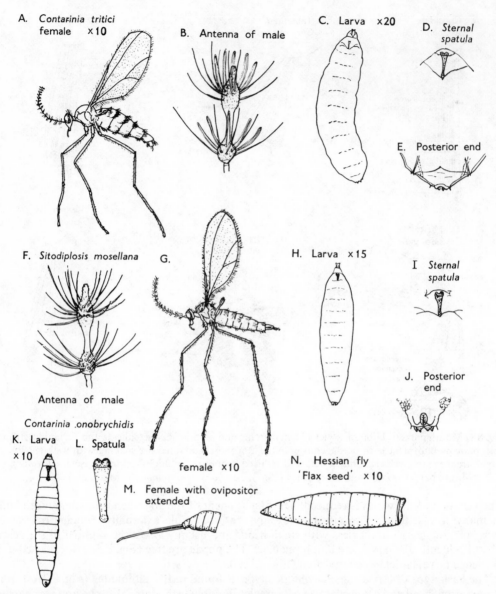

Fig. 8.2 Wheat midges. A–E, *Contarinia tritici*; F–J, *Sitodiplosis mosellana*. Notice the looped hairs on the antenna of the male *C. tritici*; K–M, *C. onobrychidis*; N, *Mayetiola destructor*, Hessian fly 'flax seed' stage. (A–J after Mesnil in Balachowsky and Mesnil (1935–36), *Les insectes nuisibles aux plantes cultivées*, Busson.)

On the mid-ventral surface of the thorax, they possess a characteristic structure, the sternal spatula (anchor process), an elongate sclerite of uncertain function (Fig. 8.3). The larvae of *Contarinia* spp. are able to leap; the body is arched, the anal crotchets locked on the spatula and, when tension is suddenly released, the larva is projected several centimetres into the air. Some larvae reproduce parthenogenetically while conditions are favourable. This is known as paedogenesis (p. 25). Paedogenetic reproduction ceases when conditions become adverse and sexual reproduction occurs again.

The number of larval instars varies from four to eight. In the lower Diptera the last larval skin is cast at pupation, but in the Cyclorrapha the last larval skin hardens to become the puparium (Fig. 2.18) in which the remains of the larval mouth hooks and spiracles can be seen.

Classification

The Diptera is divided into three suborders:

Nematocera. Gnats, mosquitoes, gall midges, crane flies and bibionid flies

These are nearly all gnat-like flies with long legs, long antennae (often plumose) and in mosquitoes and some gnats piercing and sucking mouthparts. Larval forms vary and the pupae are free. The adult emerges through a T-shaped slit in the pupal skin.

Brachycera. Horse flies, clegs, robber flies

These are robust flies with variable, but generally three-segmented antennae. The larvae have small heads with vertically biting mandibles. The pupae are usually free and often spiny. As in the Nematocera, the adult emerges through a T-shaped slit in the pupal integument.

The Nematocera and Brachycera together form the lower Diptera or **Orthorrhapha.**

Cyclorrhapha or higher Diptera contains most of the dipterous pests

The larvae are typical maggots, acephalous, apodous and amphipneustic. The pupa is coarctate, that is free within the last larval skin which is retained as a puparium. The adults have 3-jointed antennae bearing a dorsal arista, a frontal lunule, a ptilinum and 1-jointed maxillary palpae.

Suborder Nematocera

Family Cecidomyiidae

Gall midges (Figs 8.2 and 8.3) are minute, delicate flies with hairy wings and long moniliform antennae bearing conspicuous whorls of hairs. Wing venation is reduced. This important family contains many species with plant feeding larvae that attack all parts of plants giving rise to a wide range of symptoms, such as distortion, leaf curl, galls, bloated flowers and distorted pods. Other larvae are predaceous, some are scavengers and some are fungus feeders. Many phytophagous larvae are brightly coloured and the sternal spatula, of value in identifying larvae, is variously shaped (Fig. 8.3). For detailed accounts see Barnes (1946–56).

Grass seed and seed of leguminous forage crops are damaged when gall midge larvae feed on the flowers and prevent perfect seed formation.

Grass seed midges

Grass is the most important agricultural crop in Britain. It is also dried for the extraction of carotin, used for lawns and for binding loose sand particles, aiding in the consolidation of banks and the protection of the seashore from erosion. Some varieties are used for ornamental purposes. The production of good grass seed is therefore important.

There are six important genera of gall midges the larvae of which affect grass production. Their life histories are similar.

***Contarinia merceri* (Barnes)** (Fig. 8.3)　Up to fifteen golden yellow larvae are found in single florets of meadow foxtail grass (*Alopecurus pratensis*). There is one generation a year and the adults are short lived.

***Contarinia geniculati* (Reuter)**　The solitary, salmon-buff larva is found in the florets of meadow foxtail grass and cocksfoot grass and has no sternal spatula (anchor process). There are usually two generations a year. The lemon-yellow larvae of *C. dactylidis* (H. Loew) and the pink larvae of *Dasineura dactylidis* (Metcalfe) (Fig. 8.3) are found in the seeds of cocksfoot grass; both have one generation a year.

***Dasineura alopecuri* (Reuter)** (Fig. 8.3)　The orange or brick-red larvae are found in the florets of meadow foxtail and black grass (*Alopecurus myosuroides* (Huds.)). There is one generation a year and usually only one larva per floret. The larvae pupate in the florets and do not migrate to the soil as do most midge larvae.

***Sitodiplosis dactylidis* (Barnes)**　The orange or carroty-red larvae are found in the florets of cocksfoot grass (*Dactylis glomerata* (L.)) and rarely become pests. They are like *Sitodiplosis mosellana* (p. 14) in structure. There is one generation a year.

The adults appear in late May or early June when there is one generation a year, or earlier, in April and May when there is a second later generation in July or August. The adults usually appear about the same time as the grass hosts are in flower and live for 2–4 days. They are very delicate and diptera the right conditions of temperature, moisture and wind velocity for ovipositing, and fly for a brief period in the morning or evening. The eggs hatch in about 7–8 days depending on temperature, and the larvae feed on plant juices. When fully grown the larvae (not *D. alopecuri*) pass into the top layers of the soil and overwinter there in a cocoon, entering diapause in the last larval stage. They pupate the following year a week or two before emergence.

Damaging populations of midge larvae tend to build up if a stand of grass is left down for several years (D. P. Jones, 1945). The percentage of infested florets increases with the age of the crop. Examination of florets in the first and second year may show little or no infestation, but by the sixth year the infestation can reach 36–60%. As the crop gets older, midges move in from neighbouring grass stands and from the hedgerows. The accidental introduction of *C. geniculati* and *D. alopecuri* into New Zealand ruined foxtail seed production, lessening germination of the seed produced to the uneconomical level of 10%.

The damage from gall midge larvae is decreased by crop management. Old stands of grasses should be ploughed up periodically and new ones planted well away from the old. Stands of pedigree seed should be grown in isolation, choosing exposed and windswept sites rather than sheltered valley fields which favour oviposition. Sites adjacent to areas of land largely under the plough are less likely to have reserves of flies. Clean seed, free from cocoons, should always be used. Cutting or grazing sometimes delays flowering until after peak emergence of the midges but may itself also reduce the seed yield. A year's seed production may be sacrificed by taking a crop of hay, preferably on the early side to desiccate any larvae present in the young flower heads. Several cuts of short grass should also be taken from neighbouring hay fields and hedgerows to prevent flowering and lessen midge populations. To control *Contarinia lolii* (Metc.), rye grass can be sown with white clover and managed so as to take alternate crops of grass and clover seed. If the sward is heavily grazed until June in the years when the grass seed is taken, midge attack is further reduced. When the larvae remain in the seed (*D. alopecuri*), infested seed should be rejected, dry heated to 60°C for 30 min, or possibly it might be fumigated with methyl bromide. As with all pests which remain in the seed *D. alopecuri* is difficult to detect and is a problem in export and quarantine.

Leguminous crop midges

***Contarinia pisi* (Winn)** (Fig. 8.3).   The pea midge attacks peas (*Pisum sativum*) often at the same time as the pea moth (*Cydia nigricana*). Although terminal blossoms and shoots may be attacked, pod infection is most common. Attacked blossoms are swollen with short pedicels. They die without forming seeds. Attacked pods are malformed and contain numerous white 'jumping larvae'. Pea midge is widespread in Europe and all varieties of pea are attacked.

Eggs are laid in May or June in batches of twenty or more inside the flowers, on leaves or on the young shoots. After 4 days, the larvae hatch and live at the base of the ovaries, in the clustered leaves of the terminal shoot or in the pods where 20–40 larvae may feed. Up to 100 larvae may be found in one swollen flower. After 10 days, larvae of the first generation mature. They jump to the ground and spin a cocoon in the soil in which they pupate. There are two types of cocoons, one round in which the larva goes into diapause and from which the flies emerge in spring, and one oval in which the larva pupates at once (Vallotton, 1969). After about 11 days adults emerge, mate and oviposit again. Larvae of the second generation overwinter in cocoons in the soil. As for some other midges, there have been years of exceptional abundance, and relative scarcity.

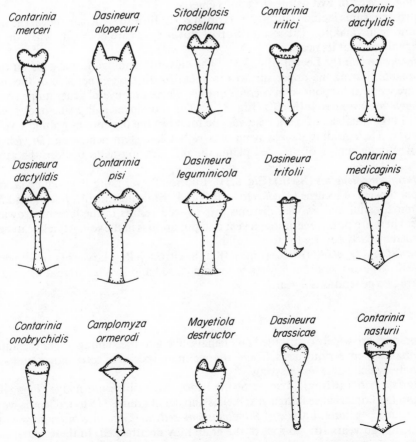

Fig. 8.3 Anchor processes or sternal spatulae of various cecidomyiid larvae. (Mostly after K. M. Smith (1948), *A Textbook of Agricultural Entomology*, Cambridge Univ. Press.)

Useful cultural methods of control include early spring sowing, the selection of early varieties, the burning of all infested haulms after harvest, crop rotation, deep ploughing to bury the overwintering larvae.

Where attacks occur frequently, routine spraying is advisable. Elsewhere crops approaching the susceptible stage should be examined; when there is one or more adult midges per leading shoot in 15% of the plants spraying is required. The onset of the susceptible stage is about 7 days before flowering when the oldest green flower buds are about 6 mm long. If growth is slow and thrips numerous, a second spray is necessary some 6–10 days later. Approved materials include azinphos-methyl/demeton-methyl sulphone mixture, demeton-S-methyl, dimethoate, dimethoate/triazophos mixture and fenitrothion. (See Leaflet No. 594.)

Three scelionids and a pteromalid (*Pirene graminae* (Hal.)) parasitize pea midge, and the bright red larvae of the midge *Lestodiplosis pisi* (Barnes) feed on the larvae.

Dasineura leguminicola (Lint.)   Larvae of the clover seed midge attack the inflorescences of red and white clover causing irregular, brown flowers and loss of seed.

The adults appear when the red clover is in blossom and many eggs are laid on each flower head. They hatch in 2–6 days and the pale yellow larvae feed at first on the developing seeds, gradually becoming salmon coloured. After 5 to 6 weeks, they drop to the ground, spin cocoons in the soil and pupate. After a few days the adults emerge to start a second generation. Larvae of the second generation overwinter in diapause in cocoons in the soil and pupate in the spring, a week or two before the adults emerge.

Control measures include cutting the crop in early June to desiccate the larvae in the flower heads during haymaking. The second crop of clover heads are not ready when the population of flying midges is at its peak.

Dasineura trifolii (F. Loew.) (Fig. 8.3)   The clover leaf midge forms galls on the leaves of clover plants causing the upper surfaces to stick together on either side of the mid rib. The larvae are white at first but soon become orange. There are several generations a year.

Campylomyza ormerodi (Kieff.) (Fig. 8.3)   The red clover gall gnat attacks red clover in Britain. Large numbers of red larvae may be found in the tap root at ground level and in the plant apex. Their feeding causes symptoms rather like stem nematode (*Ditylenchus dipsaci* (Kühn)). Young shoots of diseased plants die off. Grazing helps to keep the attacks under control.

Contarinia medicaginis (Kieff.) (Fig. 8.3)   Lucerne flower midge has yellow larvae found in the upper flowers of lucerne (*Medicago sativa*) and yellow lucerne (*M. falcata* and *M. arabica*) which remain shut and swollen. Serious loss of seed results in the lucerne growing areas of Europe. There are three generations a year and pupation is in the soil. Attacks can be prevented or lessened by early cutting for fodder and hay.

Contarinia onobrychidis (Kieff.) (Fig. 8.3)   The citron-yellow larvae of sainfoin flower midge (Fig. 8.2K–M) cause sainfoin flowers to remain closed and swollen so that no seed is formed. There are two generations a year.

Wheat midges

Wheat midges are widely distributed and behave like grass seed midges, but the importance of cereals makes them serious pests. There are two main species of seed midges and another, the Hessian fly, which damages the straw.

Contarinia tritici (Kirby)   Larvae of the lemon wheat blossom midge (Figs 8.2A–E, 8.3) feeding in the florets cause flattened spikelets and small grains. Up to ten lemon-yellow larvae may be found per floret. Larvae of *Sitodiplosis mosellana* (Géh.) are often present in the same spikelets. In bad years 10% to 55% of the grain may be attacked. In these years, clouds of *C. tritici* may be seen drifting across wheat fields when the weather is favourable. The main flight occurs when ears of wheat are appearing and eggs are deposited in the florets. The larvae hatch in 7–10 days and feed in the developing grain. After about three weeks, usually during damp weather, they jump to the ground and make cocoons in the soil, where they overwinter. They pupate in the spring, a week or so before the adults emerge. Some larvae, however, pupate as soon as they enter the soil, giving rise to late summer adults which oviposit in couch grass. The resulting larvae feed rapidly and form over-wintering cocoons in the soil. There may, therefore, be two generations a year.

Sitodiplosis mosellana (**Géh.**)   The larvae of the orange wheat blossom midge (Figs 8.2F–J, 8.3) can be distinguished from those of *C. tritici* by their orange colour and the occurrence of only one larva per floret. They do not jump and, in dry weather, are often immobile, enclosed in a transparent skin. The life cycle is similar to that of *C. tritici* although the adults fly one or two weeks later. Red eggs are laid singly in florets and the larvae feed on the sap and do not enter the grain. Larval feeding lessens the number of well-formed, viable grains. After 3 weeks feeding larvae become immobile, drop to the ground after rain, form cocoons and pupate in the following April. There is no second generation.

Larvae of *C. tritici*, *S. mosellana* and thrips are often present together and can be controlled by sprays of fenitrothion or chlorpyrifos applies during the main period of adult midge activity. Control measures against midges alone are rarely practised because outbreaks are sporadic and the damage done is usually slight. (See Leaflet No. 788.)

Several parasites have been bred from a number of stages of *C. tritici* and *S. mosellana*; these include two scelionid (Proctotrupoidea) egg parasites and fifteen larval parasites, all hymenopterous, of which the best known is the scelionid, *Leptacis tipulae* (Kirby). Predators play an insignificant part in checking the numbers of wheat midges. *Empis* sp. feeds on egg laying females as do some thrips, mites and spiders.

Haplodiplosis marginata (**van Roser**)   The saddle gall midge is common in northern Europe and lately has become more troublesome in Britain, perhaps because cereals are grown more intensively now. The midges emerge from mid-May to mid-June and lay their red eggs in reft-like groups on the wheat leaves. After hatching, the larvae move down the stem and into a leaf sheath where a cushion-like, saddle-shaped gall is formed. The larvae are blood red and 4–5 mm long. They enter the soil at the end of July, hibernate in cocoons and pupate next May, emerging soon after. Attacked shoots stay greener longer than unattacked. Attacked straws are weakened and tend to break; some yield is lost. 47%, 39% and 60% spring barley in S.E. England was attacked by saddle gall midge in 1967, 1968 and 1969 (Woodville, 1970). Fenitrothion spray applied 10 days after the peak flight of adult midges in early to mid-June is an effective control measure. (See Leaflet No. 657.) Two hymenopterous parasites attack the larvae.

Mayetiola destructor (**Say.**) (Figs 8.2N, 8.3)   The Hessian fly causes widespread damage in the wheat-growing areas of Africa, the United States, New Zealand and part of Europe but is less serious in Great Britain. It attacks the straw and not the grain. There are two main generations. The adults oviposit on spring and autumn sown wheat, and the young wheat plants soon show symptoms of attack. At first, the leaves are erect, but are bluish-green in colour and often shorter than those of uninfested plants. The central shoot may be stunted or missing and tillering lessened or stopped. Later in the summer, the straw lodges just above a node and the ear is shrivelled. When the leaf sheaths at the point of lodging are examined, white larvae or brown ensheathed larvae, often called 'flax seeds', are found. The larvae feed superficially on sap from the stem tissues. When conditions are favourable, there may be an additional summer and autumn generation, so that the tillers themselves become infested.

The female lays reddish eggs on the upper surface of cereal leaves, chiefly wheat and, after hatching, the red larvae move to the growing point or to the stem above a node, moult and start to feed. They are attached firmly to their food by their mouth parts and continue to feed for 2–3 weeks in the spring, and the second generation feeds for 5–8 weeks in the autumn. The diapausing larva passes the winter in the 'flax seed', pupae are formed in the spring and adults emerge when the wheat is at the seedling stage. All the progeny of one mating tend to be predominantly of one sex.

Normally spread is by flight, but artificial spread by 'flax seeds' in infested straw is also possible. Hessian fly was introduced into the United States from Europe in the straw bedding used by the Hessian soldiers during the American Revolution and was first found in Long Island near the camp they occupied in 1779.

'Fly free' dates for sowing autumn and spring wheat have been calculated and charted for the whole of the wheat belt of North America, where wheat is often grown continuously. Farmers are encouraged to consult these before planting. The flies live only a few days, and the dates suggested have enough safety margin to ensure that all adults will have emerged and died before the wheat comes through the ground. In the U.S.A. varieties of wheat have been bred that resist Hessian fly and their deployment has revealed that there is a gene-for-gene relationship between major genes for resistance in the host and those that regulate ability to circumvent resistance in the parasite (Hatchett & Gallun, 1970). This kind of relationship exists in rust fungi and in nematodes (p. 301).

Brassica midges

***Dasineura brassicae* (Winn.)** (Fig. 8.3)   The brassica or bladder pod midge causes swollen, prematurely ripening siliquas of cabbage, swede, turnip, rape and radish; the so-called 'bladder pod' condition which prevents normal seed development. Pod midge is an important pest in swede growing areas of Romney Marsh and Lincolnshire and is widely distributed throughout Europe and has greatly increased in importance with the large expansion of the areas under oil seed rape.

There are three or four generations a year but the first two only are important. The female uses the snout punctures in the pod made by the swede seed weevil (*Ceutorhynchus assimilis* (Payk.), p. 123), lesions caused by fungi, or other damaged spots to insert her eggs and the white larvae are said to reproduce paedogenetically (p. 25) in the siliqua, and feed on the developing seeds and the walls of the pod. After about 4 weeks the pods split open prematurely and normal, fully grown larvae descend to the soil, form cocoons and pupate. Adults appear about a fortnight later, except for the last generation which overwinters in the cocoon and pupates the following spring. Families of midges bred from pods may be unisexual.

Oil seed rape and brassica seed crops should be planted 0.5 km or more away from the previous year's crops. The adult midges are weak fliers and this distance is a fairly effective barrier. Immigrant midges lead to heavier infestations around the crop perimeter which may be misleading. (See Leaflet No. 780.)

***Contarinia nasturtii* (Kieff.)** (Fig. 8.3)   Swede midge is important in districts where swedes are grown for seed. It attacks many brassicas (cabbage, cauliflower, rape, etc.) and leads to swollen flowers, crinkled leaves ('crumpled leaf') and often kills the main shoot, so that the side shoots grow out causing a 'many necked' plant. The symptoms of 'crumpled leaf' and 'many neck' are also known as 'cabbage top' (Plate XIVD).

The main damage is caused by the June brood of yellow larvae feeding on the growing point which often leads to 'bacterial rot'. Swede midge numbers fluctuate with outbreaks in years of high humidity and temperature.

In the spring, each female lays 66–124 eggs in batches of 15–20 on the youngest parts of the plant. The larvae hatch after 3–9 days and feed chiefly in the growing point. After a few days, symptoms of attack appear. The stalks become swollen, the leaves crumple and unfolding of the terminal bud is delayed. When fully grown the larvae jump to the ground, form cocoons in the soil and pupate. After 1–3 weeks, depending on temperature, the pupa works its way out of the cocoon and moves to the surface of the soil until the anterior end protrudes; the adult then emerges. There are usually three generations a year; the larva of the third generation over-winters in the soil and pupates in the spring before the emergence of the adults.

Crop rotation accompanied by clean cultivation, is effective in keeping midge numbers down. Early sown plants are more tolerant to Swede midge attack than late (Bardner *et al.*, 1971). Old infested swede tops should be destroyed and young plants showing injury should be burnt to kill the larvae. In gardens, early larval stages can be washed off plants with a strong jet of water. The old treatment with nicotine and soft soap was so successful in Germany that spray calendars giving the dates of spraying for each generation were drawn up (Roesler, 1937).

Family TIPULIDAE

Craneflies or daddy long legs (Fig. 8.4) are well-known insects, with long legs, thin bodies and many-jointed antennae. The front of the head is prolonged forwards (Fig. 8.4B) and, in a few genera, may form a proboscis. There are no ocelli on the head and on the mesonotum there is a V-shaped suture. The adults are commonly found in September. The larvae are leatherjackets and are usually found on damp grassland but also occur on arable land especially in the year following ploughing up of low-lying grassland or ley. Leatherjackets feed on many crops and damage is greatest in the wetter districts of Britain and Ireland. Several species are injurious, the commonest being:

***Tipula paludosa* (Meig.)** (Figs 8.4A–D, 8.5A–C)   The marsh or common cranefly, which lays over 200 eggs per female in grassland either forced between the soil particles or dropped

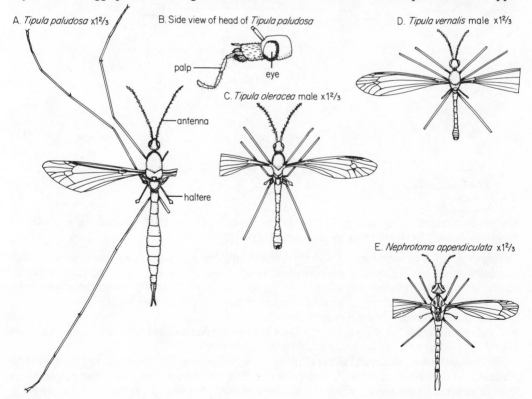

Fig. 8.4 Adult crane flies to show the long legs, antennae and halteres; B, shows the forwardly elongated head and well-developed, pendulous maxillary palps.

in flight. They hatch after about 10 days to produce tiny brown or grey larvae, rubbery to the touch. The larvae are well adapted to life in wet situations, and feed vigorously in March or April of the following year, continuing until they are fully grown in July when they are 4 cm long. They pupate in the soil and move to the surface just before the adults emerge in late August or September (Plate XVI).

Leatherjackets (Fig. 8.5) feed primarily on grass roots, so causing the grass to die and pull out easily. Heavily infested pasture loses its 'keep' and serious injury can be caused to the turf of golf greens, fairways, lawns and cricket pitches. When arable crops are attacked, plants are severed at ground level and there is ragged feeding on the leaves just above the surface of the

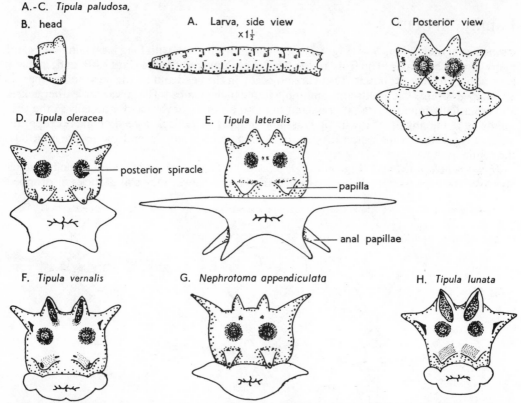

A.-C. *Tipula paludosa,*

B. head   A.   Larva, side view   C.   Posterior view
                      ×1½

D. *Tipula oleracea*   E. *Tipula lateralis*

posterior spiracle

papilla

anal papillae

F. *Tipula vernalis*   G. *Nephrotoma appendiculata*   H. *Tipula lunata*

Fig. 8.5 Structure of leatherjackets. B, note the hemicephalous head; C–H, posterior view of the last abdominal segment of six common leatherjackets. (After Brindle, 1960.)

soil. Spring sown oats and newly sown grass are liable to damage. Sugar beet crops are not normally grown in areas where leatherjackets flourish but damage to seedlings can be locally severe, irregular segments being eaten from the leaves and the growing points destroyed (Fig. 14.1, p. 284). Brassicas, flax and linseed also suffer damage (Plate XVIA–C).

Other species damaging grass crops include:

Tipula vernalis **(Meig.)** (Figs 8.4D, 8.5F)   Common in coastal and moorland areas with two periods of emergence, one in spring and one in September.

Nephrotoma appendiculata **(Pierre)** (Fig. 8.4E, 8.5G) and *N. flaxescens* **(L.)** are common in gardens and sometimes in fields. Adults emerge in May or June but the eggs remain unhatched until September so the life cycle is brought into line with that of *T. paludosa*.

Tipula oleracea **(L.)** (Figs 8.4C, 8.5D) has two periods of emergence, one in the spring and one in September but it is uncertain whether there are two generations a year.

Tipula lunata (**L.**) (Fig. 8.5H) and *Tipula lateralis* (**Meig.**) (Fig. 8.5E) occur in marshy fields and damp meadows.

Leatherjacket populations in grassland vary with the time of the year, the greatest numbers being found in the spring. In March, up to 7 500 000 per ha occur in marshland, 2 500 000 being common. The numbers in arable land, golf greens and lawns is usually less than 250 000 per ha. To estimate populations an irritant mixture, that causes the leatherjackets to rise to the surface, is watered on the sward or soil. An emulsion, St Ives Exterminator (10 l of *o*-dichloro-benzene, 2.5 l of 10% solution of sodium oleate and 2.5 l of Jeyes fluid) is diluted, 1 part in 1820 l of water and applied to the turf at 50 l per square metre with a watering can. The material is applied to random squares and the leatherjackets that come to the surface are collected, counted and identified.

Grassland may support large numbers of leatherjackets without appreciable damage, then suddenly, owing to changed climatic conditions which put additional stress on the plants, the grass may fail and bare patches appear. They are usually most numerous after prolonged damp weather in late summer and early autumn when there is enough moisture for the newly hatched larvae. When this occurs, the pasture should be ploughed up and later reseeded with seed mixtures containing deeply rooted grasses such as cocksfoot. Leatherjacket areas should be closely grazed in September, as craneflies accumulate on the long growth and oviposit there. Adequate liming and drainage of affected fields is important.

Leatherjacket damage to cereal crops occurs at populations of 250 000 per ha or less. Ploughing out from grass in fields liable to attack should be done before September (i.e. before the main oviposition period) as trouble usually follows spring ploughing, but if ploughing is too early a wheat bulb fly attack may result. Gulls, lapwings, pheasants, rooks and starlings kill many leatherjackets, which are also attacked by several parasites and may also become infested with the *Tipula* irridescent virus.

The feeding habits of leatherjackets are similar to those of cutworms (p. 83) with which they are often confused, so they are also controlled by broadcasting poison bait containing gamma-HCH or fenitrothion, preferably in the evening. Spraying with chlorpyrifos, etrimfos, HCH, quinalphos, triazophos is also effective. Treatment sometimes greatly increases grass yield from good hill fields but is uneconomic unless there are more than 2 500 000 larvae per ha in early spring. It is cheaper to apply extra nitrogen fertilizer if there are fewer than a million. (See Leaflet No. 179.)

Family BIBIONIDAE

These are robust flies, often pubescent with shorter legs and wings than other Nematocera. The antennae consist of 8–16 bead-like segments and are usually shorter than the thorax. Certain species exhibit dimorphism with reddish brown females and black males. The eyes of the males may be holoptic occupying nearly the whole of the head. The anterior veins of the wings are more strongly marked than the posterior (Fig. 8.6A). Few species are of agricultural importance. Some larvae live gregariously in meadows, on grassy hillsides and on decaying vegetation. They feed on the roots of grasses, cereals, sugar beet, hops and leaf mould. Large numbers of feeding larvae of *Bibio hortulanus* (L.), the March fly may cause slight damage to wheat in February and March. Structurally the larvae are the most primitive of all Diptera, with a well-developed head and twelve segments (Fig. 8.7). Each segment bears a band of fleshy protruberances and ten pairs of open spiracles are present. Larvae pupate in the soil and adults emerge the following April. Control is not usually required. Sprays used against leather jackets would probably work but not poison baits.

Bibio marci (**L.**) is called St Mark's fly as it appears in large numbers about St Mark's day, 25th April.

Suborder Cyclorrhapha

Family SYRPHIDAE

Hover flies (Fig. 8.6B) are brightly coloured and may be frequently observed hovering above flowers with their wings vibrating. They may be striped, spotted or banded with yellow on a blue, black or a metallic background. Many of them are beneficial, as their larvae feed on aphids, but others are phytophagous, saprophagous or scavengers in the nests of ants, termites, bees or wasps (Plate XIIIC).

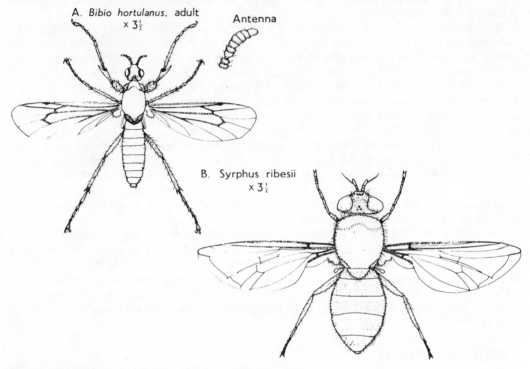

Fig. 8.6 Adult bibionid, *Bibio hortulanus*, and adult syrphid, *Syrphus ribesii*.

Syrphus ribesii (L.) and *Scaeva pyrastri* (L.) are large flies with aphidivorous larvae, which are rather slug-like with flattened ventral surfaces, pointed anterior ends and short posterior respiratory tubes. The larva of *S. ribesii* (Fig. 8.7B) is pink, and that of *S. pyrastri* green with a white mid-dorsal stripe. Long white eggs are laid in aphid colonies and the larvae destroy many aphids. They pupate near their habitat, often on leaves, and the caudal segments are cemented to a leaf or other support. The puparium is somewhat inflated. Syrphid and ladybird larvae together with parasites like *Aphidius* keep down the numbers of aphids and so help to terminate epidemics.

The larvae of *Merodon equestris* (F.), the large narcissus fly, is mainly a pest of narcissus. The lesser bulb flies, *Eumerus strigatus* (Fall.) and *E. tuberculatus* (Rond.), feed in rotten bulbs of narcissus, amaryllis, hyacinth, scilla and lily, and also sometimes in damaged carrots, parsnips, potatoes and onions (See Leaflet No. 183.)

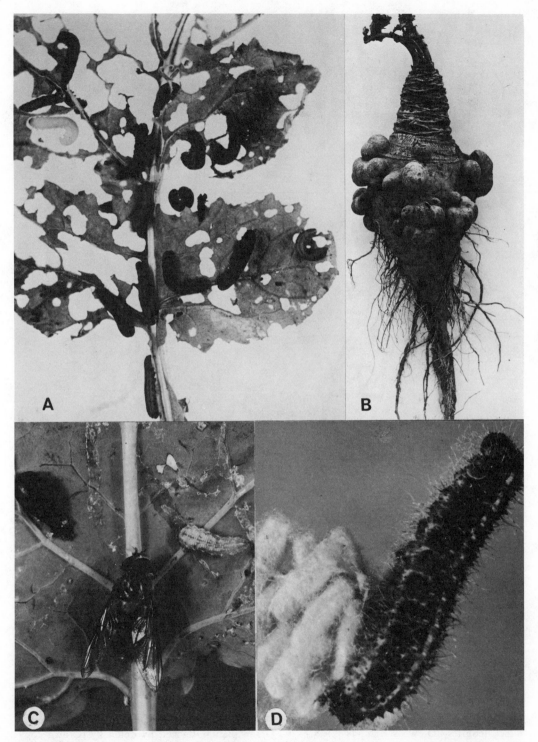

XIII **A.** Turnip leaf skeletonized by turnip sawfly caterpillars (Crown copyright). **B.** Swede galled by turnip gall weevil (courtesy W. E. Dant). **C.** Hover fly adult, larva and pupa (Crown copyright). **D.** Caterpillar of large white butterfly with cocoons of *Apanteles* sp. (courtesy H. C. Woodville).

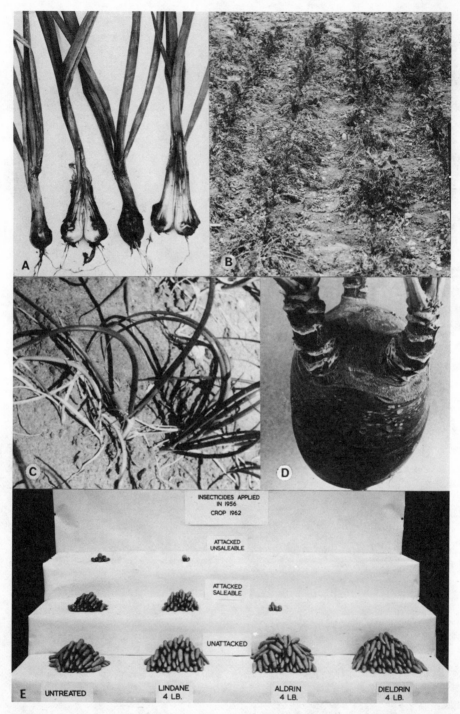

XIV **A.** Onions attacked by onion fly (courtesy W. E. Dant). **B.** Carrots wilting after heavy carrot fly attack (courtesy D. W. Wright, N.V.R.S.). **C.** Onion plants attacked by onion fly (Crown copyright). **D.** 'Many necked' swede after attack by swede midge (courtesy H. C. Woodville). **E.** The relative persistence of lindane (gamma-HCH), aldrin and dieldrin as shown by control of carrot fly six years after application (courtesy D. W. Wright, N.V.R.S.).

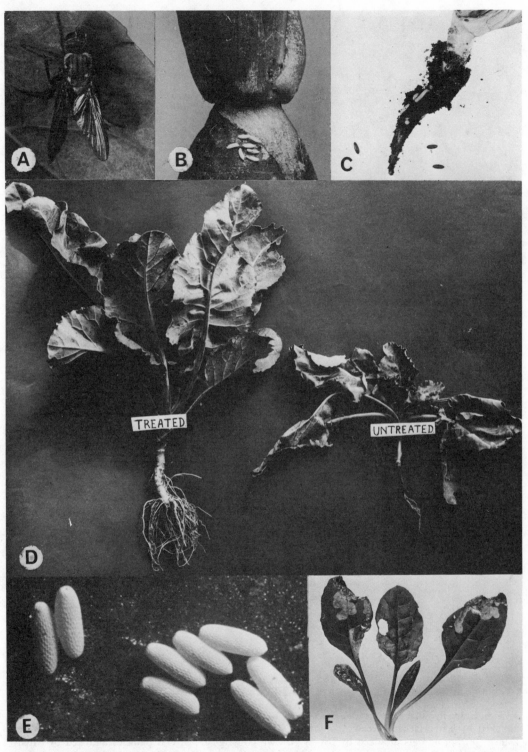

XV **A.** Cabbage root fly (courtesy H. C. Woodville).   **B.** Eggs of cabbage root fly on collar of rape (courtesy H. C. Woodville).   **C.** Cabbage root fly larvae and pupae (courtesy T. H. Coaker, N.V.R.S.).   **D.** Cabbage root fly: the effect of treatment with insecticide on cauliflower (courtesy D. W. Wright, N.V.R.S.).   **E.** Eggs of mangold fly (courtesy R. A. Dunning, Brooms Barn).   **F.** Beet leaves mined by larvae of mangold fly (Crown copyright).

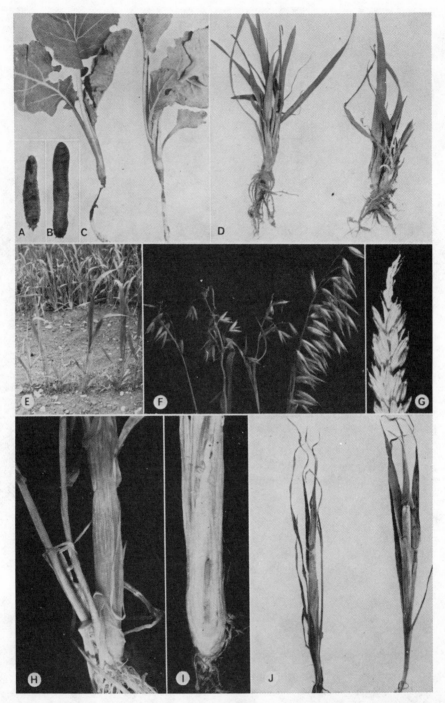

XVI **A.** Crane fly pupa (courtesy W. E. Dant). **B.** Crane fly larva (courtesy W. E. Dant). **C.** Brassica attacked by leatherjackets (Crown copyright). **D.** Spring oats attacked by frit fly (courtesy W. E. Dant). **E.** Spring oats attacked by frit fly (Crown copyright). **F.** Oat panicles attacked by frit fly (courtesy W. E. Dant). **G.** Wheat attacked by frit fly (courtesy W. E. Dant). **H.** Barley attacked by gout fly (courtesy W. E. Dant). **I.** Section of barley with gout fly larva *in situ* (courtesy H. C. Woodville). **J.** Barley. Right, attacked by gout fly; left, healthy (courtesy H. C. Woodville).

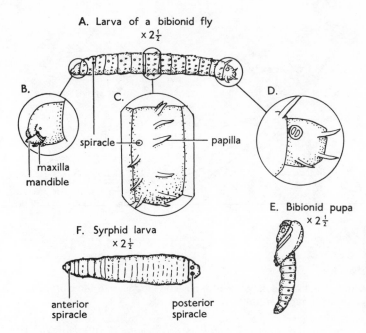

A. Larva of a bibionid fly
$\times 2\frac{1}{2}$

B.

C.

spiracle — papilla

maxilla
mandible

D.

E. Bibionid pupa
$\times 2\frac{1}{2}$

F. Syrphid larva
$\times 2\frac{1}{2}$

anterior
spiracle

posterior
spiracle

Fig. 8.7 A–E, structure of larva and pupa of a bibionid, *Bibio* sp. ($\times 2\frac{1}{2}$). Note the papillae and spiracles; F, *Syrphus ribesii* larva.

Family CHLOROPIDAE

These are small, light coloured flies, common among herbage. The arista of the antenna is bare. The larvae are mostly phytophagous although one species feeds on aphids.

Oscinella frit (L.) Although the frit fly (Figs 8.8, 8.9) is primarily an inhabitant of grassland it is a common pest of cereals. The flies are black, 1.5–2 mm long, with a globular thorax and five-segmented abdomen covered with fine hairs. The frons is wider than the eye and the frontal triangle is shining black. The black antenna has an enlarged basal joint with a short hairy arista. Dark hairs on the thorax are arranged in rows running lengthways and the halteres are yellow. The trochanters and part of the tibiae and tarsi are yellow. The claspers of the male often protrude from the end of the abdomen and on the pointed end of the female is a pair of palp-like organs at the tip of the 3-jointed, retracted ovipositor. The egg is white, slightly curved and sculptured. The larva (Fig. 8.8A) is elongated (1.5–5 mm long), white, tapering towards the head end where there is a 2-jointed antenna. The mouth hooks are short and strong and the anterior spiracles are fan-shaped with 3–7 branches, the posterior spiracles bear four tufts of branched, recurved hairs. The puparium is 2.5 mm long, yellow at first, later becoming red brown. The mouth hooks and posterior spiracles are clearly visible.

For many crop pests, the crop or crops concerned provide a far greater density of hosts than occurs in nature but, for insect pests of cereals such as frit fly, for which the natural habitat is grassland, this does not hold. Even in predominantly arable areas much grass remains in hedgerows, dyke sides, road margins and waste places. Many dipterous stem borers occur in grassland but frit fly is the only one attacking oats (Southwood & Jepson, 1961). Populations in grass are relatively stable and are somewhat higher in winter than in summer. Reasons advanced to explain this include natural enemies, interspecific competition, the limited numbers of tillers at the right stage of development and the scarcity of large seeded grasses suitable for grain feeding.

Adult frit flies are mobile and may be carried in air currents like aphids so that free interchange between populations of grassland and oat fields is possible. Flies derived from grass reach oat fields in April and May. They oviposit behind the coleoptiles of young plants,

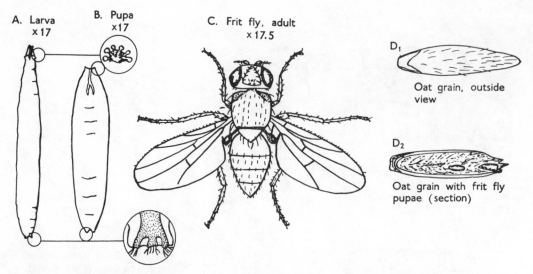

A. Larva
×17

B. Pupa
×17

C. Frit fly, adult
×17.5

D₁

Oat grain, outside view

D₂

Oat grain with frit fly pupae (section)

Fig. 8.8 *Oscinella frit*. Larva, pupa and imago with enlargement of posterior spiracles to show branched bristles. D₁ and D₂, attacked oat grains with pupae *in situ*.

the eggs occurring earlier than the similar ones of *Elachiptera cornuta* (Fall.) (Jepson & Southwood, 1960), the larvae burrow into the central shoot, pupate and emerge as adult flies after about 34 days. These flies are normally abundant in July and oviposit in the spikelets of the oat panicle, the larvae feeding upon and pupating in the seed. Flies that emerge early may oviposit on backward plants and cause damage to the emerging ear. Flies emerging from the panicle in August are 200 times more numerous than those invading the crop in spring and far more abundant than in summer grass (Southwood & Jepson, 1961) and if the weather in the autumn is warm and sunny a fourth generation of adults emerges. Eggs are laid on self-sown cereals, rye grass and early sown winter cereals. Appearance and persistence of the flies is probably affected by weather which also influences the growth of available hosts and varies with the incidence of direct attack on autumn cereals. During the winter, the larva is found mostly in grass stems. Pupation occurs in spring and the flies emerge in April.

Greatest damage is done to spring oat seedlings attacked before the four-leaf stage by the first generation of larvae. The tiny larva bores into the central shoot leaving no obvious track and mines there so that the tip of the shoot becomes completely or partly severed from the base. The central shoot yellows and dies ('dead heart' condition). If, however, the plant is somewhat older and has tillered, some tillers may escape injury and enable the plant to survive. Occasionally, many tillers are thrown out and attacked so that the plant may look like tufted grass. Second generation larvae feed on the florets and destroy the seed kernels. Dissection of plants with dead central shoots or attacked grain reveals the small white larvae or yellow-brown puparia which may be recognized by the characters of the pharyngeal sclerites and anterior and posterior spiracles (Fig. 8.8A) (Plate XVID–G).

Damage to winter oats is usually much less than to spring oats. Autumn attacks on winter wheat, barley or oats may arise from direct oviposition on early sown crops but severe attacks are uncommon. Severe attacks may, however, occur when wheat follows a rye grass ley especially in Norfolk and the Eastern Counties. Larvae hatching from eggs laid on the rye grass migrate to the wheat seedlings and destroy them. The injury from stem mining in wheat, barley and rye grass is like that in oats and produces the same outward symptoms, namely dead central shoots. Because similar symptoms are caused by wheat bulb fly, the winter generation of the gout fly and some other Diptera, larvae should be dissected out and identified

(Fig. 8.12) to determine the species responsible. Maize and sweet corn are also attacked by frit fly, especially if sown in early or mid-May, but early sowing is impracticable because they are susceptible to frost.

A measure of control can be obtained by sowing spring oats early, before mid March, so that the plants have passed the four-leaf stage when the population of flies from the overwintering generation is at its peak.

Autumn attacks on winter cereals can be lessened by avoiding over-early sowing (not before early October), and attacks after rye grass leys are controlled by breaking the leys in late July or early August, before oviposition by the flies arising mainly from the oat panicles. Oviposition may also be lessened by mowing or close grazing.

Second generation larvae feed on the seed kernel and destroy the grain, but again early sown crops suffer less than later ones and thick stands less than thin. Resistance to second generation attack is unrelated to resistance to the first.

To protect a late-sown oat crop, it should be sprayed between full emergence and the 2-leaf stage with chlorpyrifos or triazophos; in practice insecticides are not used. Maize and sweet corn may be protected by carbofuran or phorate granules applied in well-dispersed band along the rows during sowing. Chlorfenvinphos granules may be applied over the rows when the crop emerges. When fully emerged chlorpyrifos, fenitrothion, pirimphos-methyl or triazophos sprays may be used. (See Leaflet No. 110.)

Six species of Hymenoptera parasitize the spring generation of *O. frit* in England and other species have been recorded from Europe. A nematode, *Tylenchinema oscinellae* (Goodey), parasitizes the adult fly and sterilizes it. Because of the protected situation of the eggs behind the coleoptile, ground predators such as *Bembidion* spp. do not lessen the larval population as in cabbage root fly (M.G. Jones, 1969).

Chlorops pumilionis **(Bjerk.)** Larvae of the gout fly (Fig. 8.9, 8.10A, B) attack barley and wheat, especially in the south-west of England, and some wild grasses including couch grass (*Agropyron repens* (Beauv.)).

The adult females are about 4–5 mm long and the males a little smaller. The head is yellow, the basal joint of the thorax is enlarged and on the yellow thorax are three broad dark chocolate-brown bands. Dark bands are present on the dorsal surface of the brown abdomen. The female has a greenish metallic tint. There are two generations a year. Flies from the overwintering generation appear in May and lay eggs singly on the upper surface of the leaves of barley or occasionally on wheat. The eggs hatch after 8 days and the tiny larvae (Fig. 8.10B) make their way to the central shoot. When fully fed and 6–8 mm long, the larva turns so that its head is directed upwards, moves to a suitable place and pupates. After 36 days, flies of the second generation emerge in August and September. These oviposit on grasses or sometimes on early sown winter barley in October. Again the larva makes its way to the central shoot and feeds slowly during winter. In March it reverses its position and pupates. As with frit fly, the flies that oviposit on spring barley are mainly derived from grasses.

Damage to barley varies with the stage of growth of the plant. Seedling plants in autumn exhibit the usual 'dead heart' symptoms associated with dipterous miners but affected tillers are usually swollen and the central shoot is sometimes completely absent. The leaves are broader than usual and a deeper shade of green. Plants attacked in May when the ear is forming but not escaped from the ensheathing leaves, also become swollen and distorted. The larva burrows down the side of the ear rachis (stem) and weakens it so that the ear remains imprisoned and cannot escape, the whole plant rarely exceeding 20 cm in height. The bloated appearance of the plant suggested the name 'gout fly'. Damage is greatest to backward crops. When the ears are more completely developed at the time of attack, they escape from the ensheathing leaves but may be twisted and fail to ripen. If the eggs are laid on leaves below the ripening ear, the larvae that hatch migrate downwards, fail to enter it and cause little harm (Plate XVIH–J).

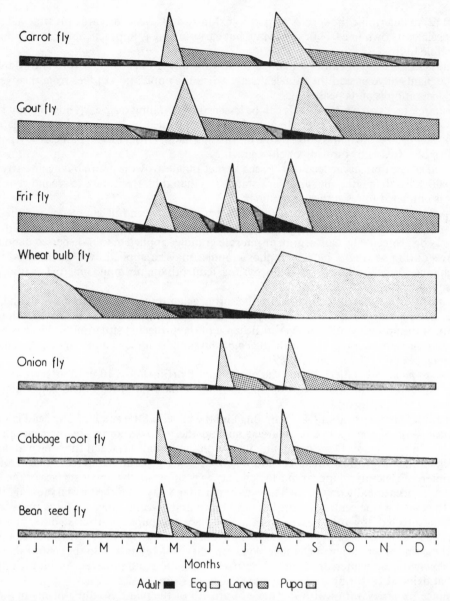

Fig. 8.9 Schematic life cycle diagrams of several species of dipterous pests, partly after T. Coaker.

Little attention has been paid to the control of gout fly as only late sown barley is liable to damaging attacks, and then only on farms in gout areas. In districts where damage to winter wheat or barley occurs it is advisable to delay sowing until after mid-October to avoid oviposition by adults which are about in September and sometimes persist into early October. (See Leaflet No. 174.)

Family OPOMYZIDAE

This is a family of small flies with larvae that are essentially grass feeders; those of *Opomyza*

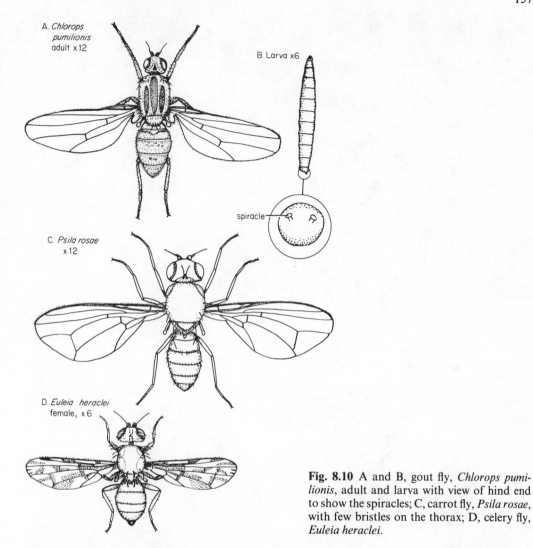

A. *Chlorops pumilionis* adult x12

B. Larva x6

spiracle

C. *Psila rosae* x12

D. *Euleia heraclei* female, x6

Fig. 8.10 A and B, gout fly, *Chlorops pumilionis*, adult and larva with view of hind end to show the spiracles; C, carrot fly, *Psila rosae*, with few bristles on the thorax; D, celery fly, *Euleia heraclei*.

florum (F.) and *O. germinationis* (L.), grass and cereal flies are sometimes found on wheat, and those of *Geomyza tripunctata* (Fall.), although normally feeding on rye grass, have been recorded from wheat.

Opomyza florum (F.) (Fig. 8.11)   The grass and cereal fly, has increased in importance with the practice of sowing winter cereals early. Adult flies appear in June–July, when the females are immature, and reappear in September–October, when they are gravid with eggs. Apparently there is only one generation a year and between appearances the flies are believed to shelter in woodlands and hedgerows where they survive predators and parasitic fungi. Females probably also feed to mature their eggs. These are laid in September–October and occasionally in November in mild weather; damage to winter cereals is somewhat greater in these circumstances. Eggs are deposited on the soil near host plants where they remain over winter hatching from January to April in the following year. The larva crawls up the plant and then makes its way between ensheathing leaves to the growing point. Larvae seem to prefer tillers to main shoots and do not migrate from shoot to shoot as do those of the wheat bulb fly. Whereas wheat bulb fly larvae chew a large ragged hole in the side of the shoot, *Opomyza* larvae make

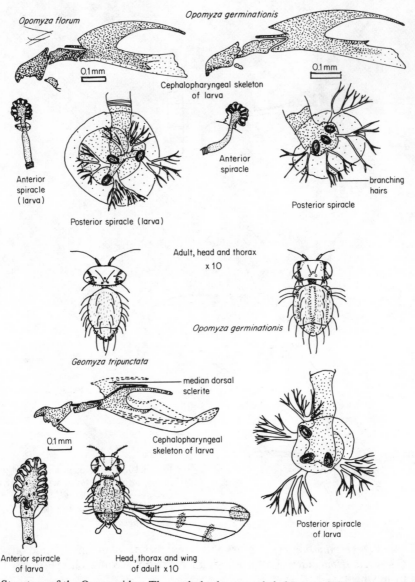

Fig. 8.11 Structure of the Opomyzidae. The cephalopharyngeal skeleton and the anterior and posterior spiracles of *Opomyza florum*, *O. germinationis* and *Geomyza tripunctata*. Note the digitate processes of the anterior spiracles and the branching hairs of the posterior spiracles. Head and thorax of the adults showing bristles.

an incision in the side of the shoot or just above the first node which is seen as a brown line encircling or spiraling around it. Otherwise the 'dead heart' symptoms produced by both species and by *Phorbia securis* (Tien.) are the same. *Opomyza* larvae pupate in May and emerge as flies in June. A single chlorfenvinphos, chlorpyriphos, cypermethrin, fonofos, triazophos or omethoate spray, correctly timed is effective. Crops sown in October or later usually escape attack.

O. germinationis is a common fly; the larvae usually feed on grasses, rarely on wheat, oats and barley. They are often present in company with those of frit fly, gout fly or wheat bulb fly.

Characters by which dipterous larvae found in cereals may be distinguished are in Figs 8.11 and 8.13.

Family PSILIDAE

Psila rosae **(F.)** (Figs 8.9, 8.10C)   The carrot fly is a well-known pest that occurs in most fields, allotments and gardens. Injury is caused by the larvae which mine into the tap roots and render them unsaleable. It is usually necessary to pull up the carrot to find the damage but a bad attack causes the leaves to turn yellow and wilt, a reaction to several kinds of injury. In gardens and small plots larvae may be sufficiently numerous to destroy the plants completely. The crowns of celery and the basal stems of celery plants are also mined. Parsnips may also be attacked and their marketability greatly reduced. In celery, heavily attacked plants are severely stunted (Plate XIVB).

The adults are black shining flies 4–5 mm long, with large yellowish brown heads. The third antennal joint is part black and part yellow, so differentiating it from the closely allied *P. nigricornis* (Meig.), the chrysanthemum stool miner, which has an all black third antennal joint. A black spot on the vertex of the head bears two bristles and is surrounded by three ocelli. The black thorax bears two pairs of dorsoventral bristles and the legs are pale yellow. The wings have brownish yellow nervures and the sub-costal vein is absent. The abdomen is dark and shining, bearing a few pale hairs, and is longer and more tapering in the female than the male. The female lays white, elongated eggs in the soil surface near carrot or other umbelliferous plants, and these hatch in about 12 days; the first stage larvae feed first on the fibrous roots and then enter the tap root and tunnel upward into the cortex where they remain, often with the posterior end protruding. The first stage larva is white or yellow and possesses two large chitinous hooks which are lost in subsequent stages. The third stage is slender (Fig. 8.13K) and shiny and often yellow because of carrot tissue in the intestine. It is amphipneustic, and on the ventral surface of each segment are transverse bands of spicules, and around each posterior segment are five to six finger-like processes. The puparium is cylindrical and pale yellow with darker ends.

Adults of the first brood emerge from mid-May to the end of June and the second brood from the end of July until the end of September (Fig. 8.12). This second brood lays more eggs in the soil in the vicinity of carrot plants, and the larvae from them again mine the tap roots and often remain in them during the winter. They pupate at various times during the winter, but by April pupation is usually finished. Pupae go into diapause at 15°C or lower.

The spring emergence of flies is short in the Eastern Counties and shows a peak in late May or early June. This first brood flies in from old carrot fields, the dispersal of flies being aided by

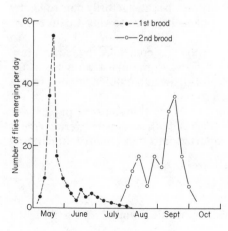

Fig. 8.12 Graph to show the first and second broods of carrot flies.

slight winds. The flies spend most of their time sheltering around the edges of the fields and in the hedgerows, dyke sides or belts of trees and only move into the carrot fields to oviposit between certain temperature ranges, usually in the early morning or late evening. Egg laying takes place chiefly around the edges of the fields, only a small number penetrating to the middle.

The second brood is on the wing for a longer period and is more damaging, especially to carrots left in the ground during winter, which become completely mined (Fig. 8.12). If carrots are lifted in October and November and clamped, 80% of the larvae are left behind in the soil while the carrots in the clamp, provided they are mature and not heavily damaged, deteriorate only slightly. Unfortunately clamping is too expensive an operation. Farmers find it cheaper to leave the carrots in the ground until needed for marketing despite the fact that larvae continue to feed during winter.

Carrots sown in mid-May are attacked by larvae arising from the tail end of the first brood of flies only, while those planted in June miss the first brood altogether and are attacked at a later stage by fewer of the second brood because they have to fly in from neighbouring fields.

The relative position of former carrot crops and the availability of shelter such as hedgerows and ditches affects the intensity of attack. Fields near woods and high river banks should be avoided. Small plots and narrow strips suffer most from edge effects. However, farmers rarely take these factors into consideration. Carrots are planted where it is most convenient and control measures are applied as necessary.

Early chemical control measures were based on the flies' habit of sheltering. The sides of the fields, ditches, hedgerows and shelter belts were sprayed as soon as the flies could be swept from the field sides. Early sprays contained sodium fluoride (0.8%) mixed with molasses and had to be applied weekly, but soon this was superseded by 0.5% DDT emulsion. Two or three applications were made for the first brood of flies and three to four for the second, toxicity lasting for about 3 weeks after application.

Twelve years after aldrin, dieldrin and gamma-HCH were first used to control carrot fly in East Anglia, populations became resistant and these insecticides failed to control the pest (Wright & Coaker, 1968). So, organophosphorus and carbamate insecticides are used instead. Where still effective, gamma-HCH seed dressings can be used but there is a slight risk of taint. Other measures to protect against the first generation include spraying with chlorpyriphos before sowing, granular insecticides applies at sowing or later (carbofuran, chlorfenvinphos, diazinon, disulfoton, phorate) and sprays against the second and partial third generation (chlorpyriphos, diazinon, etrimfos, primiphos-methyl, quianalphos, triazophos). Carbofuran granules are suitable for mineral soils only. (See Leaflet No. 68.) Chlorfenvinphos granules are most effective when placed 10 cm deep, 7.5 cm away from the carrot rows seven weeks after sowing (Wheatley, 1970). Attack by the second brood may also be satisfactorily controlled by incorporating phorate granules into the top soil to supplement a pre-sowing treatment or gamma-HCH seed treatment.

Carrot fly also attacks celery and parsnips. In these crops it is controlled by carbofuran, disulfoton or phorate granules applied broadcast or by bow-wave granular treatments as for carrots at sowing. Carbofuran can be applied in the seed furrow and chlofenvinphos sprays may be used mid-season to supplement control on parsnips only.

At least three braconids parasitize the pupae of *P. rosae*. Many thousands of pupae were collected in East Anglia and sent by air to Canada. *Chorebus* and other parasites were bred out and released for the biological control of carrot fly imported into that country, apparently without success.

Family TRYPETIDAE

The flies in this family are characterized by marbled or mottled wings, and a flattened ovipositor. The larvae are phytophagous and barrel shaped when full grown. Many live in fruits and, the Mediterranean fruit fly, *Ceratitis capitata* (Wied.) attacks succulent fruits in many sub-tropical areas of the world, except North America, causing much damage. Other larvae live in flower heads of Compositae and some are leaf miners.

Euleia heraclei (L.) (Fig. 8.10D)   Celery fly is commonly found in Great Britain, especially in the main celery- and parsnip-growing districts. The larva mines the leaves of celery and parsnips causing large blotchy blisters filled with moisture and excrement. Many umbelliferous plants are also attacked. Sometimes damage is so bad that the celery leaves become brown and curled and give the plant a scorched appearance. Plants are often attacked in the seedling beds before they are set out.

The first brood flies appear in late April, May and early June. They are shining brown with smoky wings, the eyes are deep green sometimes tinged with red and the antennae and legs are yellow. The abdomen is slightly broader than the thorax and the female is a little larger than the male (5 mm long). The eggs are laid on the lower surface of the leaves, larvae hatch in 6–10 days and mine between the upper and lower epidermes. They are white with a greenish tinge and live gregariously. On the first segment above the mouth is a pair of sensory organs, and all the body segments bear pairs of locomotory spines. The posterior spiracle opens ventrally by three lozenge-shaped slits. After 3 weeks the larvae, full grown and 7 mm long, pupate either in the leaf mine or in the soil in oval, light yellow puparia with wrinkled surfaces. In summer the pupal period lasts 25–30 days and the second brood of flies appears in August. The eggs laid by these flies do less harm, as the celery and parsnip plants are by now large with a greater leaf surface. The life cycle may be repeated and a few third generation flies appear in September, but most of the second generation overwinter in the soil as puparia, emerging in April of the following year. Celery and parsnips should not be grown on or near land heavily infested the previous year. Infested leaves should be burned rather than composted. Proximity to previous celery crops should be avoided. Celery plants can be sprayed with trichlorphon or malathion to kill the larvae before transplanting. Larvae mining leaves of mature plants can lower the value of the crop if numerous. They can be killed with malathion, trichlorphon or quinalphos sprays, but this is rarely necessary. Disulfoton granules applied against carrot fly keep populations in check.

On early sown parships, treatment is usually unnecessary as the damage caused is not great. (See Leaflet No. 87.)

Family AGROMYZIDAE

This is another family of small flies the larvae of which are leaf and stem miners with a wide range of food plants, some even mining in the cambium of trees. The larvae of *Phytomyza rufipes* (Meig.) (Fig. 8.13) mine young shoots and stems of brassicas and are particularly damaging on calabrese (Italian sprouting broccoli) (Carden, 1962). Each larva forms several tunnels and makes an exit hole in the epidermis through which it passes to pupate in the soil. There are at least two generations a year. Treatments applied against cabbage root fly and caterpillars give some protection. Trichlorphon, azinphos-methyl or demeton-S-methyl sprays are effective.

Cryptochaetum iceryae (**Williston**), with larvae endoparasitic in certain scale insects, has a remarkable life cycle. There are four larval instars, the first is an embryo-like sac without tracheae or mouth parts and an open digestive tract, but with a pair of finger-like tail processes. In successive instars, when the mouth parts and digestive tract are functioning, these caudal or tail processes increase in length and become filamentous and longer than the whole body. They appear to be respiratory in function (Thorpe, 1931).

162

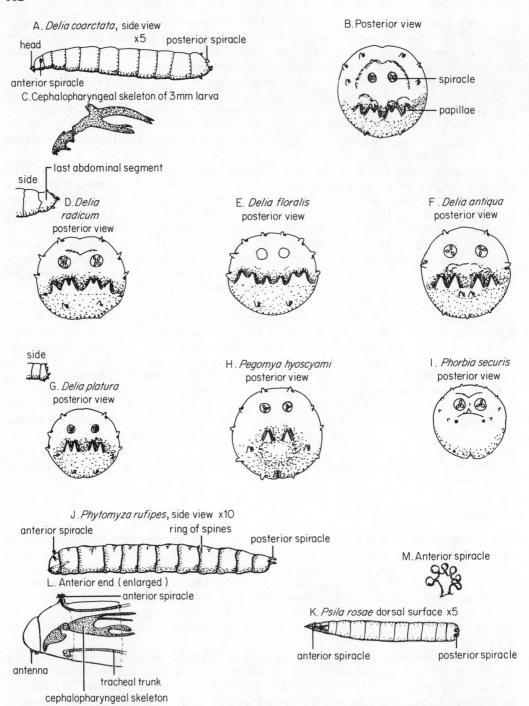

A. *Delia coarctata*, side view
head   x5   posterior spiracle
anterior spiracle
C. Cephalopharyngeal skeleton of 3mm larva

B. Posterior view
spiracle
papillae

last abdominal segment
side
D. *Delia radicum*
posterior view

E. *Delia floralis*
posterior view

F. *Delia antiqua*
posterior view

side
G. *Delia platura*
posterior view

H. *Pegomya hyoscyami*
posterior view

I. *Phorbia securis*
posterior view

J. *Phytomyza rufipes*, side view x10
anterior spiracle   ring of spines   posterior spiracle

L. Anterior end (enlarged)
anterior spiracle
antenna
tracheal trunk
cephalopharyngeal skeleton

M. Anterior spiracle

K. *Psila rosae* dorsal surface x5
anterior spiracle   posterior spiracle

Fig. 8.13 Some dipterous larvae. A side view of wheat bulb fly larva; B–I, posterior view to show caudal corona formed by papillae and the spiracles; J, side view of *Phytomyza rufipes*; K–M, larva of carrot fly. (J after K. M. Smith (1948), *A Textbook of Agricultural Entomology*, Cambridge Univ. Press.)

Family ANTHOMYIIDAE

This family contains several important crop pests.

Delia coarctata (Fall.) (Figs 8.9, 8.13A, B, 8.14A, Plate XX) Wheat bulb fly is widely distributed in the United Kingdom and is prevalent in the principal wheat-growing areas of north Europe. The larvae attack winter wheat and occasionally winter barley and rye, but not oats. The chief wild host is *Agropyron repens* (couch grass), but larvae have been found on other species of grass.

The damage is noticeable in March or April for the larva feeds inside the central shoot killing the growing point. The central shoot turns yellow and dies, and if the plants are at the single shoot stage, they are killed, and the larva migrates to another plant or tiller so that the area of damage enlarges (Plates XVII, XVIII).

The small, grey, hairy flies appear in June and July. In the male the frons is narrower than in the female, and the legs are black whereas the female has brown femora and tibiae. The wings are yellowish and the average length about 6–8 mm. Each female lays one or two batches of creamy elongated eggs in bare or sparsely covered soil in July and August. They are not laid on or near the host plant as is usual for Diptera. Dense standing crops such as cereals are avoided. Most eggs remain dormant until February. Eggs and early larvae are adapted to low

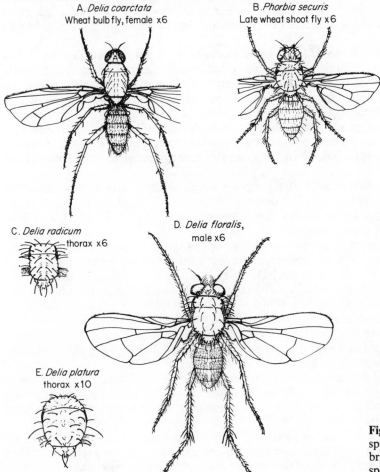

A. *Delia coarctata*
Wheat bulb fly, female x6

B. *Phorbia securis*
Late wheat shoot fly x6

C. *Delia radicum*
thorax x6

D. *Delia floralis*,
male x6

E. *Delia platura*
thorax x10

Fig. 8.14 Adult flies, *Delia* spp., dorsal view to show bristles on thorax of five species. Note hairy legs.

temperatures: eggs hatch little above 0°C and the basal development temperature for larvae is 0.5°C. After hatching, the larva moves through the soil until it finds a wheat plant, bores its way into the centre and mines the central shoot. If wheat, or other suitable host plant, is not present the larva dies soon after hatching. The fly, therefore, in ovipositing on bare soil 'takes a chance' that its host plant will be available. The newly hatched larva differs from the later instars by having two pairs of unserrated cephalic hooks, one pair above the other. The later stage is a typical muscid larva with one pair of serrated cephalic hooks. Around the anterior spiracle are 7–8 papillae, and the caudal end is truncated with posterior spiracles surrounded by a ring of spines and two bilobed papillae (Fig. 8.13B). The larva feeds on the central shoot until April or May when it pupates in the soil. The puparium is an elongated barrel with well-marked posterior spiracles. After about a month the adult emerges. Adult flies remain in and around cereal crops feeding on aphid honeydew and fungal spores, chiefly *Septomyxa* sp., and yeasts during the preoviposition period lasting about four weeks. Females mate when 10–14 days old and leave the crop about ten days later to lay their first eggs, usually within 400 to 800 m of where they emerged. Adult flies may be parasitized by fungus, chiefly *Entomophthora muscae*, and if this occurs early in July many females die before they lay their first batch of eggs (M.G. Jones, 1971). Whether attacked or not few flies lay more than one batch of eggs.

The winter wheat area has increased greatly in recent years and in some areas the crop is planted continuously. It is also planted earlier mid-September onwards, often after minimum cultivation of stubbles. Early planting is an advantage as this gives well established plants at the end of January with 2–4 tillers instead of single shoots. Spring wheat sown after 1st March is not attacked.

On heavy land, attacks are most likely after a bastard or full fallow, but these are rarely taken nowadays. On lighter land, attacks usually follow potatoes, sugar-beet or an early harvested or patchy crop that leaves bare land in late July and August. On the whole continuous cereal cropping is unfavourable for oviposition. Serious attacks tend to be local even within the areas where the pest is most common, i.e. Yorkshire, the East Midlands and East Anglia. The dead heart symptoms typical of wheat bulb fly resemble those of ley pests and, less closely, those due to wireworms but the timing and circumstance make confusion unlikely. However, *Opomyza florum* and *Phorbia securis* (Tien.) produce the same symptoms as wheat bulb fly but somewhat later.

Occasionally, seasons when wheat bulb fly eggs are exceptionally abundant have coincided with adverse weather in autumn that has delayed sowing and retarded development of the crop. Widespread and devastating attacks have followed in the spring, as in 1953, and these stimulated much work on insecticidal control.

Early autumn sown crops need protection only when egg laying has been heavy. Entomologists in the Agricultural and Development and Advisory Service monitor egg populations in areas where wheat bulb fly is troublesome. However, soil sampling and egg extraction are laborious and time-consuming, so advice is often too late to make decisions about insecticidal seed dressings when ordering wheat seed. Bowden & Jones (1979) showed that the number of gravid females caught in light traps was correlated with the number of eggs subsequently laid and that 90% of eggs were laid by 21st August. Egg populations in East Anglia were well correlated with light trap catches to 21st August so the possibility of using light traps for forecasting is being explored.

Dieldrin seed treatment was an excellent protective measure (Plate XVIID). The amount of insecticide applied per hectare was small, so there was no residue problem. Unfortunately widespread and indiscriminate use of dieldrin seed treatment when it first became available caused many deaths among grain-eating birds. The outcry from those interested in the preservation of wild life led to its withdrawal. Chlorfenvinphos and carbophenothion are organophosphorus substitutes for dieldrin in seed treatments which give fairly effective control of wheat bulb fly (Maskell, 1970). They should be used only in areas where wheat bulb fly is

XVII **A.** Wheat bulb fly (a 'Shell' photograph).   **B.** Larva of wheat bulb fly, *in situ* (Crown copyright).   **C.** Larva of wheat bulb fly (Crown copyright).   **D.** Wheat bulb fly control in winter wheat. Left, untreated, right, seed treated with dieldrin (courtesy F. D. Cowland, Rothamsted).   **E.** Eggs of wheat bulb fly (a 'Shell' photograph).

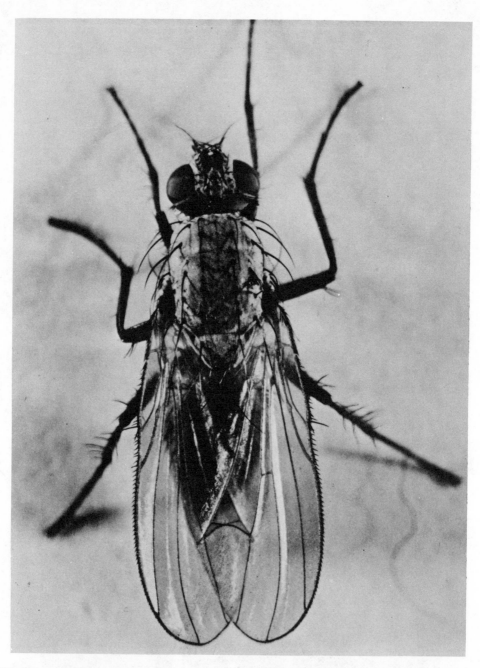

XVIII Photograph of an anthomyid fly, *Delia sp.*, ×25. (Courtesy Shell Photographic Unit.)

prevalent, in years when egg deposits are heavy, after fallow or for sowings before 1st January. Gamma-HCH treatments give some control on very late sown crops. When an attack is under way, as an emergency measure, plants may be sprayed with dimethoate, omethoate, or formothion, but such treatments are not as effective as protective measures taken before the crop was sown. (See Leaflet No. 177.)

Delia radicum (L.) (Figs 8.9, 8.13, 8.14) Cabbage root fly is a widespread pest in the northern Hemisphere, and large populations are built up in areas where brassicas are grown intensively. It is also common in market gardens, allotments and gardens, early cauliflowers and summer cabbage suffering most (Plate XVA–C).

The start of emergence of the first brood of flies in the spring (end of April–May) coincides with the beginning of flowering of the common hedge parsley (*Anthriscus sylvestris* (L.)) and is earlier in the south than in the north. The adults are 5–7 mm long, grey-black in colour, the male being darker than the female. The eyes of the male meet on the dorsal surface of the head (holoptic), but are widely separate in the female (dioptic). There are three well-marked black bands on the grey thorax, and the tapering abdomen bears a dark median and three transverse black lines. The wings are grey with brown veins and the legs are black in the male and brown in the female. At the base of the hind femora the male bears a tuft of bristles which are absent in the female. Flies feed on nectar from hedgerow flowers to develop the first batch of 40 to 60 eggs but require protein for subsequent batches (Finch, 1971).

The small white eggs are laid in the soil around the stem of young brassica plants in the seed beds or after transplanting in May. Direct drilled crops are not usually attacked until two true leaves have developed. Later broods of flies may lay their eggs on Brussels sprout buttons. The eggs hatch in 3–7 days and the larvae burrow down into the soil and feed collectively on the tap root 2–3 cm below the soil surface. The larva (Fig. 8.13D) has a smooth, white shining integument tapering anteriorly and truncated posteriorly. The anterior spiracles are fan-shaped and the caudal corona consists of six pairs of tubercles with a seventh bifid ventral pair near the anus. The posterior spiracles are knoblike and open by three slits. The first-stage larva differs from the others by having no anterior spiracles and only two slits on the posterior spiracle. After about 23 days' feeding and when they are about 8 mm long, the larvae leave the root and pupate in the soil near the plant. The puparium is red or dark brown and barrel shaped, and the pupal stage lasts from 15–35 days in summer or throughout the winter. The second generation flies begin to appear at the end of June and their numbers rise to a peak in July. Some of the pupae formed by this generation emerge in September, forming a partial third generation but others overwinter in diapause.

Many larvae may be present on the root system of a plant. They kill young plants by severing the lateral roots and leave only a 'rat's tail' of tap root (Plate XVC). Larval feeding causes early cauliflowers to curd prematurely or 'button' so that they are useless for market. Broccoli and Brussels sprout which have been attacked are stunted and may topple over in a strong wind. The leaves become limp, flaccid, blue and finally yellow. In a bad attack the plants may die. Cabbages fail to heart properly, radishes are rendered unmarketable, and damage to kale, swede and turnip may be followed by 'soft rot'. They may also attack the crowns and young sprout (buttons) of Brussels sprout plants. Even if the plant survives attack, the root action is impaired and, under dry conditions, the roots are unable to provide the plant with an adequate supply of water. When injury is less severe plants fail to make full use of fertilizer applied to the soil because of their impaired root system. The general level of infestation can be assessed by the proportion of dead or wilting plants. A method for sampling the numbers of pupae has been devised by taking known volumes of soil at known distances from the plant (5, 10 or 15 cm) and removing the pupae by flotation or by sieving (Hughes, 1960).

The adults fly intermittently, and the larvae are concealed so that early attempts to control this pest were not very successful. Various methods were tried, including tarred discs fitting closely around the stem, soil treatments and poison baits. Discs of carpet underlay work well in gardens. The first effective chemical method was 4% calomel dust (mercurous chloride) applied to transplant beds at the end of April before the flies appeared or to the soil around the base of the plants after setting out. But there are objections to mercury residues on plants and in soil and calomel treatment was discontinued. Organochlorine insecticides were used extensively but populations in many areas became resistant. However, they are being replaced increasingly by a range of other materials (Table 8.1). Early sown or transplanted crops must be protected from the end of April onwards, when the first brood of flies is beginning to oviposit, and late sown crops need protection at planting or within a few days afterwards. The attack on Brussels sprout buttons by late broods of flies is difficult to counter. It is difficult to anticipate and requires high volume sprays of materials that do not persist more than a week.

Table 8.1 Materials used to control cabbage root fly.

Materials	Methods			
	Overall	Band	Spot	Peat blocks
Granules				
Carbofuran	+	+	+	
Chlorfenvinphos	+	+	+	
Chlorpyrifos	+	+	+	+
Chlorpyrifos plus disolfoton		+		+
Diazinon			+	
Fonofos		+	+	
Fonofos plus disolfoton		+	+	
Drench				
Chlorfenvinphos			+	
Chlorpyriphos			+	
Diazinon			+	
Spray				
Chlorfenvinphos	+			
Chlorpyriphos	+			
High Volume spray for Brussels sprout buttons				
Iodofenphos	+			
Trichlorphon	+			

Once established within the buttons the larvae are not so easily killed and the marketability of the sprouts greatly decreased. (See Leaflet No. 18.)

Cabbage root flies are preyed upon by many natural enemies (Hughes & Salter, 1959; Hughes, 1959). More than 30 species of beetles chiefly belonging to the Carabidae and the Staphylinidae and two species of mites eat the eggs, larvae and pupae but only about ten are known to be important. The wingless *Bembidium lampros* (Herbst) and *Trechus quadristriatus* (Er.) occur in large numbers around brassica plants and consume many eggs. Other beetles, *B. quadrimaculatum* (L.), *Harpalus aeneus* (F.), *H. rufipes* (Deg.) and *Pterostichus melanarius* (Ill.), are also common and constitute over 80% of the total number of carabids caught on experimental plots at the National Vegetable Research Station, Wellesbourne (Wright, Hughes, Salter & Worrall, 1960; Wright, Hughes & Worrall, 1960; Coaker & Worrall, 1961). Other natural enemies parasitize the larvae and pupae. The Cynipid, *Idiomorpha rapae* (Westw.), and the larvae of the staphylinids, *Aleochara bilineata* (Gyll.) and *A. bipustulata* (L.) (Fig. 6.3), can destroy up to 65% of the pupae. Other hymenopterous parasites have also been recorded as well as a predatory anthomyiid larva. (See also p. 294 and Plate XXXIIIF.).

DDT, aldrin and gamma-HCH broadcast on the surface of the soil and rotavated into the top 7.5 cm before transplanting cabbages, caused worse root fly damage than on the plots without insecticide and the population of predatory beetles was greatly lessened. Doses too small to kill cabbage root fly larvae nevertheless were harmful to predatory ground beetles. The greatest cabbage root fly egg loss occurs on brassica plots when beetle population are numerous (see Plate XXXIII, C, D, E). (See also pages 47 and 95.)

***Delia floralis* (Fall.)** (Fig. 8.14D)   Turnip root fly is common in Scandinavia and in Scotland, where the larva attacks turnips. The wounds caused pave the way for pathogens which often cause the swollen root to rot. The caudal corona of the larva can be distinguished from that of *E. brassicae*, for the fifth and not the sixth papillae are paired (Fig. 8.13E). There are one or two generations a year: the flies are said to have a long oviposition period.

***Delia antiqua* (Meig.)**   The onion fly (Figs 8.9 and 8.15A) is a well-known pest of onions in gardens and market gardens especially in the Midlands and East Anglia. Damage is caused by the larva tunnelling into the onion bulb and is worst in June and July. The leaves become soft and flaccid, turn yellow and the bulbs rot. The fly also attacks young onions and leeks which are killed outright (Plate XIVA).

The life cycle is similar to that of *D. radicum*, and there are three generations a year, the last often incomplete. Flies appear at the end of April or at the beginning of May depending on latitude. They are grey with black bristles especially noticeable on the thorax where there are four rows and also some indistinct brown bands. There are dark triangular marks on the abdomen. The wings are yellowish and the female abdomen is more pointed than that of the male.

The white elongated eggs are laid in the soil close to the host plants. On hatching, the larva works down to the base of the onion and tunnels into it. As many as 30 larvae may be found in one large bulb and larvae migrate from one bulb to another. The larva is white and tapering

A. *Delia antiqua* Onion fly, female x6

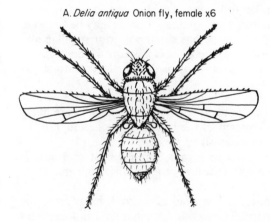

B. *Pegomya hyoscyami* Mangold fly, male x 6

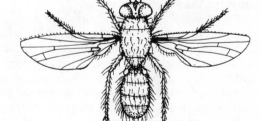

Fig. 8.15 A, *Delia antiqua*, female onion fly; B, *Pegomya hyoscyami*, male, beet or mangold fly.

with a caudal corona of tubercles, differing from *D. radicum* in that the ventral tubercles are not bifid (Fig. 8.13F). Pupation is in the soil and the puparium is dark brown and oval. Second generation flies appear in July and a partial third generation in August and September. Most puparia from the second and partial third generations overwinter in the soil in diapause.

Early sowing under glass, pricking out and planting out in April produces large plants better able to withstand attack. Small sets kept during the winter and planted out are also less liable to attack than plants raised from seed.

The first successful control was 4% calomel dust, applied to the soil of the onion bed after germination or technically pure calomel used as a seed treatment but there are objections to mercury residues in plants and soils.

Bulb onions may be protected by aldicarb or carbofuran granules applied in the furrow in moving soil at planting. Seed treatments of bromophos or iodofenphos plus a fungicide may be used for bulb and salad onions. These can be supplemented with a bromophos spray immediately after planting. The spray treatment alone gives inadequate protection. So far as possible onions should be planted well away from previously infested fields. (See Leaflet No. 163.)

In Great Britain, the chief enemies of onion fly are the parasite *Aphaereta cephalotes* (Hal.) (Branconidae) and the predator *Aleochara bilineata* (Gyllh.) (Staphylinidae). In Italy and North America other parasites occur, and in Denmark one of the most important is a nematode, *Heterotylenchus aberrans* (Jørgensen, 1955).

***Delia platura* (Meig.)** (Figs 8.9, 8.14E)   The bean seed fly is a common pest of germinating maize and beans in North America, where it was introduced from Europe and is known as the 'seed corn maggot'. In Britain it is common and attacks many crops including French and runner beans, broad beans, onions, brassicas and lettuce. *D. platura* (and *D. florilega*) sometimes attack winter cereals in October–November following the ploughing in of rape or mustard. The fly is attracted to decaying plant material and to dung. Larvae may be found in cut potato sets which have not healed properly. The larva has small posterior papillae and three lobed posterior spiracles (Fig. 8.13G). In America, the larva is a vector of the bacterial disease of potatoes known as 'black leg'. Seeds can be protected from larval attack by insecticidal seed treatment. (See Leaflet No. 760.)

***Pegomya hyoscyami* (Panz.)**   Beet or mangold fly, or beet leaf miner (Fig. 8.15), is widely distributed in Europe. It undergoes cycles of abundance and scarcity and epidemics are likely to occur when the temperature favours the development of the fly, but is cool enough to be unfavourable for an equally rapid development of its parasites.

Larvae mine the leaves between the upper and lower epidermis, forming blisters. Injury is greatest to late sown beet which has little leaf area when the first brood of flies appears. (Plate XVF).

The flies appear in early April and are 5–6 mm long. The head of the male is grey and the eyes are separated by a reddish yellow frontal stripe. The second antennal joint is sometimes red, while the thorax is pale and the abdomen reddish. The wings are pale yellow and the femora and tibiae of the legs are yellow but the tarsi are black. The female differs little from the male but the abdomen is greyer.

Eggs are laid on the lower surface of mangold or beet leaves either singly or in batches of 6–12, often in parallel rows. They are white, spindle shaped, and hatch in 3–5 days, the resulting larvae bore into the leaves and feed on the mesophyll. Larvae are dull yellow or cream, with 8-lobed anterior spiracles and a caudal corona with seven pairs of tubercles arranged in a ring around the posterior spiracles which have three apertures (Fig. 8.13H). When fully grown, the larvae drop to the ground and pupate in the soil. The puparium is dark brown with a wrinkled surface and is about 4.7 mm long. The flies emerge after 18–34 days, and there may be three generations a year and the winter is passed in the puparium in diapause.

Beet plants should be examined in the second and third weeks of May. If there are more than 30 hatched or unhatched eggs per plant at the 6–8 leaf stage or fewer if the plant is smaller, the crop should be sprayed, but it is uneconomical to spray older plants. Larvae in mines are easily killed by sprays, e.g. dimethoate, formothion, pirimiphos-methyl, trichlorphon. The last is best avoided as it is harmful to coccinellid predators of aphids and their removal may lead to an increase in the spread of yellow viruses. The mangold fly seems to have become scarcer since insecticides were used to control aphids and other pests. (See Leaflet No. 91.)

Forty-four natural enemies are listed by Sorauer (1953). Seventeen braconids, including eight species of the genus *Opius*, two chalcids, two ichneumonids, one cynipid and one tachinid parasitize *P. hyoscyami*. A fungus, *Empusa muscae*, destroys the adults and a staphylinid beetle, *Aleochara bilineata* (Gyll.), feeds on the eggs. *Biosteres carbonarius* (Nees.) a braconid, is the main parasite in East Anglia.

Phorbia securis (**Tien.**) (Fig. 8.14B)   The late wheat shoot fly occurs widely in Europe and extends to Siberia. Although common in eastern districts of Great Britain it is rarely economically important. It attacks wheat chiefly, causing the central part of the tiller to turn bright yellow, but other cereals are sometimes attacked.

The adults appear in April, are 3–5 mm long, dark and pubescent except for the last anal segment which is black and rather shiny. The larvae appear in the shoots in May and become fully grown in June, later than wheat bulb fly. The larva is white with pale yellow anterior spiracles which are bifurcated and bear many papillae so that each looks like a bunch of grapes (Fig. 8.13I). The posterior spiracles are also yellow and project backwards. There is no caudal corona. On each segment is a swelling bearing a patch of dorsal and ventral spicules. When fully grown the larva is 6–7 mm long and forms a light brown puparium in the soil. The injury may be confused with that of other dipterous miners such as wheat bulb fly, but is unimportant: control is unnecessary.

9
Arthropod pests other than insects

CLASS ARACHNIDA

The Arachnida contains scorpions, spiders, mites and ticks. Spiders are carnivorous and beneficial for they help to keep down insect populations. Some mites attack crops and stored products, others are predatory and still others are parasites. Ticks are pests of live stock. Table 9.1 shows the principal differences between spiders, mites and insects.

Table 9.1 Characters which distinguish spiders and mites from insects.

Character	Spiders	Mites	Insects
Body regions	Two; cephalothorax or prosoma and opisthosoma (abdomen)	Sac-like body – may be divided by furrow into anterior proterosoma and posterior hysterosoma	Three; head, thorax and abdomen
Head	No antennae, simple eyes (2–8 ocelli)	No antennae	Pair of antennae Pair of compound eyes and three simple eyes
	2-segmented chelicerae and pedipalps	3-segmented chelicerae and pedipalps	Three pairs of mouth appendages; mandibles maxillae, labium
Walking legs	Four pairs of 7-segmented legs	Four pairs in adult three pairs in larva	Three pairs on thorax
Wings	None	None	One or two pairs except in Apterygota
Respiration	By two pairs of lung books or by a single pair and tracheae	By tracheae opening to the exterior by paired stigmata or through the cuticle	By spiracles and tracheae

Subclass ACARI (mites)

Mites are a large group of arthropods that diverged from the main line of evolution at an early stage in the evolution of the phylum. They are mostly small, very abundant and occur in many habitats such as meadows, hedgerows and woodland, in decaying matter, in fungi, in buds and leaves, or externally on stems and foliage and also in a great range of food and stored products. Some are animal parasites and some inhabit nests, dens and dwellings. Although mostly unobserved they are, like nematodes, extremely numerous and as widely distributed on land and in fresh and salt water as insects.

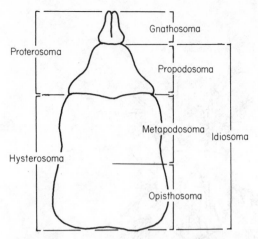

Fig. 9.1 Segmentation terminology of the Acari.

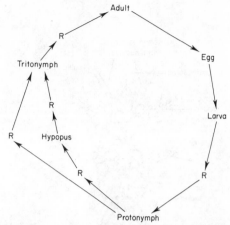

Fig. 9.2 Stages in the life cycle of the mite. R represents a possible resting stage. A moult occurs between each stage.

The body is sac-like and, because the arthropod division and underlying segmentation are obscure (Plate XIX), a rather cumbersome terminology has been devised to describe them (Fig. 9.1). The chelicerae are essentially feeding organs and are often chelate (Fig. 9.3). The pedipalps are a pair of sensory organs situated behind the mouth and arising from the gnathosoma.

Mites are able to reproduce rapidly. The sexes are separate and the male may differ greatly from the female. The egg hatches as a larva which lacks the fourth pair of legs and, after a resting stage and a moult becomes the protonymph (Fig. 9.2) which has four pairs of legs and a rudiment of the genital opening. Usually the next stage is the tritonymph which resembles the adult but is sexually immature. Alternatively there may be an intermediate stage, the hypopus which may be yellow-brown with a hard cuticle and able to move freely so that it can become attached to an arthropod or mammal for dispersal. Sometimes the hypopus is inert and relies on air currents for dispersal or persists until favourable conditions return. What triggers off the development of a hypopus or terminates its period of rest is largely unknown. Sometimes egg laying is delayed until the larva, protonymph or even the hypopus is formed. Some mites are carried on air currents helped by gossamer thread produced by silk glands in the pedipalps.

Classification and identification are difficult and are based on detailed morphology, the position of the spiracles or stigmata, and on the morphology of limbs and the distribution and form of setae (chaetotaxy). Only the barest outline is possible. For further details see Hughes (1976) and Evans *et al.* (1961).

Order Prostigmata

In tracheate forms the stigmata open at the base of the gnathosoma and associated with a groove. Includes many diverse forms that vary greatly in structure and habits. Size ranges from 1 to 1.6 mm long. Usually weakly sclerotized. The body is usually oval or flattened and covered with a smooth, shining, cuticle. *Acarapis woodi* (Rennie) infests the tracheae of honey bees causing Isle of Wight or acarine disease. (See Leaflet No. 330.)

Family TETRANYCHIDAE

Red spider mites are imported pests of trees and bushes. During dry weather, the rate of reproduction rivals that of aphids. Harvest mites belong to the Trombiculidae.

Panonychus ulmi (**Koch.**) (Fig. 9.3C) Fruit-tree red spider mite attacks apples, plums and

172

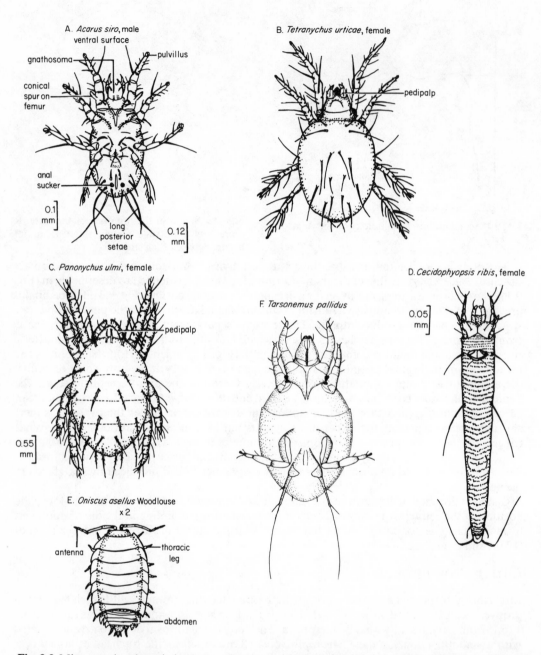

A. *Acarus siro*, male
ventral surface

gnathosoma

pulvillus

conical
spur on
femur

anal
sucker

0.1
mm

long
posterior
setae

0.12
mm

B. *Tetranychus urticae*, female

pedipalp

C. *Panonychus ulmi*, female

pedipalp

0.55
mm

E. *Oniscus asellus* Woodlouse
x 2

antenna

thoracic
leg

abdomen

F. *Tarsonemus pallidus*

D. *Cecidophyopsis ribis*, female

0.05
mm

Fig. 9.3 Mites, much enlarged. A, *Acarus siro*, the grain mite (after Hughes (1976), *The Mites of Stored Food* by permission of H.M.S.O.); B, *Tetranychus urticae*, the red spider mite; C, *Panonychus ulmi*, the fruit tree red spider mite; D, *Cecidophyopsis ribis*, the black currant gall mite; E, dorsal surface of a woodlouse to show the general structure; F, *Tarsonemus pallidus*, strawberry mite.

damsons as well as some soft fruit, except raspberries. Large numbers of mites feed on the foliage so that the leaves become bronzed and may fall, leading to loss of vigour of the tree as a whole (Evans, Sheals & Macfarlane, 1961).

P. ulmi is 0.3–0.7 mm long and reddish with a short rounded body (Fig. 9.3C). The integument is strongly striated. Winter is passed in the egg until the second half of April, when the germ band develops. This is the sensitive stage. The egg hatches in May. The first stage larvae with six legs make their way to the young foliage leaves where they feed. They grow, moult three times and become adult. The females lay 35–90 summer eggs, which are orange and smaller than the winter eggs, on the lower surfaces of the leaves and these hatch immediately. There are several generations until the red winter eggs are laid on the wood in autumn when the mites die off. Winter eggs enter diapause at the blastoderm stage which can be broken by chilling (Lees, 1953). In Britain there are usually five generations a year. The females of the last two generations produce fewer eggs. Eggs are carried from fruit tree nurseries on the wood, and so transported to new sites. In the summer, mites are carried from tree to tree by wind and air currents aided by gossamer threads of silk.

This mite is kept in check mainly by capsid (chiefly the black-kneed capsid, *Blepharidopterus angulatus* (Fall.)) by predacious typhlodromid mites (Phytoseiidae, see later), by the anthocorid bug *Anthocoris nemorum* (L.) and by the ladybird *Stethorus punctillum* (Wei). The fruit tree red spider was not troublesome before winter spraying with tar oil was introduced in the 1920s and still gives little trouble in neglected unsprayed orchards where it still occurs in comparatively small numbers. In fact, the problem has largely been created by spraying. (See Leaflet No. 10.) When spraying is discontinued it takes about two years to build up sufficient numbers of predators to control the mite.

***Tetranychus urtica* (Koch.)** (Fig. 9.3B)   The red glasshouse spider mite occurs in glasshouses on tomatoes, cucumbers and other plants and in the open on strawberries, sugar beet, hop, violet, apple and pear. The host range is wide and the adult females overwinter. Mites feed along the veins on the lower surface of the leaves and produce webbing from a pair of glands on the pedipalps. Secondary damage to the plant occurs when the webs are spun all over the leaves. Infestation of beet leaves sometimes occurs towards the end of long, hot, dry summers when the mites and their webs are found on the undersides of the leaves in shallow depressions between the veins. These areas become yellow giving a characteristic mottle to the leaf. The infestation usually appears around the edges of fields against hedges and ditches and may be confused with yellowing from sugar beet yellows viruses but is otherwise unimportant (Speyer, 1928).

In the glasshouse, *T. urticae* on cucumbers can be controlled biologically by a predaceous mite, *Phytoseiulus persimilis* (Athias-Henriot) (Parr & Hussey, 1967; Gould, 1970). The same predaceous mite also controls *T. urticae* on strawberries under cloches and outdoors (Simmonds, 1971).

To control *T. urticae* on outdoor strawberries a range of organophosphorus and pyrethroid insecticides is available plus the organotin compound cyhexatin. Endosulphan is effective against heavy infestations. Spraying is undesirable and usually unnecessary before harvest. When plants are burnt or cut off immediately after harvest, spraying should be delayed until new growth appears. Aldicarb granules may be applied to established crops but the right moisture conditions are needed for uptake and translocation. The usual source of mites is infested planting material. Only certified runners should be used. (See Leaflet Nos 224 and 226.)

***Bryobia ribis* (Thomas)**   The gooseberry red spider mite feeds on the leaves which turn pale and may drop off. (See Leaflet No. 305.)

Control of red spider mites with pesticides is difficult. The organochlorine compound, dicophol, kills all stages, eggs, young and adults, and tetradifon kills eggs and young, whereas organophosphorus compounds tend to kill only active stages, i.e. young and adults. Some fungicides are also acaricidal, e.g. bina-pacryl and quino-methionate, but these too affect mainly the active stages, so treatment may need to be repeated when surviving eggs hatch. Most kinds of red spider mites pass several generations a year and perhaps for this reason develop resistance within about five years if one kind of pesticide is used repeatedly. Some *Bryobia* mites, which pass only one generation a year, seem to develop resistance more slowly, but such resistance is not unknown. It is therefore desirable to ring the changes on the pesticides used, preferably substituting one from a chemical group different from the last used, for resistance to one member of a group (e.g. an organophosphorous compound) may confer resistance to others in the same group. These difficulties have led to the attempts at control of red spider mites biologically which are very successful in glasshouses, where population control and climatic control is easier than outdoors.

Family TARSONEMIDAE

This family includes several species which are pests of horticultural crops.

Tarsonemus pallidus **(Banks)** is a serious pest of strawberries and ornamental plants (Fig. 9.3F) (Plate XXB). While *Steneotarsonemus laticeps* (Halbt) is a major pest of narcissus and other bulbs in Britain.

Steneotarsonemus spirifex **(Marchal.)**   In Europe and America, gall-like swellings are caused on oats by large numbers of these mites feeding under the ensheathing leaves, between the plant and the stalk. The plant appears red, is shorter than usual and the rachis may be twisted. The mites overwinter in the straw, stubble and soil. Crop rotation probably prevents the build up of large populations.

Family PYEMOTIDAE

Siteroptes graminum **(Reuter)** feeds on various grasses and cereals causing 'silver tops' in oats and barley, and leading to a premature ripening of the ear. If in June, the ensheathing leaves near the ear are pulled back, the enlarged female full of eggs can be found.

Family ERIOPHYIDAE

Gall mites are minute with a reduced number of legs. They live within buds, between the scale leaves, or within leaves and their feeding often evokes the formation of galls.

Cecidophyopsis ribis **(Westw.)** (Fig. 9.1D)   Black currant gall mite commonly causes 'big bud' in black currants and attacks all varieties of currant and gooseberry. On red currants the buds do not enlarge but dry up and become blind (Plate XXA).

The black currant gall mite is 0.025 mm long, with no distinct divisions into propodosoma and hysterosoma. It has only two pairs of legs and is white and almost semi-transparent. Bristles are present on various parts of the body and may help in locomotion. There are no eyes and the abdomen terminates in a sucker-like apparatus which enables it to cling to leaves and to twigs when crawling.

The minute, semi-transparent eggs are laid within the buds in July and August. They hatch immediately and the nymphs feed on scale leaves in the bud which begins to swell and is noticeably enlarged at the end of August. The mites overwinter within the buds, passing through a diapause, and become active when the temperature rises and hours of daylight increase. They emerge as soon as the buds begin to open and are very susceptible to desiccation, moving away from places of low humidity. Migration is by crawling. Mites are also able to leap up to 5 cm in sheltered conditions and are carried further when aided by moderate winds. They travel greater distances when attached to insects, chiefly aphids, and other small invertebrates. *C. ribis* is a carrier of the organism, possibly a mycoplasm, causing black currant

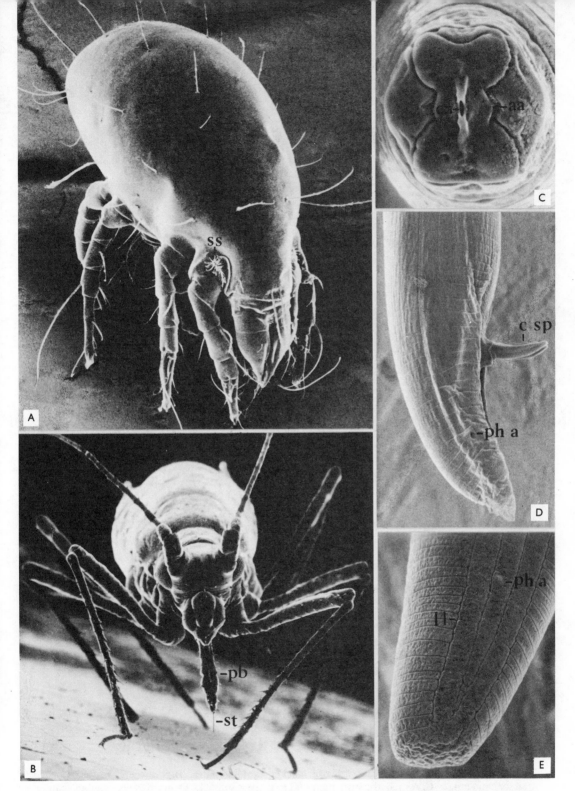

XIX Stereoscan photomicrographs. **A** *Acarus siro* the flour mite. SS, supracoxal seta, × 200. **B** A wingless aphid probing a leaf, pb, proboscis; st, stylets; × 50. **C, D** and **E** *En face* view of head, male tail and female tail of *Pratylenchus* sp., root lesion nematode. C and E, × 4000; D, × 2,000 approx. aa, amphid aperture; Csp, copulatory spicules; oa, oral aperture; ph, a phasmid aperture. Amphids and phasmids are sense organs in head and tail respectively. (**A**, Crown Copyright; **B**, courtesy Pamela Evans, **C, D, E** courtesy Sybil Clark, Rothamsted Experimental Station.)

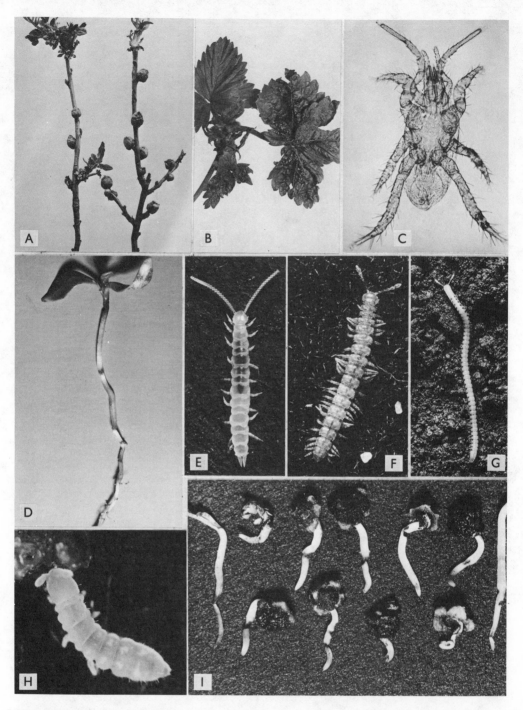

XX **A.** Big bud on black currant caused by the blackcurrant gall mite. **B.** Strawberry leaves damaged by tarsonemid mite. **C.** A predatory mesotigmatid mite from soil. **D.** Damage to beet seedling caused by the symphylid *Scutigerella immaculata*. **E.** *Scutigerella immaculata* ×8. **F.** The millipede *Brachydesmus superus* ×5. **G.** The millipede *Blaniulus guttulatus* ×3. **H.** Collembolan, *Onychiurus* feeding on white, immature females of the brassica cyst-nematode. **I.** Sugar-beet seedlings damaged by the millipede *Boreoiulus tenuis* ×2. (**A, B** courtesy A. N. Massey, East Malling, **C, H** courtesy C. C. Doncaster, Rothamsted, **D, E, F, G** and **I** courtesy Broom's Barn Experimental Station.)

reversion which, among other symptoms, reduces the number of points to the leaves, and lowers yield.

The best method of control was, and still is, to spray the bushes with $\frac{1}{2}$–1% lime sulphur spray when the flower buds open but before the inflorescences appear, i.e. while the mites are migrating as lime sulphur does not reach mites within buds. Some varieties of black currants are 'sulphur shy'. Endosulphan may also be used. (See Leaflet No. 277.)

Eriophes pyri (**Pgstr**)   Pear leaf blister mite, another microscopic mite, feeds on the tissues of pear leaves and occasionally apples, from April to August causing greenish yellow blisters, which later turn red or brown. Frequently the whole leaf becomes black and falls in late summer. It is of minor importance in Britain. (See Leaflet No. 35.)

Abaracus hystrix (**Nal.**)   The cereal rust mite is found on rye grass and meadow grasses and also on oats. It is a vector of a rye grass mosaic virus which is damaging to oats and troublesome in Italian rye grass stands.

Order **Astigmata**

Feebly sclerotized species with a soft, pallid cuticle. Stigmata and tracheae absent. Free-living species occur in a wide range of habitats including soil, surface litter, decaying plant remains, grassland, pasture, stacks of grain and hay. Many species occur in houses, dens and nests.

Family ACARIDAE

This family contains minute mites found in flour and cheese and in all forms of decaying vegetable matter, and also the common soil dwelling bulb mites. *Rhizoglyphus callae* (Oud.) and *R. robini* (Clap.) which feed on damaged narcissus bulbs.

Acarus siro (**L.**) (Plate XIX)   The flour mite occurs in all kinds of dry farinaceous products such as linseed, barley, flour, and also in cheese, in disused beehives and grain stacks. Mites infesting damp grain consume the embryos and ruin germination. Cast skins and excreta taint the grain and render it musty and unfit for consumption. Dockers and storemen handling heavily infested food stuffs occasionally develop a dermatitis, and contaminated food may cause digestive disorders.

The flour mite is 0.4–0.6 mm long, has a creamy white truncated body with trailing hairs. The legs terminate in a pulvillus and a single claw. The femur of the first leg of the male bears a conical spur while that of the female does not (Fig. 9.3A).

The egg hatches into a six-legged larva which moults into an eight-legged protonymph. The next stage or deutonymph is similar but larger and moults into the adult. Occasionally an extra stage, the hypopus occurs between the protonymph and the deutonymph. The hypopus does not feed and may be found attached to insects by its posterior ventral suckers. The life cycle, including the hypopial stage is completed in 2–4 weeks according to the temperature and relative humidity, optimum conditions being 87% relative humidity at 23°C.

Preventive spraying of the floors, walls and ceilings of storehouses and the sides of bins before use, kills mites. If the grain becomes infested, fumigate or treat as in Chapter 13. The general measures against grain pests must be followed.

Tyrophagus dimidiatus (**Herm.**) is a plant feeding species found in pastures and has been recorded from *Phlox* sp., spinach, salsify, mullein and lily. *Tryophagus palmarum* (Oud.) is a common grassland species. *Tyrophagus casei* (Oud.) the cheese mite is commonly found in stored food, cheese, grain and damp flour. *Tryophagus longior* (Gerr.) is found in grain stacks and in fields.

Order **Mesostigmata**

The stigmata open on the lateral or ventral surfaces of the body between the bases of legs 2 and 3 or 3 and 4. Again members of this order are adapted to many habitats. Many are predators feeding on other more sluggish arthropods. The family Phytoseiidae contains predators of red spider mites, e.g. *Amblyseius finlandicus* (Oud.) and *Typhlodromus* spp.

Order **Cryptostigmata**

Small, dark, well sclerotized mites with a well developed tracheal system. Primarily inhabitants of the upper layers and surface litter of soils where they play a part in maintaining fertility.

Class Crustacea

Order **Isopoda** woodlice

Most Crustacea are marine organisms but a few species occur on land. Geological evidence suggests that woodlice are amongst the most recent organisms to have moved from water to land. They are incompletely adapted to land life and are found concealed under bark, under stones or under litter from vegetation, where the relative humidity is high. Despite their 'recent' invasion of the land they are widely distributed, probably partly by human agencies, and, in favourable surroundings, populations may reach 2 500 000 per ha.

Woodlice have few natural enemies, possibly because of the possession of repugnatorial glands. Birds, ants, spiders and wasps do not molest them greatly and they have few parasites.

They occur in damp places where earthworms are absent and perform the same function. At Wicken Fen much of the soil appears to have passed through the bodies of the two common species of woodlice living there.

Agriculturally woodlice are rarely important, but they may be troublesome in glasshouses, in the vicinity of buildings and in gardens. Plants are injured near ground level and the irregular bite marks are rather like those of cutworms and leatherjackets.

The three common species are: *Oniscus asellus* (L.), *Porcellio scaber* (Latr.) and *Armadillidium vulgare* (Latr.). *A. vulgare* is common on chalk soils and if disturbed rolls itself into a ball (Fig. 9.3E).

The segmented body is that of a typical crustacean. On the head are a pair of compound eyes, two pairs of antennae, a pair of mandibles and two pairs of maxillae. There are eight pairs of appendages on the thorax, the first pair aids the mouth parts and forms the maxillipeds, leaving seven pairs of walking legs. Oostegites, outgrowths from the base of the legs of the female, form a brood pouch for the eggs, and the abdominal appendages or pleopods have developed a system of branched air-filled tubules or pseudotracheae.

The female produces eggs during the summer months and these are retained in the pouch where they hatch after about two months. The young woodlice are light in colour with fewer legs. They mature in the autumn.

Woodlice are easily killed. Poison baits of Paris green and dried blood were used, but treatment with HCH is more effective. Glasshouses may be fumigated with HCH smoke. (See Leaflet No. 623.)

Class Chilopoda

Centipedes

Centipedes are soil living organisms with a horny exoskeleton, external segmentation and the same basic arrangement of internal organs as insects. They also have a tracheal breathing system, biting mouthparts and pair of antennae. They differ from insects in having a many segmented body behind the head, each segment bearing appendages, and in having no division of the body into thorax and abdomen. The egg develops directly without a larval stage, and the additional segments are added behind as the animal grows. Centipedes inhabit the soil surface but rarely come above ground. They are also found in moist, sheltered places under stones, vegetation and litter, but like woodlice they avoid dry situations.

Centipedes are elongated, flattened arthropods, with alternate large and small segments, each bearing a pair of legs, the first pair being modified to form poison claws. Centipedes move rapidly when disturbed, are predaceous and therefore beneficial as they feed on insects and other small soil organisms. The two common genera are *Lithobius*, which is large and dark brown, and *Geophilus*, which is long, narrow and yellow and more slow moving than *Lithobius* (Fig. 9.4A).

Class Diplopoda

Millepedes

Millepedes also inhabit the soil surface where they live on vegetable matter such as decaying leaves. Occasionally they cause damage to germinating peas, runner beans and sugar-beet and to potato tubers and carrots. When damage occurs, other animals are present as well, such as slugs on potatoes and pigmy mangold beetle on sugar beet seedlings. Whether the millepedes are the primary agents of damage or not, they cause considerable additional damage once they have assembled in numbers.

Millepedes differ from centipedes in several ways; the body, except for the POLYDESMIDAE, is cylindrical and not flattened, there are two pairs of limbs to each apparent segment, poison claws are absent and they are vegetarian and relatively sluggish. Millepedes breed in the spring and summer months. Eggs are laid in a nest made of soil particles. The young are white and sluggish, have three pairs of legs and resemble *Onychiurus* springtails. Growth is accompanied by an increasing number of body segments, the full numbers not being attained for 2–3 years. The body appears to lack efficient waterproofing so all stages seem dependent on soil moisture. Soils with appreciable clay, that are moist, have an open texture and contain ploughed in stubble or other organic matter seem to favour them.

Blaniulus guttulatus (**Bosc.**) (Fig. 9.4C–F, Plate XXG)   The spotted or snake millepede is 2 cm long, has up to 60 segments and is recognized by its rather pale colour and by a row of reddish spots along the sides like port holes of a ship. It is widely distributed and is found in large numbers in open fields. Populations of 16 000 000 per ha have been recorded.

Brachydesmus superus (**Latz.**)   The flat millepede is smaller and more flattened than *B. guttulatus*. It is the most common injurious species in fields, gardens and glasshouses. The adult has 19 body segments, 29 pairs of legs in the male and 28 in the female. The eggs, about 50 per female, are laid in spring and summer in a dome-shaped nest and there are 6 or 7 moults before the adult stage is reached. The life cycle lasts 12 months.

When numerous both species and occasionally others (e.g. *Boreoiulus tenuis* (Bigler) and *Archiboreoiulus pallidus* (Brade-Birks)), severely damage young beet seedlings, causing wilting

178

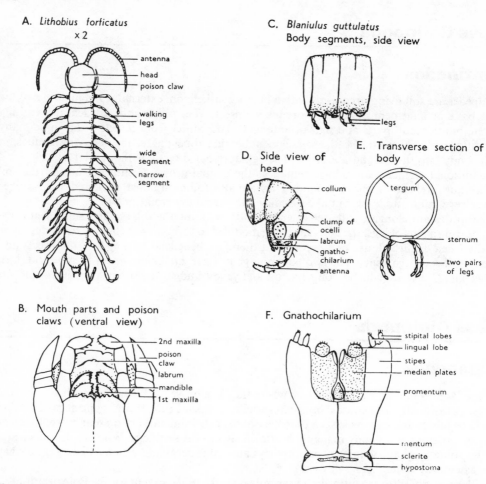

A. *Lithobius forficatus*
x 2

antenna
head
poison claw
walking legs
wide segment
narrow segment

C. *Blaniulus guttulatus*
Body segments, side view

legs

D. Side view of head

collum
clump of ocelli
labrum
gnatho-chilarium
antenna

E. Transverse section of body

tergum
sternum
two pairs of legs

B. Mouth parts and poison claws (ventral view)

2nd maxilla
poison claw
labrum
mandible
1st maxilla

F. Gnathochilarium

stipital lobes
lingual lobe
stipes
median plates
promentum
mentum
sclerite
hypostoma

Fig. 9.4 Centipede (A and B) and millepede (C–F). A and B, structure of body and mouth parts of *Lithobius forficatus* (L.) to show the flattened shape, alternate wide and narrow segments and poison claws; C–F, *Blaniulus guttulatus*; C, body segments to show double pair of legs; D, side view of head to show clump of ocelli, antenna, gnathochilarium and first segment, the collum; E, T.S. to show circular section and limbs; F, gnathochilarium for comparison with an insect labium.

and death; older seedlings are stunted but not killed. Spotted snake millepedes congregate around the base of the hypocotyl at or below seed level, injure the skin and expose underlying tissues which become discoloured (Plate XXG). Flat millepedes usually feed above seed level. Millepedes also feed on carrots, potato tubers and seedlings of other plants.

As damage by millepedes is sporadic and unpredictable, satisfactory control is difficult. Injury is usually done before counter measures can be taken. Gamma-HCH affords some protection when applied as a spray and should be worked into the soil before sowing or applied as a band spray to attacked seedlings. However, there is a risk that a following crop of potatoes or carrots will be tainted. Seed treatments containing methiocarb appear to be effective in controlling attacks on germinating seedlings. Aldicarb granules drilled with the seed also kill millepedes. On fields where attacks recur, straw should be burnt and no dung applied. A firm tilth decreases the space available to the millepedes and limits their aggregation around seedlings. (See Leaflet No. 150.)

Class Symphyla

Symphylids are white, active arthropods with a pair of long white antennae. When fully grown they are about 0.5 cm long and have 12 pairs of legs; the young have only 6 pairs. They move in cracks and major soil pores, particularly in moist clay and limestone soils. Usually they are found deeper than seeds and feed on root hairs and small roots. The lesions they cause may let in fungi and other pathogens (Plate XXD, E). For British symphylids see Edwards (1959).

Scutigerella immaculata **(Newp.)**(Plate XXF, G)   The glasshouse symphylid is a common species in glasshouses and outdoors. It is delicate, white and active. Eggs are laid in the upper soil and hatch in one to three weeks, producing miniature adults with six pairs of legs. The first moult occurs within a few days and subsequent ones take place at intervals of from two to six weeks, an additional pair of legs being added at each moult until the adult number (twelve pairs) is reached. Eggs and young may occur all the year round but are most numerous in spring. The life cycle takes three months or more but adults can be kept alive for four years.

Sugar-beet seedlings, lettuces, brassicas, beans, peas and potatoes may be attacked and young plants are particularly susceptible and readily wilt. If poorly growing plants are placed in a bucket of water and the soil stirred the symphylids float to the surface. As damage in fields is sporadic and cannot be forecast, and as the pest cannot be reached by most chemical treatments once it takes place, control is difficult. (See Leaflet No. 484.)

Millepedes, symphylids, springtails and, in beet crops, the pygmy mangold beetle usually occur together as a complex. For comments on control see p. 283 and Chapter 14.

10

Slugs and snails – Mollusca

Molluscs are soft bodied, unsegmented animals, able to secrete slime and having one or more protective shells secreted by a specialized region of the visceral mass. Many are marine, but the Class Gastropoda subclass Pulmonata in which the visceral mass is rotated through 180° relative to the foot, contains terrestrial species including slugs and snails which are of agricultural importance.

Class Gastropoda

Slugs and snails cause serious damage to cultivated plants in gardens, market gardens, allotments and on farms. They feed on and damage cereals, peas, clover, swedes and root crops, especially potato tubers. They are widely distributed but are more numerous in the wetter, milder regions of the west than in the north and east.

The head is well developed and bears two pairs of retractable tentacles, the anterior pair are sensitive to smell and perhaps taste, and the posterior ones, each with an eye at the tip, to light including infra-red (Newell & Newell, 1968), and to smell. The foot is elongated and is the organ of locomotion. The shell is conspicuous in snails, but rudimentary and often internal in slugs, and encloses the visceral hump. In snails, the whole 'head-foot' part of the body can be retracted within the shell by the action of the powerful columella muscle. At the junction of the visceral hump and the rest of the body, is the thickened collar, the edge of the mantle which secretes the principal layers of the shell. It is fused with the head except at the aperture of the mantle cavity or lung. The mouth is a transverse slit just ventral to the second pair of tentacles and contains a horny rasping tongue, the radula, with many rows of recurved teeth which work against a small transverse bar, the jaw, sawing plant tissue held by the mouth. As the radula wears, it is continuously replaced from its base. When a slug moves, alternating light and dark bands indicating muscular contraction move along the foot and a slime trail is left. Two types of slime or mucus are secreted, the one from the unicellular mucus glands of the foot is thin and watery and the other from the pedal gland, opening just below the mouth, is sticky and thick and enables the slug to move. The slime is also protective and when copiously produced removes irritants from contact with the skin. Slugs and snails are hermaphrodite, possessing both male and female reproductive organs. The opening of the genital atrium is on the right side of the body behind the head. Cross fertilization is the rule and is usually reciprocal. Eggs, laid in summer and autumn in holes in damp soil, hatch in about 32 days.

Slugs and snails feed on plants above and below ground, chiefly at night when humidity is raised and temperature lowered. They have no waterproof cover and are much affected by changes in relative humidity. During the day they seek shelter and avoid direct sunlight, behaviour which is important in determining their distribution. When crop cover is thin, inadequate or absent they retreat beneath stones, clods or into crevices and cracks in the soil. Like insects and other arthropods they are cold blooded and so activity slows down and finally ceases when the temperatures fall in the late autumn and winter.

Snails

Snails are rarely troublesome on a field scale, but may be a nuisance in gardens. The shells are well developed and hardened by calcium salts. If the animal is disturbed, the head and foot are withdrawn into the safety of the shell. Snails hibernate during the winter, often in groups, under leaves or in some shelter, when the head and foot are withdrawn into the shell and the open end is covered with an operculum leaving a small hole for breathing. Respiratory movements are slowed and heart beats sink to 4–6 per minute.

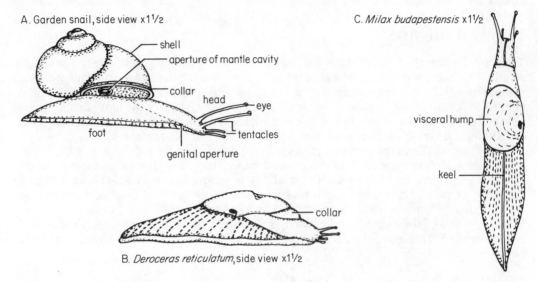

Fig. 10.1 The structure of a slug and a snail. Note the soft body, visceral mass, breathing pore, collar and tentacles.

Family HELICIDAE

Helix aspersa **(Müll.)** (Fig. 10.1A)   The garden snail is 3.5–5 cm long with a grey brown shell and pale markings. It is common all over Britain especially in gardens.

Trichia striolata **(Pfeiff)**   The strawberry snail is seldom more than 13 mm long and is grey, reddish brown or brown. It is commonly found in strawberry beds, among violets, iris and other garden plants.

Cepaea nemoralis **(L.)** and *C. hortensis* **(Müll.)**   Banded snails are common on chalky soils in leguminous forage crops such as clover, sainfoin and lucerne where they may be rather troublesome. The shell may be white, grey, pale yellow, pink or brown with one to five darker bands, spirally arranged. They can be collected in large numbers in a sweeping net when the weather is cool and moist, presumably because they have ascended the stems to feed. On dry days fewer can be collected. (See Leaflet No. 115.)

Control is rarely necessary but baits containing methiocarb or metaldehyde could be tried.

Slugs

Slugs, unlike snails, are streamlined and have no spirally wound shell, so they can move through crevices and squeeze through relatively small holes. The absence of a shell also enables slugs to tolerate smaller amounts of calcium in their environment than snails. European slugs have been introduced into the U.S.A. and are now widespread there but only one native species harms American crops. For British slugs see Quick (1960).

Family TESTACELLIDAE

A small shell is present at the rear end of the body protecting the viscera, a pair of grooves runs along the side of the body, and the upper tentacles are pointed. All these slugs are carnivorous and eat earthworms, soil invertebrates and other slugs. *Testacella* spp. is found on cultivated ground.

Family ARIONIDAE

The mantle is near the front of the body and the respiratory opening on the right front of the mantle. The foot is wider than the rest of the body and the upper tentacles have rounded ends.

Arion ater **(L.)**   is a large black slug up to 15 cm long and 2 cm wide. The tough skin is marked by longitudinally arranged coarse tubercles, and the fringe of the foot has alternating wide and narrow stripes. *A. ater* produces much mucus. It is common in gardens and hedgerows and although herbivorous does not injure cultivated plants. (Colour Plate A.)

Arion hortensis **(Fer.),** the garden slug, is small, slender, 2–4 cm long, dark grey or black with a yellow foot (Colour Plate A). It breeds in midsummer, the young grow in the autumn, winter and spring and mature within a year. The small, white opaque eggs are laid in clutches of 50 at 2–3 week intervals. The garden slug is found in gardens, hedges and fields and can burrow underground. When numerous it injures crops such as wheat, potatoes and brassicas.

Arion fasciatus **(Nilsson)** and *A. circumscriptus* **(Johnst.),** the white-soled slugs (Colour Plate A), are small, 3–5 cm long, variable in colour with a white foot. They are common in fields, hedges and woods.

Family LIMACIDAE

Limacid slugs have a small calcareous shell enclosed by a mantle and a distinct keel on the dorsal surface which may be restricted to the posterior end. The respiratory opening is behind the middle of the right margin of the mantle. The central and lateral teeth of the radula are tricuspid. *Milax* spp. are known as keeled slugs.

Milax budapestensis **(Hazay)** is slender, dark coloured and 5–10 cm long (Colour Plate B, Fig. 10.1C). The sole has a dark central zone with lighter areas on either side. When contracted, the slug is sickle-shaped. Eggs are laid in the autumn or in the spring and the newly hatched slugs are white. *M. budapestensis* is common in fields and gardens. It burrows and is especially damaging to potatoes.

Milax sowerbyi **(Fér.)** is brown with darker patches and speckles, with a paler keel and a cream or yellow sole (Colour Plate B), and produces yellow, sticky slime. It is 7–10 cm long when extended and hemispherical when contracted. It occurs in fields and gardens but is rarely a pest.

Milax gagates **(Drap.)** is smaller than *M. budapestensis* and lives in gardens (Colour Plate B).

Deroceras reticulatum **(Müll.).**   The netted slug, formerly called the grey field slug (Colour Plate B, Fig. 10.1B), is the commonest and most important slug from the farmer's point of view, damaging cereals and clover especially. Its colour varies being pale grey when young and usually mottled grey with a reddish or yellow tinge when adult. The foot is cream with a darker central zone and the rear is truncated. When extended it reaches 3–4 cm; when contracted it is hemispherical. In northern Europe, where the life cycle takes 12–15 months, it breeds throughout the year. There are two overlapping generations usually with peaks of egg laying in March–

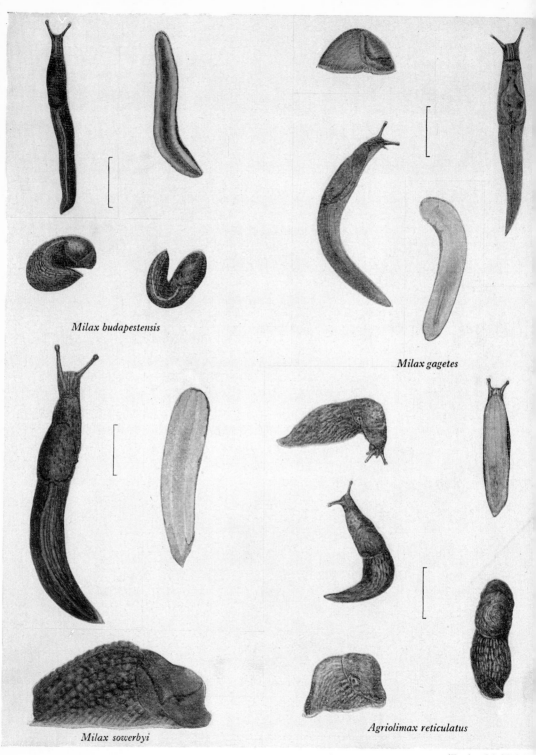

Milax budapestensis

Milax gagetes

Milax sowerbyi

Agriolimax reticulatus

(Evelyn M. Tuke)

Plate B. The grey field slug and some keeled slugs (*Milax* spp.) (p. 182). (Scale line = 1 cm)
(Courtesy Miss E. M. Tuke and *J. Anim. Ecol.*, **14,** 1945.)

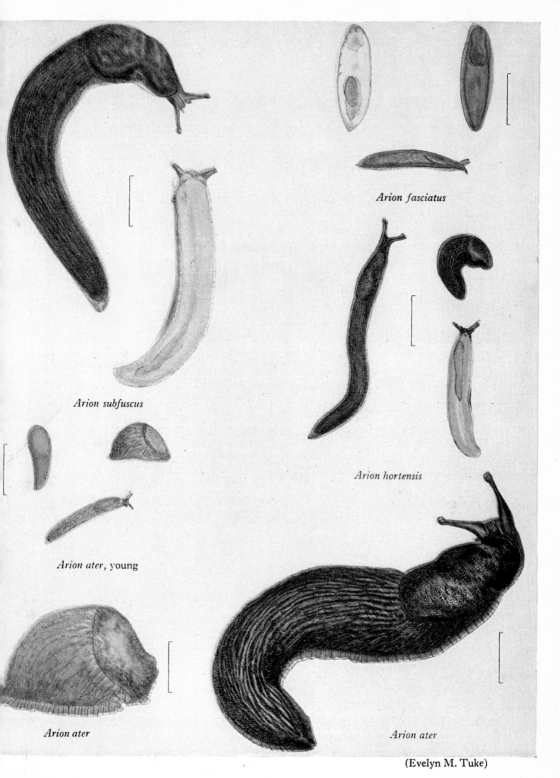

Arion fasciatus

Arion subfuscus

Arion hortensis

Arion ater, young

Arion ater

Arion ater

(Evelyn M. Tuke)

Plate A. Some common species of slug, genus *Arion*. *A. circumscriptus* is a synonym of
A. fasciatus (p. 182). (Scale line = 1 cm)
(Courtesy Miss E. M. Tuke and *J. Anim. Ecol.*, **14,** 1945.)

April and in September–October. About 300 eggs are laid in batches of 10–30 in crevices in the soil and are able to withstand freezing (Runham & Hunter, 1970). (Plate XXIC.)

Slugs are most active in June and September–October, least active in January and February when soil surface temperatures are near freezing and during long periods of drought which are thought to kill many (Webley, 1962). Slugs are said to react to darkness and to a drop in temperature (with its accompanying rise in relative humidity) and emerge to feed in the evening or after a shower of rain. In dry surroundings they are able to survive more than a 50% loss of body weight which they regain when placed on a damp surface (Dainton, 1954a, b). Organic matter favours slugs by improving the water holding capacity of soil and providing alternative food when plants are unavailable. Slugs are therefore numerous in fields which receive much farmyard manure, green manure or farm refuse. Applying dung to wheat stubble in autumn for the following potato crop favours the breeding of slugs. Ploughing in the long straw left by combine harvesters probably encourages slugs by making the soil more open so providing shelter.

Slugs spend most of their time sheltering underground. They can be extracted from soil by flotation (Salt & Hollick, 1944, p. 100), by submerging turfs (South, 1965), by placing soil samples in bowls with perforated bases and removing the slugs that crawl out, or by hand picking from dry soils. Slugs can be trapped under pieces of damp sacking, wooden boards, tiles or boxes. Their numbers can also be estimated by means of poisoned bait or by searching marked areas of ground at night. Trapping, poisoning or searching does not give numbers comparable with those found by soil sampling and are less accurate (Hunter, 1968; Runham & Hunter, 1970). Numbers can also be assessed by placing wheat grain or potato tubers on the soil surface and noting the extent of feeding, or by marking slugs with a dye or radioactive phosphorus, and recapturing them (Newell, 1966). None of these methods are very satisfactory and population estimates are subject to large errors. In one experiment 97% of *D. reticulatum* were in the top 7.5 cm of soil compared with 81% of *A. hortensis* and 72% of *M. budapestensis*. Except in December, January and February 60% of all slugs were in the top 7.5 cm. Slugs are commonest in heavy soils where there are relatively large soil aggregates with wide spaces between them. In summer they are commonest under such crops as clover, peas, beans and rape which provide much shelter by mid summer compared with crops which provide sparse cover during the drier months of the year. *D. reticulatum* is commonest in grass and cereals whereas *M. budapestensis* and *A. hortensis* are commonest in root crops and potatoes. Slugs are important pests of potatoes and are especially injurious in the Scottish Lowlands, causing a loss of 36 000 tonnes a year (Strickland, 1965). The slugs bore into the tuber and form chambers (Plate XXID). The main pest species, *A. hortensis* and *M. budapestensis*, are more numerous in the superficial layers of the soil between the ridges than in the ridges probably because the ridges are drier (Stephenson, 1967). Damage to winter wheat occurs in autumn, especially in November after wet summers on clay loams and silty clay loams. It is worse when cereals are sown directly (sod-seeded) than when drilled into ploughed land. The grains are hollowed out, the germ eaten and germination prevented so that redrilling may be necessary. Slugs also eat shoots and young leaves which are shredded when they appear above ground (Plate XXIA, B, E). Underground damage can be decreased if the field is consolidated immediately after sowing (Gould, 1961) which decreases the number and size of spaces available for shelter. Spring wheat is less susceptible possibly because the soil soon dries. Slugs may damage newly established grass leys and occasionally spring sown barley and oats. Damage to sugar-beet has increased with the practice of drilling monogerm seed to a stand. Because slugs are active at night, the crop 'goes backwards' without the culprit being seen. Realization that slugs are the cause comes too late. On land prone to slug attack test baiting with methiocarb pellets under pieces of wood or tiles before crop emergence is desirable. Then control measures can be taken in good time. Slug damage to garden and horticultural crops, e.g. lettuces, summer beans (*Phaseolus*), celery, Brussels sprout and peas, is widespread.

Little is known about the effects of predators, parasites and diseases on slug populations but natural enemies seem rarely able to keep them in check. Rooks, starlings, gulls, blackbirds, thrushes and ducks eat slugs and may also eat snails. Small mammals such as hedgehogs, moles and shrews feed on slugs as do carabid and staphylinid beetles. Slugs are often infested with flukes, nematodes, protozoa and fungi, and in certain areas with the larvae of sciomyzid flies (Stephenson & Knutson, 1966).

Populations of the three main pest species, *D. reticulatum*, *A. hortensis* and *M. budapestensis*, can be decreased to a third or a quarter by rotavating to a fine tilth (Hunter, 1967). In one experiment cultivation followed by compaction and the establishment of a grass sward subsequently mown almost eradicated the three species, probably because slugs were unable to move through the soil and there were few cracks and crevices for shelter.

Damage by slugs can be lessened by avoiding cultivars susceptible to attack. New Zealand white clover apparently contains substances that make it resistant. Maincrop potatoes especially susceptible are Maris Piper and Ulster Glade, followed by King Edward, Record and Pentland Crown. Less susceptible are Majestic and Pentland Dell and least Pentland Falcon (Winfield, Wardlow & Smith, 1967; Gould, 1965). Lifting of main crop potatoes early also lessens damage. In areas where potatoes are prone to slug damage, the tops should be examined towards the end of the season, and if many slugs are feeding on the foliage and the haulms, control measures should be applied before they die down, otherwise the slugs may go below ground and attack the tubers. Once below ground, they are more difficult to kill.

The first reasonably successful chemical method of destroying slugs was with a poison bait of metaldehyde mixed with bran (400 g metaldehyde to 10 kg bran, beet pulp or tea leaves) which was broadcast thinly or placed in small heaps in the evening when the soil is moist (Barnes & Weil, 1942). The bait attracts slugs and the metaldehyde stimulates production of slime, causes water loss or inability to move, so increases exposure to drying and to enemies. Metaldehyde and bran baits can be made into biscuits or pellets which are more effective. Adding the carbamate, carbaryl improves the kill by 15% (Webley, 1962) and small pellets are better than large (Hunter & Symonds, 1970). The carbamate methiocarb is better than metaldehyde under some conditions (Gerrard, 1971). Phorate (an organophosphorus insecticide) is claimed to kill slugs in the U.S.A. but seems ineffective in the U.K. Aldicarb is a good molluscicide. Beer and cider are attractive to slugs: apparently some fraction or fractions in yeast are responsible and immersion in a few percent alcohol for a matter of hours is lethal. Small straight-sided vessels containing beer or cider might therefore be useful in gardens and glasshouses. The beer or cider needs changing every 48 hours. Synthetic and natural product molluscicides effective against water snails that transmit liver flukes and *Bilharzia* are ineffective against slugs. To be effective these and other contact poisons for slugs have to be able to penetrate the copious watery slime which the integument secretes actively from an external glandular layer more like the lining of a typical intestine than the hypodermis of arthropods or the dermis of mammals.

Poison baits kill only those slugs actively feeding above ground and are least effective in drying conditions when many may be below. Where slug damage can be anticipated, a fine firm seedbed ensures that later on, if baiting becomes necessary, slugs are unable to penetrate so deeply and that most remain on or near the surface. Heavy rain and long exposure decreases the effectiveness of baits. In these conditions pellets based on bran or other cereal products soon break up and become unattractive. Coating the pellets with gelatine containing a fungicide prolongs their life because they do not disintegrate during rain. Hardened gelatine discs, containing a palatable additive such as bran and a molluscicide, last for many days on the soil surface and kill slugs effectively in laboratory tests (Stephenson, 1971). The molluscicides in pellets are stomach poisons. How effective soil fumigants are in killing slugs is unknown.

None of the methods outlined above are entirely satisfactory. Poison baits are essentially curative treatments applied after damage has begun and some harm done. Little attention has been paid to preventive methods. Slugs, like soil nematodes, are relatively immobile and, were their numbers greatly decreased by control measure taken in advance of planting, would take some time to recolonize or recover their numbers. Unfortunately most of the measures currently used against them kill too few. Many deep in the soil appear to survive and are able to recolonize the upper layers within a relatively short time. Direct drilling has resulted in more damage from slugs and the increase in the area under rape has led to problems in following crops partly because of the shelter afforded and partly because of the masses of plant debris left behind after harvesting.

Little research has been done on slugs. More needs to be known of their population dynamics, relationship to spaces in soil, their behaviour and physiology. A study of the integument and slime and its production might be rewarding.

11

Plant Parasitic Nematodes – Nematoda

Nematodes have long been known to attack crops but have been studied less than insects because they are minute, difficult to identify and tedious to extract from plants or soil. Although many species have been described in Europe and the United States of America, the nematode fauna of agricultural soils is imperfectly known and many harmful species await discovery. Most plant-feeding nematodes live in association with roots or underground structures as ecto- or endo-parasites. Where the soil is open and roots penetrate deeply, plant feeding species descend well below plough depth but most are in the top 15–20 cm. Bacterial feeders are common in decaying plant remains, in wounds and in lesions caused by pests or diseases, and are important secondary agents in the degradation of soil organic matter. These species can be distinguished by the absence of mouth spears or stylets, for all known plant parasites possess these structures, although they are also found in species feeding on mosses, algae, or fungi, or when predatory on small animals. Other predatory nematodes have buccal capsules armed with teeth (Fig. 11.7).

The average size of soil and plant nematodes is about 1 mm (1000 μm). Juveniles and small species may be less than 100 μm long: exceptionally large species exceed 1 cm. Animals so small are unable to force their way through soil and are confined to the existing labyrinth of passages between soil particles. Worm-like shape and undulating motion are well adapted to this kind of habitat as well as to the spaces between folded leaves in buds, between ensheathing leaves and stems, or within the air spaces of plant tissues.

Movement and activity of nematodes in soil and indeed of all small soil animals are much affected by soil structure, aeration and moisture (Wallace, 1963; Jones, 1975; Southey, 1978). Although nematodes may squeeze through pore necks with diameters somewhat smaller than their own, the diameter of the necks determine whether passage is possible. The mean free path that can be traversed before encountering ends or impassable necks is probably also important especially for long nematodes. The cross sectional diameters of most common nematode species in soil range from about 30 to 50 μm; those of juveniles and small species range from about 20 μm upwards. Therefore, except in coarse sandy soil, nematodes cannot penetrate the spaces between particles but are confined to macropores between aggregates (see Plate XXVIC–F). Fine particle size and close packing, as in dense clays, are unfavourable to nematodes. Open texture, provided sufficient moisture is retained, is favourable. The forces binding water in soil are best represented by the suction pressure, also called hydrostatic suction pressure deficiency or matric potential, and are usually measured in cm of water on a logarithmic scale (pF scale). The moisture percentage of a given soil can be related to suction pressure by a curve called the moisture characteristic (Fig. 11.1). When the soil is saturated with water the pores are blocked and oxygenation prevented. Little movement of nematodes occurs in such soils. As the soil drains, pores empty, aeration increases and the moisture characteristic inflects. Movement and activity of nematodes increases, reaches a maximum and then begins to decline as the forces retaining water increase. Thin, discontinuous, water films at relatively small suctions probably bind the nematodes to the soil particles. Substances dissolved in soil water have little effect on nematode movement until their concentration is excessive and approaches

Distribution of water in pores

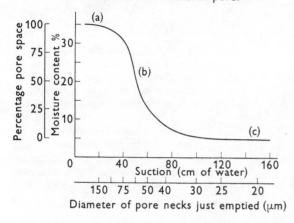

Fig. 11.1 The relationship between suction pressure and moisture percentage in graded sand (a) pores full; (b) emptying; (c) almost empty. This curve is called the moisture characteristic. For the moisture characteristics of some real soils see Jones (1975).

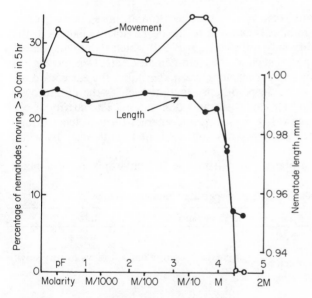

Fig. 11.2 Relationship between suction forces withdrawing water from second-stage juveniles of *Ditylenchus dipsaci*, their length as a measure of turgor and ability to move. Suction forces (osmotic) are from solutions of urea of different molarity. Turgor is lost around pF 4.0 ($\equiv$ 10 atmospheres) when ability to move is also lost. Nematodes are killed by $\geqslant 1$ M urea. Data from Blake (1962).

pF 3.5 (Fig. 11.2). At these high suction pressures nematode body turgor collapses and movement is no longer possible.

These considerations apply to all plant parasitic nematodes, for all have a soil phase. Some, however, find their hosts in the soil surface, invade them as seedlings and are carried up above the soil during growth. Others such as *Aphelenchoides ritzemabosi* (Schwartz) climb plants when they are covered by water films or droplets. The relative humidity of soil spaces under the dusty surface rarely falls below 98% (Fig. 11.3), which is equivalent to 4.45 on the pF scale. Moisture tensions greater than pF 2.2 prevent movement and, as already stated, suctions greater than about 4.0 cause collapse. Some deep soil species may avoid desiccation by migrating downwards and other possess quiescent stages that can endure a degree of desicca-

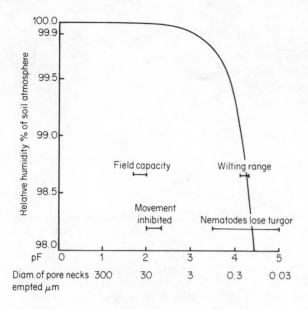

Fig. 11.3 Relationship between the relative humidity of the soil atmosphere and suction of water generated by drainage and evapotranspiration; calculated from the equation $pF = 6.51 + \log_{10}(2 - \log_{10}RH)$ where RH is the relative humidity % (Schofield & da Costa, 1935). The diameter of soil pore necks just emptied is given by $D = 3000/h$, where D is pore diameter in μm and h is the suction in cm of water (Wallace, 1963).

tion. Soil surface forms, however, are subjected to severe desiccation at times and to greater extremes of temperature because their habitat is exposed to the more violent changes in the aerial climate where the relative humidity may fall to 50% (pF 6.0) which represents a large desiccating power. Many soil surface species such as the stem nematode (*Ditylenchus dipsaci* (Kühn)), the chrysanthemum nematode (*Aphelenchoides ritzemabosi*) and the ear cockle nematode (*Anguina tritici* (Stein.)), and some deep soil species also, such as the potato cyst-nematode (*Globodera rostochiensis* (Woll.)), can withstand desiccation. This attribute obviously has survival value for those species that possess it and enables them to live through adverse conditions and to be transported readily in dry soil, seeds and plant trash, by wind, animals and man (Table 11.1).

Species that become endoparasitic in plants move from the soil environment to a plant

Table 11.1 Types, methods, distances moved and rates of dispersal of nematodes (Jones, 1980).

Types	Method	Distance moved	Rate
Self-dispersal	adults and juveniles of most species 2nd stage juveniles of species with sedentary females }	immediate vicinity only	slow or very slow
Assisted dispersal			
Natural agencies	rain splash	immediate vicinity	slow
	water currents	local	sometimes rapid
	wind	mostly local	sometimes rapid
	seed animals }	medium and long	mostly slow
Human agencies	agricultural operations	local	rapid
	seeds, planting material	medium and long	immediate
	commerce unrelated to agriculture }	long	rapid
	tourism	long	fortuitous

environment; such species may make only one journey through the soil to the host. Root ectoparasitic species however live wholly in the soil which is their sole environment. When they are able to feed, water stress can be relieved by ingesting plant sap, but it is unlikely that they can drink from the thin water films when soil dries. Except when the soil is very wet it is untrue to say that they live in water films which are usually much thinner than their cross sectional diameters. Many plant invading species appear rather unselective and often invade plants or varieties that are unsuitable hosts (Jones, 1960). Host suitability may mean not just the provision of food but the correct response to feeding and sometimes the induction of special cells (transfer cells) to nourish the parasite (M. G. K. Jones, 1981). In fact all grades of interrelationship exist from facultative and polyphagous external parasites to obligate and host specific internal parasites. In the cyst-forming nematodes (Heteroderidae), integration between host and parasite has gone furthest. Here the encysted eggs tend to lie dormant until stimulated by a host root growing near. The root gives out a specific hatching factor that diffuses at great dilution over a few cm and reaches the cysts. The eggs then hatch and juveniles find and invade the root. In natural settings a gene-for-gene relationship (p. 301) probably adjusts the balance between host resistance and parasite virulence (Jones, Parrott & Perry, 1981).

Compared with adult insects and most, though not all insect larvae, nematodes lack size, power, sensory organisation and mobility. Therefore they cannot be expected to find their hosts with the same precision and alacrity as do some insects. The distances covered unaided are of the order of 10 cm per year. Wind, water currents and animals assist dispersal. Agricultural operations cause mass movements of soil, manures, seeds and produce, so disseminating nematodes far and wide (Table 11.1).

Soil, the medium in which most nematodes live, imposes limits on movement (Plate XXVIC–F) and sensory perception. Hatching factors, sex attractants and other substances that integrate

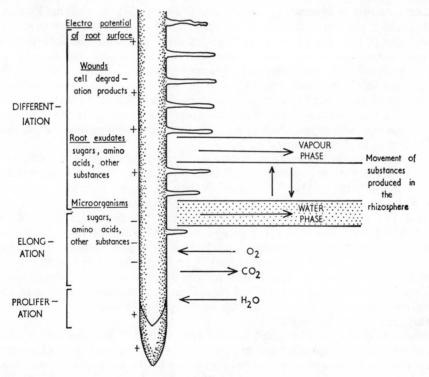

Fig. 11.4 Factors affecting the host finding of plant-feeding nematodes in soil.

the behaviour of the nematodes with each other and with their hosts appear all to be relatively inactive, scarcely volatile, water-soluble substances with many OH groups and difficult to purify and characterize. Some of the factors affecting host finding are summarized in Fig. 11.4. Soil nematodes respond to gradients of moisture, carbon dioxide and unknown substances given out by plant stems and roots (Jones, 1960; Wallace, 1961a). Experimentally they respond to a few acidic and reducing compounds, to redox potentials and to gradients of electrical potential. Some of the stimili given out by plant roots and stems in soil must be common to many plants but some must be specific to a given species or even to a limited part of the plant (e.g. attraction of *Heterodera* and *Meloidogyne* to root tips (p. 225), and *Ditylenchus dipsaci* to pseudostems of oat seedlings (p. 209)). Movements of soil nematodes are probably random at distances greater than a few millimetres from host stems or roots until they come under the influence of gradients of substances in the rhizosphere. No one has yet succeeded in identifying the substances, measuring their gradients of concentration or determining the threshold of response of nematode sense organs.

Many soil nematodes are polyphagous, but even the species with limited host ranges probably test or invade plants rather indiscriminately and may be trapped in unsuitable hosts. The observed host specificity of some nematodes may depend more on the failure to reproduce in unsatisfactory hosts than on host selection. Overspecialization in host finding, feeding or parasitic habits might lessen chances of survival in a changing environment, whereas 'trial feeding' on a range of plants would favour survival of aberrant individuals able to reproduce in the milieu provided by new hosts.

General structure

The general structure of a nematode is illustrated in Figs 11.5, 11.6, 11.7 and Plate XIXC–E. The body is bilaterally symmetrical with tripartite, radial symmetry in the region of the head and oesophagus (Fig. 11.6B, C) and consists essentially of two tubes; the outer formed by the cuticle, hypodermis and neuro-muscular cells, and the inner by the gut. In between lie the gonads, also tubular, but there is no true body cavity because the space is filled by a few large, vacuolated cells. Inside the cuticle and body wall the contents are under turgor pressure and all orifices are closed by valves rather than sphincters. The cuticle, body wall and pressurized contents form a hydrostatic skeleton, which interacts with the musculature. The oesophagus or pharynx is a muscular organ which pumps any food ingested into the hind gut against the turgor pressure of the body contents.

The cuticle is in layers, some of which contain fibres. That in the Tylenchida usually has three layers, the inner being structured. Additional layers are added in cyst-nematodes (Shepherd, Clark & Dart, 1972). The homologies betweeen the Tylenchida and other groups are obscure. Although the cuticle is extensible, it is cast periodically, four moults and five stages being usual. The moulting process and the way in which it is controlled have not been fully elucidated (Lee & Atkinson, 1976). The cuticle invades all body apertures and the linings of these are cast with each moult. Segmentation is lacking but the cuticle is annulated, sometimes strongly, and there are lateral folds or incisures along both sides. As in insects, the cuticle is a barrier to the ingress of substances and helps to limit water loss. It is also flexible and opposes the turgor pressure of the body contents and the muscular contractions that cause charácteristic serpentine movements. These are in the dorso-ventral plane; nematodes seen moving on slides or in dishes are lying on their sides.

The musculature consists of a single layer of cells applied to the hypodermis with only a portion of the cytoplasm modified into muscle fibres. A process from each cell extends to the longitudinal nerve. Muscle cells are divided into banks by the nerve chords and are arranged into rows of two, four, eight or more cells.

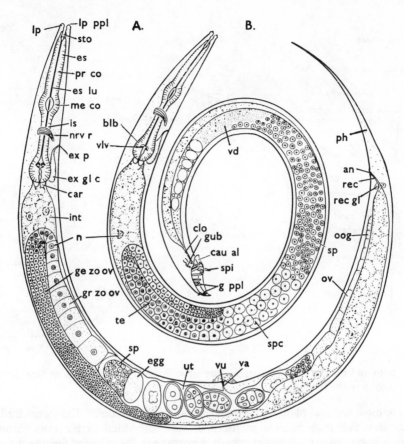

Fig. 11.5 The structure of a nematode, *Rhabditis* sp. A, female; B, male. (Courtesy H. Hirschmann, N. Carolina State College, from Sasser and Jenkins (1960), *Nematology*, Univ. N. Carolina Press.)

an	anus	gub	gubernaculum	rec	rectum
blb	bulb of esophagus	is	isthmus of esophagus	rec gl	rectal gland
car	esophago-intestinal valve	int	intestine	sp	sperm
cau al	caudal alae	lp	lip region	spc	spermatocyte
clo	cloaca	lp ppl	lip papilla	spi	spicule
egg	egg	me co	metacorpus of esophagus	sto	stoma
es	esophagus	n	nucleus	te	testis
es lu	esophagus lumen	nrv r	nerve ring	ut	uterus
ex gl c	excretory glar d cell	oog	oogonium	va	vagina
ex p	excretory pore	ov	ovary	v d	vas deferens
g ppl	genital papillae	ph	phasmid	vu	vulva
ge zo ov	germinal zone of ovary	pr co	procorpus of esophagus		

The gut is a simple tube without muscles stretching from oesophagus to rectum. Its wall is a single layer of cuboidal cells with a conspicuous bacillary layer, believed to consist of modified cilia but more probably microvilli concerned with absorption. When healthy, the intestinal cells are filled with granules and globules of reserve food.

Head and oesophagus

The head may be offset from the body or confluent with it and possesses an internal supporting skeleton (Fig. 11.6A). The terminal mouth (Fig. 11.6A) is surrounded by six lips which may be fused or modified in various ways. Around the mouth are sixteen papillae (Fig. 11.6B), a pair of inner and outer ones on each lip and four on the head, but their number and position varies

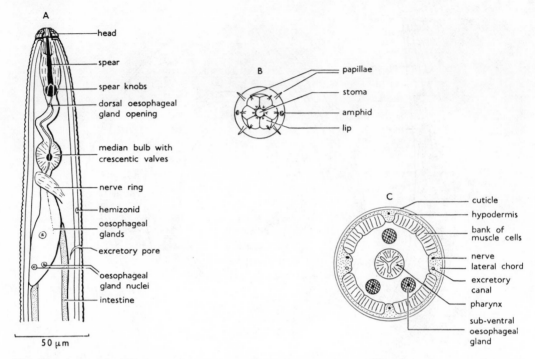

Fig. 11.6 A, head and pharynx of a nematode, side view; B, head of nematode, *en face* view; C, cross-section in the region of the pharynx.

greatly and in most soil and plant nematodes they are inconspicuous. The stoma leads into the buccal capsule which may be free or more or less embedded in the musculature of the oesophagus. The buccal capsule varies greatly according to the mode of feeding. Plant feeding forms possess a spear or stylet.

The muscular oesophagus is sometimes a simple tube but more often has one or more enlargements or bulbs (Fig. 11.7). Normally there are three oesophagal glands, one dorsal and two subventral, situated towards the rear end of the oesophagus. Often these glands are embedded in the oesophagus but sometimes they are free or form a lobe that projects backwards beyond the oesophago-intestinal junction. Ducts from the glands may lead directly into the oesophageal lumen or run forward some distance. Ultrastructural studies have revealed an elaborate system of sophisticated valves and other structures associated with feeding. These vary from group to group (e.g. Shepherd, Clark & Hooper, 1980).

Nervous system and sense organs

A single nerve ring or occasionally two nerve rings (e.g. *Xiphinema*, *Longidorus*) embrace the oesophagus about halfway along its length (Fig. 11.5). Normally the nerve ring consists of two principal masses of nerve cells, the paired dorso-lateral and ventral ganglia which resemble the supra- and sub-oesophageal ganglia of other invertebrates. These ganglia are united by nerve tracts around the oesophagus, and from this ring arise dorsal, lateral and ventral chords as well as nerves stretching forward to innervate the cephalic papillae and amphids. The paired amphids on the head and phasmids on the tail are thought to be chemoreceptors. Each consists of an external pore, a pouch and a group of sensilla-like cells including cilia. The head papillae may be tactile as also may be the various papillae on other regions of the body. Eye spots are absent except in a few marine forms and in larvae of the insect-parasitic mermithids.

Excretory and reproductive organs

The excretory system is a glandular organ, the renette (Fig. 11.5 ex.gl.c.), composed of one or more cells, or a combined glandular and tubular organ opening by a common excretory pore near the nerve ring. How this system functions is unknown.

The gonads are a pair of tubes each ending in an apical cell below which lies a terminal, proliferative cell. Males with two testes are described as diorchic, but often one testis is suppressed and they are then monorchic. The vasa deferentia open with the anus into a common cloaca (Fig. 11.5). Paired copulatory spicules lie in outgrowths of the cloacal wall and vary greatly in shape and size. Accessory copulatory organs, the gubernaculum and telemon, may also be present. When the ovaries are paired (didelphic) the vulva is usually about halfway along the body and the ovaries stretch fore and aft (amphidelphic). When the ovary is single (monodelphic) the vulva is usually further back and the ovary stretches forward (prodelphic). The suppressed ovary may be represented by a post-uterine sac (Fig. 11.11). If the ovaries are long, their tips are often reflexed once or twice.

The developing eggs usually lie in a single line (Fig. 11.5) but the oöcytes of *Anguina* and a few other species are in multiple rows around a central rachis (Fig. 11.17). The ovaries of highly specialised plant parasites are often greatly developed. The uterus prolapses in a few insect parasites such as *Sphaerularia* spp. and *Sphaerulariopsis* spp., and uterus and ovaries grow enormously and greatly exceed the size of the parent body.

Nematodes are normally bisexual with approximately equal numbers of males and females but often males are rare or unknown and parthenogenesis is presumed to occur. Parthenogenesis may also occur when males are numerous (e.g. some species of *Meloidogyne*). The clover cyst-nematode (*Heterodera trifolii* (Goffart)) is apparently triploid, which may account for the absence of males and the occurrence of parthenogenesis in this species. The relative number of males and females may be determined by environmental conditions which sometimes cause sex reversal. In *Mermis*, an insect parasite, females only are produced when one to three parasites are present in a host, mixed sexes when there are three to twenty-three, and all males when there are twenty-four or more present (Christie, 1929). Isolated juveniles of *Meloidogyne incognita* in tomato roots usually become females, multiple infections produce mainly males (Triantaphyllou, 1971). Unsuitable hosts may produce many males and suitable ones many females. The proportion of males of *Globodera rostochiensis* also increases when juveniles crowd into the roots or the nutrition of the plant is poor (Trudgill, 1967).

Eggs and juveniles

Before fertilization by the amoeboid sperm in the oviduct, eggs are enclosed in a delicate vitelline membrane. After fusion with a sperm, a fertilization membrane arises which thickens to form a shell containing chitin. Soon after fertilization an inner lipoid membrane develops and an outer, protein layer may be added by uterine glands. This is thick in some animal parasites but poorly developed in plant and soil forms. The gelatinous egg sac prominent in *Heterodera* and *Meloidogyne* females is a tanned, secreted collagen produced by rectal glands.

Eggs of plant and soil nematodes have a characteristic ovoid shape (Fig. 11.5). Development usually begins at once and may proceed beyond the first moult before hatching. Hatching may be delayed if circumstances are adverse or, in *Heterodera* and *Globodera* spp., require the stimulus of a hatching factor emanating from host plant roots.

Juveniles are usually called larvae but the term is a misnomer because the juveniles differ little from adults except in the state of development of the gonads. In *Criconemoides* (Fig. 11.9) the juvenile is a true larva possessing a spiny cuticle which is cast at the last moult to give an adult with a smooth cuticle.

Life cycle

The life cycles of plant-feeding nematodes are mostly simple and direct, even among species such as *Ditylenchus dipsaci* that are obligate internal parasites. In the more highly adapted parasites the female tends to become sedentary, to lose power of movement and to degenerate into little more than an egg laying machine. Examples of semi-sedentary and fully sedentary female nematodes are in Figs 11.12, 11.18. Males generally remain worm-like, but sometimes cease to feed when adult and may have degenerate mouth parts (e.g. Criconematidae).

The infective stage in the life cycle varies. In *Heterodera* it is the second stage juvenile newly hatched from the egg, in *Anguina* the second stage juvenile expelled from the gall, in *Ditylenchus dipsaci* the pre-adult juvenile and in *Aphelenchoides* the adult. The examples quoted are also the overwintering stages and the ones able to withstand desiccation (for *Heterodera* within the cyst and for *Anguina* within the seed gall).

Less is known about the life cycles of migratory plant-feeding species and as a rule no one stage can be defined as the infective stage or as that adapted to withstand adverse conditions. Wallace (1963) describes their biology and Zuckerman, Mai & Rohde (1972), Zuckerman & Rhode (1981), and Southey (1978) contain useful reviews of several subjects.

Plant feeding and the Injury it causes

When feeding on the cells of the host plant, a spear-bearing nematode presses its lips against the cell wall and makes repeated thrusts with its spear until penetration is achieved. With the spear extended it remains motionless while saliva is injected into the cell. Initially the nematode may be force fed until the turgor pressure of the cell eases. The juices are sucked back by periodic pulsations of the oesophaegeal bulb or by contractions of the whole oesophagus (Doncaster, 1976). Some semi-sedentary parasites remain feeding for long periods. In sedentary parasites like *Meloidogyne* the neck is mobile and the spear is thrust in turn into the transfer cells which feeding induces around the head. Females of *Heterodera* and *Globodera* rupture the root cortex and remain with the head embedded. Feeding is similar to that of *Meloidogyne* but the head lacks mobility and the globules of sticky effusion found around the head may possibly anchor the nematodes in position, or alternatively plug cracks in the external cuticle of the neck region (Shepherd & Clark, 1978).

Feeding resembles that of the Hemiptera (Insecta, p. 36) and produces similar symptoms but on a smaller scale. *Helicotylenchus*, *Radopholus* and *Pratylenchus* produce necrotic lesions, *Trichodorus* stops the growth of root tips. *Heterodera*, *Meloidogyne*, *Nacobbus* and other spp. induce enlarged multinucleate transfer cells which may be giant cells or syncytia (M. G. K. Jones, 1981) (Plate XXVIIID) which act as nectaries upon which they feed. The last three also provoke cell division and cause galls. Those caused by *Meloidogyne* and *Nacobbus* on roots are nodular but those caused by *Anguina* are more highly organized. *Xiphinema* feeding produces small curved, clavate galls on root tips. *Xiphinema*, *Longidorus* and *Trichodorus* also transmit the so-called soil-borne viruses; another resemblance to the feeding of Hemiptera. *Ditylenchus* causes soft galls and stunted misshapen growth when feeding on stems, leaves and flowers whereas cankers develop in harder tissues such as potato tubers or sugar-beet crowns. *Aphelenchoides* feeding as an ectoparasite in buds causes surface lesions that lead to distortion as leaves expand. When feeding as an internal parasite of more mature leaves, typical blotch symptoms are caused, delimited by major veins which for a time act as barriers to spread.

Field symptoms usually are patches of unthrifty, stunted plants. Typical galling is sufficiently characteristic for easy diagnosis, but root-feeding species produce only symptoms of general starvation and debility that are easily passed off as waterlogging, bad soil conditions, mineral deficiencies or virus disease. Endoparasitic nematodes can be demonstrated by dissection, or by suitable staining or extraction methods, but migratory ectoparasites are lost if roots are washed. These must be extracted from soil by methods outlined in Southey (1984). Nematodes are sometimes associated with other organisms in disease complexes caused by viruses, bacteria and fungi (Raski & Hewitt, 1963). For viruses certain species of nematode are the vectors (p. 203). For bacteria and fungi, nematodes may merely pave the way by causing wounds that break the epidermis, create cavities or modify cell metabolism and lower resistance. Sometimes, however, they may actively transport spores. Russian workers list the whole complex of nematodes associated with diseased plants and claim that some non-spear bearing forms occurring within plant tissues are injurious (Paramonov, 1968). *Rhadinaphelenchus cocophilus* (Cobb) which causes red ring disease of coconuts in the Caribbean area and *Bursaphelenchus xylophilus* (Steiner and Buhrer) which causes a similar condition in pines in Japan and the U.S.A., are both transmitted by wood-feeding beetles. *B. xylophilus* is a potential pest of conifers in Europe.

Classification

The system of classification followed below is that of Hooper in Southey (1978) which, together with Thorne (1961) and Mai & Lyon (1962), is useful for identification. Body length, other dimensions and ratios of these are used to characterize nematodes:

L = length in millimetres
a = length/greatest diameter
b = length/distance from tip of head to end of oesophagus
c = length/tail length
$$V = \frac{\text{distance from tip of head to vulva} \times 100}{\text{length}}$$

The details necessary to describe a species are in Goodey (1959a), while methods of extracting, counting, fixing staining and mounting are in Southey (1978).

The Nematoda are spilt into two subclasses on details of the excretory system. Plant-feeding nematodes occur in the Orders Dorylaimida and Tylenchida but many forms common in soil, around plant roots and in decaying plant, material belong to the Order Rhabditida. Insect parasites occur in the Rhabditida, Tylenchida and Trichosyringida, and predatory nematodes are found in the Dorylaimida, Rhabditida and Tylenchida. An abbreviated list of Orders, Sub-orders, families and genera is set out in Table 11.2 followed by brief notes on important Orders and families.

Notes on important groups

Order Dorylaimida

Superfamily DORYLAIMOIDEA Long nematodes. Females usually with two relatively short reflexed ovaries. Males usually with two testes and a row of genital papillae along the ventral side, the bursa characteristic of many Rhabditida and Tylenchida is lacking, copulatory spicules usually rather massive. Common in soils about plant roots, some feed on higher plants, some on algae, some on fungi, some predatory.

Family LONGIDORIDAE Adult nematodes 2 to 10 mm long, spear long, slender and needle-like with an extension and one or more guiding rings. Amphids conspicuous. Posterior oesophagus a wide muscular cylinder in which the salivary glands are embedded (Figs 11.7F, G, 11.8, Plate XXXA).

Table 11.2 Classification of nematodes.

Order	Sub-order	Super-family	Family	Genus
Sub-class ADENOPHORA				
Dorylaimida	Dorylaimina	Dorylaimoidea	Dorylaimidae, Longidoridae	Dorylaimus, Xiphinema, Longidorus
Enoplida	Diphtherophorina	Diphtherophoroidea	Trichodoridae	Trichodorus, Paratrichodorus
Mononchida	Mononchina	Mononchoidea	Mononchidae	Mononchus
Sub-class SECERNENTIA				
Rhabditida	Rhabditina	Rhabditoidea	Rhabditidae	Rhabditis
			Panagrolaimidae	Panagrolaimus, Turbatrix
			Cephalobidae	Cephalobus
			Steinernematidae	Neoaplectana
Tylenchida	Tylenchina	Tylenchoidea	Hoplolaimidae	Hoplolaimus, Scutellonema, Rotylenchus, Helicotylenchus, Rotylenchulus
			Pratylenchidae	Pratylenchus, Radopholus, Nacobbus, Hirschmanniella
			Tylenchidae	Ditylenchus, Anguina
			Tylenchorhynchidae	Tylenchorhynchus
			Allantonematidae	Allantonema
		Heteroderoidea	Heteroderidae	Heterodera, Globodera
			Meloidogynidae	Meloidogyne
		Criconematoidea	Tylenchulidae	Tylenchulus
			Criconematidae	Criconema, Criconemoides, Hemicriconemoides
			Hemicycliophoridae	Hemicycliophora
			Paratylenchidae	Paratylenchus
	Aphelenchina	Aphelenchoidea	Aphelenchidae	Aphelenchus
			Aphelenchoididae	Aphelenchoides, Rhadinaphelenchus, Bursaphelenchus
			Sphaerulariidae	Sphaerularia
Trichosyringida	Trichosyringina	Mermithoidea	Mermithidae	Mermis

Order Enoplida

Super-family DIPHTHEROPHOROIDEA　Usually shorter than 2 mm, conspicuous wine-glass shaped amphids, spear tooth-like of complicated structure. Oesophagus a simple tube with elongate conoid terminal bulb.

　　Family TRICHODORIDAE　Plump nematodes, 1 mm long or shorter, characteristically curved spear without basal knobs, blunt tail. Cuticle thick and loosely fitting (Fig. 11.8). Males with or without a bursa.

Order Mononchida

Super-family MONOCHOIDEA and family MONONCHIDAE   Large, thick-bodied nematodes with capacious stomas armed with cuticular teeth (Fig. 11.7B). Carnivorous feeding on protozoa, nematodes, tardigrades and small oligochaetes. Nematodes are swallowed whole or after cuticle has been ruptured and body contents sucked out. *Mononchus* spp. and related genera have been suggested as possible agents for biological control of plant parasitic nematodes.

Order Rhabditida

Mouth lacks spear, varies in shape according to mode of feeding. Oesophagus with pro-corpus (with or without median bulb), isthmus and terminal bulb. Valve-like structures may be present in the median or terminal bulb or in both. Male tail often has a bursa, usually supported by rays (Fig. 11.5).

The order contains many forms common in soil and in decaying plants. The oesophagus of two, *Rhabditis* and *Panagrolaimus*, are in Fig. 11.7C, D. Those with narrow stomas are usually bacterial feeders, and those with wide stomas armed with teeth are usually predatory. Some species are transported by insects as dauer larvae attached to the insect cuticle. The Steiner-nematidae are insect parasites.

Order Tylenchida

Mouth with hollow, protrusible spear, oesophagus with a procorpus, median bulb, isthmus and terminal glandular region (Fig. 11.7E, *Tylenchorhynchus*). Crescentic valve-like structures are present in the median bulb only. In one sub-order, the TYLENCHINA, the dorsal oesophageal gland duct opens just behind the base of the spear, the median bulb is usually not prominent and the males often have a bursa without rays. In the other, suborder, the APHELENCHINA, the duct opens in the prominent median bulb just in front of the crescentic valves and the males rarely have a bursa.

The Tylenchida contains most of the known plant parasitic nematodes. Some species feed on algae and fungi; some may possibly feed on small animals. The Allantonematidae and Sphaerulariidae are insect parasites. Females of Sphaerulariidae evert the female reproductive system through the vulva into the haemocoel of their host insect (Poinar & van der Laan, 1972). For additional information on nematode systematics see Stone *et al.* (1983).

Super-family TYLENCHOIDEA   Medium sized nematodes averaging 1 mm long. Cuticular annulations distinct but never pronounced. Head skeleton often sclerotized, mouth spear moderate to well developed, median oesophageal bulb distinct, offset ovate, usually not prominent. Glandular portion of terminal bulb may be separate and overlap intestine. Mostly vermiform but females may become swollen and sedentary, e.g. *Nacobbus*, *Rotylenchulus*, one or two ovaries, with median vulva or one ovary with a post-uterine sac and posterior vulva. Males more or less normal, rarely degenerate. Bursa present.

Family HOPLOLAIMIDAE   Robust nematodes with sclerotized head skeletons and massive mouth spears. Terminal bulb overlaps intestine, two ovaries, median vulva, position of phasmids, which may be large, varies. Feeding habits mainly external on roots, but sometimes invade them (Fig. 11.10).

Family PRATYLENCHIDAE   Head low, flattened, spear prominent. Oesophageal glands in lobe overlapping intestine. One or two ovaries. Phasmids post anal. Bursa terminal or subterminal. Internal parasites of roots. Vermiform except females of *Nacobbus* (Fig. 11.1, Plate XIXC–E).

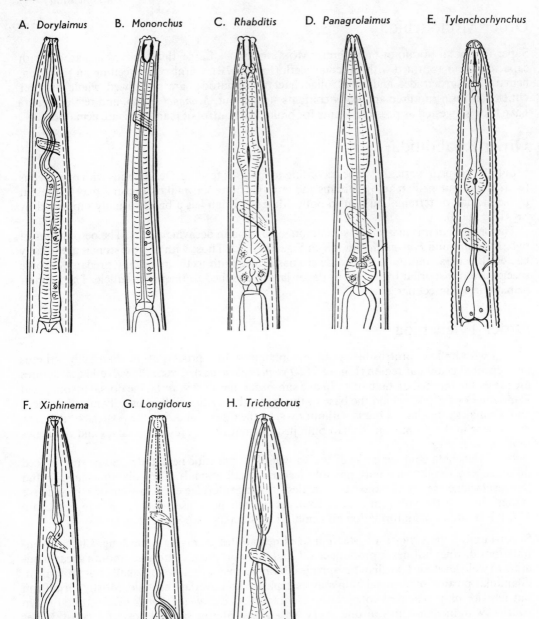

A. *Dorylaimus*   B. *Mononchus*   C. *Rhabditis*   D. *Panagrolaimus*   E. *Tylenchorhynchus*

F. *Xiphinema*   G. *Longidorus*   H. *Trichodorus*

Fig. 11.7 The head and pharynx of some common soil and plant nematodes.

Family TYLENCHIDAE   Cuticle thin; annulations, head skeleton, spear, median bulb and phasmids all inconspicuous. Salivary glands embedded in terminal oesophageal bulb not or only slightly overlapping intestine. Usually one ovary and post-uterine sac. Feeding habits various ranging from fungi, external browsing on roots to internal parasitism, of stems leaves and flowers. Some produce well-organized galls on leaves and flowers.

Family TYLENCHORHYNCHIDAE   Rounded head, head skeleton slightly sclerotized, spear well developed, glandular terminal bulb does not overlap intestines. Usually two ovaries with median vulva. Bursa surrounding tail tip. Phasmids post-anal, conspicuous. Soil dwellers browsing on root tips, epidermal cells, root hairs, etc (Fig. 11.7E).

Superfamily HETERODEROIDEA   Nematodes with marked sexual dimorphism. Oesophageal bulb well-developed with prominent crescentic valves. Oesophageal glands overlap intestine ventrally. Adult males without a bursa. Females swollen to sub-spherical. Endoparasites that usually become semi-endoparasitic by enlargement of both sexes.

Family HETERODERIDAE   Spear and head skeleton well developed. Second-stage juveniles invade roots, both sexes enlarge, males regain worm-like form, females enlarge greatly and when dead become cysts enclosing some or all of eggs. Induce syncytial transfer cells in host tissues. Egg sac formed in most genera. (Figs 11.18, 11.19, Plates XXVII, XXVIII, XIXA–C.)

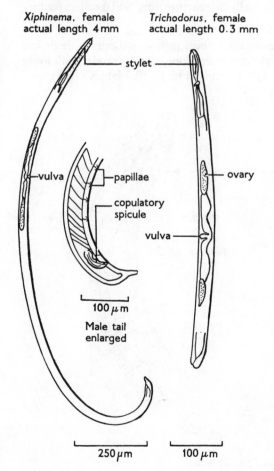

Xiphinema, female
actual length 4 mm

Trichodorus, female
actual length 0.3 mm

— stylet —

vulva

papillae

ovary

copulatory
spicule

vulva

100 μm

Male tail
enlarged

250 μm   100 μm

Fig. 11.8 Migratory ectoparasitic nematodes, *Xiphinema* and *Trichodorus*.

Family MELOIDOGYNIDAE   As for Heteroderidae but with a less well-developed spear. Induce giant cells in host tissues and galls in which females and males may be more or less embedded. Female does not form a cyst, all eggs laid externally in an egg sac (Figs 11.24, 11.25, Plate XXX).

Superfamily CRICONEMATOIDEA   Except for *Hemicycliophora*, which is about 1 mm long, most are 0.5 mm or less. Cuticular annulations pronounced. Head skeleton lightly sclerotized, mouth spear usually elongate especially in the female, median oesophageal bulb large and muscular, running into anterior part of oesophagus and with pronounced crescentic valves, isthmus short, terminal bulb rounded or spatulate. Intestine probably always lacks lumen, rectum and anus poorly developed. Vulva posterior, single ovary, no post-uterine sac. Females may become swollen and sedentary. Males degenerate with reduced mouth spear and oesophagus, bursa absent or poorly developed.

Family TYLENCHULIDAE   Adult females swollen, procorpus long, spear short. Gelatinous matrix of egg sac produced by excretory system, opening of which more than a third of body length behind the head. Root ectoparasites with head embedded in host tissues (Fig. 11.12).

Family CRICONEMATIDAE   Cuticle with pronounced, retrose annules especially in *Criconema* and *Criconemoides*, which progress atypically like earthworms. Soil dwellers feeding externally on roots (Fig. 11.9).

Family HEMICYCLIOPHORIDAE   Annules marked but never retrose, isthmus more or less amalgamated with terminal bulb to form cylinder. Spear long. Adult females with loose cuticular sheath. Root tip feeders inserting stylets deeply (Fig. 11.10, Plate XXIIIE).

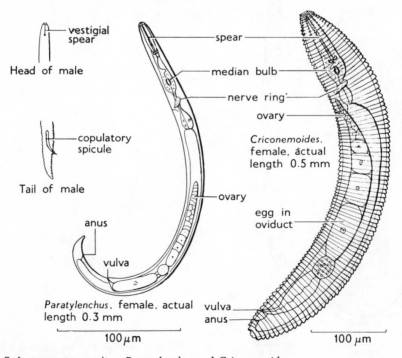

Fig. 11.9   Sedentary ectoparasites, *Paratylenchus* and *Criconemoides*.

Family PARATYLENCHIDAE. Small, plump nematodes, annules plain, fine isthmus slender. Soil dwellers that feed externally on roots, often very numerous (Fig. 11.9).

Superfamily APHELENCHOIDEA Spear weak, with or without basal thickenings, rarely with knobs. Median oesophageal bulb prominent. Females with single outstretched ovary, post-uterine sac well developed in bisexual species.

Family APHELENCHIDAE Spear lacks basal thickenings, oesophageal glands in terminal lobe overlapping the intestine. Males with bursa, copulatory spicules curved. Common in soil around plants within plants and in decaying plant tissue. Usually feed on fungi (Fig. 11.16).

Family APHELENCHOIDIDAE Oesophageal glands as separate lobes extending dorsally over the intestine. Intestine joins oesophagus immediately behind the median bulb. Female tails usually conoid with one or more short terminal bristles. Male copulatory spicules massive, thorn-like, bursa small or absent. Male tail curved ventrally. Some are unspecialized plant feeders with hosts embracing algae, fungi, lower and higher plants (Figs 11.13, 11.16). Some are associated with insects and spread by them.

Order Trichosyringida

Oesophageal glands form a separate organ, the stichosome lying outside the confines of the oesophagus. The Mermithidae are insect parasites. The eggs have peculiar branched filaments by which they adhere to the surfaces of plants on which they are laid. The larvae have a spear with which they pierce the gut of their insect host and invade the haemocoel. The adult intestine has no lumen and functions as a fat body.

Ecological classification

Because the characters used in zoological classification require greater resolving power than that of the average student microscope, nematode genera and species of economic importance are considered in groups based on the type of parasitism (Winslow, 1960).

Ectoparasites

These fall into two groups:
Migratory ectoparasites
    Xiphinema spp.   Dagger nematodes (Figs. 11.7F, 11.8, Plate XXIIIA).
    Longidorus spp.   Needle nematodes (Fig. 11.7G).
    Trichodorus spp.   }
    Paratrichodorus spp. } Stubby root nematodes (Figs. 11.7H, 11.8).
    Tylenchorhynchus spp.   Stunt nematodes (Fig. 11.7E).

Sedentary ectoparasites
    Paratylenchus spp.   Pin nematodes (Fig. 11.9).
    Criconemoides spp.   Ring nematodes (Fig. 11.9).
    Hemicriconemoides spp. Sheath nematodes
    Hemicycliophora spp.   „   (Fig. 11.10, Plate XXIIIE).

***Xiphinema diversicaudatum* (Micol.)** The main species of dagger nematode in Britain averages 4.5 mm long, feeds on young root tips by inserting its spear deeply and causing galls and distortions (Plate XXIIIA–D). A wide range of woody and herbaceous plants are attacked including rose, raspberry, strawberry, celery and several weeds. *X. diversicaudatum* is frequently associated with clay soils, hedgerows and uncultivated woodlands and is rare in dry sandy soils.

Longidorus elongatus **(de Man)** and *L. attenuatus* **(Hooper)**, needle nematodes, occur in open textured soils. The first extends further north than *X. diversicaudatum* and is common in medium loams in eastern Scotland and north-east England. It also occurs in sandy and fen peat soils in England. *L. attenuatus* is frequent in sandy soils in East Anglia. Both species feed on root tips and cause galling (Plate XXIIB, C). *L. elongatus* attacks many plants including grasses, cereals, vegetable and root crops, raspberry, strawberry and weeds. *L. attenuatus* also appears to have a wide host range. Both species stunt sugar beet seedlings and cause fangy roots, up to 2000 nematodes/litre soil being recorded around plants growing poorly and many fewer around healthy ones (Whitehead, Dunning & Cooke, 1971; Jones & Dunning, 1972).

Trichodorus and *Paratrichodorus* **spp.**   Stunt nematodes are common in coarse sandy soils where they feed on the roots of many plant species, sucking out the contents of root hairs and epidermal cells. Large numbers aggregate around actively growing root tips causing browning and some thickening, and stopping elongation. Heavily attacked roots have the appearance shown in Plate XXIID and known as the stubby root condition. Sugar-beet seems especially susceptible but heavy infestations also stunt the growth of ryegrass, winter wheat, brassicas and other vegetables. Many Scottish soils infested with *Trichodorus* are deficient in copper: 16 ppm of copper sulphate are toxic *in vitro* (Cooper, 1971).

Tylenchorhynchus **spp.**   Stunt nematodes are common in sandy soils and attack most plants. Some feed on root hairs, some on epidermal cells and some feed and burrow among the cap cells of root tips. Root symptoms are usually inconspicuous but browning and the death of root tip does occur. *T. dubius* (Bütschli) is a common species in British light soils. Numbers tend to increase under grass and, when numerous enough injure ryegrass, wheat, potato, turnip and pea (Sharma, 1971). Field beans and Sitka spruce seedlings are also injured (Whitehead & Fraser, 1972).

Docking disorder

The patchy stunting of sugar beet seedlings, so-called Docking disorder, which occurs on coarse sandy soils in Eastern England, the Midlands and parts of Yorkshire, is caused by the feeding of needle and stubby root nematodes. Viruses sometimes carried by the nematodes play little or no part in the stunting due to feeding which, although different in the two kinds of nematodes, has the same general effect on the seedlings. Attacked seedlings make little growth and fail to develop effective root systems, so the plants remain small and sickly with symptoms of nitrogen and magnesium deficiency apparent in the leaves. In sandy soils shallow, stunted root systems cannot reach nitrogen leached down the soil profile by spring rains. The severity of the stunting varies with May rainfall. A dry spell early in May inactivates the nematodes and enables roots to grow whereas recurrent rainfall enables nematodes to continue feeding (Jones, Larbey & Parrott, 1969). As the soil dries in June, feeding ceases and plants may grow, but root systems remain small and fanged with feeder roots tending to run horizontally. Little can be done to cure stunted beet crops in May but, in fields with a past history of patchy stunting, seedlings may be protected and enabled to develop effective root systems by injecting small amounts of a dichloropropene nematicide (DD or Telone) 15 to 20 cm deep in the row positions a few days before drilling (drilling at the same time is hazardous because the nematicide vapour is phytotoxic). This treatment has been superceded by the use of granular aldicarb, oxamyl or carbofuran placed in the seed rows during drilling. After harmful nematodes are removed, the plants with healthy root systems require far less nitrogen to give maximum yields (Cooke & Hull, 1972).

Because there are several different species with overlapping host ranges amongst crops and

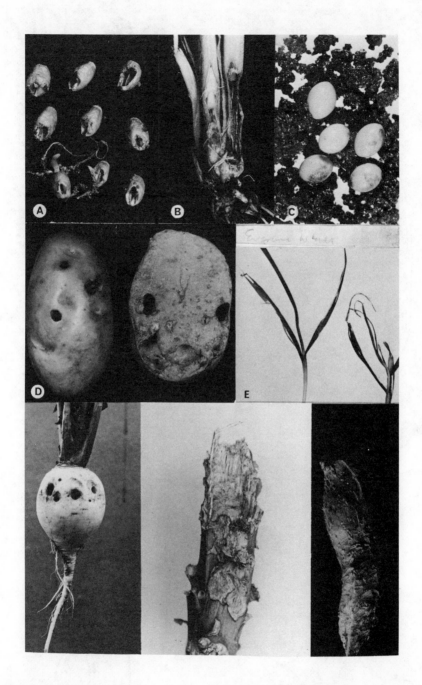

XXI **A.** Wheat grains hollowed by slugs (Crown copyright). **B.** Base of barely plant rasped away by slugs (Crown copyright). **C.** Slugs eggs in soil (courtesy V. Stansfield, Rothamsted. **D.** Slug holes in potatoes (Crown copyright). **E.** Cereal leaves with interveinal areas rasped away by slugs (Crown copyright). **F.** Slug holes in turnip (Crown copyright). **G.** Rat damage to kale stem (Crown copyright). **H.** Base of apple root stock, gnawed away by rodents (courtesy H. C. Woodville).

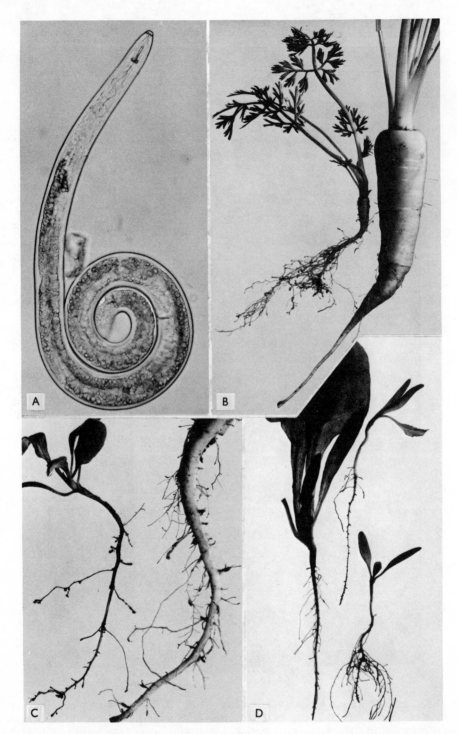

XXII **A.** Microphotograph of a spiral nematode, *Helicotylenchus* sp.   **B.** Healthy and stunted carrot plant. Stunted plant attacked by *Longidorus elongatus*. Note root tip galls.   **C.** Roots of sugar-beet plants attacked by *Longidorus attenuatus*. Note root tip galls and marked stunting of left hand plant.   **D.** One healthy and two stunted sugar-beet seedlings. The stunted seedlings show some thickening and darkening of the main and of lateral roots and the suppression of lateral roots caused by the feeding of *Paratrichodorus teres* (all courtesy C. C. Doncaster, Rothamsted).

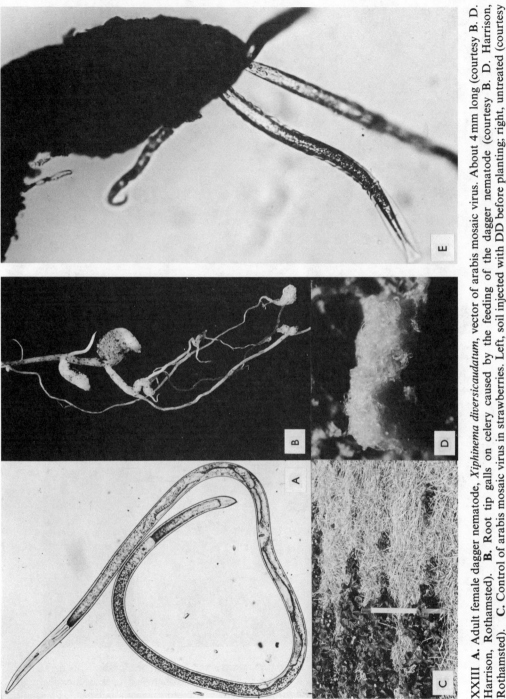

XXIII **A.** Adult female dagger nematode, *Xiphinema diversicaudatum*, vector of arabis mosaic virus. About 4 mm long (courtesy B. D. Harrison, Rothamsted). **B.** Root tip galls on celery caused by the feeding of the dagger nematode (courtesy B. D. Harrison, Rothamsted). **C.** Control of arabis mosaic virus in strawberries. Left, soil injected with DD before planting; right, untreated (courtesy F. D. Cowland, Rothamsted). **D.** Eelworm wool formed by preadult stage of stem nematode on the base of an infested narcissus bulb (courtesy C. C. Doncaster, Rothamsted). **E.** Swollen root tip of sugar-beet being fed upon by sheath nematodes, which have inserted their stylets deeply (courtesy C. C. Doncaster, Rothamsted).

XXIV **A.** Section through tissue of a susceptible oat plant attacked by stem nematode. Note nematode head and the separation of surrounding cells (courtesy C. D. Blake, Rothamsted). **B.** Section through tissue of a resistant oat plant attacked by stem nematode. Tissue around the nematode head is normal; cells have not separated (courtesy C. D. Blake, Rothamsted). **C.** Stem nematode attack on oats in the field (courtesy F. D. Cowland, Rothamsted). **D.** Oat plant stunted and thickened by stem nematode attacked, note the twisted tillers issuing from the base (courtesy F. D. Cowland, Rothamsted).

weeds, crop rotation is useless against root ectoparasitic nematodes. Grass crops seem to favour several species, e.g. *Longidorus elongatus* and *Tylenchorhynchus dubius*. On low lying peat soil in East Anglia, grass reseeded in spring after previous grass is sometimes heavily attacked by *L. elongatus*. However, when reseeding is delayed until autumn, damage is slight. Marling coarse sandy soils, if economically feasible, would change their texture and structure, improve their fertility and make them less suitable media for root ectoparasitic nematodes. (See Leaflet No. 582.)

Virus transmission by nematodes

The migratory ectoparasites, *Xiphinema*, *Longidorus*, *Trichodorus* and *Paratrichodorus* are vectors of soil-borne viruses (Table 11.3). The polyphagous habits of the nematodes and the wide host ranges of the viruses fit each other well. Moreover, the related polyhedron shaped viruses are transmitted by the closely related nematodes *Xiphinema* and *Longidorus*, and the rod-shaped viruses by the more distant genera *Trichodorus* and *Paratrichodorus* (Taylor in Southey, 1978).

Table 11.3 Some vectors of plant viruses. See Taylor in Southey (1978).

Vector	Virus strain	Plants affected include
Polyhedron-shaped viruses (Nepoviruses)		
Xiphinema diversicaudatum (Micol.)	Arabis mosaic	Cherry, cucumber, grapevine, raspberry, rhubarb
	Brome grass mosaic†	Cereals and grasses
	Strawberry latent ringspot	Blackcurrant, cherry, celery, peach, plum, raspberry, rose
X. index	Grapevine fanleaf	Grapevine
Longidorus attenuatus (Hooper)	Tomato blackring English strain	Celery, lettuce, peach potato, raspberry, strawberry, sugar-beet, tomato
L. elongatus (de Man)	Raspberry ringspot English strain	Blackberry, raspberry redcurrant, strawberry
	Scottish strain	Blackberry, raspberry
	Tomato blackring Scottish strain	As for English strain above
L. macrosoma (Hooper)	Raspberry ringspot English strain	As above
	Brome grass mosaic†	As above
Rod-shaped viruses (Tobarviruses)		
Trichodorus primitivus (de Man) *T. viruliferus* (Hooper) *Paratrichodorus anemones* (Loof)	Pea early browning English strain	Pea, lucerne
P. pachydermus (Sein.) *P. teres* (Hooper)	Dutch strain	Pea, lucerne
T. primitivus (de Man) *T. similis* (Sein.) *T. viruliferus* (Hooper) *P. anemones* (Loof) *P. nanus* (Allen) *P. pachydermus* (Sein.) *P. minor* (Colbran)	Tobacco rattle European strains	Potato, tobacco bulbous ornamentals

† Not a Nepovirus.

These nematodes transmit viruses to many woody and herbaceous plants. Those viruses with polyhedral particles have similar biological and physical properties but are serologically distinct. Several viruses are transmitted by the same vector species but serologically distant strains appear to be transmitted by different species in the same genus. In the laboratory specificity can be broken down when rapid changes of vector are made, perhaps because transmission occurs while mouthparts are contaminated with particles which are not absorbed on a retaining surface. In fields, however, specific associations between virus and vector are evident. Serologically different strains of rod-shaped viruses can be transmitted by different *Trichodorus* and *Paratrichodorus* species.

In transmitting viruses of both kinds, adults and juveniles of all three genera seem equally efficient. In *Xiphinema* virus particles are retained on the cuticular lining of the rear part of the stylet (stylet extension) and of the oesophagus, in *Longidorus* on the inner surface of the guiding sheath around the forepart of the stylet and in one species on the cuticular lining of the spear cavity. In *Trichodorus* and *Paratrichodorus* virus particles are retained on the cuticular lining of the oesophagus from just behind the stoma to the junction with the intestine. Nematodes become viruliferous after being in the presence of infected hosts for fewer than 24 hours and the actual feeding times required range from 15 to 60 min. The virus does not pass through the egg stage and is lost at each moult, presumably because the cuticular linings on which it is retained are sloughed. Once acquired, the virus can persist in its vector nematode for long periods without access to infective hosts, sometimes for several months, but *Longidorus* retains viruses for a shorter period than *Xiphinema*, *Trichodorus* or *Paratrichodorus* possibly because the particles lodge in the stylet only.

Weed seeds are an important reservoir of raspberry ring spot and tomato black ring viruses in which they overwinter, for these viruses persist no more than two months in their vector (*L. elongatus*). Arabis mosaic is not commonly found in weed seeds but persists eight or more months in its vector (*X. diversicaudatum*). Tobacco rattle virus causes a form of spraing (concentric corky rings) in potato tubers (another is caused by mop-top virus which is vectored by the fungus *Spongospora subterranea* (Wallop.)). Tobacco rattle virus is widespread in many weed and crop hosts. However, it is usually localised in roots near nematode feeding sites and rarely becomes systemic in the host plant. Pea early browning virus behaves similarly.

The vector species multiply at different rates on different hosts, usually rather slowly. The life cycles of *Xiphinema* and *Longidorus* are long, one generation occupying about a year and adults possibly living longer. *Trichodorus* and *Paratrichodorus* have a shorter generation time but it is doubtful whether there are more than two generations a year on most crops in the field in the U.K. Several species of *Longidorus*, *Paratrichodorus* and *Trichodorus* sometimes occur together but usually one is dominant. The distribution of virus in crops is usually correlated with that of the appropriate vector and, as for aphid and other vectors, large populations are not required. But virus may be spread without the vector by plants propagated vegetatively, especially if raised from aerial cuttings. For further details see Southey (1978).

The most likely source of most virus infestations transmitted by nematodes is infested planting material and possibly in some cases viruliferous weed seeds. Viruliferous nematodes spread only a matter of cms per generation (Jones, 1980). Grass leys, orchards, woodland and hedgerows are important reservoirs of *Xiphinema* and *Longidorus* spp.

Difficulties from the nematode-transmitted viruses can be avoided by not planting susceptible crops on land infested with virus-carrying nematodes and by keeping down weed hosts. Whether the appropriate vector is present in dangerous numbers can be estimated by soil sampling and whether the virus is being carried by tests with suitable bait plants. About 40% of some 600 soil samples examined by Taylor (1972) contained *X. diversicaudatum*, *Longidorus* spp. or both but only 1% were infective, all with viruses transmitted by the former. In Scotland, Cooper (1971) found 68% of *Trichodorus* populations carried tobacco rattle virus.

Although nematicides do not eliminate the nematodes, especially those deep in the soil, they give long lasting and effective control of the virus diseases. Dichloropropene nematicide (D-D or Telone) kills more than 95% of *Trichodorus* and *Paratrichodorus* and decreases spraing in potatoes by about 90% (Cooper & Thomas, 1971). D-D and dazomet control virus spread by *Xiphinema* and *Longidorus* in strawberry crops (Taylor & Gordon, 1970).

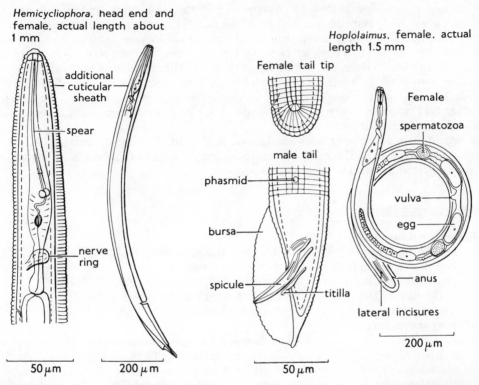

Fig. 11.10 Sedentary ectoparasite, *Hemicycliophora* spp. Migratory semi-endoparasite *Hoplolaimus* spp.

The sedentary ectoparasites are all members of the Criconematidae. Ring and sheath nematodes are uncommon in British soils but pin nematodes (*Paratylenchus*) are common and often overlooked because they are small (300–500 μm) (Fig. 11.9). In some Dutch soils they increase after rye, peas, and potatoes or beet followed by cereals, and it is claimed that rye is injured by dense populations in sandy soil. Winslow (1964) regards *Paratylenchus* as a migratory species.

Semi-endoparasites

These fall into two groups:

Migratory semi-endoparasites
Hoplolaimus spp.	Lance nematodes (Fig. 11.10).
Rotylenchus spp.	
Helicotylenchus spp. }	Spiral nematodes (Fig. 11.11, Plate XXIIA).
Scutellonema spp.	
Tylenchorhynchus spp.	Stunt nematodes (Fig. 11.7E).

Sedentary semi-endoparasites
  Rotylenchulus reniformis
    (L. and O.)   Reniform nematode (Fig. 11.12).
  Tylenchulus semipene-
    trans (Cobb)   Citrus nematode (Fig. 11.12).

Migratory semi-endoparasites are a cosmopolitan group. They normally feed with only the front part of the body embedded in host roots but there is a tendency with some species and hosts to become endoparasitic. Lance, spiral and stunt nematodes are common in some British soils where they are part of a complex of organisms dependent on plant roots. *Helicotylenchus vulgaris* (Yuen), which occurs in a wide range of crops without causing noticeable harm, is sometimes associated with patchy stunting of sugar-beet on calcareous soils (Spaull, 1982) *Hoplolaimus uniformis* (Thorne) attacks Sitka spruce seedlings in England, and peas, carrots and other field crops on sandy soils in Holland. The role of most species as agents of plant disease requires examination.

No sedentary semi-endoparasites are known in Britain. The citrus nematode has been transported around the world on citrus root stocks and the reniform nematode, a polyphagous root parasite, is widespread in hot countries.

Endoparasites

Migratory endoparasites
  Pratylenchus spp.   Root lesion nematodes (Fig. 11.11, Plate XIXC–E).
  Radopholus similis (Cobb)   Burrowing nematode.
  Ditylenchus dipsaci (Kühn)   Stem and bulb nematode (Figs 11.13, 11.15).
  D. destructor (Thorne)   Potato tuber rot nematode (Fig. 11.15).
  *Aphelenchoides spp.   Bud and leaf nematodes (Figs 11.13, 11.16).
Sedentary endoparasites
  Anguina spp.   Flower and leaf-gall nematodes (Fig. 11.17).
  Heterodera spp.   Cyst-nematodes (Figs 11.18, 11.19).
  Meloidogyne spp.   Root-knot nematodes (Figs 11.24, 11.25).
  Nacobbus spp.   False root-knot nematodes (Fig. 11.24).
  * often ectoparasitic in buds.

In the sedentary endoparasites the life cycle is modified, and the female loses mobility or becomes saccate. Females of *Heterodera* and *Globodera* spp. become secondarily ectoparasitic as they enlarge and rupture the root cortex. Migratory endoparasites have simple life cycles, no obvious morphological adaptations to parasitism and all stages occur together in parasitized plant tissues.

***Pratylenchus* spp.** Root lesion nematodes are cosmopolitan and normally attack roots but sometimes damage other underground plant structures.

Root lesion nematodes are ubiquitous in British soils and can be found within the roots of weeds, crop plants, ornamentals, soft and top fruits. In cereal fields root lesion nematodes are usually the commonest plant parasitic nematodes. Here and elsewhere, the species present vary with soil type. In roots and other organs below ground, lesion nematodes cause lesions and cavities and multiply *in situ*. The lesions are invaded by secondary pathogens and, when roots are heavily attacked, the root cortex over appreciable lengths may be completely destroyed. Because the degree of harm which different species do to host roots varies, accurate identification is essential. Grass weeds appear to play an important part in the overwintering of harmful populations of some species. No economic control measures yet exist for field crops. Because most fields contain species mixtures effective crop rotations are difficult to devise.

XXV **A.** Lucerne seedling. Left, with swellings caused by stem nematode; right, unattacked (Crown copyright).   **B.** Lucerne plants. Left, healthy; right, stunted by stem nematode (Crown copyright).   **C.** Stem nematode patch in a second year lucerne crop. The patch, probably started from seed-borne infection, has spread downhill in surface drainage water (Crown copyright).   **D.** Potato tubers rotted by tuber nematode (courtesy W. E. Dant).   **E.** Onion plant attacked by stem nematode (courtesy C. C. Doncaster, Rothamsted).   **F.** Onion tissue with all stages of stem nematode, stained in osmic acid (courtesy late T. Goodey, Rothamsted).   **G.** Onion flower with stem nematodes: the nematodes may become seed-borne (courtesy late T. Goodey, Rothamsted).

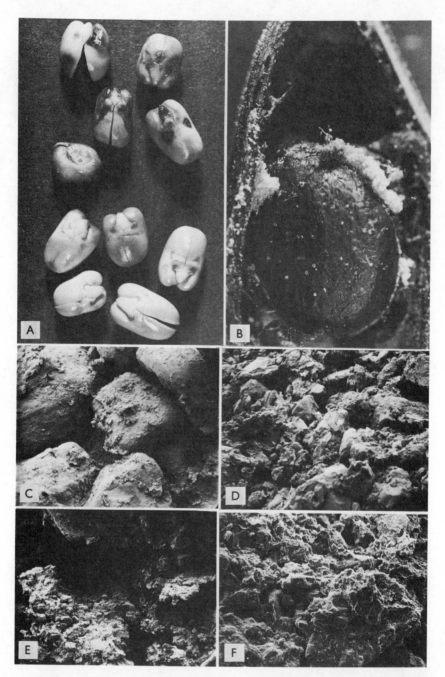

XXVI **A.** Healthy and infested seeds of field bean with testas removed. The lesions on the five upper seeds contain *Ditylenchus dipsaci*. **B.** Eelworm wool (*Ditylenchus dipsaci*) on field bean seed within ripe pod. **C** to **F.** Stereoscan microscope pictures of the surface of soil peds to show the available space to nematodes and other small organisms. Each picture is 400 μm wide, nematodes are mostly 20–40 μm wide. **C.** Coarse sand. **D.** Soil with range of particle sizes and relatively little space between them. **E.** Clay soil with cracks and hole probably made by root. **F.** Dense clay soil, roots and organisms confined to cracks (all courtesy Rothamsted Experimental Station).

XXVII **A.** The near spherical females of the golden nematode on potato roots. These are white when young, turn golden yellow and finally brown (courtesy C. C. Doncaster, Rothamsted). **B.** Head of a cyst-nematode juvenile much enlarged (courtesy C. C. Doncaster, Rothamsted). **C.** A patch of potato plants devastated by the golden nematode (courtesy W. E. Dant).

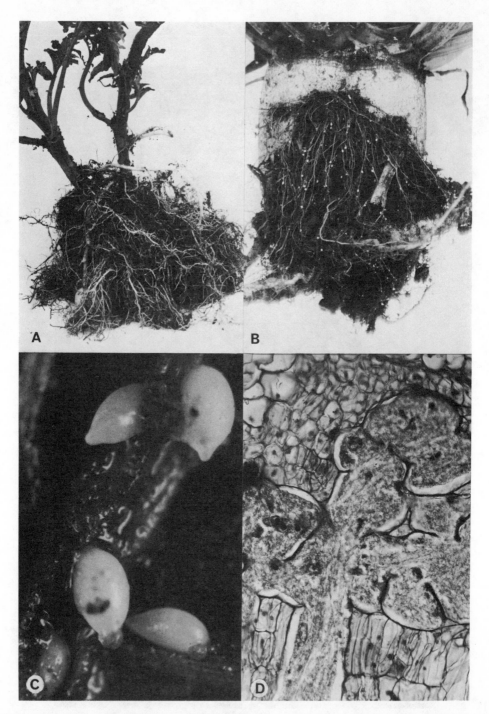

XXVIII **A.** Potato plant attacked by the golden nematode: note the excessive production of fibrous roots (courtesy W. E. Dant). **B.** Sugar-beet plant attacked by beet cyst-nematode: note the absence of a tap root and the abundance of fibrous roots bearing white female nematodes (courtesy W. E. Dant). **C.** Swollen white females of the clover cyst-nematode on clover roots. Note the lemon shape and the gelatinous egg sac at the vulval end of some individuals (courtesy C. C. Doncaster, Rothamsted). **D.** Syncytial transfer cells formed by cyst-nematode feeding in the roots of wild beet (courtesy A. M. Shephed, Rothamsted).

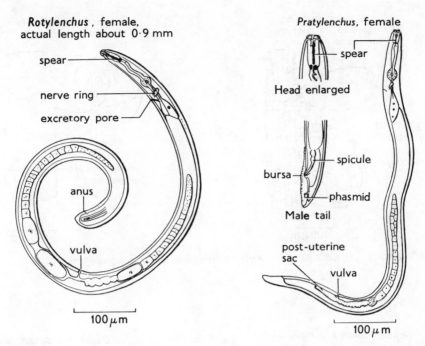

Fig. 11.11 Migratory semi-endoparasite, *Rotylenchus*, a spiral nematode. Migratory endoparasite, *Pratylenchus*.

Although they might conceivably be used to diminish numbers and perhaps change the dominant species to less harmful ones.

P. penetrans (Cobb) injures bulbs, ornamental Ranunculaceae, scabious, roses and lettuces. In gardens *Tagetes* spp. may be planted before, or interplanted with, susceptible plants. The roots of *Tagetes* spp. contain nematicidal principles (e.g. α-terthionyl) which kill nematodes that invade them. *P. penetrans* is the single most destructive nematode in north-east U.S.A. It may well be more injurious in Britain than is realized. In Britain *P. fallax* (Seinhorst) is pathogenic to cereals and other crops (Corbett, 1972) and causes patchiness in cereals on Bunter sands. *P. thornei* (Sher and Allen) injures wheat in the U.S.A. *P. coffeae* (Zimmermann) severely damages coffee in Indonesia and the Caribbean area, and citrus in India and the United States. *P. penetrans* and *P. pratensis* (de Man) are common in European fields. In reclaimed polder soils in Holland populations increase under wheat, barley and oats whereas they decrease under rye, flax and sugar-beet.

The burrowing nematode *Radopholus similis* (Cobb) causes lesions in roots like those caused by *Pratylenchus*, and these become infected by other pathogenic soil organisms which play a part in root injury. A biological race of *R. similis* causes spreading decline of citrus in Florida, U.S.A. (Christie, 1959) which in the deep moist coral soils spread by approximately one tree space a year (Jones, 1980). Attempts to arrest the spread of the disease and eradicate the nematode by grubbing up trees over an area greater than that affected, burning the roots, applying a heavy dose of nematicide and trying to keep the land weed free for two years before replanting have not been entirely successful; policies that depend on total eradication rarely are. Another biological race of *R. similis* causes black-head toppling disease of banana in several parts of the world. Black pepper is also severely damaged in the East Indies. This nematode like many others is readily transported on root stocks: it occurs in Britain only in tropical houses.

208

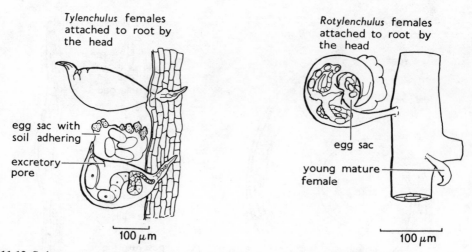

Fig. 11.12 Sedentary semi-endoparasites, *Rotylenchulus* spp. and *Tylenchulus* spp.

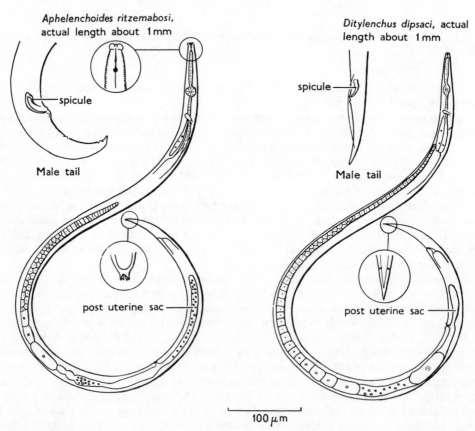

Fig. 11.13 Migratory endoparasites, *Aphelenchoides ritzemabosi* and *Ditylenchus dipsaci*.

Nacobbus aberrans **(Thorne)** (Fig. 11.24)   The false root-knot nematode has only the tip of its body exposed. It parasitizes a wide range of crops and weeds in the Andes, including potatoes, sugar-beet in sandy soils in western U.S.A. and rarely on glasshouse tomatoes in Britain. The root galls that it causes, although slightly different, may be confused with those of *Meloidogyne* spp. It is difficult to understand why *Nacobbus* has been less successful in colonizing potatoes in Europe and the U.S.A. than *Globodera*. Perhaps it lacks a stage able to withstand desiccation. Sometimes regarded as a potential pest in the U.K.

Ditylenchus, Aphelenchoides, Heterodera, Globodera and *Meloidogyne* are important economic genera in Britain and *Anguina* is especially interesting. These genera are considered separately.

Stem and bulb nematodes

Ditylenchus dipsaci **(Kühn)**   The stem and bulb nematode (Figs 11.13, 11.15) is an important pest in cool, moist regions. These include Scandinavia, north-west Europe, Britain, Swiss mountain valleys, Canada, the northern parts of U.S.A. and irrigated valleys further south in the U.S.A. It occurs in a number of biological races, morphologically indistinguishable from each other, but having distinct host ranges.

The life cycle of all races is simple and is complete in 19–23 days on onions at room temperature (much above soil temperature). Mating is essential, females produce 200–500 eggs and begin to lay 4 days after the last moult (Yuksel, 1960). Except for the giant bean race, which is tetraploid, all races appear to interbreed. Pre-adult juveniles and young adults withstand desiccation and are the infective stages that overwinter in the soil, moving downwards in winter and upwards in spring. Field populations (Fig. 11.14) are very great at the end

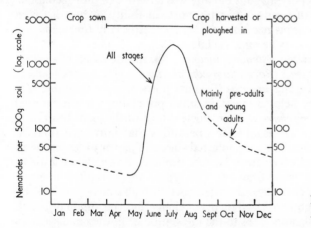

Fig. 11.14   Annual population cycle of *D. dipsaci* on a favourable host crop.

of summer after an infested crop, for a single plant may harbour 50 000 or more nematodes (of the order 2500/500 g soil). Mortality in autumn and winter is great and the number of survivors in spring is few, rarely exceeding 100 per 500 g of soil (Seinhorst, 1956, 1957). In sandy and light soils in Holland numbers decrease more than in heavy soils, where they tend to stabilize around 50 nematodes per 500 g. As few as 10 nematodes in 500 g cause serious injury to susceptible crops like onions, and at 25 per 500 g no healthy plant is left in a field.

Stem and bulb nematode persistence probably is largely due to weeds and marginal host plants in which some reproduction is possible red and white dead nettles (*Lamium purpureum* (L.) and *L. album* (L.)) are important hosts of the giant race. Although the pre-adult stage can survive dry for a long time, dry conditions rarely last long in the field. Consequently it is unlikely that adults or immature stages can survive without food. There is no convincing evidence that fungi are fed upon as by *D. destructor*.

When conditions in the soil surface are favourable, plants are invaded rather indiscriminately. If feeding kills the cells and causes necrotic lesions accompanied by some distortion of growth, the plant is unsuitable as a host and reproduction cannot occur. If cells are not killed, a slow rate of reproduction may be possible but, for maximum reproduction, the pectinase enzymes of the nematode must dissolve the middle lamellae of parenchymatous cells so that they separate, become rounded and allow cavities to form between them. In this way the nematodes create an environment in which they move freely and reproduce rapidly (Plate XXIVA, B). As numbers increase, feeding causes cells to collapse, the tissue becomes disorganized and the environment less favourable. Nematodes return to the soil or are desiccated as the plant remains dry out.

Although little is known of the genetic and physiological interrelationships of biological races, specific tissue reactions probably determine which plants are suitable hosts of a given race and which are very susceptible to injury. Heavy invasions of suitable or unsuitable host plants may kill seedlings outright, but light invasion by a few nematodes leaves unsuitable hosts little affected. Only in suitable hosts can numbers increase until the plants are crippled. Races other than the giant bean race appear to be able to interbreed freely on common host plants (Sturhan, 1970), yet new races able to multiply freely on resistant cultivars of affected crops seem to arise rarely.

Most of the symptoms observed in suitable hosts (Table 11.4) are a result of dissolution of the middle lamellae (bloating, blistering and light green colour), but some are separate or secondary effects of salivary stimulation by the nematodes (inhibition of elongation, excessive tillering, outgrowth of lateral buds, curvature, twisting and distortion).

Of the nine races known in Britain, the oat race is the most polyphagous (Plate XXIVC, D). The races on clovers and lucerne are less polyphagous but all races have one or more economic plant as susceptible hosts and a range of weeds of varying host suitability. A host list is in Table 11.5, but much work on host ranges is needed; host records, especially for weeds, are often doubtful because field populations may be mixed (Plate XXVA–C, E–G).

Many common weeds such as chickweed (*Stellaria media* (L.)), mouse-eared chickweed (*Cerastium vulgatum* (L.)), black bindweed (*Polygonum convolvulus* (L.)), cleavers (*Galium aparine* (L.)) and scarlet pimpernel (*Anagallis arvensis* (L.)) are hosts of more than one race and, although not always suitable, they help to maintain small populations in the interval between susceptible crops. *Lamium purpureum* (L.) is an efficient host of the giant bean race.

Stem nematodes are dispersed by all the usual farm operations that move soil or farm produce. The carting of foliage for hay or silage from infested stands of clover or lucerne is an important mode of spread. In these and other crops surface drainage down slopes leads to elongation of infested patches in the direction of drainage. This type of spread is often observed in 2-year-old stands of lucerne. Stem nematodes are often carried high above soil level by elongation of the flowering shoots of slightly infested plants of clover, lucerne, onion or teasel. Movement of the nematodes within the plants leads to the infestation of flowers and seeds so that pre-adult nematodes are desiccated and adhere to the seed coats and to cavings. Thus the nematodes become seed-borne on crop and weed seeds (Green & Sime, 1979) (Plate XXVIB). Seed-borne infestions are usually, though not always, light and give rise to patches of stunted plants of limited size in the first year where soil surface temperature is low and moisture adequate. Both the oat and the giant races of *D. dipsaci* infest field beans (*Vicia faba* (L.)) and may become seed-borne. Infestations begin at the base of the stem, which turns reddish brown, and may spread upwards to the pod-bearing regions. Ripe infested pods contain many coiled juveniles attached to the seed, especially in the slit in the hilum. When seed testas are removed, necrotic patches containing nematodes are seen in depressions on the cotyledons (Plate XXVIA). Infested seeds are a potent means of spreading the nematodes (Hooper, 1971).

Chemical control of stem nematodes is difficult because the nematodes that survive, even when few, multiply fast enough to become damaging by autumn. Oxime carbamates (e.g. aldicarb, oxamyl) and the fumigant powder nematicide, dazomet protect young plants from being heavily attacked in spring. Survivors spread and multiply little during midsummer, when the soil is usually rather dry, but increase rapidly as autumn advances and the moisture returns.

Table 11.4 Stem nematode symptoms in different crops.

Crop	Swelling	Lateral bud development	Dwarfing	Distortion	Canker or other symptoms
Oats Rye Maize	Bases of seedlings thickened and puffy	Excessive production of tillers	Affected plants do not run up to ear	Tillers distorted	Base of stem becomes necrotic
Onion Leek Shallot	Tissues bloated, bulb-necks thick and puffy		Plants remain small	Leaves twisted and misshapen	
Beans (*Phaseolus*) Beans (*Vicia*) Peas Potato (haulms)	Affected parts swollen and puffy, sometimes with blisters on stems		Some stunting	Stems and foliage twisted	Discoloration of base of stem
Sugar beet Mangold	Leaf and petiole galls sometimes formed on young plants	Growing points of seedlings killed, leading to development of multiple crowns	Seedlings stunted	Leaves of small plants irregular in shape	Crown canker develops in autumn
Carrot Parsnip					Crown canker in autumn with splitting of leaf bases
Red clover White clover Vetches Lucerne Trefoil	Thickening of stem bases, tissues spongy. In seedlings characteristic galls are produced on the hypocotyl	Some development of basal buds in lucerne	Plants severely stunted	Some distortion of leaves, petioles and stems	
Teasel	Cabbagy plants		Severe stunting		
Bulbs	Small round; pale swellings on leaves (spikels)		Leaves and flower-stems stunted	Leaves curved or twisted, flowers malformed	Brown rings in infested bulbs
Strawberry	Petioles thickened, spongy	Weak secondary growths	Plants dwarfed	Leaves crinkled, margins turned down. Petioles twisted	

Table 11.5 Host ranges of some British races of stem nematode.

Host plant	Oat race	Red clover race	White clover race	Lucerne race	Giant bean race	Narcissus race	Tulip race	Hyacinth race	Phlox race
Barley	−								
Oat	++++								
Rye	++								
Wheat	−								
Maize							+++		
Beans (*Phaseolus*)	++++				+++				
Beans (*Vicia*)	++++				+		++		
Peas	++++						+++		
Vetches	++++				−		++		
Potatoes (haulm)	++								
Sugar beet	++++				+		++		
Mangold	++						++		
Alsike	+	++	+++		−				
Kidney vetch	++++	+++		+	−				
Lucerne	+	+++	++	+++					
Red clover	+++	+++	++	−					
Sainfoin	++	++		−	+	+	++		
Trefoil	+	+			−				
White clover	+		+++	−	−				
Carrot	+++						++		
Parsley	+++++								
Parsnip	++								
Turnip	++						++		
Rape	++								
Rhubarb	+++++								
Teasel	+++								
Strawberry	++++	+++				++	+++++	++	
Bluebell						++	+++++	+++	
Hyacinth						+++	+++	++++	
Narcissus						++++	+++	++	
Onion	++++						++		
Tulip		++					+++++		
Phlox							++		++++ (and other ornamental plants)

−   No invasion after repeated tests, immune.
+   Invaded; plant sometimes produces necrotic lesions, seedling may be killed if invasion is heavy. Unsuitable hosts.
++ Invaded, larvae may mature and a few eggs may be laid, no marked symptoms. Poor host, highly resistant.
+++ Invaded, larvae mature and reproduce at slow rate, symptoms not pronounced. Moderate host, resistant.
++++ Invaded, middle lamellae dissolved, reproduction rapid, symptoms pronounced. Good host, susceptible.

So, although a crop as sensitive as onions can be established in heavily infested land, many of the bulbs harvested contain nematodes and these bulbs eventually rot in store. A second application of nematicide to control later generations is impractical on food crops but would be possible on bulbs and teasle crops. Pre-treatment with dichloropropene nematicide of land to be planted with strawberries, which are relatively poor hosts, gives some control of *D. dipsaci* and *Aphelenchoides ritzemabosi*. It also controls *Xiphinema diversicaudatum* and *Longidorus elongatus* and the viruses they transmit. In lucerne crops infested from seeds, the patches are about 1 metre across at the end of the first year, spread markedly in the second year and may engulf the whole stand in the third or fourth year, again providing that soil surface temperature and moisture are favourable for nematode migration assisted by other means of dispersal (Atkinson & Sykes, 1981). Infections arising from the soil, crop residues or weeds are usually more serious and more widespread in the first year than those arising from seed unless it is exceptionally heavily infested. Stem nematodes are also spread in bulbs and in strawberry transplants. Straw from infested oat crops, the crowns of infested sugar beet and all infested crop residues are potent inocula for the ensuing year.

Lucerne, onion and clover seed may be disinfested by fumigation with methyl bromide, and this is now standard practice for lucerne seed imported from the Continent. Seed is usually treated by seed merchants in fumigation chambers. The concentration of gas, exposure time and moisture content of the seed must be carefully regulated or germination is impaired.

Strawberry runners, onion sets and narcissus bulbs may be disinfested by hot water treatment. Strawberry runners require 46°C for 10 minutes and onion sets and narcissus bulbs 44°C for 2 or 3 hours respectively. Strawberry runners must be immediately cooled by plunging in cold water (Southey, 1978). The design and management of baths, the physiological state of the bulbs and their pre-treatment are important considerations. Unless infested bulb residues are carefully segregated and destroyed, the benefits of treatment may be negated. Hot water treatment saved the daffodil and narcissus bulb industry from disaster and continues to do so. Treatment is never perfect and infested bulbs are often on sale. Tulip bulbs may also be disinfested by hot water treatment but flowering and growth are adversely affected in the following year. Alternatively they may be dipped in a solution of thionazin at room temperature 18.5°C for 2 hours, the solution being agitated continuously. Subsequently the bulbs should be drained and dried in a well ventilated store. Thionazin is sometimes phytotoxic as a soil treatment. Only small amounts are sold and it has become increasingly uneconomic to manufacture.

Resistant varieties against some races of stem nematode have been successfully bred, e.g. the winter oat varieties Peniarth, Pennal and Panema. The spring varieties Manod and Early miller are tolerant rather than resistant in Britain against the oat race; the red clover varieties Merkur, Resistenta and Ulva in Sweden against the red clover race; the rye variety Heetvelder in the Netherlands against a race that attacks rye; the lucerne varieities Nemastan, Talent and Lahontan in the U.S.A. against the lucerne race. Foreign varieties, although resistant, are not always satisfactory yielders in Britain, but from the well-adapted varieties may be bred. Resistance is also sometimes found in local seedstocks. E. B. Brown found such a stock of red clover resistant to the clover race in Hertfordshire which has now been multiplied and sold under the name Pearce's broad red. (See Leaflet Nos 175, 440, 449, 460, 461.)

The potato tuber-rot nematode

Ditylenchus destructor **(Thorne)** which is illustrated in Fig. 11.15 is morphologically close to *D. dipsaci* from which it differs in having a relatively thicker body, a shorter oesophagus with the terminal glandular lobe overlapping the intestine by about one body-width, the possession of six and not four lateral incisures and a shorter, blunter tail. In females the post uterine sac

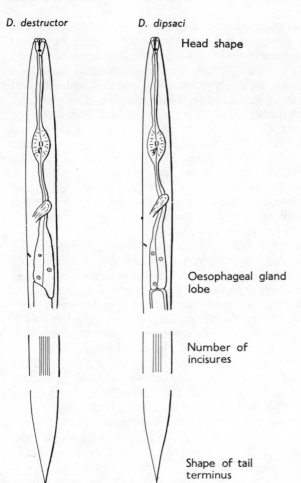

D. destructor   D. dipsaci

Head shape

Oesophageal gland lobe

Number of incisures

Shape of tail terminus

Fig. 11.15 Distinguishing features of *Ditylenchus dipsaci* and *D. destructor*.

extends two-thirds of the distance to the anus (halfway in *D. dipsaci*) and the eggs are twice as long as wide (three times in *D. dipsaci*). *D. destructor* is a deep soil nematode not a soil surface dweller, and has no infestive stage able to withstand desiccation.

Experimentally *D. destructor* may be cultured on a range of plants and on many fungi. Its body dimensions vary considerably according to the host on which it develops. In the field, natural infections occur on potato tubers, bulbous irises, *Tigridia*, hop roots, lilac and occasionally beet (Plate XXVD). Underground stems of corn mint (*Mentha arvensis* (L.)) and creeping sowthistle (*Sonchus arvensis* (L.)) are also attacked. The earliest external symptoms on potato tubers are discoloration beneath the periderm (outer skin). Lesions increase in size and coalesce, the periderm becomes papery and eventually cracks to expose the diseased tissue beneath. Early internal symptoms are pearly white spots which develop into dark, loose and woolly lesions. As rot progresses and the lesions are invaded by soil fungi on which nematodes may also feed, the tissue acquires a brown matted-wool appearance. Outbreaks occur most commonly in the Fenland, and often there is no obvious injury when the tubers are clamped but rot, once under way, progresses rapidly until large sections of the clamp are affected and the yield of saleable potatoes greatly lessened. (See Leaflet No. 372.)

On fields and farms where tuber-rot nematode occurs, weed hosts should be destroyed.

Infested tubers should be sold at once, not clamped and none should be used as seed which might spread the nematode to new areas.

D. radicicola (Greeff) causes small twisted galls on the roots of grasses including barley, oats, wheat and rye. In Britain it is found mostly on *Poa* spp. but is of no economic importance. *D. angustus* (Butler) causes 'ufra' disease of rice in Bangladesh. *D. myceliophagus* (J. B. Goodey) is closely related to *D. destructor* and attacks mushroom mycelium.

Bud and leaf nematodes

Aphelenchoides spp. (Fig. 11.16).
Noteworthy species of *Aphelenchoides* are listed below:

Species	Habits
A. ritzemabosi (Schwarz)	Feeds on chrysanthemum, strawberry, black currant, tobacco and over 100 other plants.
A. fragariae (Ritzema Bos)	Feeds on strawberry and over 250 other plants, mainly ornamentals, including many ferns.
A. blastophthorus (Franklin)	Feeds on scabious and other ornamentals.
A. subtenuis (Cobb)	Feeds on narcissus, other hosts unknown.
A. besseyi (Christie)	Feeds on strawberries, rice and ornamentals – requires warmer conditions than other species.
A. parietinus (Bastian)	Common in decaying plants. Probably feeds on fungi.
A. limberi (Steiner) A. composticola (Franklin) A. saprophilus (Franklin) A. sacchari (Hooper) A. dactylocerus (Hooper)	Occur in mushroom compost. Injurious to mushroom mycelium. Probably also feed on other fungi.

The first five species are called bud and leaf nematodes. They are soil surface forms like *D. dipsaci* but exhibit much less host specialization. Some climb plants readily when surface films of moisture are present. Whether they feed as ectoparasites in buds or as endoparasites in leaves seems to depend on the plant concerned. They are facultative plant feeders not needing specific tissue reactions. Work on chrysanthemums infested by *A. ritzemabosi* suggests that the more resistant varieties are those in which the browning of tissues is most pronounced (Wallace, 1961b). Within leaves, which are entered by the stomata, the spread of the nematodes and the spread of injury is limited to some extent by veins so producing typical leaf blotch symptoms. The stage resisting desiccation is mainly the adult, but persistence in soil is not long when host crop plants or host weeds are absent. Dispersal is mainly by infested plant stocks. *A. besseyi* is seed-borne in rice and *A. ritzemabosi* in *Callistephus chinensis* (Nees).

Aphelenchoides ritzemabosi and *A. fragariae* are important on strawberries in Britain feeding as ectoparasites within buds, rarely as endoparasites within leaves. The leaves of infested plants range from rudiments to almost normal size. The edges are crinkled, less hairy and become reddish in some varieties. Runners are shortened, flowers misshapen and yield greatly lessened. Pale feeding areas, sometimes with colonies of nematodes, may be seen near the bases of leaf midribs of unopened leaves. The symptoms are known as 'spring dwarf' or 'crimp' in the U.S.A. and are most pronounced in spring after a wet autumn.

Propagating stocks of strawberry may be hot-water treated at 46.1°C for 10 minutes (which also controls *Ditylenchus dipsaci* and insect pests) but is near the tolerance limits of the plants. They should be plunged into cold water immediately after treatment, planted under good conditions and well-firmed. Hot-water treatment of chrysanthemum stools at 43.4°C for

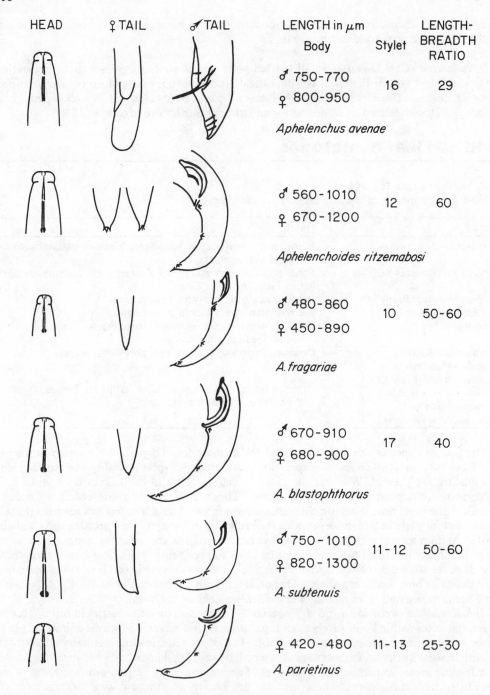

HEAD	♀ TAIL	♂ TAIL	LENGTH in μm		LENGTH-BREADTH RATIO
			Body	Stylet	
			♂ 750-770	16	29
			♀ 800-950		

Aphelenchus avenae

			♂ 560-1010	12	60
			♀ 670-1200		

Aphelenchoides ritzemabosi

			♂ 480-860	10	50-60
			♀ 450-890		

A. fragariae

			♂ 670-910	17	40
			♀ 680-900		

A. blastophthorus

			♂ 750-1010	11-12	50-60
			♀ 820-1300		

A. subtenuis

			♀ 420-480	11-13	25-30

A. parietinus

Fig. 11.16 Distinguishing features of species of *Aphelenchus* and *Aphelenchoides*.

XXIX **A.** Pea plant stunted by cyst-nematode attack on root system; note lack of root proliferation and of root nodules (courtesy F. D. Cowland, Rothamsted).   **B.** Cyst-nematode larvae (*Heterodera* spp.) expressed from root tissue (courtesy C. C. Doncaster, Rothamsted).   **C.** Enlarged portion of root system of pea plant infested with pea cyst-nematode to show immature white females that will later become brown cysts (courtesy F. D. Cowland, Rothamsted).   **D.** Wheat ears in which many of the grains are replaced by galls or cockles of the ear cockle nematode (courtesy C. C. Doncaster, Rothamsted).

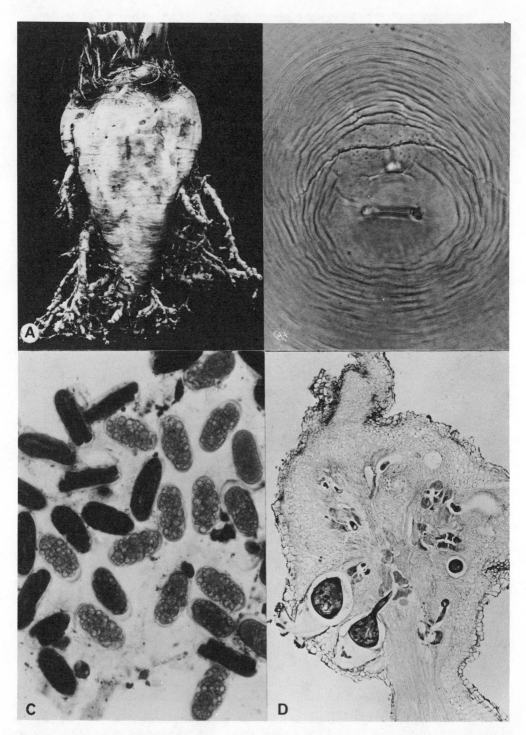

XXX **A.** Sugar-beet plant attacked by the root-knot nematode (courtesy W. E. Dant). **B.** The perianal pattern of a female of the northern root-knot nematode, *Meloidogync hapla* (courtesy M. T. Franklin, Rothamsted). **C.** Eggs of a root-knot nematode stained with New Blue R stain to differentiate living from dead (courtesy A. M. Shepherd, Rothamsted). **D.** Section through a root-knot nematode gall to show nematodes *in situ* and the development of giant cells and the hypertrophy of surrounding tissues (courtesy C. C. Doncaster, Rothamsted).

20 minutes or 46.1°C for 5½ min is desirable, according to variety. Dazomet or aldicarb granules may be applied and worked into the soil before planting when there is a history of *A. ritzemabosi* infestation. Thionazin drenches or aldicarb granules control the nematode in infested chrysanthemum beds. Because of residue problems aldicarb should not be used on established strawberry crops or for spring planting (see Southey, 1978). (See Leaflet No. 337.)

Aphelenchus avenae **(Bast.)** (Fig. 11.16, Plate XXXIIIF) is a cosmopolitan nematode frequently found in mushroom beds and diseased plants where it feeds in fungi. *Rhadinaphelenchus cocophilus* (Cobb) causes red ring disease of coconut palms and also attacks oil palm. Nematodes gain entry via soil and roots or are carried by the palm weevil. Eventually a cylinder of affected, reddish tissue extends from the base to the apex of the bole. *Bursaphelenchus xylophilus* (Steiner and Buhrer) attacks pines, cedars and larches in Japan and the U.S.A.

Flower and leaf-gall nematodes

Anguina spp. (Fig. 11.17).

Species	Plants attacked
A. tritici (Steiner)	Flower galls on wheat, rye, emmer, spelt.
A. agrostis (Steiner)	Flower galls on many species of grass.
A. graminis (Hardy)	Leaf galls on many species of grass.

A. tritici causes 'ear cockles' in wheat. Once common it has virtually disappeared from Britain but is serious where farming is primitive as in south-east Europe, India and elsewhere. Larvae escape from galls in soil and find wheat seedlings from distances up to 30 cm, either from the side or from below. Conditions favouring invasion are similar to those for *Ditylenchus dipsaci* and seedlings and tillers are susceptible as long as the growing points are near the base of the plant. Larvae make their way amongst the ensheathing leaf-bases to the growing point. Heavily attacked seedlings may be killed or, if they survive, have buckled, twisted and rolled leaves and the ears if formed are short and misshapen. The ectoparasitic juveniles are carried up with the ear as the plant grows (Plate XXIXD). Occasionally galls are formed on leaves but usually floral parts are invaded by seven or eight pairs of nematodes which rapidly become adult, mate and lay many thousands of eggs. The eggs soon hatch into first-stage juveniles and these moult to second-stage larvae which are infective and resist desiccation. The endoparasitic nematodes induce a gall or cockle that replaces the grain in an infected floret. Early on, a cavity forms surrounded by granular cells, the so-called nutritive layer, on which the nematodes feed. Later, as the ear ripens, the wall of the gall hardens and darkens. Because the size and shape of the gall is similar to wheat grains, the galls are harvested with the grain and can be separated from it only by thorough cleaning. Apart from the death of seedlings, and stunting and malformation of plants, grain yield is lessened, sometimes by as much as 60%.

Wheat, rye, emmer and spelt are all susceptible; barley and oats, though invaded, are nearly immune and fodder grasses unaffected. Control of the disease is easily achieved by seed cleaning coupled with a single year's rest from a susceptible crop, for although the nematodes remain viable in dry galls for many years, they do not survive a full year in soil when a host crop is withheld. Galls that fall from the ear liberate their juveniles into the soil as soon as they become moist: there seems to be no mechanism to delay liberation until a wheat crop is available. (See Leaflet No. 172.)

A. agrostis forms flower galls on many grasses including species of *Agrostis*, *Poa* and *Festuca*. Its life cycle is similar to that of *A. tritici*, but the juveniles do not escape from the galls in soil by their own movements. The gall contains mucoproteins that imbibe water, burst the gall and

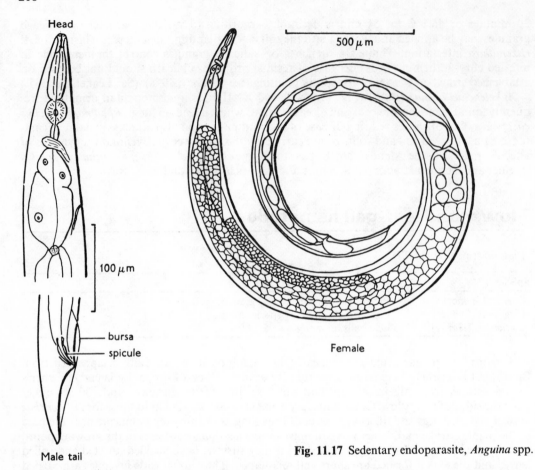

Head

500 μm

100 μm

bursa
spicule

Female

Male tail

Fig. 11.17 Sedentary endoparasite, *Anguina* spp.

expel the larval mass; the optimum conditions for expulsion being those also suitable for grass seed germination. Young infested plants show no obvious symptoms but flower symptoms are striking. Galled florets have glumes two or three times longer than normal, and heavily galled panicles of *Agrostis stolonifera* have on this account been described as a different grass species, *A. sylvatica*. The nematode is sometimes a serious pest where bent grass is grown for seed. Control is as for *A. tritici*. Galls in seed can be killed by soaking in hot water at 52.2°C for 15 minutes followed by 23.9°C for 24 hours.

Anguina graminis is similar to the two preceding species but galls are formed towards the bases of the leaves. They are yellowish at first but later turn deep purplish brown. The disease seems to be confined to Europe.

Cyst-nematodes

Heterodera and *Globodera* **spp.** (Figs 11.18, 11.19). The outstanding features of the cyst-nematodes are enlargement of the female which assumes a lemon or sub-spherical shape, and the tanning of the body wall after death to produce a horny cyst enclosing all or most of the eggs. In many species too, eggs do not hatch freely unless stimulated by substances that exude from the roots. Second-stage juveniles, the sex of which appears to be determined environmen-

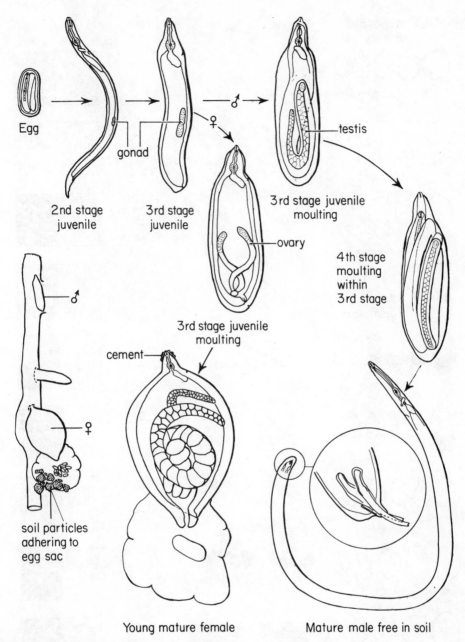

Fig. 11.18 Life cycle of a cyst-nematode, *Heterodera* sp.

tally, induce large syncytial transfer cells in host roots. The following species of cyst-nematodes commonly occur in field soils:

G. rostochiensis (Woll.)	Yellow potato cyst-nematode, golden nematode
G. pallida (Stone)	White potato cyst-nematode
H. schachtii (Schm.)	Beet cyst-nematode
H. avenae (Woll.)	Cereal cyst-nematode

cont.

220

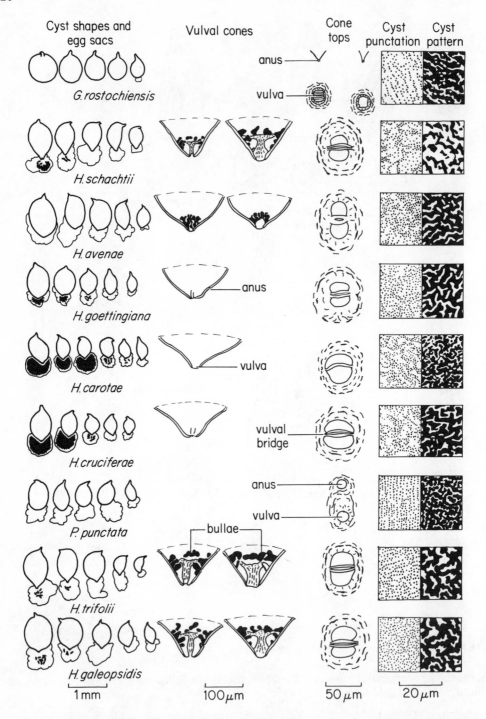

Fig. 11.19 The distinguishing characters of some common species of *Globodera*, *Heterodera* and *Punctodera*.

	♂ tail	Spicule tip	Cyst length μm	Cyst breadth μm	Male length μm	Larval length μm	Colour of large ♀♀	Colour of cysts	Subcrystalline layer
rostochiensis			1000–400	860–180	1100	471	golden	chestnut	No
schachtii			1100–450	750–200	1500	459	white	brown	Yes
avenae			1000–450	800–150	1400	582	white	dark brown	marked
goettingiana			940–400	780–250	1300	474	white	brown	No
carotae			760–300	600–120	1200	451	white	brown	Yes
cruciferae			830–400	570–180	1200	414	white	brown	Yes
punctata			830–470	450–230	1000	brown	white	brown	Yes
trifolii	Absent		1000–450	750–200	–	502	cream	brown	Yes
galeopsidis	Absent		1000–450	850–220		518	cream	brown	Yes

SD 5% for larvae

Table 11.6 *Heterodera*, *Globodera* and *Punctodera* spp., host plants and hatching responses.

Species	Name	Host families and crops	Hatching responses
rostochiensis (Woll.) *pallida* (Stone)	Potato cyst-nematodes	*Solanaceae.* Potato, tomato, egg plant	Water hatch usually low, responds to host root diffusates and those of some other Solanaceae
schachtii (Schm.)	Beet cyst-nematode	*Chenopodiaceae.* Sugar beet, fodder beet, mangold, red beet, spinach beet *Cruciferae.* All types of brassica, mustard, cress, radish, etc. *Polygonaceae.* Rhubarb *Caryophyllaceae Amaranthaceae, Labiatae Portulacaceae, Scrophulariaceae* (*Leguminoseae* and *Solanaceae* in U.S.A.)	Water hatch high, responds to host diffusates, those of some non-hosts and some simple chemicals
avenae (Woll.)	Cereal cyst-nematode	*Graminae.* Oats, barley, wheat, some grasses	Hatches in water in spring cereal root diffusate
goettingiana (Liebs.)	Pea cyst-nematode	*Leguminoseae.* Pea, field bean, broad bean, vetch	Water hatch trivial, hatches in diffusate from host roots of a definite age
carotae (Jones)	Carrot cyst-nematode	*Umbelliferae.* Carrot	Water hatch low, carrot root diffusate
cruciferae (Franklin)	Brassica cyst-nematode	*Cruciferae.* All types of brassicas and many other plants	Water hatch low, host root-diffusates and some other Cruciferae
punctata (Thorne)	Grass cyst-nematode	*Gramineae.* Some grasses, wheat in U.S.A.	Unknown
humuli (Fil. & Sch. Stek.)	Hop cyst-nematode	*Urticaceae.* Hops, hemp, nettles	Water hatch low, host root diffusates
trifolii (Goffart)	Clover cyst-nematode	*Leguminoseae.* Clovers, Chenopodiaceae *Caryophyllaceae, Polygonaceae, Labiatae, Scrophulariaceae*, etc.	Water hatch high, responds to some host and non-host root diffusates
galeopsidis (Goffart)	Galeopsis cyst-nematode	As *H. trifolii*	As *H. trifolii*

H. goettingiana (Liebs.)	Pea cyst-nematode
H. cruciferae (Franklin)	Brassica cyst-nematode
H. carotae (Jones)	Carrot cyst-nematode
H. trifolii (Goff.)	Clover cyst-nematode
H. galeopsidis (Goff.)	Galeopsis cyst-nematode
Punctodera punctata (Thorne)	Grass cyst-nematode

H. humuli (Fil. and Sch. Stek.), the hop cyst-nematode, occurs in hop gardens and occasionally on annual nettles elsewhere. Host ranges of all the above are summarized in Table 11.6 and the diagnostic features of all but *H. humuli* are in Fig. 11.19 (see also Hesling in Southey, 1978).

Until recently cyst-nematodes were regarded as serious pests in temperate regions only, but although most crop losses are in temperate regions, especially Europe and the U.S.A., cyst-nematode problems have been recognized in many countries and new ones are frequently described.

G. rostochiensis and *G. pallida* (Table 11.7) are important pests of potatoes in Britain and Europe generally, and were probably introduced from the Andes plateau of S. America (the

Table 11.7 Some differences between *G. rostochiensis* and *G. pallida*.

	G. rostochiensis	*G. pallida*
Juveniles		
length, body	465 μm	485 μm
stylet	22 μm	24 μm
tail	44 μm	51 μm
stylet knobs	Backward sloping shoulders	Forward pointing shoulders
Females		
stylet length	23 μm	27 μm
anus to vulva	60 μm	44 μm
*Granek's ratio	>3.5	<2.0
ridges between anus and vulva	21	12
colour	golden	white or cream

* Distance between anus and edge of vulval basin divided by mean diameter of vulva.

original home of the potato) some time after 1850, and have since spread and threaten to become coincident with potato growing (Fig. 11.20). Overcropping before and during two world wars so increased soil infestions in the principal potato growing areas of Britain that losses towards the end and immediately after World War II were severe (Jones, 1969b, 1973b). *G. rostochiensis* and *G. pallida* are sibling species which do not interbreed and are distinguished most easily by the internal colour of the females, golden yellow and white or pale cream respectively. (See Leaflet No. 284.)

H. schachtii The beet sugar industry was founded in Germany about 1800 and it was the practice to grow the crop repeatedly on the same land around the small factories. After some years the yields declined and factories were forced to close. The disease was called 'rubenmuedigheit' (beet weariness), but as early as 1859 beet cyst-nematode was recognized as the cause and described as *H. schachtii* in 1871. Although the cause was known and it was established early that brassicas were important hosts, a similar pattern of overcropping with host plants was followed in other European countries and in the U.S.A. Mangolds and brassicas were grown in Britain long before sugar-beet became established in 1925, but there is no evidence that *H. schachtii* was prevalent before that year, although the pest was probably native except in the Fens, which until recent times were flooded. The Continental pattern began to be repeated in Britain where competitive factories sprang up in East Anglia, especially in the

224

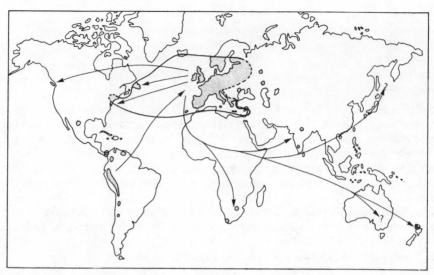

Fig. 11.20 Map to illustrate the combined spread of *G. rostochiensis* and *G. pallida* from the Andes.

favourable Fenland areas inland from the Wash. The first evidence of trouble came when a few infested fields were found in 1935. Crop rotation was encouraged from 1935 onwards and later enforced through factory contracts with growers and through the more effective Sugar Beet Eelworm Order, first issued in 1943, subsequently revised on several occasions (Petherbridge & Jones, 1944; Jones, 1951) and revoked in 1977 (see p. 232). Profiting from Continental experience, no serious losses have been incurred, but the pest has continued to spread, especially in the black fens and in a few other areas (Southey *et al.*, 1982; Whiteway *et al.*, 1982). (Fig. 11.23, Plate XXVIIIB.) (See Leaflet No. 233.)

H. avenae is probably native or long naturalized in Britain, and occurs everywhere in land cropped with cereals. In fields where cereals have been grown frequently, populations show little tendency to increase and little if any yield is lost. Following the great expansion of the cereal acreage after World War II, outbreaks in areas new to cereal growing were frequent, especially on light soils, but have become less frequent. Soils where cereals are grown frequently contain the fungi, *Nematophthora gynophila* (Kerry & Crump) and *Verticillium chlamydosporium* (Goddard) (Kerry *et. al.*, 1982) that limit population increase. They are killed by drenching with formalin but not by most nematicides. Oats are susceptible to injury, spring wheat less and spring barley least (Jones, 1972). (See Leaflet No. 421.)

H. goettingiana is also probably native. In some areas (e.g. Suffolk) considerable acreage of peas and beans were grown in the nineteenth century and populations have since been maintained on field beans (*Vicia faba*) which are little injured. With the increase in pea growing and the rise of factories for quick freezing, canning and packeting as well as acreages for picking green, the existence of these field populations has become apparent. Intensive pea growing has not so far led to a marked increase in populations. Nevertheless this nematode is a serious threat to the industry (Plate XXIXA, C). (See Leaflet No. 462.)

H. cruciferae, also native and widespread, sometimes causes poor patches in brassica crops, especially drilled crops of spring cabbage on land where many brassica crops have been grown. Recently it has been found on oilseed rape. *H. carotae* is more limited in distribution but occasionally stunts carrots on fields where carrot crops are taken frequently in the black fens and on light soils in East Anglia. *Punctodera punctata* (Thorne), *H. trifolii* and *H. galeopsidis* occur in fields but are not associated with crop damage in the U.K. However, a race of *H. trifolii* known as the yellow beet cyst-nematode occurs in sandy soil in the South Netherlands where beet is grown frequently. Evidently it has been established for some years as beet crops are severely stunted (Maas & Hijbrock, 1983). Their presence and that of some other species may complicate the estimation of soil populations because recognition of cysts, especially lemon-shaped ones, is difficult and tedious. Round-cyst nematodes similar to *G. pallida* and *G. rostochiensis*, attacks tobacco and other Solanaeace in some parts of the U.S.A. *H. glycines* attacks soya bean in the U.S.A., Egypt and Japan. For additional information on British species of *Heterodera* and *Globodera* see Hesling and Williams in Southey (1978).

Life cycle

The life cycle of a typical cyst-nematode is detailed in Fig. 11.18. After hatching, the second stage juveniles invade the roots of host plants just behind the root tip, near the points of emergence of the endogenous lateral roots or, sometimes, where lateral roots have ruptured the cortex of the main root. At first they move through the cells, usually upwards but mostly downwards to the root tip in *H. avenae*, leaving a trail of ruptured end walls. Soon, however, they settle with their heads near the endodermis opposite a phloem group and commence feeding. Their saliva induces the formation of enlarged multinucleate syncytia with densely granular contents, which are transfer cells (Plate XXVIIID). The juveniles thicken, shorten a little, and moult three times before becoming adult. The sex of juveniles becomes obvious in the third stage and from then on males and females diverge. The fourth-stage male re-elongates within the cuticle of the third stage which by now has either ruptured the cortex of the root to which it is attached or lies just beneath the epidermis. After the fourth moult, the fully formed male, a little over 1 mm long, escapes into the soil. Fourth-stage females remain saccate and enlarge further until they rupture the cortex completely, remaining attached with the head buried in the root and seeming to be held in place by a sticky effusion produced in the neck region. The ovaries enlarge greatly, the gut distends and, after moulting, the characteristic gelatinous egg sac to which soil particles adhere is secreted (absent in round-cyst nematodes). At about this stage males congregate around the females and fertilize them. Sometimes several males occur inside the egg sac. Although typically root parasites, heteroderid juveniles also invade buried leaves underground stems and tubers (e.g. potatoes) and form cysts on them.

After fertilization, the female produces many eggs, 100–500 or more, some of which may be deposited in the egg sac. The colour of the immature females is white at first but may become cream or golden yellow. After death, the body wall hardens, turns brown and forms a protective envelope for the eggs retained in the body. It is pervious to water and ingress is helped by apertures which develop at the head and vulval ends. Through these apertures hatched juveniles emerge.

The number of generations per year is governed by soil temperature, soil moisture and the length of time the host crop occupies the ground. Development is slow in spring, speeds up in midsummer, declines in autumn and is stationary in winter in the U.K. It can be modelled by accumulated temperatures above a threshold of about 10°C (Jones in Southey, 1978; Jones, 1983). Soil moisture is adequate in most months, but in light open soils it may be limiting in the spring for invasion, in June for males and in July and August for a second generation of *H. schachtii*. Sugar-beet is vegetative from March to September and allows completion of two or two and a partial third generations of *H. schachtii*. In hot, irrigated Californian valleys where beet is a winter crop, five generations are possible. *H. avenae*, *G. rostochiensis*, *G. pallida* and *H. goettingiana* are limited to one generation by the relatively short vegetative lives of their host crops. A partial second generation of *G. rostochiensis* may sometimes occur in potatoes that continue growing into the autumn. *H. carotae*, *H. cruciferae* and *H. trifolii* probably pass at least two generations a year.

Early potatoes (e.g. Epicure) and quick maturing varieties of pea (e.g. Meteor, Kelvedon Wonder) are not in the ground long enough for one generation of *G. rostochiensis* and

H. goettingiana, respectively, to be completed fully and do not support such heavy soil populations as main crop and late maturing varieties.

Dispersal

Dispersal of cyst-nematodes in soil, or in soil adhering to plant produce, occurs in many ways. Cysts of *G. rostochiensis* and *G. pallida* are carried by seed tubers raised on infested land, in

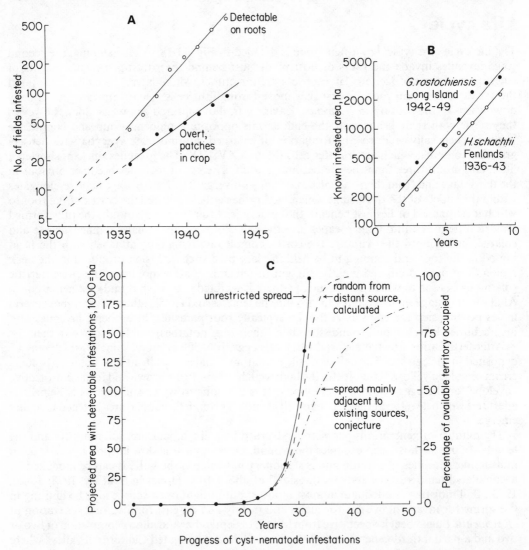

Fig. 11.21 A. Increase in the number of beet fields found infested with *Heterodera schachtii* by examining root systems between mid-June and September in the East Anglian Fens.   **B.** Increase in the area infested with *H. schachtii* in the East Anglian Fens compared with the area infested with *Globodera rostochiensis* in Long Island, U.S.A.   **C.** Colonization of a large territory (200 000 ha). Continuous line, unrestricted spread at the average rate of the curves in B. The shape of the curves in C would be similar for colonization of smaller territories but the time scale shorter. (From Jones, 1980.)

soil adhering to tubers, especially around the eyes, and in soil loose in seed bags. Cysts are sometimes formed on tubers and therefore may occasionally be embedded in the skin. Thus host and parasite are dispersed together. Similarly other species may be dispersed by their hosts in lumps of soil equal in size to the seed. Dispersal by wind and water are probably the natural means of spread often accentuated by cultivation which exposes light surface soil (e.g. the Fenlands, the Brecklands) and by drainage and its accompanying floods (e.g. the Fenlands). Machinery is an important means of spread within farms and there is little point in trying to limit it by cleaning after a heavy infestation has occurred. Transplants and stecklings carry nematodes in adhering soil or in their roots, e.g. *H. schachtii*, *H. cruciferae*, *G. pallida* and *G. rostochiensis* are frequently spread by brassica transplants. Except where massive quantities of soil are moved by high winds, floods or excavations, spread in the ways described above produces soil populations far too sparse to detect by sampling and some time must elapse under suitable host crops before soil populations reach damaging concentrations. Even where relatively large amounts of soil are carried on transplants the process would have to be repeated many times to create sizeable populations. Soil derived from many sources accumulates at sugar-beet factories, chicory factories and vegetable processing plants. This soil contains cysts of all kinds and care is necessary in disposing of it. In the U.S.A. beet factory 'dump dirt' accelerated the spread of beet nematodes in many areas.

The rate of spread from field to field of *H. schachtii* in the East Anglian Fenlands and of *G. rostochiensis* in Long Island, New York State, U.S.A. turns out to be similar (Fig. 11.21). It is easy to demonstrate the importance of soil movements, especially harvesting, in the within field spread of these species. Assuming that the movements of second stage juveniles extend an infestation by 10 cm per generation, then from a cyst introduced into the centre (most favourable position) of a square field of area 1 ha, it would take 500 generations by self-spread alone to reach the sides and 707 to reach the corners; if placed in one corner (the least favourable position) it would take 1414 generations to reach the opposite corner. It is common knowledge that fields become almost totally infested in fewer than 10 years (i.e. ten generations of *G. rostochiensis* and about 20 full generations of *H. schachtii*). New infestations started by one or two cysts in a small lump of soil would take at least three generations to colonize the initial focus fully during which time there would be too few cysts per g soil to ensure dispersal, i.e. effectively there would be a latent period before harvesting and other operations began the process of dissemination. Subsequent spread would tend to proceed in waves at the end of each successive latent period. These considerations tend to explain why the initial spread of cyst-nematodes goes on insidiously and unseen. Not until the infestation is already well-dispersed, possibly beyond the confines of the original field, do crop symptoms appear as stunted patches. As many as five potato crops may be grown on slightly infested land before the infestation is patently obvious. In individual fields and over larger tracts the later stages of spread are explosive. This stage has recently been reached by the soya bean cyst-nematode, *H. glycines* (Ichinohe), in the U.S.A.

Hatching

The eggs of *H. avenae* hatch freely in water in early spring but more rapidly in exudates from cereal roots. Few eggs of *G. pallida*, *G. rostochiensis*, *H. carotae* and *H. humuli* hatch in water, but they hatch freely in leachates from pots of their respective host plants. *H. goettingiana* can be induced to hatch only in leachates from pea roots 4–6 weeks old and bean roots 6–8 weeks old. *H. schachtii*, *H. trifolii* and *H. galeopsidis* are intermediate. Many juveniles hatch from eggs in water but many more hatch in root exudate.

Normally, vigorous hatching occurs only in the presence of root exudates from host plants but sometimes related marginal hosts or non-hosts stimulate hatching and therefore serve to trap juveniles in the soil. This is true of most races of *Solanum nigrum* which is resistant to

G. rostochiensis and *Coronopus squamatus* (Forsk.), *Hesperis matronalis* (L.) and *Beta patellaris* (Moq.) which are resistant to *H. schachtii*. Occasionally hosts fail to evoke a massive hatch, e.g. clover and *H. trifolii* (Winslow, 1955).

Although the hatching reaction probably has survival value and helps to ensure that the cyst-nematodes remain dormant except when host roots are near, 30–50% of the eggs of most species of *Heterodera* and *Globodera* hatch spontaneously in spring in fallow land or non-host crops.

A partial characterization of the chemical structure of the hatching factor for *H. schachtii* is known (Masamune, 1982). Several factors can be extracted from potato roots. All seem to be partially unsaturated, long chain fatty acids, water soluble, non-volatile, unstable and active at great dilutions. Many dyes and simple organic substances also evoke hatch. The second stage juveniles in heteroderid eggs are bathed in a solution of the sugar trehalose which keeps them in a partly dehydrated state and prevents them from moving. Hatching factors seem to operate by changing the permeability of the egg shell. This allows the sugar to escape, the juvenile to become fully hydrated and to begin movements which lead to eclosion. Probably the innermost of the three layers of the egg shell which is made of lipoproteins is the one affected by the hatching factors (Perry, Wharton & Clarke, 1982; Perry & Clarke, 1981).

Crop injury

The above ground symptoms of cyst-nematode attack are patches of stunted plants. The patches resemble those caused by waterlogging, acidity, poor soil conditions and lack of major or minor plant nutrients. Foliage becomes pale and later yellows, and weeds often grow up and smother plants in the worst affected patches. This is because the root systems are impaired and the plants are unable to obtain essential water and minerals (Table 11.8). Affected plants wilt easily, and the severity of the symptoms is aggravated by environmental stress such as high temperature, bright sunlight and drought. When plants are lifted, a common indication of cyst-nematode attack is excessive production of lateral roots (Plate XXVIII). Peas and field beans are exceptional in not exhibiting root proliferation when attacked by *H. goettingiana*. Instead the root systems are smaller and lack the characteristic nodules of leguminous plants. In cereals *H. avenae* juveniles inhibit the growth of root tips. Many lateral rootlets grow out just above the points where juveniles have settled so that the root system is dwarfed, shallow

Table 11.8 Effects of cyst-nematodes on root systems (after Jones, 1976a).

Roots	Smaller, more branched
	Search smaller volume of soil
	Take up water and nutrients less efficiently
	Water and nutrients shunted to female nematodes
	Formation of nitrogen fixing nodules may be inhibited
Haulms	Fewer, shorter stems
	Usually the same number of smaller leaves per stem
	Total N, P, K, Mg, Ca, Na in haulms decreased
	Concentration of: N little changed
	P, Mg decreased
	K much decreased
	Ca, Na increased
	Dry matter increased
Water usage	Less efficient
Other pathogens	May be assisted or encouraged

and matted. Where soil populations are heavy stunting is severe and may affect whole fields, leading to premature death of the plants and little or no yield. Losses may also be considerable before extreme symptoms of attack appear. When the range of population density is large enough an S-shaped curve is usually found by plotting yield against log population density, i.e. there is a maximum yield at small population densities and a minimum yield at large densities (see p. 292). In pots and small plots, there is sometimes a significant increase in yield between uninfested and slightly infested ones. It seems that small densities cause branching of the root system which slightly increases the efficiency of water and nutrient uptake. The principal effects of infestation are summarized in Table 11.8.

Cyst-nematode attack is readily diagnosed by the abundance of immature females adhering to the roots. Paradoxically, when the attack is excessively heavy, few cysts are formed but roots may be stained with osmic acid or lacto-phenol-cotton blue and cleared to demonstrate the nematodes within them (Southey, 1984).

Like other soil nematodes, the injury caused by cyst-nematodes may pave the way for attacks by other pathogens. Verticillium wilt fungus (*Verticillium dahliae*) appears to gain entry not when juveniles of *G. rostochiensis* invade potato root systems but when males and females rupture the root cortex. The disease appears earlier and is more damaging when large numbers of *G. rostochiensis* are present in infested soil (Corbett & Hide, 1971). There may also be an association between this nematode and the fungus *Rhizoctonia solani* (Kühn). Damage by *H. schachtii* to seedling sugar-beet is also enhanced by certain soil fungi.

Soil populations

Anscombe (1950), Jones (1955) and Fenwick (1961) discuss the principles of population estimation. Methods are described in Southey (1984). Brown (1969) using small field plots in fields infested with *G. rostochiensis*, found the diminution in yield could be related almost equally well to the initial population level or the logarithm of the initial population level. Experiments with the same and other species of *Heterodera* in pots and microplots covering a wider range of populations suggest that the relationship between yield loss best fits the logarithm of the initial population except at very low population densities. This kind of relationship would be expected on theoretical grounds because of intraspecific competition within infested roots which are both habitat and food (Seinhorst, 1965). A curious feature of some laboratory experiments, which may or may not occur in the field, is stimulation of growth

Table 11.9 Populations, yield lost and crop frequency: a summary of field experience for *G. rostochiensis* (Jones, 1973b).

Preplanting nos. Eggs/g soil	Post harvest nos. Eggs/g soil	Crude multiplication	Crop lost (approx.)	Years between successive potato crops to	
				restore pre-planting nos.	avoid crop loss
1	30	× 30	None	9	1–2
10	100	× 10	Little	6	4
50	200	× 4	Quarter	4	6
100	300	× 3	Half	3	7
200	250	× 1.25	Failure	1	6–7
300	210	× 0.7	Most	0	6
†Equilibrium numbers:					
230	230	× 1.0	Most		6–7

†Equilibrium density varies with soil, season and host cultivar and is often less than 200 eggs/g.

by small populations. Probably the complete curve would be sigmoid with its slope position varying according to the degree of environmental stress imposed upon the plants.

Populations of new cysts (females) at the end of the season are usually correlated with yield. As the initial density increases from one which has little effect on yield, the number of new cysts increases, reaches a peak as root growth is affected and then declines. New cyst formation is therefore density dependent (Table 11.9 and Fig. 1.3).

Populations left after a host crop consist partly of old eggs and cysts and partly of new eggs and cysts which cannot be distinguished in soil samples. When a host crop is grown, root exudates stimulate most but not all of the eggs to hatch. After harvest the old and the new population together are related to preplanting numbers by the type of relationship shown for *G. rostochiensis* in Fig. 1.3. Increase is clearly density-dependent, whereas decrease under fallow or non-host crops is usually density-independent suggesting that the juvenile within the egg and cyst is well protected and little affected by enemies. The annual cycle of *G. rostochiensis* and *G. pallida* in a potato field is of the type shown in Fig. 11.22. That of *H. schachtii* with its two generations would be more complicated. Populations of potato cyst-nematodes behave rather consistently probably because the soil climate, especially the soil temperature, is similar from year to year during the period when juveniles are establishing themselves and developing in potato roots (Jones, 1973a). The juveniles invade the roots and thereafter are largely independent of soil moisture. The crop is planted into soil in much the same climatic state because the sowing date varies with the earliness or otherwise of the season and because approximately the same weight of seed tubers is sown from year to year and about the same amount of fertilizer is applied. The same soil temperature relationships must hold for many soil nematodes and other soil pests in the early part of the season, but those living entirely in the soil are much affected by variations in soil moisture as seems to be the case for *Trichodorus*, *Paratrichodorus* and *Longidorus* spp. (see page 187, also Figs 11.1–11.3 and Jones, 1983).

Crop rotation

Crop rotation remains an important method of controlling cyst-nematodes in Britain. For most species a four-to-six-course rotation is usually sufficient to ensure that the population left after a susceptible crop has fallen to a safe level before the next susceptible crop is grown. For potato cyst-nematodes, only potatoes need to be considered, but for beet, brassica, cereal and pea cyst-nematodes more than one crop is concerned. Brassicas and other cruciferous seed crops such as oilseed rape, turnip seed and cress seed, are all efficient hosts of *H. schachtii* but are less susceptible to injury than sugar-beet, mangold, fodder beet and red beet. Similarly, barley is an efficient host of cereal cyst-nematode, less susceptible to injury than oats, and field bean is an efficient host of pea cyst-nematode, less susceptible to injury than peas. For brassica cyst-nematode also, oil seed rape and other cruciferous crops are hosts.

Control by crop rotation, although generally useful, has limitations. The population density after a susceptible crop has been grown in rotation is determined by several factors. Paradoxically, the small density left at the end of a rotation favours rapid increase largely because root systems are less damaged and can support more nematodes. Heavy soils are less favourable than light for cyst-nematode hatch and migration, and they usually hold more water so that plants are better able to withstand stress. Moist seasons and moist situations, such as the Fenland area, river valleys and irrigated areas favour cyst-nematodes. Efficient host crops, such as rape seed and cress seed for *H. schachtii*, encourage development of greater populations. When these factors combine to give exceptionally high post-harvest populations, longer than average periods of rest from susceptible crops are needed before the next one can be grown safely.

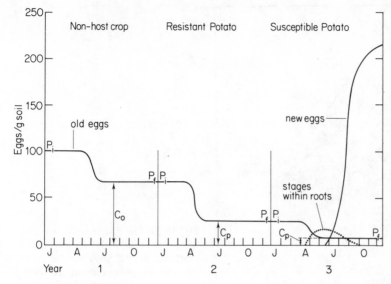

Fig. 11.22 Population changes in the cyst-nematode *G. rosto-chiensis* when a non-host crop is grown for 2 years followed by a host crop. Note the slow decline under the non-host, mostly due to spontaneous hatch, and the rapid increase when a host crop is grown.

Crop rotation may be supplemented by estimates of soil populations. These are easily measured for potato cyst-nematodes with their distinctive spherical cysts and less easily for the species with lemon-shaped cysts. For a fee the Agricultural Development and Advisory Service will estimate soil populations each year. On the basis of cyst and egg counts fields can be placed in four categories: 'absent, safe, borderline and unsafe'. More categories are often used but, because of errors in field sampling, it is doubtful whether any more are justified. Samples containing no cysts do not indicate freedom from infection because the size of the sample relative to the great weight of top soil per ha is so small. Nil counts merely indicate that no cysts were found in the sample of that particular size. For potato, beet, cereal and pea cyst-nematodes, the economic threshold (p. 2) lies between 10 and 50 eggs per g of air-dried soil (10–50 cysts with contents per 100 g). The figure for potatoes and *G. rostochiensis* and *G. pallida* is 20–30 eggs per g, for sugar-beet and *H. schachtii*, 10 eggs per g, for oats and *H. avenae* about 10 eggs per g and for barley about 30 eggs per g. The comparable figure for stem nematode damage to onions is less than 10 nematodes in 500 g soil, but in that case multipli-cation must occur to produce serious injury. Potatoes stand rather more nematodes than other crops, but the levels vary and Advisory Entomologists vary their recommendations on the basis of past experience in a particular area. The cyst-nematode thresholds turned into popula-tions per ha are of the order of 2.5×10^{10}.

For beet cyst-nematode, crop rotation was enforced by the rotational clause in factory contracts and, in the Fenlands and one or two other places, by the Beet Eelworm Orders 1943 to 1962 (Fig. 11.23). Field surveys were made and fields of sugar beet or mangolds found infested were placed in two categories: 'sick' and 'infested'. A sick field is one in which patches of stunted plants make up an appreciable part of the field area. An infested field is one in which females are found on the plants; small patches of stunted plants may also be present. Sick fields had to be rested from susceptible crops for 5 years and then be sampled to ensure that the soil population had fallen to a safe level before sugar beet, mangolds and brassicas could be grown. Infested fields had to be rested from susceptible crops for 3 years. Within areas scheduled under the Beet Eelworm Order, a field not known to be infested could be planted with beet and other susceptible crops once every 3 years. Control of cropping was by a licensing system. This system was difficult to operate and became more so as more and more land was found infested. The Beet Cyst Nematode Order 1977 revoked earlier Orders but allowed

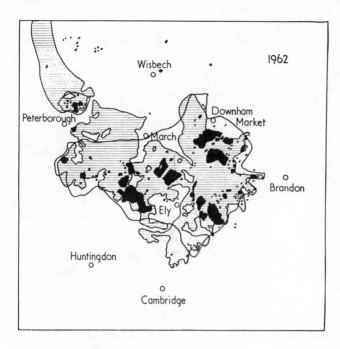

Fig. 11.23 The distribution of beet cyst-nematode in the Fenlands in 1962 showing the areas of fen peat soil and the boundaries of the areas scheduled under the Beet Eelworm Orders.

statutory controls to be re-introduced if necessary. The rotational clause has also gone from growers controls with the sugar factories. Yet the rotational clause in factory contracts and the Orders have prevented the serious losses which were incurred on the Continent and in the U.S.A.

Attempts to control cyst-nematodes by trap-cropping, that is by growing a susceptible crop for a period of about 6 weeks and lifting or destroying it before egg formation begins, have not met with success. Because roots do not have time to fill the soil and root diffusates do not reach all cysts, the lowering of the soil population is disappointing and there is always the risk that if unforeseen circumstances prevent destruction of the crop at the correct time, a population increase may occur. A substitute for rape, turnip, and mustard which are grown as catch crops after early potatoes, early carrots or peas in areas infested with beet cyst-nematode is fodder raddish. If left in the ground for more than about 6 weeks, other cruciferous catch crops may increase soil populations.

Resistant varieties

Ellenby (1952) first found evidence of resistance to potato cyst-nematode in some lines of *Solanum tuberosum* ssp. *andigena* and *S. vernei* (*balsii*), subsequently resistance genes were found in several other tuberous solanums. Attempts to breed for resistance began almost simultaneously in Britain, the Netherlands and the U.S.A. By chance all the early breeders used populations of the nematode that were *G. rostochiensis* pathotype Ro1 (Table 11.10). However, when resistant potato hybrids were tested against more populations it soon became clear that others behaved differently. It is now known that there are two species of potato cyst-nematodes not one (Table 11.7), that they do not interbreed freely, and that each has its own pathotypes or sub-races.

G. rostochiensis pathotype Ro1 is dominant in North and South Ireland, in Scotland and East and southern England. *G. pallida* pathotype Pa3 is dominant in the Humber basin embracing parts of Yorkshire, Nottinghamshire and Lincolnshire, and in the Channel Islands.

Table 11.10 Some pathotypes of potato cyst-nematodes, European nomenclature.

Differential *Solanum* hosts	Resistance genes	*G. rostochiensis* yellow females					*G. pallida* cream females		white females	
		Ro1	Ro2	Ro3	Ro4	Ro5	Pa1	Pa2	Pa3	Pa4
S. tuberosum	None	+	+	+	+	+	+	+	+	+
ex *andigena*	H$_1$	−	+	+	−	+	+	+	+	+
ex *kurtzianum*	K$_1$K$_2$	−	−	+	+	+				
ex *vernei* 58.1642/4	?	−	−	−	+	+				
ex *vernei* 62.33.3	?	−	−	−	−	+	−	−	+	−
ex *vernei* 65.346/19	?	−	−	−	−	−				
ex *multidissectum*	H$_2$	+	+	+	+	+	−	+	+	+
ex *andigena* × *multidissectum*	†(H$_1$H$_2$H$_3$)						−	−	−	+
ex *andigena* × *multidissectum*	†(H$_1$H$_2$H$_3$ +?)						−	−	−	−

+ able to reproduce freely, − scarcely able to reproduce,
† contain genes analogous or homologous to H$_1$H$_2$H$_3$ and possibly some additional minor genes.

Elsewhere in England and Wales fields with both types of population are intermingled in various proportions. Fields infested with *G. pallida* Pa1 are scattered or in pockets in areas of Britain infested with either Ro1 or Pa3; Pa2 and Pa4 also occur. Whereas *G. rostochiensis* populations in Britain are all nearly homozygous for Ro1, populations of *G. pallida* other than Pa1 are rather heterozygous mixtures of pathotypes. Other pathotypes of both species occur in the Andes and on the Continent. As the pathotypes were present before resistant varieties were grown, it seems both species and their sub-races existed previously and were introduced from the Andes (Fig.11.20). Their distribution patterns probably arose from the original introductions of cysts and from their subsequent dispersal in commerce, especially that in seed potatoes. So far as is known, pathotypes of *G. rostochiensis* other than Ro1 do not occur in the U.K., but pathotypes of *G. pallida* are widespread in fields where Ro1 is dominant and repeated cropping with varieties possessing resistance gene H$_1$ (e.g. Maris Piper) leads to replacement by pathotype *G. pallida*. In North America related species of round-cyst nematodes, some of which reproduce on potato, occur on tobacco, other solanaceous crops and weeds (for a discussion of pathotypes see Anon, 1972a).

Resistant plants exude hatching factors from their roots, are invaded by juveniles as heavily as are susceptible plants and produce many males but very few females. Apparently most of the juveniles fail to evoke the correct tissue response in the roots and fully functional, long-lasting syncytial transfer cells are not formed. As sex appears to be determined by nutrition, these inadequate transfer cells are unable to support females which ultimately require 500–1000 times more food. When resistance is from a single dominant major gene the total number of adults produced is the same as from a susceptible variety. When two genes are present even fewer juveniles become female and males are fewer and smaller. Additional major or minor resistance genes suppress the production of males still further. Resistance from one gene increases tolerance to root invasion slightly if at all, but additional genes may possibly improve tolerance. So far there has been no attempt to breed for tolerance as such, although some susceptible varieties are known to do better than others in infested land. For example, Pentland Crown does better than Pentland Dell.

Transfer of resistance from *S. tuberosum* ssp. *andigena* to cultivated potato is relatively easy because both plants are tetraploid. Most other species of *Solanum* with genes for resistance are diploid and a few triploid. They also possess many undesirable botanical and biochemical characteristics that make breeding more difficult. Varieties with a measure of resistance to *G. pallida* pathotypes are in British proving trials and should soon be avilable. Dutch breeders have produced many varieties with resistance (see Jones, 1976b).

Until acceptable varieties fully resistant to most populations of both species are available, those with resistance from gene H_1 only should be planted rationally. In introducing them to a field for the first time, they should be planted only when the soil population has decreased to a safe level, otherwise yield may be lost from juveniles invading the roots. In late June and July not fewer than 20 plants should be lifted from parts of the field and the roots examined for females (cysts). If no females are seen, either the infestation is slight or the population is one unable for the time being to multiply on the chosen cultivar. If many white females are seen the species is *G. pallida* and the cultivar is unsuitable. If no females were seen on the potato roots the field may be planted again with the resistant cultivar three or four years later when the roots should be examined as before. Proceeding cautiously in this way, on suitable fields, the resistant cultivar may remain usable for 12–20 years by which time more effective ones may have been bred. Unless they are, it will be necessary to revert to a five- or six-course rotation or to use a nematicide.

Sources of resistance in barley, wheat and oats have been found to *H. avenae* and breeding for resistance has produced some varieties of spring barley with monogenic resistance derived from barley No. 191 (Table 11.11), for example Ansgar, Tintern, Sabarlis and Tyra. Similarly the resistant oats Nelson, Trafalgar and Panema have been derived from *Avena sativa* (L.) or *A. sterilis* (L.) but do not tolerate root infestation well. There are as yet no acceptable resistant wheats. Tests with resistant hybrids against populations of *H. avenae* has revealed a situation similar to that in potato cyst-nematodes. Pathotype 2 is common in Britain whereas in the Netherlands, Denmark, Sweden and Germany pathotype 1 predominates. In Britain pathotype 1 is often mixed with 2, and 3 is rare. All races of *H. avenae* have hosts among grasses but grass weeds do not appear to be efficient hosts (Cook, 1982): rye is susceptible to most races but few females are formed on its roots. As with resistance to *G. rostochiensis*, that against *H. avenae* appears to be derived from dominant major genes which interfere with the function of syncytial transfer cells in roots and shift the sex ratio in favour of males.

Breeding for resistance to the more polyphagous *H. schachtii* is more difficult. There is some evidence that selected families of *Beta vulgaris* produced fewer cysts, but pronounced resistance has been found only in the wild species. These do not interbreed freely with sugar-beet to produce viable seed. With much difficulty, a few successful crosses that will grow on their own roots have been made in the U.S.A. and are in trials. (See also Chapter 15 under Resistant varieties.)

Quarantine

Because cyst-nematodes are serious pests most countries have quarantine regulations which aim at preventing entry. Potato-growing countries where *G. rostochiensis* and *G. pallida* are either not known, or strictly confined, exclude potatoes completely or refuse to accept seed potato tubers or planting material with adhering soil if raised on infested land.

Seed potatoes exported from the U.K. require phytosanitary certificates which include a statement that the consignment is free or believed to be free from cysts of *G. rostochiensis* or *G. pallida*. Fields where it is intended to grow seed are sampled at the rate of 100 or more small cores per 4 ha (10 acres) to give a bulk sample of 500 ml soil. The sampling tool is of the cheese-corer type on the end of a short stick and the sampler pursues a W-shaped path up and down the field. The whole sample is extracted and the extract searched (Southey, 1984). If any cysts are found the field is barred from seed production; if none are found the growing crop is inspected in July for trueness to type and for virus infestation (see p. 61). Twenty plants are lifted and the roots examined for females: this is too few to detect any but gross infestations. Tuber samples are also examined at harvest but usually not specifically for nematode cysts nor is soil in the bottoms of seed bags tested, the main reliance is on the pre-crop soil sample although if missed the nematodes may have increased perhaps a hundred-fold on the intervening crop. It would be more sensible to sample the soil beneath the seed dressing machines.

Table 11.11 Reactions of some pathotypes of *H. avenae* and of some related species to sources of resistance (adapted from Cook & York, 1982; Andersen & Andersen, 1982).

Notation	H. avenae							H. hordecalis	H. bifenestra
UK	1		2			3	G		
Dutch	A	B	C	D	E		G		
Danish	Ha11	Ha51	Ha12	Ha21	Ha61	Ha23/33	Ha13		
Occurrence									
Australia									
Britain	†P		P			P	P		
Scandinavia	P	P	P	P	P	P	P		
Western Europe	P	P	P	P	P		P	P	P
Barley cvs									
Varde	+	+	+	+	+	+	+	+	+
Emir/Herta	+	−	+	+	+	+	+	+	+
Drost/Ortolan RHa1	−	+	+	−	−	+	+	+	+
Ansgar/Sabarlis/Tintern RHa2	−	−	−	−	+	+	−	−	+
Morocco RHa3	−	−	−	−	−	+	−	−	+
Moroccaine	−	−	−	−	−	±	−	−	+
Wheat cvs									
Most European	+	+	+	+	+	+	+	−	+
Loros/AUS108941	−	·	−	−	±	+	±	−	−
Oats cvs									
Most European	+	−	+	−	+	+	+	·	+
*Panema 1	−	−	−	·	−	+	+	·	+
**Nelson/Trafalga 1	−	·	−	·	−	+	−	·	+
Avena sterilis 2/3	−	−	−	−	−	−	−	−	−

†P indicates presence, + susceptibility, − resistance, · not tested.

* ex *A. sterilis*, ** ex *A. sativa*

Note: There are three or more pathotypes in India one of which is probably D(Ha21).

The problems associated with soil sampling for statutory or certification purposes are considerable and are not always clearly understood by growers, seeds merchants, exporters or quarantine authorities at home and abroad. There are 2 million litres of topsoil to the hectare down to a plough depth of 20 cm, i.e. 8×10^9 ml per 4 ha sampled. Assuming that cysts are distributed randomly, the numbers found in repeated 500 ml samples (or any other sampling unit) should fit the Poisson distribution from which the chances of detection can be calculated. These are shown in Table 11.12 together with the equivalent number of cysts in the topsoil of 4 ha. From this it can be seen that one cyst in a single sample represents a population of anything from 0 to 16 million cysts per 4 ha. Further a 50:50 chance of detection is not achieved until the population density is 11 million cysts per 4 ha and a 95 to 99% chance of detection is not reached until the density reaches or exceeds 48 million per 4 ha. In practice the chances of detection are smaller because cysts are not randomly distributed but aggregated in pockets

Table 11.12 Number of cysts found in 500 ml soil sample, equivalent number in 4 ha of top soil and percentage chances of detection.

Mean no. of cysts in 500 ml soil	Equivalent number in 4 ha of top soil (millions)	Percentage chances of:	
		detection	failure to detect
0.05	0.8	5	95
0.1	1.6	10	90
0.5	8	39	61
0.7	11	50	50
1	16	63	37
2	32	85	15
3	48	95	5
4	64	98	2
5	80	99	1

which may be missed, sampling is not random and extraction and collection are imperfect. For example, at 1 cyst per sample the chances of detection are reduced to 33% or fewer. What a 'certificate of freedom' means is that a sample of a given kind was examined, no cysts were detected in it and it is presumed that all the field top soil is similar, a presumption which may not be justified. Increasing the quantity of soil collected or decreasing it to half and reducing the area sampled to say 0.5 ha (as for *Globodera* spp. in the Netherlands) would improve the chances of detection only marginally, for the amount of soil and the number of samples it is practical to process is limited.

Sometimes juveniles of *Globodera* spp. invade developing tubers and eventually produce females that become cysts so that there is always a risk that the tubers themselves will be infested. Usually however, any accompanying cysts will be in soil attached to tubers that tends to fall to the bottom of seed bags which foreign quarantine inspectors are likely to examine. If any are found, very large consignments of seed may be rejected with resultant losses which may be as much as £1 million on occasion. Efforts have been made to disinfest tubers by washing (Mabbott, 1960) or by dipping them in sodium hypochlorite solution (Wood & Foot, 1975). For various reasons, mainly cost and the encouragement of bacterial disease when pressure hoses are used, these methods have not caught on. Growing in coarse sandy soils (as in the Netherlands) produces tubers with the minimum of adhering soil is helpful but not always possible.

The principles behind quarantine sampling of large consignments of produce or planting stocks for the exclusion of mites or insects are similar to those for soil-borne nematodes but not so clearly demonstrable (Jones, 1969a).

Nematicides

When successful, nematicides offer an immediate and effective method of increasing yield and controlling cyst-nematode populations. Most trials in the UK have been against potato cyst-nematodes (Whitehead, 1975). Three types of nematicides are available which kill or immobilize the nematodes: soil fumigants, organophosphorus compounds and oxime carbamates. Fumigants include dichloropropene, chloropicrin, methyl bromide or compounds which release the gas methyl isothiocyanate in the soil such as a Vorlex (20% methyl isothiocyanate in DD), Trapex (20% methyl isothiocyanate in xylol), dazomet and methamsodium. All are toxic to plants so they must be applied to the soil from one week to several months before planting. Methyl bromide applied under polythene sheeting to retain the gas acts quickly and soon disappears, so crops can be planted within days of treatment. This treatment is too expensive outdoors but is economic for glasshouse tomato crops. As methyl bromide gas is very poisonous strict safety precautions are necessary in applying it and there is the possibility that tomatoes and other produce planted later may contain appreciable bromine residues. Chloropicrin is effective but expensive and unpleasant to apply.

DD applied to cool, moist, well-drained sandy soils kills about 80% of the eggs within cysts in the topsoil but fewer in soil 2–4 cm deep. DD is less effective in clay or peat than in very sandy soil and its performance is influenced more by soil moisture than by soil temperature. Diffusion of the vapour is greatly decreased if the soil pores are blocked by water but is rapid and easily escapes from soil surface if the soil is too dry. Although almost insoluble in water, the right amount of moisture in soil helps to retain the vapour long enough to ensure an efficient kill.

Dazomet prill (a coarse powder) incorporated into the topsoil by rotary cultivation decreases the number of juveniles invading the roots by 98% or more. Most eggs are killed but some hatch later and may succeed in reproducing. Dazomet is effective in sandy soils and in soils containing appreciable clay but not always effective in peat soils. These may also retain the vapour for a long time and cause damage to the crop after planting. To obtain a better kill the fumigant may be applied in two doses, one before and one after ploughing, or a liquid fumigant may be injected deeply and followed by a granular nematicide like dazomet applied to the topsoil subsequently. Fumigants often cause some slight phytotoxicity which is masked by the growth improvement resulting from the control of nematodes. DD sometimes taints potato tubers and occasionally causes malformed ears in wheat.

Applying a fumigant to soil in the autumn creates difficulties. Before application the field must be worked to seedbed condition and all plant residues removed, and in this state many soils deteriorate over winter and must be ploughed or otherwise cultivated to attempt to provide proper structure for planting in the spring. For this reason non-phytotoxic granular materials are preferred which may be applied to the seedbed in the spring. Among the most effective are oxime carbamates such as aldicarb and oxamyl, which when thoroughly incorporated into the top soil before planting prevent injury to crops and also stop the nematodes multiplying so that at the end of the growing season fewer usually remain than before planting (Whitehead in Southey, 1978).

Integrated control

Two criteria can be used to assess whether control measures are successful or not. The first relies on yield increase and the second on only a small population remaining after harvest. A good yield increase can often be obtained with a kill that is inadequate to control the post-harvest population. A nematicide, a resistant variety or a short rotation all leave survivors that multiply greatly and bring populations to near peak numbers. In effect they aggravate the problem of control so that, for example, a nematicide used one year leaves an increased number of nematodes at the beginning of the next whereas growing potatoes continuously would produce populations that oscillated over a range well below the numbers following the use of a nematicide (and give a smaller yield) (see Fig. 15.9). How large a kill is needed to give diminished populations after taking a susceptible crop depends on the maximum reproductive rate the nematode can achieve. Thus the original population is restored at harvest after a kill of 80% by a reproductive rate of five times, 90% kill by a rate of 10 times and after a 99% kill by a rate of 100 times. To control a cyst-nematode like *G. rostochiensis*, that can multiply by perhaps as many as 70 times in the season, a kill approaching 99% is required. A standard dose of DD, a resistant potato variety grown for one year, or other crops grown for four years, all decrease the nematode population by no more than about 80%, leaving a residue of 20% of the numbers before planting. With this rate of kill a multiplication rate of only 5 times brings the populations to its pre-planting numbers. Combining any two of these methods gives a kill of 96%, a 4% residue and requires a reproductive rate of 25 times to restore preplanting numbers. If however all three methods are combined, the kill is 99.2%, the residue 0.8% and the balancing reproductive rate must be 125 times. For *G. rostochiensis* any two of these methods barely gives adequate control but all three combined require a balancing reproductive rate greater than the nematode can achieve, leaving fewer after harvest than before planting. Some nematicides, e.g. aldicarb, give kills approaching 99% and may give effective population control. The effect of integrating control measures in this way can be calculated from the following equation

$$\log_{10} F = \log_{10} I + n_1 \log_{10} \left(\frac{100-d}{100} \right) + n_2 \log_{10} \left(\frac{100-k}{100} \right) + n_3 \log_{10} \left(\frac{100-r}{100} \right)$$

where F is the final population at the end of the sequence of treatments, I is the starting population before they were applied, d is the average annual percentage decrease when crops other than potatoes are grown, n_1 the number of years for which they are grown, k is the percentage kill achieved by the nematicide and n_2 the number of times it is applied during the period, r is the percentage of eggs hatched by the resistant variety and n_3 the number of times the resistant variety is planted (see also Jones, 1969a, 1973b Whitehead, 1980). Effects of selection by the resistant variety are ignored in this calculation.

Root-knot nematodes

Meloidogyne spp. (Figs 11.24, 11.25)

M. naasi (Franklin)	Cereal root-knot nematode
M. hapla (Chitwood)	Northern root-knot nematode
M. artiellia (Franklin)	British root-knot nematode
M. incognita (K & W)	Southern root-knot nematode
M. javanica (Treub)	Javanese root-knot nematode
M. arenaria (Neal)	Pea-nut root-knot nematode

Meloidogyne naasi occurs in Europe and the U.S.A. It is widely scattered from south-west England through Wales and the Midlands to Yorkshire and Lancashire and seems to be increasing. In Belgium, *M. naasi* is a pest of sugar-beet. It attacks plants in six families including the Gramineae, Cruciferae, Leguminoseae and Chenopodiacea. In Britain wheat increases numbers most, barley is a good host, rye relatively poor and oats resistant. Italian and perennial ryegrass are good hosts. It is noticed most commonly after several barley crops in succession. Poor patches are most often observed in late sown crops in dry weather and galls are evident on the roots when plants are lifted.

M. hapla and **M. artiellia** are also native in Britain. The first occurs in widely scattered

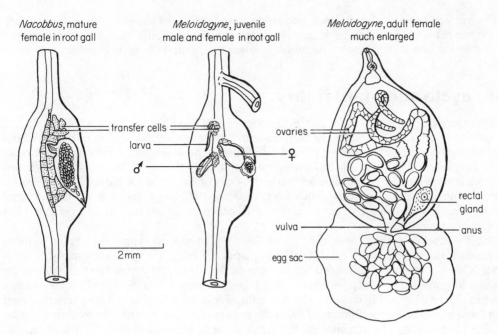

Fig. 11.24 Sedentary endoparasites, *Meloidogyne* spp. and *Nacobbus* spp.

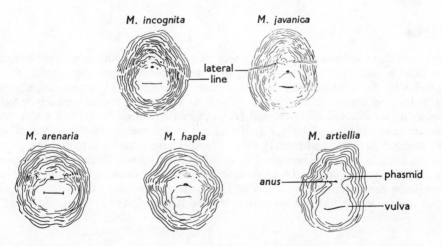

Fig. 11.25 Perianal patterns of some common *Meloidogyne* spp. (Courtesy M. T. Franklin.)

localities mostly on light soils but occasionally causes severe damage to crops such as carrots, mangolds, sugar beet, kale and scabious (*Scabiosa caucasica* Bieb.) in the Breckland area of Suffolk, to ornamentals in Norfolk, Suffolk and Kent, and to parsnips in Wiltshire.

The other species listed above are aliens, native to warmer countries where they and other species occur on coffee, tea, soya bean and other plants, and take the place of cyst-nematodes as root parasites. In Britain the alien species are found only in glasshouses, especially on tomatoes and cucumbers. When infested glasshouse soil is spread on fields, injury to crops of potatoes, beet, carrots, etc., may occur but rarely persists for more than a year, probably because soil temperatures are too low.

The various species of root-knot nematode have wide host ranges, although there are differences between species and evidence of variation between populations of the same species. For further information consult Franklin in Southey (1978) and Goodey, Franklin & Hooper (1965).

Life cycle and plant injury

The life cycle is similar to that of cyst-nematodes (Fig. 11.18) but the saccate females do not form cysts, all eggs are laid in the egg sac, and both female and egg sac tend to be buried within galls induced by feeding. The extent of root galling varies, however, and when small the females are mostly external like females of *Heterodera*. Apparently the sex of juveniles is determined environmentally and large multinucleate transfer cells (giant cells) are induced in host roots. Males of most species seem to be functionless (Santos, 1972). Eggs hatch freely in water and are only feebly stimulated by root exudates. Persistence in the absence of hosts is much less than in cyst-nematodes.

Heavily attacked plants are stunted and exhibit above ground symptoms similar to those described for cyst-nematodes (p. 228). When plants are lifted the galls are usually obvious (Plate XXXA). Juveniles, males and females can be demonstrated within the root systems by staining in osmic acid or lacto-phenol-cotton blue (Southey, 1984). Saccate females and egg sacs can also be dissected from galls. The root galls of *Nacobbus* have sometimes been confused with those of *Meloidogyne*. Species of *Meloidogyne* are distinguished mainly by their perianal patterns (Fig. 11.25). The juveniles are similar to those of *Heterodera* except in minor details. The third stage juvenile has a pointed not rounded tail. For other differences see Franklin in Southey (1978).

Control

Because of the many crops attacked, control of root-knot nematodes in warmer countries by crop rotation is difficult. Non-host crops such as Bermuda grass are helpful. Fallowing accompanied by frequent cultivation to desiccate the nematodes is sometimes practised. Because soil temperatures are higher and soils on which trouble occurs are usually light, soil injection with DD is often effective, and DD and methyl bromide, applied under gas tight sheeting, are valuable for seed beds, e.g. of tobacco, coffee and tea seedlings. Steaming, DD, metham sodium, dazomet and methyl isothiocyanate have been used to sterilize glasshouse soils. In the UK, *M. naasi* is controlled by crop rotation with oats, potatoes and brassicas. Rape is a relatively poor host and a single crop of oats or potatoes decreases numbers of eggs and juveniles by as much as 95% (York, 1980). There is some evidence of resistance in wild and cultivated barleys which has not yet been exploited (Cook & York, 1982). (See Leaflet No. 307.)

Enemies of nematodes

The groups of organisms containing enemies of nematodes are shown in Table 11.13. Apart from work on predacious fungi, knowledge of other groups is scanty.

Table 11.13 Enemies of plant parasitic nematodes.

	Predators	Parasites
Amoebae	Nematodes	Bacteria
Ciliates	Enchytraeids	Sporozoans
Tardigrades	Collembolans	Endoparasitic fungi
Turbellarians	Mites	Nematode capturing

Except for amoebae and some nematodes, all the predatory animals have larger cross sections than their prey and so would be unable to follow them through narrow pore necks in soil. In clay soils or those with stable aggregates and a good structure, roots, nematodes and their enemies tend to occupy the same soil spaces. There the enemies might be more effective. Gilmore (1970), comparing the numbers of collembola in soil with the number of nematodes, concluded that, at the rates of feeding observed in laboratory cultures, the collembola were potentially able to exert a controlling influence. However, in his cultures, the collembola had no choice of food and the nematodes no means of escape. In field soils, alternative food is abundant and most nematodes except sedentary or semi-sedentary forms can move through pores that collembola cannot penetrate. There is little concrete evidence as yet that predators exert an appreciable effect on soil populations.

Little is known of bacterial or protozoan parasites in Britain but attempts have been made to control cyst-nematodes by nematode-capturing fungi. About fifty species are known and some are widely distributed in field soils without exerting much controlling influence. Active nematodes are trapped by sticky hyphae or contractile rings, and their integuments pierced by trophic hyphae which ramify through their bodies absorbing the contents. Fungi have been added to infested soil as spores, cultures or mycelium, alone or with plant material and dung. The quantities of plant material added, dry or green, are usually unrealistic and the lessening of the population negligible (Jones, 1976a). Why naturally occurring predacious fungi are ineffective is unknown as also are the factors that would stimulate their activity (Cooke, 1962). It seems that they are not totally dependent on nematodes and perhaps use them only as a source of extra nitrogen.

Under continuous cereal culture, numbers of the cereal cyst-nematode, *Heterodera avenae*, decrease. This is largely due to two species of endoparasitic fungi, *Nematophthora gynophila* (Kerry and Crump) and *Verticillium chlamydosporium* (Goddard) (Kerry & Crump, 1982). Unlike the nematode-trapping fungi that attack second-stage juveniles that tend to remove individuals surplus to the carrying capacity of roots and exert little influence on multiplication, these attack females and destroy eggs or entirely prevent their formation. Since females are only a small fraction of the original juvenile population and egg production is essential to replace mortality at all stages, their destruction has a pronounced effect on population density. The spores of both species are common in fields where cereals are grown regularly. *N. gynophila* has motile zoospores which *V. chlamydosporium* does not; both require moisture for the infective process and possibly for this reason are less effective in well-drained light soils. These fungi also attack other *Heterodera* spp. but not *Globodera* spp. This may be why *H. schachtii* has spread less rapidly and been less troublesome in East Anglia in mineral soils where cereals are grown than on peat fen soils where root crops predominate.

Parthenogenesis

Parthenogenesis, unknown in mammals and rare in birds, is common in nematodes, mites and insects. It seems to be associated with the exploitation of temporary resources in habitats that are discontinuous in time and space: a characteristic of agricultural crops especially when not grown as monocultures. It is also associated with small size, short generation times, reduced mobility of females, polyploidy, polymorphism, neotony and paedogenesis. Sometimes it seems to be a solution to polyploidy which has produced an unbalanced chromosome set.

In ephemeral environments or restrictive ones like the soil, parthenogenesis has some advantages. Resources are not wasted on producing males, the uncertainty of mating when populations are sparse is avoided and a successful genome can be exploited with maximum speed from a single survivor. Parthenogenesis is said to be an evolutionary blind alley because gene flow is restricted. Whether this is entirely true is doubtful for some parthenogenetic animals have speciated.

It is possible that ephemeral and unstable habitats have always been available and that agriculture has merely increased their number. On this view species with reproductive adaptations that enable them to colonize such habitats have evolved in evolutionary rather than agricultural time. Many with unusual reproductive habits, including for example slugs and snails which are outbreeding hermaphrodites, have been and remain as successful as obligate bisexual species (Clark, 1973). Aphids, with their alternation of bisexual reproduction to pass the adverse conditions of winter and parthenogenetic reproduction during summer for rapid exploitation of summer plants, seem to have the best of all worlds (Clark & Lowe, 1973).

It is worth pointing out in passing, that although parthenogenesis is commoner in some genera than others (e.g. root-knot nematodes, *Meloidogyne* spp.) it often occurs as an exception in a related group of organisms which occupy a particular kind of niche (e.g. flowers, seed, growing points of stems or leaves, the root cortex). These genera are sometimes called affinity groups and are important in breeding for resistance (see Jones, 1979 and p. 302).

12

Birds and small mammals – Vertebrata

Since birds and mammals are warm blooded, their activity is not limited by low winter temperatures to the same extent as insects and other invertebrates. Therefore, they may cause much loss at all times of the year. Small mammals are active, mobile creatures and birds are amongst the most mobile creatures that exist. Their power of flight is far superior to that of insects, and is more purposeful and more independent of air currents. Birds and mammals possess well-developed senses; sight being especially acute in birds and smell in mammals. Because of the greater development of the central nervous system, behaviour is also more complicated, and is often varied in the light of experience.

Losses caused by birds and mammals have been the subject of much controversy. Attitudes towards warm-blooded pests may be coloured by sentimentality or social habits. Uncertainties can be resolved only by careful and objective study of behaviour, feeding habits and population densities, and by assessment of the crop damage caused. Local or national sentiment may sometimes hinder control even when the animal is clearly harmful, for public clamour is aroused by wholesale destruction or methods considered cruel. No such nice feelings are aroused by the destruction of invertebrates.

As pointed out in Chapter 1, farming has greatly modified the habitat of soil animals. Equally it has modified the habitat of above ground animals, including mammals and birds. The rabbit, for example, appears to be favoured by large areas of close-cropped grass and nutritious young cereals. In areas climatically suitable, it has tended to follow or even precede man into forest or bush (e.g. Australia, New Zealand). The woodpigeon may originally have been an inhabitant of woodland and hedgerows and these provide nesting sites from where its powerful flight enables it to reach out to pastures, leys, root crops, pulse crops which, except in hard winters when there is heavy snow cover, provide an abundance of food the year round. The accumulation of harvested produce in barns provides an abundance of food for rodents, and farm buildings and grain stores provide winter shelter. Birds like the sparrow have become largely dependent on man. Farms, villages and urban areas furnish them with shelter, nesting sites and food.

The birds and small mammals that are pests, like many invertebrate pests, appear to lack effective predators or other natural enemies that might limit their numbers. Although vertebrate predators, parasites and disease, sometimes limit population densities, usually they remove only animals surplus to the carrying capacity of the environment. This may vary from year to year, and in adverse years a shortage of prey may lessen the breeding success of predators. Man is able to exterminate large mammals but action against birds and small mammals usually succeeds in depressing numbers for a short time only. This has the effect of reducing competition and increasing breeding success, so that the population returns quickly to its former level. As with invertebrate pests, the chances of solving a pest problem once and for all are remote. Control measures must be continued indefinitely and aim at limiting population densities to acceptable levels (Thompson, 1963). The traditional methods of shooting and trapping often used against mammalian and avian pests merely effect minor local improvements which are soon nullified by immigration and breeding. Traditional methods coupled with cheap cartridge schemes or with a bounty payment for pelt, head or tail, are not, as a rule, very effective and are sometimes open to abuse. Organized and coordinated campaigns need careful planning and sustained leadership best supplied by a government-sponsored organization also involved in research on the habits, ecology and distribution of the animal in question. Only careful study of habits and of the animal population cycle can show when is the most effective period to apply control measures. It is tempting to shoot, trap or poison when numbers are high, but a more lasting effect may sometimes be obtained by measures taken when numbers are at their lowest annual level which is usually at the beginning of the breeding season.

The law, through the Poisons Act, 1972, regulates the types of poison, their handling and the species against which they can be used. Under the Animal (Cruel Poisons) Act, 1962, and the Animals (Cruel Poisons) Regulations, 1963, the use of poisons considered cruel may be prohibited by Orders in Council. Red squill and yellow phosphorous are two poisons prohibited in this way. A clause in the Protection of Animals (Amendment) Act, 1927 states: 'that it shall be a defence that poison was placed by the accused for the purpose of destroying insects and other invertebrates, rats, mice or other small ground vermin where such is found necessary in the interests of public health, agriculture ... and that he took all reasonable precautions to prevent injury thereby to dogs, cats, fowls or other domestic animals and wild birds'. The interpretation of the words 'small ground vermin' raises difficulties. Poison may not be used against rabbits because they are game, squirrels because they are arboreal or coypu because they are large. In effect, poison baits can be used only against rats, mice, voles, shrews and moles. However, the Agriculture (Miscellaneous Provisions) Act, 1972, section 19, gives power for an Order to be made authorizing the use of poison against grey squirrels and coypus. In the equivalent Scottish Act, only the word vermin occurs so that poisons may legally be used against a wider range of animals in Scotland. The Prevention of Damage by Rabbits Act, 1939, specifies that gassing rabbits does not contravene the Protection of Animals Act, 1911.

The Wildlife and Countryside Act, 1981 repealed the Protection of Birds Act, 1954, 1967 and the Conservation of Wild Creatures and Wild Plants Act, 1975 and amended several others, e.g. the Ground Game Act, 1948, the Protection of Animals (Scotland) Act, 1912, and the Criminal Procedure (Scotland) Act, 1975. Part I of Schedule 2 of the Wildlife and Countryside Act, 1981 lists birds that may be killed or taken outside the close season (roughly from February to March inclusive). This includes Canada, greylag, pink-footed and white-fronted geese, the last in England and Wales only. Part II of Schedule 2 contains birds which may be killed or taken at any time by authorized persons. This includes carrion crow, collared dove, feral pigeon, jackdaw, jay, magpie, rook, house sparrow, starling and woodpigeon. The landowner or occupier of farmland is an authorized person under the Act. All birds except those listed in Schedule 2 Parts I and II are fully protected throughout the year. (See Royal Society for the Protection of Birds booklet *Wild Birds and the Law*.)

The Wildlife and Countryside Act, 1981 also prohibited certain methods of killing or taking of wild birds (Section 5) and wild animals (Section 11). Section 5 includes almost every conceivable means of killing or capturing birds except shooting and cage trapping. It includes any electrical device for killing, stunning or frightening; bird lime, bows, crossbows, automatic and semi-automatic weapons, dazzling and target illuminating devices, sound recordings and various live decoys. Poisons, poisoned or stupefying baits are also banned. Section 11 is almost equally restrictive in the methods prohibited for use against small mammals. Again, except for traps and poisons permitted to be used against rats and mice and for gassing, ferreting, snares, dazzling and netting in the case of rabbits, only shooting and cage trapping are permitted for those animals which may be killed or taken by authorized persons or under certain licensed circumstances. Cage traps for birds (Plate XXXIIIA, B) or mammals must be inspected daily and any creatures caught humanely killed, usually by shooting.

Altogether more than twenty Acts affect the destruction of birds and small mammals, directly or indirectly, as well as an almost equal number of Orders, Regulations and Statutory Instruments made under the various Acts. Any control measures devised must fit within the framework of this mass of legislation, a partial list of which appears below.

Some legislation related to harmful mammals and birds

Game Act, 1831
Hares Act, 1848
Game Licences Act, 1860
Gun Licence Act, 1870
Ground Game Act, 1880
Hares Preservation Act, 1892
Ground Game (Amendment) Act, 1906
Protection of Animals Act, 1911-1927
Protection of Animals (Amendment) Act, 1927
Destructive Imported Animals Act, 1932
   Grey Squirrel (Prohibition of Importation and Keeping) Order, 1937
   The Non-indigenous Rabbit (Prohibition of Importation and Keeping) Order, 1954
   The Coypu (Importation and Keeping) Order, 1962
   The Coypu (Keeping) Order, 1982
Firearms Act, 1937, 1968
Prevention of Damage by Rabbits Act, 1939
The Agriculture Act, 1947 (Sections 98-101, 106-107)
Prevention of Damage by Pests Act, 1949
Pests Act, 1954
   The Spring Traps Approval Order, 1957
   The Small Ground Vermin Traps Order, 1968
   The Spring Traps Approval (Amendment) Order, 1966, 1968, 1970
Game Laws (Amendment) Act, 1960
Noise Abatement Act, 1960
Public Health Act, 1961 (Section 74)
Air Guns and Shot Guns Act, 1962
Animals (Cruel Poisons) Act, 1962
Animals (Cruel Poisons) Regulations, 1963
Animals (Restriction of Importation) Act, 1964
Hares (Control of Importation) Order, 1965
Theft Act, 1968
Agriculture (Miscellaneous Provisions) Act, 1972
Poisons Act, 1972
Wildlife and Countryside Act, 1981

Poison baits are used against rabbits in New Zealand, apparently without harmful effects on stock or wildlife. Before poison baits can be used in Britain against harmful birds or mammals other than rats, mice and moles, it will be necessary to find: (1) poisons relatively specific to the undesirable species; (2) bait bases highly attractive to them and not to other species; (3) a method of using bait and poison which excludes harm to humans, domestic animals and wildlife; or (4) a combination of all three. Poisons used against harmful birds and mammals are always likely to involve an element of risk to man, domestic animals and wildlife, and the risk has to be balanced against the harm done and the political pressures that both evoke. Studies of animal and bird behaviour are being made as well as studies of sensory physiology which may bear on the problem. Birds, despite the relatively few taste buds located at the base of the tongue, nevertheless seem to possess an acute sense of taste. They respond to substances in solution and, by suitable additives, it is possible to prevent them drinking but impossible to prevent them feeding. They appear to select food by sight and by using tactile sense located in the beak and tongue. In the U.S.A., in places where seed is broadcast on a dry surface, bird repellents have met with success, but as yet, little success has been had with bird repellents in Britain (Wright *et al.*, 1980).

Class Mammalia

The main external features of mammals are:
(1) The covering of hair which conserves body heat.
(2) The mammary glands of the females which produce milk for the young.
(3) The short mobile neck between head and trunk.
(4) The ears with external lobes or pinnae.
(5) The differentiation of the teeth into incisors, canines, premolars and molars.
(6) The separate anus and urinogenital openings, the penis of the male and vagina of the female.
For internal structure consult a textbook of zoology and for natural history Matthews (1968).

Order Rodentia

Rodents are numerically the largest and most widely-distributed order of mammals. They are placental, mostly small or very small, and characterized by a single pair of growing, incisor teeth in both jaws with enamel confined to the front surfaces. Canine teeth are absent, the premolars and molars may be rooted or continuously growing, and the cusps of the upper molars may wear down to form zigzag ridges. Members of unrelated groups have frequently become adapted to similar habitats, and this has led to superficial similarities in structure and appearance. This group contains rats, mice and voles (Muridae), squirrels (Sciuridae) and coypu (Capromyidae).

Order Lagomorpha

This order contains hares and rabbits (Leporidae) characterized by two pairs of incisor teeth in the upper jaw, with the smaller second pair situated behind the first. Enamel extends right round the tooth surfaces. Hares and rabbits also have short, recurved tails, long ears and elongated hind limbs.

Order Insectivora

The Insectivora is another group of small, placental mammals. Nearly all members are nocturnal and terrestrial, many are fossorial or arboreal and a few are aquatic. The premolar and molar teeth bear sharp cusps adapted primarily for feeding on insects. Like the rodents, insectivores are very prolific. Moles (Talpidae) and shrews (Soricidae) are the only British representatives of the superfamily (Soricoidea) and are closely related. The cusps on the molar teeth form a W-pattern. Many species of shrew have scent glands on the sides of the body behind the armpits. Their secretions are believed to be repugnant and few carnivorous animals molest them. Hedgehogs are placed in a separate family (Erinaceidae) and superfamily (Erinacoidea). This group is in many ways the least specialized of living insectivores.

Order Carnivora

This order includes flesh-eating, placental mammals with long piercing canine teeth for tearing flesh and toes provided with sharp claws for seizing prey. The tail is usually long and the face adorned with vibrissae. There are about 300 species varying greatly in size and not all are flesh feeders. The Order falls into three groups: the cat-like forms (Feloidea); the dog-like forms (Canoidea); and seal-like forms. The Canoidea includes foxes (Canidae), stoats, weasels, wild mink and badgers (Mustelidae).

Rats

Rattus rattus **(L.)** The roof rat.
Rattus norvegicus **(Berk.)** The Norwegian rat.

Both species of rat, like the rabbit, were introduced into Britain, the roof rat is now thought to have reached Britain before the Crusades, possibly in Roman times, and the Norwegian rat from Central Asia in the 1720s. The differences between them are shown in Table 12.1 (Plate XXXIC, D). The Norwegian rat, a burrower, at home in drains and sewers, has probably partly ousted the roof rat which is primarily a climber, but colonies of the latter exist at several sea ports. The roof rat does not appear to be able to survive outdoors in Britain but in the laboratory the heavier male survives 779 ± 225 days and the female 837 ± 227 days (Bentley & Taylor, 1965) and its numbers have recently declined, possibly because of improved methods of poisoning, improvements in the construction of buildings and the reduction of infestation in ocean-going ships (Bentley, 1964). Rats carry bubonic plague, the vector of which is the rat flea, *Xenopsylla cheopis* (Roths.) and Leptospiral jaundice (Weil's disease). Human infection occurs from wet, rat-infested surfaces and is commonest among farm labourers and sewer men. Bacteria of the *Salmonella* group which cause food poisoning are carried in excreta of rats and mice. Rats are omnivorous and cause damage to stored products such as grain and flour, wasting much by holing sacks and containers and by contaminating more with faeces. Rats also damage buildings, embankments, drains and sewers, and may gnaw through lead piping and electric cables. The population of Norwegian rats in Britain has been stated as approximately equal to the human population (50 000 000), but numbers are not accurately known.

Table 12.1 To show the structure of *R. rattus* and *R. norvegicus* after R. A. Davis (1970).

Character	R. norvegicus	R. rattus
Colour	Brownish grey or black, grey belly	Brownish grey, black. May have white, cream or grey belly
Average adult weight	250 g	Less than 250 g
Tail length	Shorter than head and body combined	As long as or longer than head and body
Ears	Thick, opaque, hairy	Thin, translucent, large, almost hairless
Snout	Blunt	Pointed
Droppings	In groups, sometimes scattered, spindle shaped	Scattered, sausage or banana shaped
Habits	Burrowing, can climb, swims, shows new object reaction. Lives indoors, outdoors and in sewers	Non-burrowing, agile climber, shows new object reaction, lives indoors
Distribution	Everywhere, town and country	Almost exclusively in ports
Litters	4–5 a year	5–7 a year

Populations pass through periods of abundance and scarcity. Norwegian rats are hardy and many live throughout the year remote from human habitations in hedgerows, coppices, and patches of waste land finding what food they can. These outlying colonies often serve as reservoirs from which re-infestation occurs after clearance.

On a Hampshire farm studied by D. C. Drummond (Anon, 1962), numbers of Norwegian rats were fewest and distribution least scattered in early summer. At this time the rats were feeding mainly on grass leaves, especially cereals. From July to October, grain was plentiful and rats spread around cornfields and stubbles. Numbers increased to a peak in October, but thereafter grain in fields became increasingly scarce and rats turned to turnips and corn ricks, most having started to feed on ricks or to move into or near farm buildings by December. From the October peak, numbers fell in fields and continued to do so until the following summer, although there was plenty of spring-sown grain for food between February and May. On most farms rats are found near obvious sources of food in winter, that is in or near houses, buildings, poultry runs, corn or straw stacks, root clamps and around the margins of kale and root fields. Plate XXXIIF illustrates feeding on winter wheat seedlings.

Mice and voles

Mus musculus **(L.)** The house or domestic mouse like the rabbit and rat is an alien, thought to have come from Asia some thousands of years ago (Plate XXXIE). Damage is of the same type as that caused by rats. The house mouse weighs about 15 g, is 7.6–10.2 cm long, including tail, has a pointed snout, black beady eyes, long sensitive brown ears and soft brownish-grey fur. It is easily distinguished from the young of the Norwegian rat by the tail which is as long as the head and body. The underparts are paler than the upper but there is much colour variation. Mice are active and are good climbers. They breed rapidly having 4–10 litters of 5–6 young a year depending on the environment. The young are born naked and blind but develop quickly. Small populations occur in fields at all times but large ones develop in buildings and anywhere where there is much food and shelter. Under favourable circumstances the population doubles in 2 months, trebles in 4 months and quadruples in 6 months.

Apodemus sylvaticus **(L.)** The wood or long-tailed field mouse is easily distinguished from the house mouse by its brown colour with light underparts, including the underside of the tail, prominent eyes and ears, and longer limbs (Plate XXXIF). When fully grown, it is 8.8–11.5 cm long, nocturnal, active and able to climb, jump and swim. It makes underground burrows with several openings and stores food for the winter when it is less active. Two or three litters of 5–9 young are born each year. Field mice are short-lived and few adults breed in more than one season. Although sometimes found in buildings, *A. sylvaticus* is equally at home and about equally as numerous in woodland, hedgerows or open fields and seems to be the only small mammal able to survive the upsets to its nesting sites that result from cultivation. Woodmice have no marked preferences for fields with different crops in winter when their main food is grain, waste sugar-beets, weed seeds and soil invertebrates. Numbers vary from around 20 per ha in late autumn to 1 or fewer in May–June but numbers vary greatly from place to place and year to year. Individuals have a range of as much as 150 m. Field mice are able to locate seeds and other food by smell. Newly sown sugar-beet is especially vulnerable if early sown, the soil is dry and some pellets are not properly covered. Once alerted by exposed pelleted seed, mice dig out seeds accurately up to 4 cm deep over lengths of row leaving tell-tale pits and pellet fragment. Feeding ceases when the seed germinates (Green, 1979). (See Leaflet No. 626.) Enemies include owl, fox, weasel, hedgehog and viper.

The slightly larger yellow-necked field mouse, *A. flavicollis* (Melchior), is common in the West Country but occurs locally all over England.

Microtus agrestis **(L.)** The short-tailed vole is a grassland animal, preferring open country meadows, parkland and young plantations to woodland. It is predominantly vegetarian and feeds on succulent grasses, little piles of which are taken into the domed chambers in the runs and tunnels which it builds beneath the grass roots.

Voles are distinguished from mice by short blunt snouts, rounded heads and short ears. The short-tailed vole is about 10 cm long when adult, has ears so short that they are almost hidden in the fur, a greyish-brown pelt, six pads on the undersurface of the foot and a tail about one-third the length of head and body. This species is diurnal and, although rarely seen, is probably the commonest of small British mammals, serving as an important item of diet for many predators, especially the short-eared owl which congregates where voles are plentiful.

Clethrionomys glareolus **(Schreb.)** The bank vole prefers banks, hedgerows and coppices. Like the short-tailed vole, it makes shallow runs. Nests are constructed of chopped, dry grass, sometimes lined with moss, wool or feathers.

The bank vole is 8.2 cm long when fully grown, reddish above with white, light grey or buff underparts and grey feet. Its ears are rather more obvious than those of the short-tailed vole. It is also more omnivorous and rather more active than the short-tailed vole. At times it climbs small trees to feed on seeds, fruits and nuts, and is a good swimmer and diver.

Arvicola amphibius **(L.)** The water vole or water rat, is 15–20 cm long when fully grown with a tail 12 cm long (Plate XXXIB). The pelt is usually brown but sometimes black and the underparts are brownish-grey. The water vole rests in burrows in the banks of streams, rivers and ponds, but sometimes lives away from water in fields, hedgerows and gardens. Most of its time is spent on land where it is active mainly by night. It is a good diver and swimmer. Although its diet is mainly vegetarian including water-plants, dandelions, grasses and bark, fresh water molluscs are also eaten.

Damage by field mice and voles

Mice and voles are mostly nocturnal and rarely seen. Field mice are said to graze winter wheat in autumn and spring causing some damage, but not on the same scale as rabbits. Field mice also dig up newly planted peas, feed on strawberries, blackberries and other soft fruits, nuts, bulbs, birds' eggs and honeycomb. Feeding on ungerminated sugar-beet seed may lead to patchy crops and appreciable field losses.

Field mice and voles enter glasshouses, probably by accident and injure plants there but may be excluded by proofing entrance points with 6.35 mm mesh wire netting. Voles may injure brassicas and anemones. The burrowing of water voles may cause considerable damage to river banks and dykes in the Fenland. The short-tailed vole sometimes gnaws the roots and bases of fruit trees causing spectacular damage and bringing about the death of trees. Mice and voles may bark trees near ground level, but the bank vole sometimes climbs higher and girdles saplings 1–2 m above ground level. All mice and voles eat tree seeds, but it is doubtful whether this is of economic significance. A feature of vole attacks on trees is a pronounced preference for one species in an area at a particular time. But the preference is strangely inconsistent, other species being preferred at different places and times. Beech is especially liable to attack but no kind of tree is immune (see Leaflet No. 49). Populations of mice and voles fluctuate rather widely. Veritable plagues of the short-tailed vole sometimes occur when numbers exceed 1250 per ha. On the Continent, plagues of a related species, *M. arvalis* (Pall.), are frequent, but vole plagues in Britain are rare. Peaks in local abundance are not synchronous from place to place, nor with peaks in populations of other small mammals; their causes are unknown, and the inevitable population crashes that follow cannot be explained. They are not thought to be due to disease, predators or lack of food but perhaps to the adverse physiological effects of overcrowding. When numbers rise, the preferred food supplies, mainly grass, are used up and they begin to feed on other plants.

Control of rats, mice and voles

A good rodenticide should be odourless, tasteless, specific to rodents, and should not induce tolerance or act so rapidly that the rodent perceives warning symptoms before a lethal dose has been eaten. No poisons meet all these requirements.

Of many traditionally used acute poisons only 2.5% zinc phosphide is considered nowadays to be suitable for the control of field rodents. It is termed an acute poison because it is lethal after a single dose and causes death within a relatively short time. A second class of poisons was introduced about 1950. These are termed chronic poisons because they are used at concentrations at which they kill only after several doses. The best known and most widely used of these is warfarin (3–(1–phenyl 2-acetylethyl)-4-hydroxy coumarin) employed at 0.025% against mice and the roof rat. This and related coumarin compounds, as well as several indanediones, are termed anticoagulants because they interfere with the mechanism of blood clotting and cause haemorrhages (Bentley & Larthe, 1959). Coumatetralyl (Greaves & Ayres, 1969) and chlorophacinone are two anticoagulants at present available. Warfarin can be obtained in four forms (1) ready to use baits, (2) water-soluble powders for poisoning drinking points, (3) water-soluble warfarin for absorption on baits and (4) poison dusts for laying on surface area over which rats and mice run. Other anticoagulants are prepared as ready-to-use baits and as mastermixes: coumatetralyl is also available as a poison dust. In the late 1970s three new, so-called 'second-generation' anticoagulants difenacoum, bromadiolone, and brodifacoum were introduced. These compounds are generally effective against strains of rodents that have become resistant to the older anticoagulants but being more toxic, they present an increased risk of poisoning for non-target animals. Baits that may be used as carriers for acute poisons include soaked wheat, damp coarse oatmeal, damp sausage rusk, damp bread rusk, cooked kibbled maize, and damp bread mash. Suitable warfarin baits for use against rats and mice are listed in Table 12.2.

Rats generally avoid strange objects and may take time to become accustomed even to a new pile of bait. At first, therefore, if given an acute poison bait immediately, they might take insufficient of it to be killed but enough to put them off feeding on the same bait or poison for several months. This 'bait shyness' is overcome by 'prebaiting'. Unpoisoned bait is supplied

Table 12.2 *Warfarin baits for rats and mice.

	Ingredient	Parts by weight	Per cent	Remarks
(1)	Pinhead oatmeal	16	85	Mix the dry ingredients first and then
	Castor sugar	2	15	add oil, paraffin or glycerine. Gly-
	0.5% warfarin mastermix	1	5	cerine produces the most attractive
	Technical white oil, liquid paraffin B.P., or refined, straw-coloured glycerine	1	5	bait. Gives 0.025% warfarin in final mix.
(2)	Medium oatmeal	19	95	Not as acceptable as bait (1).
	0.5% warfarin mastermix	1	5	Gives 0.025% warfarin in final mix.
(3)	Soaked wheat	19	95	An alternative to (1) and (2) under dry conditions.
	0.5% warfarin mastermix	1	5	Soak the wheat overnight and drain before adding warfarin. The most palatable bait. Gives 0.025% warfarin in final mix.

*Other anticoagulant rodenticides can be substituted for warfarin.

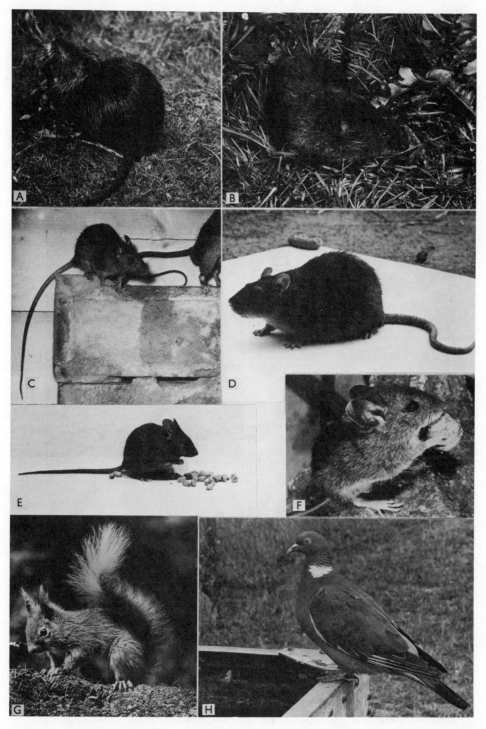

XXXI **A.** Coypu (Crown copyright).   **B.** Water vole (Crown copyright).   **C.** Roof rat (Crown copyright).   **D.** Norwegian rat (Crown copyright).   **E.** House mouse (Crown copyright).   **F.** Long-tailed field mouse (courtesy H. C. Woodville).   **G.** A red squirrell, *Sciurus vulgaris leucourus* (Kerr) (copyright M. S. Wood).   **H.** A woodpigeon, *Columba palumbus* (L.) (courtesy R. K. Murton).

XXXII A. Effect of rabbit grazing on winter wheat before myxomatosis epizootic; right, rabbits excluded. **B.** Left, oat plant grazed by rabbits; right, ungrazed. **C.** Kale, gnawed by rabbits. **D.** End of
cereal leaf grazed by a rabbit. **E.** Coypu damage to sugar-beet alongside a drain. **F.** Rat damage to
winter wheat seedlings (all Crown copyright). **G.** Woodpigeon damage to peas (courtesy F. D. Cowland,
Rothamsted). **H.** Woodpigeon damage to Brussels sprout (courtesy F. D. Cowland, Rothamsted).

XXXIII **A.** One of a group of Chardonneret bullfinch traps set in a bramble thicket on the edge of an orchard (Crown copyright). **B.** Rook, jackdaw and crow trap of the 'sheep netting' type with fine-mesh wire netting around the sides. In this version the netting top is replacing by slats. The entrance hole is on the top (Crown copyright). **C–E.** Cabbage root fly and cauliflowers (courtesy National Vegetable Research Station). **C.** Note partial stand in untreated plot in foreground. **D.** Plots with insecticidal barriers eliminating ground predators. Note the much heavier attack. **E.** Plots treated with insecticide eliminating cabbage root fly and predators. These pictures stress the importance of ground predators. **F.** *Aphelenchus avenae* feeding on mushroom mycelium (courtesy C. C. Doncaster).

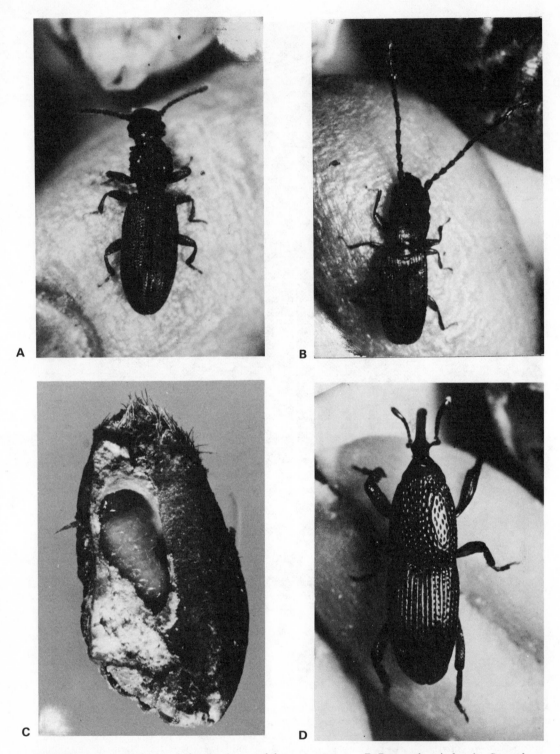

XXXIV **A.** Saw-toothed grain beetle, *Oryzaephilus surinamensis.* **B.** Rust-red grain beetle, *Cryptolestes ferrugineus.* **C.** Grain weevil larva in wheat grain. **D.** Grain weevil, *Sitophilus granarius* (all Crown copyright).

until rodents are feeding freely after which poisoned bait is supplied. Prebaiting is not essential for chronic poisons such as warfarin.

In dealing with rat infestations, the locality concerned is first reconnoitred to delimit the area and to find the best places for baiting. Rats often leave tracks running through grass or across arable land. Other signs of activity are holes, footprints, tail marks and droppings. Greasy marks on surfaces, absence of dust on ledges, and evidence of feeding, all disclose the whereabouts of rats indoors. Bait is laid in holes and along runs and where necessary is placed in baiting boxes, land-drain tiles blocked at one end, or under improvised covers to protect it from weather and against farm stock, or children. This usually delays consumption until the rats have become familiar with the objects used.

Domestic mice eat smaller amounts of food, often from several places, and live only a few feet from or even within sight of their food. Their behaviour is more variable than that of rats and they investigate new objects, so that in treating food stores with warfarin, several small baits of about 25 g need to be placed at intervals of 1–2 m over the whole of the area where mice are present. For safety, covers are often necessary and these may or may not enhance the efficiency of treatment.

Where rats and domestic mice occur together, treat first as for rats and continue the treatment for mice if any remain after rats have been killed.

When using warfarin or other anticoagulants, the baiting points should be visited and replenished every two or three days until feeding stops, which may take a fortnight or more. Rat holes should then be stopped and a few days later a search made for freshly opened holes, droppings, feeding or other signs of activity. If survivors remain, repeat the operation using a different bait base both in new and in the old positions.

Rats have become resistant to warfarin in several districts (Greaves & Ayres, 1967). Resistant rats were first found on a farm in Scotland in 1958 and have since spread widely in Scotland. In 1960 resistant rats were discovered on two farms in Powys and have now colonized parts of adjacent counties. In 1968–69 pockets of resistant rats were found in Kent and Hampshire and are now well established in these counties. In other smaller outbreaks the rats were killed by acute poisons. In Powys there is a genetical link between resistance and colour, and resistance appears to be a dominant trait. In 1966 attempts to contain resistant rats by trying to eradicate them from a 5 km perimeter with acute poisons was unsuccessful but the spread of resistant rats to outside areas was slow. Resistant rats require more vitamin K in their diet than susceptible ones and in the field there may not always be an adequate supply (Greaves, 1970). Now that resistance has appeared alternatives to warfarin are being actively sought. Norbormide is selectively toxic (Rennison, Hammond & Jones, 1968) to rats but causes 'bait-shyness'.

The Prevention of Damage by Pests Act, 1949, makes it obligatory for any non-agricultural occupier to notify the local authority if substantial numbers of rats or mice are noticed on his land. The local authority may require any owner or occupier to deal with an infestation and may itself deal with it at the owner or occupier's expense. For further details of rat and mouse control consult Drummond (1970) and Davis (1970).

Field mice and voles can be controlled by anticoagulant baits using baiting boxes or other suitable covers to protect birds and wild life, and to shelter the bait from the weather. Trapping with ordinary break-back traps is also possible but laborious and rarely used.

British Sugar Corporation field staff assess field mouse populations annually in late February–early March. However, numbers fluctuate greatly and in 1980 ranged from 0.1 to 26.1 per hectare. Nevertheless, large numbers in a given year alert farmers in areas where damage to sugar-beet has been experienced. Crops can then be examined in the first few days after drilling. It is far too late to take action once the seedlings begin to emerge because by then the damage has been done.

In practice local plagues of voles or field mice are difficult to counter. Just when the situation seems desperate, numbers decline suddenly and damage ceases.

Squirrels

Sciurus carolinensis **(Gmel.)** American grey squirrel
Sciurus vulgaris leucourus **(Kerr.)** Red squirrel (Plate XXXIG)

The red squirrel has red-brown fur, a long plumed tail, tufted ears and measures 39 cm, 18 cm of this being tail. The eyes are prominent and bright, and the ears pointed. The forefeet have four digits and a rudimentary thumb, the hindfeet have five digits all with long, curved, pointed claws, and hairy soles, enabling them to climb and grip. The hind limbs are longer than the fore limbs which hold nuts and food for the teeth to work on. The tail is held out behind and acts as a rudder when jumping from branch to branch. The grey squirrel differs in having grey fur and no tufts on the ears, and is more vigorous and adaptable than the red.

Squirrels are diurnal; the red species is less often seen on the ground and seeks more shelter than the grey. Both build nests or dreys in trees; these are bulky structures composed of twigs, strips of bark, moss and leaves. The actual breeding nest is a ball-like structure in the dreys or else inside a hollow tree. Litters of three to four blind, naked young are born in spring and again in summer.

The grey squirrel was introduced into Britain from North America in the nineteenth century. It spread slowly at first (Fig. 12.1), more rapidly from 1937–1952, and more slowly thereafter, a pattern similar to the sigmoid infestation-curve typical of a species colonizing an unoccupied terrain or one replacing another with which it competes successfully. Figure 12.2 shows the distribution of red and grey squirrels in England and Wales. Red squirrels are commonest in areas not yet colonized by grey squirrels and in areas only recently occupied by them. Squirrels are absent or rare in the fenland areas around the Wash and possibly also in parts of Lancashire. As the grey squirrel has advanced so the red squirrel has retreated and is becoming increasingly rare.

Squirrels feed chiefly on vegetable material such as seeds, nuts, fruit, young shoots and buds, but birds' eggs and nestlings are also taken. Red squirrels are destructive pests in plantations of conifers for, besides feeding on the pine cones and their seeds, they strip the bark off leading shoots thereby deforming them and ruining them for production of straight timber. Many different trees are attacked by grey squirrels, but beech and sycamore suffer most. Damage frequently occurs to tree butts that are 20–40 years old and 10–30 cm in diameter. The bark and cambium are gnawed away until the butt is completely girdled and the tree dies. The buttress roots of older trees may be attacked. Damage is also done higher up the bole at some convenient side branch that serves for standing. Damage to butts is obvious but that to main

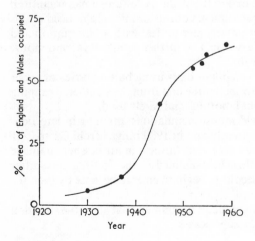

Fig. 12.1 The spread of the grey squirrel in England and Wales. (After Lloyd, 1962.)

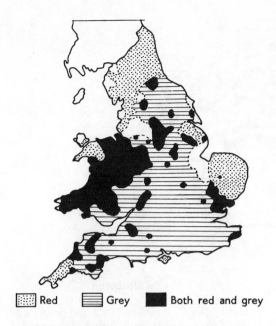

Red   Grey   Both red and grey

Fig. 12.2 Distribution of red and grey squirrels in 1959. (After Lloyd, 1962.)

stem and branches higher up often passes unnoticed. This damage lets in fungi and greatly lowers the value of the timber. Damage to agricultural and horticultural crops may also be severe locally, and is usually related to nearby wood or parkland. Altogether the damage done by squirrels is of minor importance compared with that done by rabbits and woodpigeons.

The grey squirrel is well established and unlikely to be eradicated. For this reason there seems no point in indiscriminate killing which has little effect on population levels. Organized control measures are essential only in places such as forestry plantations, where serious damage is caused. Squirrels may be shot when trees are leafless and dreys can be poked out with long poles. The best results, however, have been had by trapping just before or during the main period of damage from May to August. Woods cleared in winter are usually recolonized by early spring. Baited cage traps are most successful, but humane spring traps set in tunnels with or without bait are also used (Taylor, 1963). Warfarin baits may now be used in counties with no viable populations of red squirrels.

The coypu

Myocastor coypus **(Molina)**   The coypu (Plate XXXIA) is a large rodent measuring up to 60 cm and weighing up to 11 kg and has orange incisors when mature. It is an active swimmer with webbed hind feet and much of its waking time is spent in the water, but large shallow burrows are made in river banks and dyke sides. Road sides and road bridges in marshy areas may be undermined. The burrows of coypus are less damaging to river banks than those of the musk rat, earlier found in Britain, but now believed to have been exterminated.

Coypus are mainly nocturnal. They raid fields on sugar-beet, kale and cereals, and cause much damage when they are numerous (Plate XXXIID, E). They also gnaw the roots of trees but their main food is aquatic vegetation and molluscs including the fresh water mussel (*Anodonta cygnea* (L.)). After having their litters, females are thought to disperse. In cold wintry weather more animals are seen in daytime than normally. Populations are usually assessed by trapping and, by signs of activity such as recently used runs, fresh droppings, grazed vegetation or damaged crops.

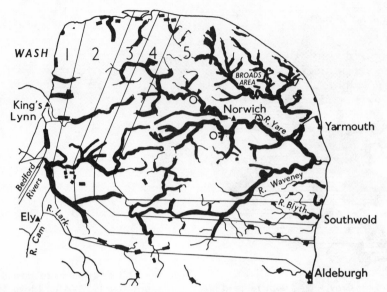

Fig. 12.3 The distribution of coypu in East Anglia, 1944–59 (after R. A. Davis, 1963), and progress of the campaign against them up to March 1963; ○—sites of former coypu farms. Area 1, outer area of scattered infestations first cleared; 2–4 areas successively surveyed, trapped and cleared; 5, trapping by the now disbanded rabbit clearance societies, river boards and individual occupiers. The main infested area now extends as far south as Colchester.

The coypu was introduced into Britain from South America for its fur about 1929. Escaped animals had established themselves firmly in East Anglia by 1944 and were well entrenched in the marshy areas of the Broads. Since then they have spread along East Anglian rivers to the Fens, and if left alone, they might well colonize the many Fenland waterways (Fig. 12.3).

A trapping campaign was launched in 1962 to exterminate the coypu from the greater part of East Anglia by working systematically in belts 6.4 km wide stretching in arcs from the Wash to the Suffolk coast. The operation was coordinated from a Coypu Control Centre in Norwich and was carried out by Ministry of Agriculture Field Officers with the help of Rabbit Clearance Societies (now disbanded) and River Boards. The aim of the campaign was to confine the coypu to the Broads where extermination may not be possible because the terrain favours the animals. The hard winter of 1963 killed many coypus and hindered the campaign by dispersing the survivors. Trapping in subsequent years was effective until a run of mild winters allowed the coypu to re-establish itself and the campaign had to be intensified. The hard winter of 1981–82 again depleted numbers. In 1981 another ten-year eradication campaign was begun. Coypu are not now trapped in Britain but caught alive and shot. The Coypus (Keeping) Orders, 1962 and 1982, prohibit the keeping of coypus except under licence and their presence on land is notifiable (see Leaflet No. 479). They are about as susceptible to warfarin as *Rattus rattus* but more resistant than *R. norvegicus*. In Russia, coypu are cropped for fur chiefly. When populations are sparser than 10–25 per ha, little trouble is caused. The animals congregate on rafts supplied with dry grain and are trapped or killed there.

Rabbit

Oryctolagus cuniculus **(L.)** The structure of rabbits is well known. The external ears are large and laterally placed. The fur is dense, grey or grey brown and the longer hairs are ringed with black. The coat thickens in winter. The forelegs are shorter than the hind and there is a coating of hair on the soles of the feet. The tail is short, upturned and lined with white. The forepaws are used for burrowing, the loosened soil being thrown back between the hind feet. Rabbits feed on many plants, especially grasses.

The rabbit originated in the western Mediterranean area and was probably brought to Britain by the Normans. Introduced from Europe to Australia and New Zealand, it thrived and became a serious pest because of an abundance of food and a lack of effective enemies. Rabbits used to be the most conspicuous small mammals in Britain, the peak population at the end of the breeding season being of the order of 100 million, but numbers fell sharply after the myxomatosis epizootics of 1954–56.

Rabbits usually live gregariously in warrens composed of many burrows made chiefly by the does. Warrens may be in banks, overgrown hedgerows, copses and shelter belts on sandy soil. Burrows are about 15 cm in diameter and intercommunicate with each other. Although some rabbits have always lived above ground in dense undergrowth, relatively more have done so since the myxomatosis epizootic.

The breeding season begins in January and ends in June but there is sporadic breeding in all other months of the year. Many litters are resorbed within 14 days of being conceived, but the remainder go to full term at 28–30 days. Males are fecund long before the onset of the main breeding season which begins sharply so that a large proportion of does become pregnant almost simultaneously. Does also become pregnant again immediately after producing litters or after their resorption, and for the duration of the breeding season about 90% of does are pregnant at any time.

The does make end-chambers or stops lined with dried grass, dried leaves and fur from their bodies in which the young are born blind, deaf, naked and helpless. The number of young born per doe varies, but average figures would be four or five per litter and 20 to 30 per year. For the first fortnight after the birth of a litter, the stop is closed or covered with earth and difficult to find. After about 18 days the young emerge but remain close to the stop at first. Although occasional individuals may wander over a mile, the movements of rabbits in a colony are rather restricted. It is doubtful whether they habitually range over an area greater than 20 ha, and the usual feeding movements are much less. Adult does tend to have their own particular patch of earth astride a burrow which they defend aggressively from other females, young rabbits and sometimes bucks. The social organization of the warren during the breeding season is based on the territorial behaviour of the does and the dominance of a few bucks. Young males tend to emigrate from the warrens in which they are born and, of those that remain, only a few survive amongst the older animals.

Two curious features of rabbit behaviour are enurination and reingestion. Enurination is a manifestation of sexual behaviour. The buck may turn his hindquarters towards the doe and eject a jet of urine backwards, but usually the act is preceded by some form of parade, circling, tail flagging or chasing. Reingestion is the habit of eating the faecal pellets. Rabbits feed mainly at dusk and at night and reingestion occurs while they are lying up during the day and is thought to be equivalent to chewing the cud in ruminants. The pellets consumed are soft, mucous-coated and quite different from the usual hard, round type.

The influence of prolonged drought on the rabbit population of the island of Skokholm was studied in the summer of 1959 by Lloyd (Anon, 1962). It greatly lessened the food supply of the 6 000 to 10 000 rabbits living there on 100 ha of land, so that in March 1960 only 103 rabbits could be counted. By October 1960 the population had risen to about 1000 and by October 1961 to 4000. Before the population crash, the breeding season was short, lasting only three months, there were few pregnancies per doe, few live young per doe, and the net rate of increase was nearly zero. After the crash, in 1960, the onset of breeding was delayed, but a second short season occured in September and October; increase was sevenfold. In 1961, the breeding season lasted four months, the number of pregnancies averaged four per doe, giving about twenty-two young per doe, but increase was only threefold. Although the Skokholm population is atypical because it lacks predators, the above observations illustrate the resilience of a rabbit population to variation in its food supply. As in other animals, the reproductive rate rose when the population density was low and declined as it increased. According to Lack (1954), the rabbit is also one of those animals adapted to withstand natural enemies by having effective means of hiding or escaping, and also by reproducing rapidly, so that when pressure is relaxed it soon becomes numerous again.

Dense colonies of rabbits deplete the food supply of their grazing grounds and greatly affect the vegetation. Other density-dependent effects that operate within the warrens are physical and physiological stresses from competition and fighting (Lloyd, 1970a). These factors and disease set upper limits to population density but usually not before the nearby vegetation or crops are severely affected. In some districts cereals, especially winter wheat, is almost the only winter food available. Before 1952, winter cereal crops were so severely grazed that some were total failures (Plate XXXIIA–D). Grazed crops tiller poorly, ripen late and yield badly. In addition much grass that should have been feeding sheep or cattle was consumed. Observations by Gough & Dunnett (1950) and surveys by Church, Jacob & Thompson (1953) and Church, Westmacott & Jacob (1956), showed that an average of 75 kg of grain was lost per acre over England and Wales and that damage was concentrated mainly on about one-third of the acreage. Probably an equal amount of damage was done to grasslands. Farmers were often surprised and impressed by the evidence of rabbit damage produced by wiring off portions of fields. Damage became negligible after populations were depleted by myxomatosis. On some farms the growth of grass became an embarrassment for a time because there was insufficient stock to consume it. Populations remained small for a while, but have since begun to recover and rabbit damage is on the increase again.

Control of rabbits

Much more attention has been given to the control of rabbits in Australia and New Zealand than in Great Britain and the European Continent. Attempts to use myxoma virus for biological control were unsuccessful at first, but in 1950, partly as a result of unusual rainfall, which appears to have provided favourable conditions for the breeding of mosquito vectors, an epizootic spread with surprising rapidity. Relatively wet seasons which favour annual epizootics followed until 1957, when drought intervened. Since introduction, two other factors have also greatly reduced the effectiveness of the virus; the appearance of attenuated strains and natural selection of the rabbit population. Virulent virus causes lesions about the head of susceptible, diseased rabbits in which there are dense concentrations of virus particles. Death follows in 10 to 12 days with mortality well over 99%. Attenuated virus causes smaller and more persistent lesions, and mortality is not so great. From the nature of the disease and its method of spread, moderately attenuated viruses seem likely to be more infectious than virulent or weak strains and appear to supplant virulent strains within a matter of months. They have been recovered from rabbit populations towards the end of epizootic initiated by virulent strains of virus. The effects of selection on rabbit populations in Australia have become evident and rabbits are now more resistant. Whereas a somewhat attenuated strain killed over 90% of susceptible rabbits caught in one area soon after the introduction of myxomatosis the same strain killed only 30% of rabbits collected there after the population had been subjected to seven annual epizootics.

Myxoma virus was introduced into France from South America via Lucerne in 1952. The disease spread to Britain but followed a course different from that in Australia. The principal vector was the rabbit flea, *Spilopsyllus cuniculi* (Dale), and not mosquitoes, although a few outbreaks in domestic rabbits may have been mosquito borne. The first outbreaks were in Kent and East Sussex in October 1953. The centres of infection were wired off with rabbit-proof netting and, as far as possible, all the rabbits within them were killed. Several further outbreaks followed in rapid succession and it soon became clear that the policy of elimination would fail. The Pests Act, 1954, made it an offence to use an infected rabbit to spread the disease among uninfected ones, but this had little impact on the course of the disease. Spread of myxomatosis in 1954–55 was patchy and gradual. Islands of susceptible rabbits escaped infection and continued to breed: a very few rabbits that caught the disease recovered. Secondary outbreaks followed in 1956, 1957 and 1958, mostly in the eastern and northern parts of the country. Less virulent strains have replaced that originally introduced and very virulent strains killing 99% of the rabbits that become infected are now uncommon. Strains killing 70–95% are commonest, possibly because rabbits carrying them are infective for the longest time. In some areas the proportion of immune rabbits is substantial and forms the bulk of winter populations. When these rabbits breed in spring, antibodies are passed to their young but disappear within two or three months. These young rabbits then become susceptible, so outbreaks of myxomatosis occur in late summer, leaving immune animals that have survived infection by weaker virus strains and those that escape infection to pass the winter. Before the epizootics 94% of agricultural holdings were infested with rabbits, about half seriously; now 60% of holdings are infested but only 2% seriously. Although outbreaks of myxomatosis still occur and populations are smaller than they were, numbers are increasing and control measures are still needed (Mead-Biggs, 1966; Lloyd, 1970b; Ross, 1972).

Rabbits should be controlled all the year round. Poison baits are illegal but shooting, snaring, ferreting, dazzling, trapping and gassing may be used. The Pests Act, 1954, stipulates that only humane traps may be used, and these must be placed in the burrows and visited every day. The procedure in gassing is first to block all holes. Within a few days the rabbits open those in active use and cyanogenic powder (NaCN) is spooned down these, 30–60 cm inside and the hole reclosed or alternatively, powder may be forced into holes with a positive-pressure foot-pump. For large-scale operation, a power-driven pump or impeller is necessary and holes need not then be blocked in advance. Dust is forced into one of the holes in a group and the other holes are sealed as soon as the white powder begins to issue. Chloropicrin is extensively used in New Zealand but not in Britain.

Gassing fails to exterminate rabbits because holes are missed, some rabbits live above ground, and immigrants soon move in from neighbouring infested areas. After gassing, survivors must be killed and re-establishment prevented by dazzling, shooting, ferreting and trapping. With few exceptions every occupier of land is required to destroy them under the Pests Act, 1954. (See Leaflet No. 534.)

Hares

Lepus europaeus occidentalis (**de Winton**) Brown hare
Lepus tumidus scoticus (**Hilz.**) Blue hare
Lepus tumidus hibernicus (**Bell**) Irish hare

Hares differ from rabbits in several ways. They do not have permanent homes or burrows, but live above ground, moving from place to place. During the day they rest in depressions in grass, between grass tussocks or in hollows in the ground. These resting places are known as 'forms'. The body and hind limbs are long and the ears are tipped with black. The fur is generally russet brown, the upper parts are tawny and the ventral fur is white. The hare is a solitary animal except during the mating season in February and March when the jack (male) hares show elaborate courtship behaviour, from which the expression 'mad as a March hare' is derived. There may be three or four litters a year and in each two to four young, the leverets, are born in the form. Unlike rabbits, the young are born with their eyes open, a short furry coat, and are able to move. When together, the leverets of a litter lie in a rough circle with their heads pointing to the middle. Each leveret makes its own little form and, when a month old, becomes independent and disperses.

The adult hare is about 61 cm long and weighs 4.7 kg. The eyes are large and prominent and have a wide range of vision. The tail is curved over the back or straight behind, and is black dorsally and white ventrally and laterally. The jack hare has a smaller body, shorter head and redder shoulders than the doe.

Hares feed on grain, bark, roots and many herbaceous plants including sugar-beet, and are particularly destructive to young plantation trees and to market gardens, where they feed avidly on carrots, lettuce and turnips.

There are three kinds of hare: the brown hare common in England, Wales and the Lowlands of Scotland; the blue or mountain or Scottish hare confined to the higher regions of Scotland, northern England and Wales; and the Irish hare found only in Ireland. The blue hare is smaller than the brown hare and does not make a form but uses cover of loose stones and rocks. In the winter its coat changes to white as does that of the larger, Irish hare. Although usually solitary, the Irish hare is gregarious at times when two or three hundred move together in droves across country.

Shooting is the only effective means of control. Poisoned baits are illegal because of the risks to game, domestic animals and wildlife. In recent years hare populations seem to have decreased and control measures are rarely necessary.

Shrews

Sorex araneus castaneus **(Jenyns)**   Common shrew. Shrews are widely distributed in Great Britain but not in Ireland. They are active insectivores and feed both day and night. They are very small (7.5 cm long, including the tail), with short, velvety, dark red brown or grey fur, with a well-whiskered, flexible, pointed snout and small, spikey hairs on the tail. The hind foot is over 12 mm long and the toes are separated. Scent glands give out an offensive smell, stronger in the male than the female; this is repellent to some mammalian predators, but owls and snakes feed on them. Shrews make runs in the hedge bottoms, copses and amongst dead leaves. In the summer they move into fields and rough pastures. They climb readily and, although able to burrow, rarely do so. They are short lived, highly active and soon die if deprived of food. The breeding season lasts from May to November and each female may have litters averaging five young. Nests are made from woven dry herbage and grass. Males fight furiously with each other.

Shrews destroy insects, but there is no evidence that the closely related mole, or other insect-feeding mammals or birds have an appreciable effect on insect populations.

Sorex minutus **(L.)**   The pigmy shrew is 5.7 cm long and the smallest British mammal. The water shrew (*Neomys fodiens bicolor* (Schr.)) occurs in the banks of streams and rivers, but has a more local distribution than the other shrews.

Moles

Talpa europaea **(L.)**   Moles reach 20 cm long, including tail, and have powerful digging forelegs, inconspicuous eyes and ears, and short dark velvety fur that does not ruffle. The mole feeds predominantly on earthworms but also kills harmful insects and it is claimed that their burrows aerate and drain soil. But moles burrowing after insects, usually under rows of seedlings, may destroy plants making campaigns against them necessary. Molehills in grass and cereals can also damage machinery and degrade pastures. Moles can be trapped in their runs or killed by earthworms baited with soluble salts of strychnine. Sometimes rags soaked in petroleum, paraffin or carbolic acid are put down the runs in an attempt to repel them. Altogether moles are of very minor importance agriculturally and methods of controlling them have been little studied. (See Leaflet No. 318.)

Class aves. birds

The main features of birds are:
1) The external covering of feathers.
2) The long mobile neck separating head and trunk.
3) The fore limbs modified as wings.
4) The hind limbs supporting the whole weight of the body.
5) The feet and lower legs covered with scales, and having four toes, three directed forwards and one backwards.
6) The facial region forming a beak covered with horny skin, with a featherless patch of skin at the base (the cere).
7) The absence of teeth.
8) The eyes with a third eyelid, the nictitating membrane, drawn across the eye from the side.
9) The ear openings covered with feathers and lacking pinnae.
10) The rectum and urinogenital systems opening in a common cloaca with a transverse slit beneath the tail.
11) A gland above the tail producing oil to preen the feathers.
12) A higher body temperature than in mammals and a system of air sacs rounding the body contours, entering the larger bones, so increasing buoyancy, aerodynamic and respiratory efficiency.
13) The pronounced keel on the breastbone to which the powerful pectoral muscles are attached.
14) The large-yolked egg protected by a shell and a complicated behaviour pattern designed to ensure the safety of the eggs and young after they have hatched.

For an account of avian anatomy and physiology see Thomson (1964) and for identifications Peterson, Mountfort & Hollam (1974) or Heinzel, Fitter & Parslow (1972). For natural history see Perrins (1974). Existing birds belong to the subclass NEORNITHES. The domestic hen, grouse, pheasant, partridge, guinea fowl and peacock all belong to the Order **Galliformes,** pigeons and doves to the Order **Columbiformes** and most common birds, including the house sparrow, starling, carrion crow, rook and their allies belong to the Order **Passeriformes.**

Feathers are of four main kinds (Fig. 12.4). Quill feathers have interlocking hooks which provide the plane surfaces for flight and occur on the wings and tail. Contour feathers are smaller than quill feathers with the interlocking devices less strongly developed. The down feathers, on the breast, have little or no interlocking and the filoplumes are short and hairlike.

In Britain a few species of birds are harmful to agriculture and horticulture, cereals, peas, brassicas, rape and sugar-beet being especially vulnerable. Whether harm is done by those that wholly or partly feed on crops depends on numbers and the season when feeding occurs. In trying to work out control measures a knowledge of the ecology of the species is important but after much effort effective control continues to elude us. Predatory birds may be regarded as beneficial because their diet often includes rabbits, rats and mice, but there are few reliable assessments of their effectiveness. Birds harmful to agricultural and horticultural crops include woodpigeon, rook, starling, house sparrow and skylark. The first three form large flocks and move from nesting sites or roosts to feeding grounds whereas the others feed locally near their nesting territories. Partridges and pheasants also consume crops and are harmful at times.

Columba palumbus **(L.)**   The woodpigeon or ringdove (Plate XXXIH) is primarily a wood-

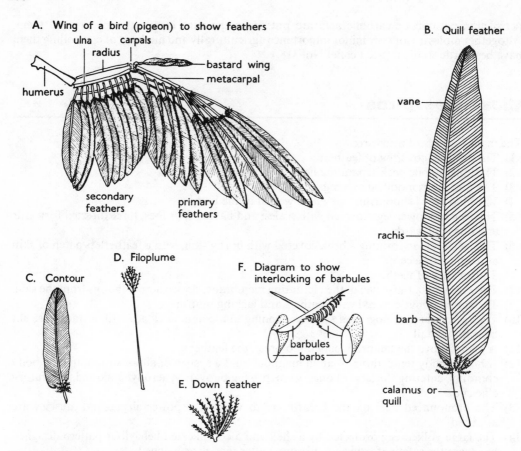

A. Wing of a bird (pigeon) to show feathers
ulna   carpals
radius
bastard wing
metacarpal
humerus
secondary feathers
primary feathers

B. Quill feather
vane
rachis
barb
calamus or quill

D. Filoplume

C. Contour

F. Diagram to show interlocking of barbules
barbules
barbs

E. Down feather

Fig. 12.4 Diagram of different kinds of feathers. A, skeleton of wing and feathers to show where the primary and secondary feathers arise; B, quill feather from wing or tail; C, contour feather from body; D, filoplume from body; E, down feathers F, diagrammatic representation of interlocking barbules of quill feathers.

land bird. The clearing of forests for agriculture and their replacement by fields with hedgerows, copses and scattered trees, and the crops planted by man, have created favourable habitats with abundant food in which the woodpigeon has multiplied and become a pest. Numbers of woodpigeons are declining generally because of the encroachment of towns, buildings, roads and people, but there are greater concentrations on the remaining arable land.

The woodpigeon is a plump, blue-grey bird about 38 cm long with glossy, green and purple neck feathers, a white collar patch on either side of the neck, and white bands on the wings which are conspicuous in flight. Juveniles are darker than adults, have brown tipped feathers and at first no white neck mark. The smaller stock dove (*Columba oenas* (L.)) lacks the white collar and wing patches and has two, short black bands on the wings. It occurs in open parkland rather than woods. The rock dove (*Columba livia* (Gmel.)), another close relative, has a white rump, a black terminal band and some white feathers on the tail, two broad black bands across the secondary wing feathers and a white underwing. This species, in its wild form, is confined to rocky coasts, in north England, Scotland and Ireland; it is the ancestor of the domestic pigeon which, in turn, has given rise to the 'feral' flocks found in towns.

Woodpigeons feed mainly on clover, cereal, brassica crops and particularly oilseed rape and are most destructive during the winter and spring months. At roosts and resting places they also consume tree buds. During frosty weather they feed on kale, cabbage and Brussels sprout, spoiling whole fields. Spring sowings of cereals and legumes suffer in late February and early March, and when these are exhausted the birds are forced back on to leys, pastures and oilseed rape. In June and July they may feed on sugar-beet leaves, leaving only midribs and main veins. When cereal crops begin to ripen in the late summer, there is an abundance of food and the pigeons apparently do less damage. If storms flatten the cereal crops, the birds tend to concentrate there as they can obtain grain easily. For some time after harvest, spilt grain gives ample food and after this the pigeons turn again to clover or rape. Crops preferred are wheat, barley, peas, beans, clover, rape, turnips and other seedlings, roughly in that order, and grain with its high protein content is taken in preference to foliage (Plate XXX). Damage may also be caused to fruit such as gooseberries, currants and cherries, and to market garden crops. Wild seeds and fruits when available are also eaten, including holly, ivy and elderberries.

It is still generally believed that large flocks of woodpigeons arrive in the autumn from Scandinavia to augment our resident birds (Colquhoun, 1951), but observations and ringing experiments indicate that most British woodpigeons are resident and the species is only a partial migrant (Murton & Ridpath, 1962), although in Europe there are pronounced migratory movements in the autumn south-westwards from Scandinavia involving the whole population (Fig. 12.5). In Britain there is a small southward movement in November consisting mainly of first-year birds and probably comparable with the long migratory movements in Europe. Juvenile birds up to one year old tend to move from places where they were reared and make longer flights than the adults, which may extend for 15, 30 or even 500 miles.

During the summer months, woodpigeon flocks disperse and the birds move to their breeding territories. Food is collected early in the morning, but in October and November, when there is less food available and shorter daylight hours, feeding is restricted to a shorter period and flocking evident. Hard weather causes further concentration and large flocks feed on vulnerable fields, giving a false impression of abundance.

The main breeding season is between July and September (Fig. 12.6) and is preceded by pronounced territorial behaviour (Murton & Isaacson, 1962), although nests can be found as early as March. The nest is a flat platform of sticks 17.5–22.5 cm in diameter placed in a fork of a tree, in ivy or other suitable cover, where there is protection from above. When cover is relatively sparse, as in a mature deciduous wood, about three occupied nests per ha is the rule.

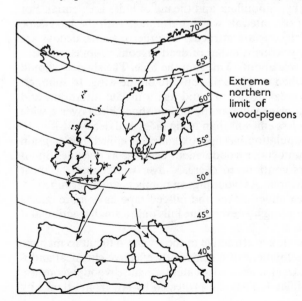

Extreme northern limit of wood-pigeons

Fig. 12.5 Map of Europe showing autumn passage of woodpigeons. The dotted arrow in Britain represents the tendency for British birds to fly south in the autumn. (After Murton & Ridpath, 1962.)

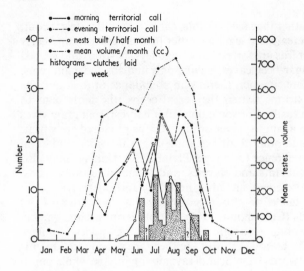

Fig. 12.6 The activity of wood-pigeons in woodland. There is a relationship between morning and evening territorial calls of males, testis size and nest-building activity. Eggs are laid mainly in July and August. (After Murton & Isaacson, 1962.)

In thickets or hedgerows the density of occupied nests is greater, as many as twelve occupied nests in every 1 500 m of hedgerow. The clutch usually consists of two eggs, but as many are taken by jays, magpies and other predators, many repeat clutches are laid, each pair laying an average of five clutches. On average, each pair rears only 2.1 young. The eggs hatch in about 18 days, both sexes share in incubation, and the young stay in the nest between 20 and 30 days. The young are fed at first on a whitish creamy substance, pigeon's milk, which is secreted from the crop wall and mixed with other food before being regurgitated. While young are being fed, much food rich in protein is required; this is available chiefly in August and September when cereal grains are abundant. Pigeons are attracted to food by other pigeons feeding, but once on the ground each seeks its own food. As birds copy each other, a flock is able to exploit the most profitable food source and flock size is related to food density where the birds are feeding (Murton, 1971).

Shooting pigeons may lessen numbers locally and the accompanying noise and activity scares birds away for a time. Concealment by camouflage and the use of hides is essential. For successful shooting, a hide should be built of materials which harmonize with the landscape and placed down or across the wind. As pigeons are attracted by other pigeons, decoys are used to bring them within range. The decoys consist either of dead pigeons propped up with stocks, or artificial ones, set up in a semicircle about 15 m from the hide. The shooter should only leave the hide to pick up dead birds when no more pigeons are on the wing. In summer, the best times for shooting are from daybreak to 9 a.m., and again in late afternoon and evening. Battue shooting or flighting takes place when many people shoot birds over a wide area as the birds return to roost, but this is less efficient than using hides and decoys.

At the end of the breeding season the population has approximately doubled, but as grain and other foods grow scarcer in autumn numbers of woodpigeons decrease as juveniles find it difficult to compete with adults. Numbers continue to decrease over winter and the most critical period is in late February or early March. As woodpigeon numbers are determined by the winter food supply, changes in the area under clover and oilseed rape have more effects than attempts to control their numbers. Shooting however done kills only a small fraction and serves only to remove an expendable surplus.

The use of substances inhibiting reproduction is attractive in theory but difficult to manage successfully in the field. It requires the development of a sufficiently effective chemical and a baiting technique that will safeguard non-target species adequately. Because woodpigeons are quick, wary and mobile, their control is difficult. They are protected by law from attacks by many possible methods, some of which might be considered cruel or would be likely to have harmful side-effects on man, domestic animals or wildlife. For these reasons bird problems cannot usually be dealt with like those posed by invertebrate animals. Detailed study of the biology of woodpigeon (Murton, 1965; Murton & Wright, 1968) have shown that it is impossible to secure widespread population reductions.

Crops attacked at an early stage may be ruined. However, depending on the timing and severity of damage, some crops make compensatory growth and recover well. After heavy snow, however, excessive numbers of birds may feed on the aerial parts of Brussels sprouts and other brassicas and so cause financial loss. The eating of peas and spring cabbage may delay maturation of these crops. This may prevent peas being harvested by viner or result in the early market for spring cabbage being missed. Both lead to heavy financial losses. Current research work is aimed at improved methods of scaring birds from known vulnerable crops. Unfortunately birds soon become habituated to scares of many kinds, especially when they are not backed by actions that lead to physical harm to some of the marauders. It seems that frequent changes of method accompanied by shooting work best, but this approach is costly and time consuming. Effective methods of scaring or repelling woodpigeons from vulnerable crops are still required. (See Leaflet No. 165.) Feral pigeons are sometimes troublesome on newly sown cereals and peas but the damage is very local. (See Leaflet No. 601.)

Passer domesticus (L.)   The house sparrow is the best known but not the commonest of British birds. It lives near buildings and survives in temperate regions wherever it finds food. The sparrow has spread spectacularly in the last century. It followed the extension of grain-growing areas across N. America and the U.S.S.R. Feeding on waste grain from horses' food it has followed armies into new areas. When a railroad was being driven across Australia from the already infested eastern region to the western region, sparrows were shot at the railroad camps, so preventing establishment in the west. Sparrows were unknown on Scottish islands until crofts were set up, and when the crofters left sparrows also deserted the islands. Although in autumn sparrows disperse to some extent into the countryside there is no migration.

Sparrows are about 14.5 cm long, with a short brown, heavy bill. The male has a dark grey crown, chestnut nape, black throat, whitish cheeks and a short, white wing bar. Females and juveniles are dull brown above and dingy white below.

Holes, thatch, walls, trees, rainwater pipes, eaves of houses, ivy and other creepers on walls, and nests of house martins are all places in which nests are built. Untidy spherical nests of straw or dried grass are built in branches of trees in some areas. Four to six greyish or white eggs, speckled with black and brown markings, are laid. Two or three broods may be raised between May and August.

Examination of house-sparrow crops has shown that adults feed largely on grain and that the quantity consumed is greatest at harvest time. Where there is much grassland, and corn-growing is unimportant, little harm is done. House sparrows are very destructive in gardens where they strip the buds from plum trees, destroy buds, blossoms and fruits of currants and gooseberries, and tear the flowers of crocuses, primroses and polyanthuses. Leaves of carnations, chrysanthemums and lettuces are pecked off and peas are attacked at all stages. The birds are often encouraged by being fed during the winter months, and much of their diet at all times consists of household waste and chicken food. Not all the house sparrow's activities are harmful, however, for the nestlings are fed mainly on a diet of insects.

Sparrows around farm buildings may be destroyed by shooting, trapping and poking out nests, but immigration from outside soon replenishes the depleted population. Regular nest destruction between May and August is essential, and all the nesting material should be pulled from each nesting site. Nets may be used under licence to catch sparrows disturbed from their roosts on hedges. Funnel traps with entrances at ground level may also be used. Whole wheat, crushed oats, or bread crumbs are used as bait, and the cages are placed at established feeding points and left open until the birds have become accustomed to them. Traps should be visited twice daily to kill any sparrows caught and left open if they cannot be attended. Trapping is usually less effective in August and September. Bird scarers may be used to drive sparrows away, but they are inefficient and also unpopular near houses because of the noise they make. Pest control operators may obtain licences to use stupefying baits in and around certain types of premises. (See Leaflet No. 169.)

Sturnus vulgaris **(L.)**   The starling is one of the commonest birds in the British Isles except on high land or in dense forests. Just over 22 cm long, the male is glossy black with metallic purple or green tints and has a relatively short tail and pointed wings. The flight and tail feathers are dark brown and tipped with white or buff. In the spring the spots gradually wear off and the beak becomes yellow. The female is duller but speckled in the winter. The juvenile is a uniform greyish brown, and, after a complete moult in late summer, it looks very like the adult.

The starling normally breeds in April, building untidy straw nests lined with feathers, in inaccessible holes in trees, or buildings. Clutches of 4 to 6 eggs are laid, incubation lasts about 11 days and the young are able to leave the nest in about 3 weeks. Breeding now occurs throughout the British Isles and Ireland, in places where the bird was formerly only a winter visitor. The nests are widely spaced but may exceed 200 ha^{-1}. Males lack mate fidelity and pairs are not strongly attached to nesting sites. This breeding strategy although different from that of many other birds seems to pay off. The starling social system also helps to minimize the time spent sampling feeding sites for suitable food sources (Verheyen in Wright *et al.*, 1980).

Starlings feed on a variety of plant and animal materials, such as grain, fruit, sprouting grain, insects, including leatherjackets, click beetles, wireworms, chafer larvae, earthworms, spiders and snails. They often associate with cattle, feeding on insects in dung or those disturbed by their grazing, and alight on the backs of sheep to remove ectoparasitic insects. In fruit-growing areas, flocks eat ripening cherries, apples, pears and soft fruit. The removal of food from silage clamps, poultry yards and cattle-feeding troughs can cause heavy financial losses.

Breeding birds usually roost near their nesting sites and the young and non-breeding birds form flocks by day and fly to communal roosts in the evening. Vast numbers of immigrants arrive from the Continent in the autumn and fly away again in the early spring, roosting for a time with the resident birds. The communal roosting habits are troublesome both in cities, where they foul buildings, and in the country, where they damage forestry plantations, breaking branches by sheer weight of numbers and contaminating the ground beneath. There are many roosts in the British Isles, in trees and thickets, on cliffs and buildings, and many are re-occupied every year. The starlings flock to these roosts along regular flight lines from distances of up to 48 km. They leave in parties at daybreak. Tens of thousands of birds may be present in a roost, but the numbers can change rapidly. Dispersal of starlings from roosts produces what are known as 'ring angels' on radar screens. The waves of birds fly out in concentric circles like the ripples caused by a stone thrown into a pool.

The large autumn and winter flocks cause most damage and nuisance, the noise of their singing being troublesome in built-up areas. They eat much poultry food and cause game birds to desert those parts of coverts where they establish roosts. A critical factor in damage to autumn sown crops of winter wheat and barley in November–December is the arrival of continental migrants and the establishment of winter roosts. Large flocks numbering 100 000 or so can do much damage in a few minutes. However, 100 birds feeding continuously for many days also cause damage. Sowing seeds less deeply as a strategy to help control wheat bulb fly when seed is treated with insecticide may enhance bird damage. Starling beaks are 3–3.5 cm long but birds can probe more deeply for seed grains which need to be 5 cm or deeper in the soil for safety. Changes in the growing of cereals (greater area sown, sown earlier and shallower in autumn) have led to the view that starlings and some corvids could become habituated to the consumption of grain (Feare in Wright *et al.*, 1980). Migrants from the

Continent have been suspected of carrying the virus of foot and mouth disease to Britain but, although it has been shown that birds are able to carry virus, no one has yet shown that they do, in fact, act as carriers.

It is quite impossible to control the enormous numbers of starlings by shooting. In the country, firing rockets into flocks as they are returning to roosts or broadcasting recordings of starling distress calls are often successful but the operations must be repeated for several nights until the birds give up. This is only a palliative, however, as the birds are driven elsewhere. On buildings in towns jelly-like materials applied to ledges often unsettle the birds. To prevent ingress of starlings, openings in animal feeding units may be fitted with curtains of plastic strips. These allow men, animals and machinery to be moved in and out but deter the birds. (See Wright *et al.*, 1980, and Leaflet No. 208).

Corvus frugilegus (L.) The rook is common throughout Britain and is especially numerous in southern and eastern Scotland. It has glossy black plumage with a slight purple tint. A patch of greyish white skin extending from the base of the beak to the eyes, and the slender, pointed greyish beak serve to distinguish it from the otherwise similar carrion crow (*Corvus corone* (L.)). The white face-patch is not developed until the end of the first year. Full-grown rooks are 45.8 cm long and are larger than jackdaws (*Corvus monedula* (L.)), which reach only 33 cm and have grey napes and ear coverts.

Rooks are gregarious when nesting and feeding. Winter roosts contain thousands of birds as well as many jackdaws and some carrion crows. Flocks sometimes disperse from the winter roosts every morning and visit their particular rookeries *en route* to their feeding ground. The winter roosts are deserted in spring when the rookeries are re-occupied, which are usually in tall trees where nests consisting of large masses of twigs woven into the uppermost branches are built. Between three and six greenish eggs are laid in March or April, they hatch in a little over a fortnight and the young leave the nest about 4 weeks later. Adult British rooks do not migrate, but a few migrants from north-west Europe arrive in eastern England in the autumn.

The rook is primarily a grassland bird. Its food is mainly small animals that are found in the soil surface but, although insects are eaten, it is doubtful whether the rook or other birds with a similar diet have any appreciable effect upon populations of noxious insects. Rooks also feed on fields of cereals, peas, beans, roots and potatoes, on stubbles and on maize grown for silage. Sugar-beet seedlings are sometimes uprooted. Since insecticidal seed treatments to control wireworms have been used routinely, rook damage to sugar-beet seedlings has diminished greatly. Lodged corn is also eaten, and it is here that the most serious losses are incurred by the farmer. Feeding grounds and rookeries are rarely more than 1.5 km apart.

It is unnecessary to take action against rooks unless their number becomes excessive for the area in which they are living. Owners on whose land rookeries exist may be required to kill them but these powers are rarely used. Trapping and shooting are the main methods of control. Traps with funnel entrances may be made of light timber and 38 mm mesh wire-netting, with a door in the side for access. The rooks enter through the funnel, the top of which opens in the centre of the cage roof (Plate XXXIIB). The whole cage is a 1.8 m cube at least, and the funnel 60 cm diameter at the top tapering to 25 cm at the bottom and from 30 to 60 cm long. An alternative funnel requiring a larger cage is 2 m across at the top, 25 cm at the bottom and ends 45 to 60 cm from the floor of the cage. Rooks are trapped in rick yards, near heaps of chaff or in other places where they feed regularly. Traps placed in fields should be near a tree or fence on which they can first perch. Baits used in the traps are grain, chaff, or bread and a hen may be used as a decoy. Trapped rooks or jackdaws which must be supplied with food and water also serve to decoy others. Any birds caught must be removed after dark and destroyed.

Shooting is usually confined to the month of May when young birds may be killed as they are about to leave the nests. Scarecrows are of little use as birds soon grow accustomed to them. (See Leaflet No. 244.)

In areas that favour it, the carrion crow (*Corvus corone* (L.)) is more numerous than any other corvid. Its diet is similar to that of the rook but includes cattle food, carrion, birds' eggs, and the young of birds and mammals, but it is most disliked for its habit of attacking the eyes of sheep and lambs. In fact the carrion crow attacks sheep only when they have rolled onto their backs and cannot get up and attacks and kills lambs that are sickly and would die from other causes. In Scotland, Ireland and the Isle of Man, the carrion crow's niche is taken over by the hooded crow (*Corvus cornix* (L.)). The jackdaw (*Corvus monedula* (L.)), magpie (*Pica pica* (L.)), and jay (*Garrulus glandarius* (L.)) feed on insects, worms, slugs, mice, wild fruits and berries. All take eggs and may destroy young game. The jackdaw sometimes damages agricultural crops, and the jay feeds on top and soft fruit, and on peas in gardens. (See Leaflet No. 242.) Control measures are shooting and, where possible, nest destruction.

Phasianus colchicus **(L.)** Pheasant and *Perdix perdix* **(L.)** Partridge. Game birds are keepered in well-wooded areas and both partridges and pheasants sometimes graze beet seedlings, especially in West Norfolk. Pheasants may damage singled beet by pecking the crown of the plant just above soil level and eventually felling it. Serious damage may occur in May and June but later in the year the plants are too large to fell. In dry years they may also dig out young potato tubers. Apparently cock birds do most damage. If it can be arranged, shooting of surplus cocks is desirable.

Alauda arvensis **(L.)**   The skylark, a passerine bird widespread in Europe and abundant in open fields, is brown with white on the tail feathers, a long posterior claw and a raisable crest on the head. Numbers appear to have increased with the spread of arable cropping and the

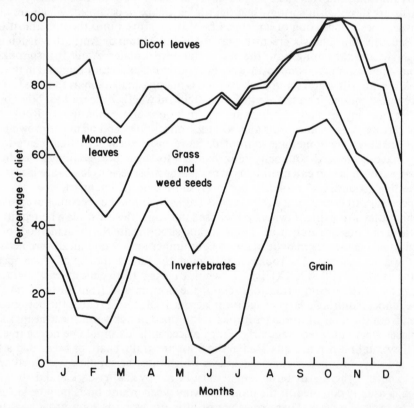

Fig. 12.7 Diet of the skylark based on data from three sites and two years (from Green, 1978). In individual fields and in different years the proportions of the constituents vary greatly and are much influenced by what is available.

removal of hedgerows and hedgerow trees. In East Anglia from December to March densities per ha ranged from about one on ploughland and ryegrass ley, to two on winter wheat and, exceptionally, to more than ten on clover ley (Green, 1978). Skylarks are only partial migrants, being displaced only in cold weather. Two broods are raised in April to July in a nest made on the ground, usually in an existing hollow and protected by a tuft of grass. Clutches usually contain 3–4 eggs. The lark feeds on a mixed diet consisting of invertebrate animals and plant material.

The proportions vary throughout the year (Fig. 12.7). Cereal grain predominates in the diet in spring and autumn. When exposed grain is unavailable, skylarks take the kernels from seedlings up to 6 cm tall. However, even when much grain is available skylarks also feed on leaves of wheat, barley, ryegrass and clover; at other times these are taken in quantity. Seasonal variations in the diet of the skylark are comparable with those of woodpigeon, rook and partridge. Sugar-beet seedlings have become vulnerable because crops are drilled to a stand and weeds are eliminated by pre- and post-emergence herbicides. Feeding, which occurs in April and May, results in the loss of all or part of the cotyledons. The growing point is not usually damaged. If defoliation continues over a period, early growth is prevented and much yield is lost. Skylarks are not exclusively territorial and the breeding pairs established in beet fields in spring tend to forage where food is most plentiful.

Although herbicides might be used to regulate the supply of weed seeds, the relationship between weed numbers and damage to beet seedlings is complex (Green, 1980). Applying herbicides to the seed row and leaving weeds between rows to be killed later is helpful. Aldicarb seed furrow treatments applied against soil inhabiting and aphid pests decrease grazing damage by skylarks but in most years would be uneconomic if applied to deter skylarks only. Presumably they detect the compound or its degradation products in the foliage which are translocated upwards from the roots and become systemic. The skylark is a protected bird under the Wildlife and Countryside Act, 1981 and may not be killed.

Geese

Anser anser (L.)	Greylag
A. fabalis brachyrhynchus (Baillon)	Pinkfooted
A. albifrons (Scop.)	Whitefronted
Branta leucopsis (Bechstein)	Barnacle
B. bernicla (O. F. Müller)	Brent
B. canadensis (L.)	Canada

The above six species of goose feed mainly on cereal crops and grassland as do some swans and ducks. The original mainly coastal habitats of the first five have largely been replaced by farm or forest land. Nevertheless the geese have tended to remain in or near their usual haunts. Evidently attachment to these habitats is strong and although relative population densities may have changed, no large scale changes in the pattern of distribution of the species seem to have occurred (Owen in Wright *et al.*, 1980). The first five species breed in the arctic in summer and migrate southwards in autumn. The once numerous migratory bean goose (*A. fabalis fabalis* (Latham)) is now rare and occurs mainly in Scotland. Originally introduced as an ornamental bird in parks, the Canada goose has increased greatly and is resident mainly in central and southern England. In fields alongside lakes and streams it sometimes strips the grain from cereal crops in summer and grazes grass and cereal crops at other times. In autumn greylag and pinkfooted geese feed on grain in stubbles and on waste potatoes. There are usually few complaints about the feeding of geese until the turn of the year but, in hard weather, swedes and carrots may be consumed. In Lancashire pinkfooted geese have been known to attack growing carrot crops whereas usually they consume only waste roots left after harvest. Late winter and spring migratory geese move from the margins of estuaries and coastal mudflats to nearby seedling cereal crops and grass. Damage to cereals is partly from grazing the foliage and partly from puddling and compacting the soil surface by treading. In Scotland in late April and early May greylag and pinkfooted geese graze the spring growth on improved pastures intended to provide an 'early bite' for stock.

Problems from migratory geese are likely to occur in coastal areas (north and south of Inverness, the island of Islay, the northern shores of the Solway Firth, the Wash, north Norfolk, East Suffolk, Essex, the Portsmouth–Chichester area). In addition to the coastal areas of the Firths of Forth and Tay, greylag and pinkfooted geese may cause damage in large areas of the central lowlands of Scotland stretching inland to Loch Lomond. The numbers of geese concerned are small on a national basis. In 1978–79 they were estimated to total about 254 000 of which some 175 000 were on arable land. Numbers vary from year to year according to survival and breeding success as does the damage done. However, these numbers are localized and concentrated enough to cause serious losses to farmers.

Scaring devices give only limited and temporary relief, and intensive harassment fails to dislodge them. Only the farmer suffering damage to his crops can decide how much time and money he can afford to spend on crop protection. The greylag goose is protected by special penalties in the close season (Part II, Schedule I of the Wildlife and Countryside Act, 1981) to safeguard small local breeding populations in the Outer Hebrides, Caithness, Sutherland and Wester Ross. Protection extends to birds, their nests and dependent young and prohibits intentional disturbance. Canada, greylag and pinkfooted geese are listed as quarry species (Part I of Schedule 2) which may be taken outside the close season. The close season is normally 1 February to 31 August or 21 February to 31 August below high water mark. Barnacle and whitefronted geese are not quarry species. The sale of dead geese is totally banned. Exceptions to acts, such as shooting in the close season, which would otherwise be offences, include killing to prevent serious damage to crops, but this does not apply to birds listed in Schedule I such as the greylag goose in certain areas. Section 16 of the Act contains provisions for licensing the shooting of geese during the close season to prevent damage to crops. The licensing authorities are the Departments of Agriculture for Scotland and Northern Ireland and the Ministry of Agriculture, Fisheries and Food for England and Wales. Before a licence can be issued the advice of the Nature Conservancy Council must be sought and those who obtain licences are required to report the numbers and dates when pinkfooted and greylag geese were shot. (See DAFS Advisory Leaflet 'Wildgeese and Scottish Agriculture'.)

Some general considerations on birds

Bird problems are not new but have perhaps been heightened by economic pressures and the greater yields now at risk. Farmland which occupies about 80% of the land surface of Britain supports about 50–100 million breeding pairs of some 130 different species. About two-thirds of these species have populations of fewer than 100 000 pairs which could be threatened by drastic and effective control measures, if available. The biggest problem from the conservation angle is winter flocks of relatively rare species which cause local damage, i.e. migratory geese. To counter these, adequate refuges and compensation to farmers seem possible measures. Other injurious birds are numerous, widespread and in no great danger of extinction especially as they are part of larger European populations. Only small fractions of these populations are responsible for damage to crops. Their diets are flexible and crop plants feature irregularly and perhaps incidentally in their diets. Despite their great mobility some harmful species are 'local' in their operations (Table 12.3).

More knowledge, forethought and well-planned control strategies are urged by ornithologists (see Wright *et al.*, 1980). Unfortunately these and other hopeful suggestions are not too helpful to farmers suffering financial loss here and now.

In discussions and debates in Brussels on the EEC Directive on the Conservation of Birds

Table 12.3 Movement of some bird species associated with damage to crops. Percentages based on birds recovered by ringing; first figure April–August, second September–March. (After Flegg in Wright *et al.*, 1980.)

	Distances moved (km)		
	0–10	11–100	>100
Woodpigeon	73–68%	26–30%	1–2%
Skylark*	72–63%	21–13%	7–25%
Rook	76–66%	24–33%	0–1%
House sparrow	95–94%	5–5%	<1–<1%

* Based on a sample size of 49, others > 100.

(with which British legislation broadly complies), members of the British delegation were surprised and dismayed by the utterly irrational determination of some delegations to achieve blanket protection at all costs or to minimize or delete from the Directive all concessions to sport or pest control. As a result of heavy legal restraints many farmers feel unduly restricted both in the species they may kill and the methods they may employ. This is especially so when the restraints apply to harmful birds in no immediate danger of extinction. Unfortunately unreasoning bird lovers greatly outnumber informed ornithologists (Colling in Wright *et al.*, 1980).

Bird behaviour is resourceful, adaptable and plastic. Little real progress has been made with scares, scaring devices alone or in conjunction with alarm signals, or with chemical bird repellents. The idea of breeding cultivars with resistance to birds is in its infancy. For further reading on these and other matters see Wright *et al.* (1980).

13

Pests of stored grain

Pests found feeding in stored grain in Britain do not attack grain in the field. Whether they are truly native or not is debatable as there have been a number of climatic changes over recent centuries and many have been present so long that they may be described as naturalized. The grain weevil was probably introduced in prehistoric times and saw-toothed grain beetle may have been native in Roman times, but *Ptinus tectus* (Bld) did not come until 1894.

Stored grain pests live in grain stores, silos and warehouses where they are favoured by shelter, temperatures somewhat higher than outdoors, and by great concentrations of food material. To succeed they must be able to multiply rapidly when conditions are good and they must be able to survive when they are bad. In farm stores, most pest species can survive the winter temperatures experienced. Stored grain usually becomes infested by contact with existing foci within farm buildings or by contact with bought feeding stuffs, especially unprocessed imported cereals and oil cakes, or grain from other farms. Endemic populations exist in barns and warehouses, often within the structure of the buildings themselves because of age or faulty construction. The number of generations a year depends on temperature, relative humidity and available food.

Oryzaephilus surinamensis (L.) (Fig. 13.1E)   The saw-toothed grain beetle (Plate XXXIVA) is now the most important source of losses in farm-stored grain, possibly because of the changed methods of handling grain following combine harvesting. The free-living larva will attack grain which has only microscopic damage but appears superficially sound and feeds preferentially on the germ. The saw-toothed grain beetle occurs all over the world, feeds on organic stored products, is about 3 mm long, slender, flattened and dark brown. On each side of the thorax are six tooth-like projections. Adults may live for six months or more and can overwinter in farm buildings, and although able to fly, they rarely do so. The eggs are laid in or near the food and hatch in 3–17 days. The flattened, active larvae have brown heads, six legs, dark chitinous terga, and each segment bears two or more long bristles. Before pupation, which lasts from 6 to 21 days, the larva forms a protective case by sticking together bits of food material. The life-cycle from egg to adult is about 80 days at 20°C but only 20 days at 32–35°C. The insect is unable to breed below 18°C (Howe, 1956 a,b), but is encouraged by warm storage, hence the desirability of ensuring that grain passed through a dryer is properly cooled. Infestations usually develop after the grain has been put into store or into infested bags.

Sitophilus granarius (L.) (Fig. 13.1A)   The grain weevil is far less important than it was but is still often found. It attacks all types of grain but not grain products, except hard material like macaroni. The larva lives entirely within grain damaging the endosperm and adversely affecting germination. *S. granarius* can breed at 30°C whereas the related *S. oryzae* (Fig. 13.1B) and *S. zeamais* (Motschulsky) sometimes introduced with feeding stuffs are tropical and although they require higher temperatures for successful breeding they are becoming more common in U.K. stored grain: they rarely survive over winter.

The adult grain weevil, which cannot fly, is dark brown to nearly black, rather shiny and covered all over with fine pits (Plate XXXIVD). Including the slender snout, it is about 3 mm long. The life cycle from egg to pupa is passed unseen within the grain. The female chews

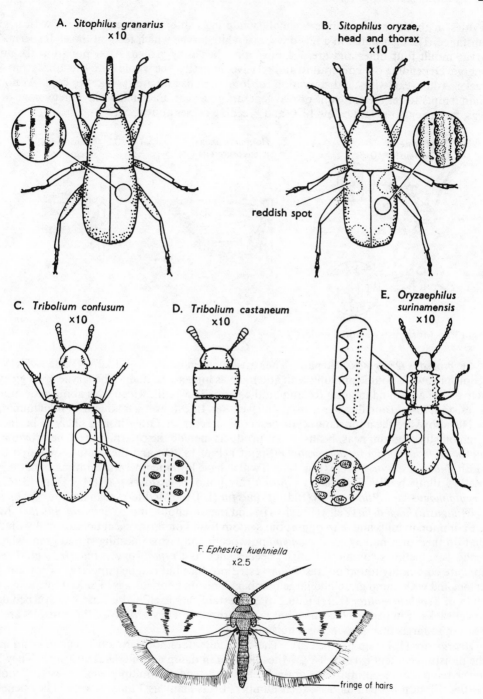

Fig. 13.1 Pests of stored grain and *E. kuehniella* which is a pest in flour mills.

cavities in the kernels and inserts small, white eggs into them, sealing them with a gluey substance. The eggs hatch in a few days into white larvae which feed on the endosperm. The larvae moult four times and are 3–4 mm long when fully grown. After pupation the adults emerge. Depending on temperature and relative humidity, the whole life cycle takes from 4–16 weeks. Adults can withstand starvation for long periods and are resistant to cold. At normal grain temperatures there are one or two generations a year. Each female lays an average of 190 eggs, but no eggs are laid below 13°C and breeding ceases above 35°C.

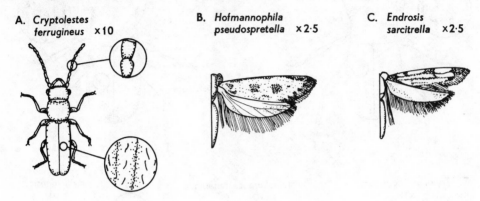

A. *Cryptolestes*
   ferrugineus ×10

B. *Hofmannophila*
   pseudospretella ×2·5

C. *Endrosis*
   sarcitrella ×2·5

Fig. 13.2   More pests of stored grain.

***Cryptolestes ferrugineus* (Steph.)**   The rust-red grain beetle (Fig. 13.2A) (Plate XXXIVB) is reddish brown, about 2 mm long with antennae as long as the body. The larvae are legged and active, and are found in grain feeding on the germ. The adult, which lives about nine months, is resistant to cold and flies in warm, sunny weather. Infestation is favoured by high humidities.

   The above are the important beetle pests of stored grain. Other insects likely to be encountered on stored grain, peas, beans or oat products include the caterpillars of the brown house moth, *Hofmannophila pseudospretella* (Staint.) (Fig. 13.2B), and the white-shouldered house moth *Endrosis sarcitrella* (L.) (Fig. 13.2C) which have lower temperature optima than the grain beetles. Flour beetles *Tribolium confusum* (Dv.) and *T. castaneum* (Herb.) (Fig. 13.1C, D), *Cryptophagus* spp., *Ptinus tectus* (Bld.), *Ptinus fur* (L.), the Mediterranean flour moth, *Ephestia* (= *Anagasta*) *kuehniella* (Zell.) (Fig. 13.1F) and the warehouse moth, *Ephestia elutella* (Hueb.), are common in mills and warehouses but seldom found on farms except occasionally on those that do their own milling. *T. castaneum* now occurs on farms infesting stored grain where *P. tectus* is sometimes common. Fungus beetles such as *Cryptophagus*, *Ahaverus* and *Typhaea* spp. are commonly found on freshly harvested produce but can be important. They are often associated with damp grain and large numbers indicate poor storage. The yellow meal-worm, larva of *Tenebrio molitor* (L.) (Fig. 6.2, p. 94), establishes itself in dust and undisturbed debris under sacks or in neglected corners but is not primarily a pest. Many of these insects are only pests in so far as their presence is a nuisance.

   ***Acarus siro* (L.)** (Fig. 6.2, p. 175)   The flour mite does not normally occur in grain unless the moisture content exceeds 14%. Mites flourish in damp and moderate warmth. They have lower temperature optima than grain beetles and tend to multiply when it is too cool for beetles. Flour mites can increase their numbers by as many as 7 times a week. They penetrate damaged grain through microscopic cracks in the testa and feed on the germ to which they impart a characteristic musty odour.

   ***A. farris* (Oudemans)** and ***Tyrophagus longior* (Gerr.)** can occur in stored grain but are confined to the 3–4-month period after harvest. Mites are sometimes excessively numerous in recently baled hay and workers may suffer skin irritations on coming into contact with mites or their excreta. Mites may be killed by passing the grain through a dryer and up to 90% may be killed by passing through a conveyor system. Mite infestations in baled hay decline with time as the fungi on which they feed die out. Infested baled hay should be isolated from stored grain or feeding stuffs. Another mite, *Lepidoglyphus destructor* (Schrank), is commonly found in company with grain mites in granaries (Griffiths, 1960). For details of mite life cycles see p. 171, and for further information MAFF leaflet No. 368 and reference booklet 409.

Ecology of masses of grain

Because grain is alive, it respires and produces heat. If properly dried and cooled, respiration is slow and heat production is at a minimum. Grain, like air, conducts heat poorly and air movement within grain masses, which might help to cool grain, is very slow. If therefore grain is put into store with a moisture content below 14%, cooled rapidly to as far below 15°C as possible and protected from damp air and leaking roofs, it keeps a long time, from harvest to May or later in the following year. (Oilseed rape, compound feeding stuffs have a much lower safe moisture content, some as low as 8%.)

The temperature of a mass of grain remains constant or changes little over long periods. Heating of bulks of grain usually starts at one or more hot spots where insects have gained a foothold and begun to breed, or where the moisture content has risen above 16% and allowed the growth of moulds. The heat produced where insects and fungi are active does not escape but warms the grain and enables the insect to become more active and to produce still more heat. In time the insects move away towards a cooler zone. Larvae inside grains cannot move, they complete their life cycle or are eventually killed. Heating caused by insects stops when the temperature reaches about 37°C in the centre of the infested spot, that caused by fungi goes on to 45°C and then bacterial heating may take the temperature to 50°C or more. Heating reaches the periphery and spreads throughout the bulk. The infestation ultimately reaches the exposed surface.

Although some water is produced by the respiration of the grain and of the insects, this does not usually affect the moisture content of the grain. Movements of moisture, which have far more serious consequences, arise from absorption of water from grain by air in the hot spots and movement of the air to cooler places, where the absorbed moisture is deposited. Free water may also appear at the grain surface or where the grain is in contact with cool, conducting surfaces. At these places, damp heating occurs (see Leaflet No. 404), and the grain may sprout. As the temperature of the grain rises, it is attacked by a succession of moulds increasingly tolerant to heat, until the temperature rises to 60-65°C. The grain eventually cakes, loses quality and germinating power and, if completely neglected, becomes worthless.

Grain should never be stored damp or mixed with dried grain. Besides encouraging fungi that generate heat, damp grain is also more easily attacked by insects. When insects invade this kind of grain there is a race between cooling caused by lower autumn temperatures and heating from the activities of insects and fungi: while the grain remains warm enough the insects multiply and the outcome can be heating or cooling depending on which proceeds faster. Heating from the activity of insects alone probably never occurs in Britain unless it is induced by grain weevils.

Preventive measures

The principles of grain storage are dealt with in MAFF Booklets 2136, 2137, 2415, 2416 and 2417. Buildings used for storage should be cool, dry, watertight and without artificial heating.

They should be kept in good repair and have concrete floors with smooth finish and few corners. There should be no dampness from leaking roofs, damaged gutters or faulty damp courses. Cracks in floors and walls should be filled. Doors should fit properly and the whole building be constructed so as to exclude birds, rats and mice. Cleanliness and tidiness are essential. Industrial vacuum cleaners can be used to remove dust, insects and their residues from floors and sacks. Heating them to 60°C for 30 minutes or to 55°C for an hour kills the most resistant species and this is essential to prevent re-infestations. Old grain residues, old sacks, unfrequented corners and nooks and crannies should be cleared out and the dust and rubbish burnt to kill living insects. Infected returned sacks often lead to rapid re-infestation of grain. After thorough cleaning, the building should be sprayed with chlorpyriphos-methyl, fenitrothion, malathion or pirimiphos-methyl. Wettable powder formulations are the most effective, especially on porous surfaces of brick, concrete or wooden bins. All this should be completed, several weeks before the new harvest. Large numbers of insects may emerge from cracks and crevices and be killed by residual films of insecticide. These should be cleaned up and killed as before: not all are necessarily dead. Sacks can be fumigated with carbon tetra-chloride mixture or methyl bromide but this should only be done by specialists.

In buildings with a previous history of infestations, admixture of grain with chlorpyrifos-methyl, fenitrothion, malathion, methacrifos or pirimiphos-methyl is a useful preventive measure. The insecticide can be applied as a dust or spray as the grain passes into storage and applicators designed for the purpose are available. The application rates recommended on the packs of insecticide should be followed precisely and, as the operation is a specialized one, expert advice should first be sought. If the insecticide applied does not have the desired effect it may be that one or more of the insects present are resistant. Again, expert advice should be sought.

Regular inspection once a week is important so that prompt action can be taken the moment insects appear. A probe type of thermometer is useful for taking temperatures. Heating of dry grain usually indicates insect activity. A cargo sampler or sample spear can be used to remove samples of grain which should be sieved and the sievings examined for insects and mites. Damp grain should never be stored in bulk. The maximum moisture content permissible is 16% for storage from harvest to January, 15% from harvest to April, and 14% from harvest to May, but lower moisture contents are advisable.

Curative measures

If inspection reveals an infestation, the grain should not be moved as this might cause spread. Turning grain, spreading or attempts at cooling are usually ineffective. It is necessary to kill the insects. It is possible to eradicate an infestation by fumigating grain *in situ* but this must be done by a recognized fumigation contractor. It is no longer considered safe for farmers and merchants to do the work themselves and the British Pest Control Association will supply a list of contractors.

The choice of fumigant depends on location and type of store, pest species present and the temperature of the grain. Methyl bromide, phosphine and the liquid fumigant mixtures are the ones most widely used in the United Kingdom.

Apart from small bulks or stacks of bagged grain, methyl bromide can only be used where circulatory facilities are installed and these are usually confined to large silos of the grain elevator type. Dosage rates range from $25-50\,\mathrm{g\,m^{-3}}$ depending on temperature and the exposure period is usually only 24 h. Bulk grain on floor stores or in silos may be fumigated with liquid fumigants consisting of mixtures of halogenated hydrocarbons. Dosage rates range from $\frac{1}{2}$ to 1 litre per tonne and exposure periods of 7-10 days are necessary. Grain must be thoroughly aired before use, particularly if being fed to animals without further processing or compounding. Phosphine generated from proprietary metallic phosphide preparations is now widely used and is particularly effective against *Oryzaephilus surinamensis*, the most common pest of grain in the U.K. Dosage rates range from $1-5\,\mathrm{g\,m^{-3}}$ with recommended exposure periods of 3–16 days depending on temperature.

The advice of servicing companies should be sought on all fumigation matters and they will be pleased to quote for such treatments. However, it should be borne in mind that, unlike contact insecticide treatments which give some residual protection, fumigation will not prevent re-infestation after treatment.

Insects in other parts of the structure not reached by fumigation may be killed with smoke, dust or spray formulations of contact insecticides. For walls, wettable powders or emulsion formulations are convenient e.g. malathion, gamma-HCH (or a mixture of both), fenitrothion, or bromophos. Insects are often able to escape insecticidal treatments, even those with residual effects, for several months by hiding in deep cracks and crevices within the structure of a granary. Smoke formulations do not fumigate; they deposit fine films of insecticide on exposed surfaces which should be free of dust. They are less effective than sprays.

Resistance to storage pests

In well-built and well-managed grain stores large quantities can be kept for long periods with suitable pesticidal treatments. Under these conditions ability of the grain to resist attack is unimportant. Smaller quantities of grain stored on farms in poor conditions are more at risk. Although storage pests increase exponentially, the rate being determined partly by the storage temperature, in absolute terms numbers remain small for several months after which increase is 'explosive'. Grain with resistance delays the onset of the 'explosive' phase. In developing countries, e.g. Malawi, farmers store only sufficient grain for their immediate needs and little is carried over to the following year. Losses are small and insecticides unnecessary: local crop cultivars and storage methods match local needs (Golob, 1981). Introduction of heavy yielding modern cultivars could change the picture and result in a need to protect grain with insecticides. Resistance can be compared by measuring the rate of increase of storage pests under standard conditions. Such tests done on triticale, barley, maize and wheat showed that triticale was about twice as susceptible as the other three (Dobie & Kilminster, 1978). Dobie *et al.* (1979) also found that the winged bean, *Psophocarpus tetragonolobus*, was immune and a cultivar of cowpea, *Vigna unguiculata*, highly resistant to attacks by seed beetles (Bruchidae, p. 119). Sometimes resistance to storage pests seems to be related to husk or integumental characteristics which provide a mainly mechanical barrier to feeding. In maize cultivars, a good husk coverage, which some 'improved' varieties lack, is related to resistance against *Sitotroga cerealella* (Oliv.) and *S. zeamais* (Mot.) (Dobie, 1977). In the winged bean and cowpea resistance to seed beetles seems to be related to large concentrations of a protease inhibitor in the cotyledons (Dobie *et al.*, 1979).

14

Crops and their pests

In this chapter crops and their pests are brought together, the information desirable before control measures are attempted is briefly discussed, and crops are considered in related groups so that their pests can be put in perspective.

In planning control measures the first essential is accurate identification of the animal concerned. The second is as complete a knowledge of its life cycle as possible, including periods of activity and quiescence, method of overwintering, type of feeding by young and adults, whether exposed or concealed, time spent feeding, flying and sheltering and, where appropriate, information about behaviour. The third essential is a knowledge of the animal's host range, and the fourth a study of population densities and the influence of climatic, edaphic and biotic factors upon them. Without some of this knowledge control measures can only be empirical, *ad hoc*, and curative as opposed to preventive. Curative measures are usually less effective than preventive measures, where these are possible, because action is not taken until some damage has been done and some or all the benefits of control lost. For many pests that feed in the concealment of soil, roots, stems, seed-pods, or seeds, only preventive measures are effective.

Information required about the crop attacked includes type, time of sowing, critical stages, plant populations and seed rate, rate of growth and time of maturation, plot size and shape.

Origin of infestations

Attacks early in the growing season by relatively immobile organisms such as nematodes, molluscs, woodlice, millepedes, symphylids, collembola or other flightless insects as well as by overwintering insect larvae or adults, all arise from populations that built up previously in the affected field. The origins of such pest problems are to be sought in the preceding year or years during which suitable host crops have been grown too frequently or conditions created that favoured increase. Attacks by immobile pests later in the year are usually the result of multiplication in the crop, e.g. races of the stem nematode, slugs in potatoes, some mites. Attacks by highly mobile insects, mammals or birds often bear little relationship to preceding events in the field where the attacked crop is growing; those by migratory birds and by many aphids may bear none at all. However not all insects, birds or mammals migrate far. The flights of many insects result only in local dispersal and populations are related to previous cropping in the immediate vicinity. Thus, carrot fly, cabbage root fly, wheat bulb fly and onion fly concentrations are greatest in the localities where their hosts are grown intensively and pigmy mangold beetle is dense only in areas where sugar-beet is grown frequently.

Type of crop

Crops fall into two groups: those raised from seeds and those raised from tubers, transplants, runners, bulbs, etc. Some brassicas, leeks, celery and onions (sets) are sown as seeds in seed beds or raised in soil blocks or containers and planted out later, and therefore fall into both groups. Most seeds are vulnerable when first sown. Tubers, transplants and bulbs are less vulnerable because they represent more plant material and so, in establishing root and shoot systems, are better able to replace roots or buds that are destroyed.

Time of sowing

Although hard and fast rules cannot be laid down, the earlier a crop is sown and the longer its vegetative life before maturation, the better it withstands pest attacks and virus infection and the more it compensates gaps where plants are killed. Sometimes sowing can be delayed so that the susceptible stage of the crop comes after the peak activity of a pest (e.g. brassicas to avoid flea beetle attack, carrots to avoid the first brood of carrot flies) but some loss of yield is then inevitable.

Critical stages

For crops raised from seed, the most critical stage is during and just after germination. If the hypocotyl or radicle is injured or severed, the plant may die. For those crops singled like sugar-beet, sown to a stand and kept free of weeds by herbicides, pests present before sowing or arriving during the seedling stage are concentrated on the plants which are the sole food supply. Where seed is harvested, the flowering period is a critical time (e.g. frit fly attacks on oat panicles, pea moth attack on peas, gall midges on grass seed). For transplanted crops, the period immediately after transplanting and before vigorous growth begins is critical.

Plant populations

Although plant populations are determined by the seed rate they are much affected by soil conditions immediately after sowing. In Table 14.1 the same weights of seed were sown per unit length of row, but the numbers of seedlings that emerged were different at the different sites. Nowadays many fewer seeds are sown, so their survival is essential.

For pests like wireworms, cutworms and leatherjackets which destroy seedlings, the relative numbers of pests and seedlings per ha is important. With a wireworm population of $5\,000\,000\ ha^{-1}$, there are 2 per plant for wheat ($2\,500\,000$ seedlings ha^{-1}), and 50–67 per plant for sugar-beet ($75\,000–100\,000$ plants ha^{-1}). The greater 'resistance' of wheat to wireworm attack is partly a reflection of the denser plant population.

Increasing the seed rate in the hope of establishing more seedlings is helpful but marginal, because it is rarely possible or economical to increase the rate more than double as this may give rise to new problems and sometimes promotes the spread of fungal disease. Similarly, sowing bait crops such as wheat between sugar-beet rows to control wireworm damage is

Table 14.1 Numbers of seedlings established from the same amounts of sugar-beet seed in six different fields. After Jones & Humphries (1954).

Field	Seedlings per 20 m of row	
	Untreated	Treated with combined insecticide and fungicidal seed treatments
1	242	631
2	330	465
3	374	602
4	374	565
5	550	715
6	330	673

unlikely to be effective unless the balance between crop and pest is close (Jones & Jones, 1947). In practice it is not so much the pest numbers per unit area but the fraction that reach and feed on plants that is important. This fraction varies with soil and season. To ensure protection, seed baits would need to be used at an excessive rate and, if left long, would compete with the crop for nutrients. Use of baits other than seeds (e.g. unwanted potatoes or carrots) is rarely effective or economic.

Rate of growth

The rate of growth of the crop determines how rapidly it passes through critical stages and how effectively losses of plants or plant parts are compensated. Seedlings remain vulnerable while they are small but become less susceptible as they grow. Increase in size is accompanied by increase in thickness of stems and roots, by increase of the number of stems in cereals and leguminous crops and by increase in bulk in other crops. Slow-growing crops are more easily overwhelmed than those growing rapidly because new plant material produced may not exceed that consumed.

Rate of growth is determined partly by season and partly by the skill of the farmer. Low temperatures slow growth and also inactivate invertebrate pests. Drought also slows growth, but is often accompanied by raised soil temperatures. The soil surface temperatures associated with fine clear weather in spring favour soil surface insects like flea beetles and pea and bean weevils while plant growth remains slow because the roots are in relatively cold soil. Feeble, slow-growing crops withstand pests badly. Often soil conditions need to be improved by draining, adjustment of the pH by liming and by making sure that no major soil nutrients (NPK) or trace elements are limiting. Soil structure may need to be improved by proper soil management and cultivation. Symptoms due to these factors may be wrongly attributed to pests incidentally present. Sometimes the correct control measure is improved crop husbandry rather than resort to pesticides.

In fields on well-run farms the growth of crops is often remarkably uniform so that the amount of plant material per unit area is relatively constant. In spring, the soil environment below the top few cm is also far more equable than above ground and seasonal and local differences are moderated by the shifts of sowing date they cause (Jones, 1973a). For a time the crop is akin to a laboratory experiment in which food is constant and the initial density of plant-feeding organisms is variable. A relationship between pest density and yield loss probably exists for all pests and is implicit but not precisely expressed by the concept of an 'economic threshold' (p. 2). Relationships between initial pest density and yield loss have been demonstrated for a number of relatively immobile pests such as wireworms and nematodes (Fig. 15.1). For more mobile pests such as the adult stages of many insects, the numbers initially present at a particular stage of the crop is important even though the numbers that cause injury may be the result of immigration and multiplication.

One feature of insect attacks on crops is that they may end abruptly because of the onset of moulting (e.g. wireworms in spring) or pupation (wheat bulb fly, cutworms). If growth has kept pace with injury, the remaining plants grow bigger when feeding ceases and lessen the ultimate effect on yield. Leatherjacket attacks terminate when the pre-pupal stage is reached. As species of leatherjacket differ in the time when this occurs (e.g. *Nephrotoma appendiculata* pupates earlier than other species), the importance of accurate identification is emphasized. Where immature forms of insects are concerned, some assessment of the stage reached is essential before advising direct control measures, and none should be applied if maturity is likely to be reached in a few days and the adult stage is harmless.

Plot size and shape

These are important in attacks by mobile pests. The initial infestations of aphids, such as the bean aphid on bean or beet seed crops, are always denser around the outsides (Taylor, 1962). This is partly because of hedgerow shelter and partly because the field margins are first encountered as the insects fly or drift in air currents. Large fields and plots suffer less from these edge effects because the ratio of periphery to area is small. Small plots or long thin plots may be all edge. This is often so in gardens, allotments and smallholdings where pests like carrot fly, onion fly, bean aphid, cabbage aphid and large white butterfly can be devastating, although here also overcropping, lack of rotation and shelter play a part.

Shelter

Distribution patterns of insects near hedgerows are similar to those about artificial windbreaks but insects originating in the hedgerow have different patterns from those flying or drifting over it from elsewhere. In calm weather hedgerows have little effect on the latter but in windy weather their numbers may increase ten times. Insects accumulate most when the wind blows perpendicularly to the hedgerow and most are deposited on the leeward side at a distance approximately 2–3 times the height of the hedge, with numbers diminishing up to 7–10 times hedge height. This effect varies with the species of insects and their flying habits (Lewis, 1968). Dykes also provide shelter of a different kind around the margins of fenland fields. In windswept places the benefits of shelter outweigh the problems it creates. Elsewhere, whether the value of shelter as a haven for beneficial insects is greater than the harm as a source of pest species is uncertain, for no one has found satisfactory ways of quantifying either.

General pests

These are those pests which have weak host preferences and attack many kinds of crop. They include woodlice, earwigs, cutworms, leatherjackets, rabbits, hares, rats and mice, all of which move over the soil surface and can be attacked by poison baits where appropriate. Wireworms, chafer grubs, millepedes and many migratory soil nematodes also fall into this class but are difficult to kill because they are protected and concealed by the soil in which they live. The silver Y moth is migratory and highly mobile, but its caterpillars feed on plants well above ground level. Although they might take baits, their habits make it unlikely that they would encounter them. As general pests, woodpigeons, rooks and starlings are in a class of their own because of their great mobility.

Pests of cereals

The critical stages of cereal crops are: (1) the seedlings up to the four-leaf stage and the formation of tillers; (2) the state of emergence of ears from the ensheathing leaves; and (3) the flowering stage. Many pests attack cereals in the seedling stage which is the most vulnerable. Of the general pests, wireworms have declined in importance following the use of insecticidal seed treatments and soil insecticides. Leatherjackets are troublesome mainly after ploughed grass in the higher rainfall areas of the north and West Country. Rabbits have declined in importance because of myxomatosis, but their numbers are increasing again. Of the cereal flies and other minor pests (Table 14.2) wheat bulb fly attacks are alleviated by insecticidal seed treatments and curative treatments are possible in late winter and early spring. Many winter wheat crops are sown early and so are better able to withstand attacks which occur after the turn of the year. However, seed treatments applied to these early sown crops may not remain effective long enough. Although it is possible to exert some control of frit fly on oats by curative

Time	Pest	Stage	Notes
BARLEY			
October–November	Frit fly	Third generation larvae	Eggs laid on early sown winter barley, cause dead central shoots, uncommon
	Gout fly	Second generation larvae	Eggs laid on early sown winter barley, cause blind and enlarged (gouted) plants. Uncommon
	Slugs	All stages	Hollow out newly sown grains, graze young shoots, rasp away leaf lamina leaving only the veins. Ends of leaves tattered. Heavy soils only
October–April	Rabbits, hares		Graze foliage, delay establishment
April–May	Wireworms		Rarely serious in barley. Produce ragged holes through ensheathing leaves
	Slugs		Hollow out newly sown grain, graze young shoots, rasp leaf lamina leaving only the veins. Ends of leaves tattered. Heavy soils
April–June	Cereal cyst-nematode	Larvae invade roots	Sometimes causes patches of stunted plants
May–July	Gout fly	Larvae injure developing ears	Plants swollen and gouted. Serious outbreaks rather uncommon, worst on late sown spring barley
July–August	Cereal aphids	Alatae migrate in and found colonies of nymphs, apterae and alatae	Infest leaves and panicles, spread viruses such as barley yellow dwarf and stunt plants
	Woodpigeons, sparrows		Consume grain, sparrows troublesome near buildings
OATS			
October–November	Frit fly	Third generation larvae	Eggs laid on early sown winter oats, uncommon. Larvae move from grass or ley as for young wheat
	Stem nematode		Plants stunted and thickened, tillers distorted. Affects winter sown oats most. On heavy soils only
	Slugs	All stages	Grains not hollowed, graze young shoots, rasp leaf lamina
October–April	Rabbits		Graze foliage, delay establishment
	Opomyza germinationis	Larvae mining central shoots	Eggs laid near plants. Overwinter as larvae mainly in grasses
	Stem nematode	Pre-adult larvae invade seedlings	Most common on moist soils with a high clay content. Plants stunted, thickened at the base, tillers twisted
April–May	Wireworms	Larvae 6 months to 4 years old	Most troublesome after ploughed up grass. Produce ragged holes through ensheathing leaves at base of stem
	Frit fly	First generation larvae from eggs laid on plants	Cause dead central shoots and excessive tillering. Worst on late sown spring oats
April–June	Cereal cyst-nematode	Larvae invade roots	Causes patches of yellowed and stunted plants. Most serious on light soils
May–July	Frit fly	Second generation larvae attack developing grain, from egg laid on the oat panicle	Lessen grain yield. Worst in late sown spring oats after first generation attack on the seedlings
July–August	Cereal aphids	Alatae migrate in and found colonies of nymphs, apterae and alatae	Infest leaves and panicles. Spread viruses and stunt plants
	Oat leaf beetle	Adults and larvae	Feed on foliage, rarely numerous enough to cause serious harm
	Woodpigeons, sparrows		Feed on grain, sparrows troublesome near buildings

Table 14.2 Pests of cereals.

Period	Pest	Stage	Remarks
WHEAT			
October–November	Frit fly	Third generation larvae killing central shoot	Eggs laid occasionally on early sown winter wheat, or larvae move to wheat plants after late ploughing of rye-grass ley. Dead central shoots
	Gout fly	Second generation larvae killing central shoot	Eggs laid occasionally on early sown winter wheat, uncommon. Central shoot sometimes completely absent, plants swollen, leek-like
	Common rustic moth	Caterpillar mines central shoot	Dead central shoots; injury in fields where wheat follows rye-grass ley on which the eggs are laid, most obvious next spring
	Slugs	All stages	Hollow out newly sown grains, graze young shoots, rasp away leaf lamina and leave only veins. Ends of leaves tattered. Heavy soils only
October–April	Rabbits, hares		Graze foliage and delay establishment
	Opomyza germinationis	Eggs laid near or on winter wheat plants in autumn	Larvae cause dead central shoots. Overwinter mainly in grasses
	Geomyza tripunctata		
February–March	Wheat shoot beetle	Adults and larvae active at low soil temperatures	Ragged wounds below, at and just above soil level. Occurs after clover-rye-grass leys on light soil over chalk
March–April	Wheat bulb fly	Eggs laid in previous July and August, most larvae hatch in February	Mine the central shoots, move from dead seedlings to new plants or tillers. Worst on late sown or backward winter wheat. Spring wheats usually escape attack
April–May	Wireworms	Larvae from 6 months to 4 years old	Serious attacks after grassland with large populations is ploughed. Produce ragged holes through ensheathing leaves at base of stem
	Opomyza florum	Eggs laid near early sown winter wheat plants, larvae hatch in spring	Larvae cause dead central shoots. Occurs mainly in grasses
May–June	Late wheat shoot fly	Eggs laid on plants	Larvae kill central shoots which turn bright yellow
	Hessian fly	Larvae feed between leaves and stem	Straws weaken and lodge, turn yellow prematurely, uncommon
May–July	Cereal aphids	Alatae migrate in and found colonies with nymphs, apterae and alatae	Infest leaves and panicles, spread virus diseases
	Wheat blossom midges	Adults oviposit in ears, larvae in florets	Extent of injury not assessed
	Cereal thrips	Adults oviposit in ears, larvae in florets	Extent of injury not assessed
	Wheat stem sawfly	Larvae burrow in base of straw	Rarely more than occasional stems affected
	Gout fly	Larvae damage developing ear	Wheat race of the gout fly, uncommon
June–July	Saddle gall midge	Larvae gall stems	Attacked shoots stay green, ears fail to open
July–August	Woodpigeons, sparrows		Feed on grain. Sparrows important only near buildings

sprays, early sowing remains the best control measure. Except on some dry soils, e.g. light soils over chalk and some coarse sands, the cereal cyst-nematode is controlled naturally by parasitic fungi provided cereals are grown with reasonable frequency. Cultivars of oats and barley that resist the commonly-occurring pathotypes are available. Cereal aphids which are abundant and injurious in some years can be controlled by aphicidal sprays applied from the ground or the air. Transmission of barley yellow dwarf virus by the bird cherry aphid can lead to serious yield losses in early sown winter wheat and barley. Aphicidal granules applied at seeding or sprays applied in November and again in the following June in Wales and S.W. England give effective control. Measures against aphids also control thrips and wheat blossom midges at the flowering stage. The intensity of attack by these insects varies from year to year and the economics of the damage they cause awaits thorough assessment.

Stem nematodes can be devastating pests of oats, especially on clay soils or in moist situations. Resistant winter oat cultivars are available (one or two of these also have separate resistance to cereal cyst-nematode) but currently there are no satisfactory cultivars for spring sowing. Some quick growing ones are tolerant rather than resistant and may leave behind populations of stem nematode damaging to other susceptible crops (e.g. beans, onions). Stem nematode resistant cultivars of rye yield poorly on uninfested land.

The expansion of the area sown with cereals does not seem to have produced any new cereal pests. However, sod-seeding, minimum cultivations and very early sowing have changed the emphasis of certain pests. Slugs are sometimes favoured by sod-seeding and early autumn sowings catch the later generations of cereal flies, common rustic moth and some aphids, notably the bird cherry aphid. Shallower drilling to improve the effectiveness of seed treatments aimed against the wheat bulb fly are thought to favour attacks on seeds and seedlings by field mice and grain-eating birds.

Pests of grassland

Established grassland provides plant food in greater quantity and variety than arable land. Because the soil surface is covered throughout the year by a dense mat of stems and roots, the microclimate is less harsh. For these reasons, grassland harbours more species and greater numbers of animals than arable crops. Populations also seem to fluctuate rather less violently and do less harm than in arable land where ploughing and frequent change of crop creates instability. If powerful insecticides like gamma-HCH are applied to grass, the yield may be earlier but the total yield is not much changed, suggesting that the many insects present do little harm in most fields.

Chafer grubs are sometimes a problem in grassland, in lawns and in sports turf on light soil near woodlands and are difficult to control except by cultivating and re-seeding. Leatherjackets may damage grassland in wet low-lying fields or where rainfall is high. The grass aphid (*Metopolophium festucae* (Theob.)) survives mild winters without egg laying and multiplies excessively the following spring. Antler moth caterpillars occasionally damage hill pasture but are spectacular rather than important.

Because of the low value of grassland, control measures other than adding lime or fertilizer to encourage growth, or ploughing and re-seeding, are rarely practicable. More can be spent on lawns and sports turf. Leatherjacket populations in grass are controlled by poison baits or more simply and effectively by spraying with gamma-HCH. In re-seeding grassland there is always the risk that the many plant-feeding insects that survive ploughing may attack the delicate seedlings which represent only a fraction of the food previously available. They may be protected by sprays or seed treatments containing gamma-HCH. The needle nematode, *Longidorus elongatus* (de Man), is troublesome where grass is grown intensively on peat soils in East Anglia. Spring re-seeding often fails in patches but autumn seedlings succeed, possibly because the soil is then drier. Valuable grass seed crops are best protected against the many grass seed midges that attack them, and cocksfoot against cocksfoot moth, by cultural measures (pp. 78, 143). Aphids are readily controlled by spraying with systemic organophosphorus compounds. Many of the pests endemic in grass attack cereals, especially when they follow grass (e.g. wireworms, leatherjackets, cereal flies, common rustic moth (Table 14.2)). The establishment of leys is sometimes hindered by frit fly larvae which can be killed by appropriate sprays. Rye grass mosaic virus is spread by the cereal rustic mite and can cause serious losses to perennial and Italian rye grass seed crops. Stands sown in spring suffer more than those sown in autumn in the first year. Heavy grazing of spring sown crops in autumn by sheep reduces the incidence of the virus: no recommended chemicals are available for mite control. (See Leaflets Nos 595, 616 and reference book 186.)

Pests of potatoes

Potatoes have no critical seedling stage because they are grown from tubers containing reserves of food adequate to ensure early and rapid growth. Injurious pests must either be very numerous (e.g. potato cyst-nematode), reproduce rapidly (e.g. buckthorn-potato aphid), be insidious, as are slugs and wireworms that feed on tubers, or have other effects such as aphids that spread viruses.

The potato aphids (p. 59) are outstandingly important. Besides directly affecting the yield of ware crops, they enforce seed production away from main crop areas, complicate the process of roguing out unwanted plants, and increase transport costs and the hazards of freezing in transit. Aphids in ware crops can be controlled by organophosphorus insecticides applied as sprays or granules. Seed potatoes in chitting houses can be protected from aphids by gamma-HCH or pirimicarb smokes.

The potato cyst-nematode is also an important pest because it enforces long rotations and prevents the full exploitation of good potato-growing land. Resistant varieties provide a partial solution. Those bearing gene H_1 are effective against *Globodera rostochiensis* which is almost wholly pathotype Ro1 in the U.K. This resistance is ineffective against populations of *G. pallida* which infests about half the U.K. potato land. *G. pallida* populations are mostly mixtures of pathotypes, and cultivars bred to resist them are only partly effective. Available granular nematicides are costly but are used increasingly. The Colorado beetle, now firmly established on the Continent, is less dangerous than previously thought because most potato-growing areas in Britain are probably near its northern climatic limit and because it is readily controlled by modern insecticides. Wireworm holes in potato tubers remain a problem because absolute freedom from wireworm holes is now demanded by supermarkets and by the housewife. Slugs, cutworms and potato-tuber nematodes are important locally and seasonally.

Pests of sugar-beet, fodder beet, red beet and mangold

Sugar-beet growing in Britain was firmly established by 1925. Mangolds have been grown much longer and most pests of beet and its allies stem from mangolds but extension of the beet acreage and the practice of drilling pelleted monogerm seed to a stand has led to an increase in old pests and the appearance of some new ones (Jones & Dunning, 1972) (Fig. 14.1).

Varieties of *Beta vulgaris* have delicate seedlings very susceptible to injury. Hence all seed is treated with methiocarb which is effective against a range of seedling pests. Under good conditions they rapidly become more resistant but lack the high population density of cereals and their power of tillering. Sowing to a stand to avoid singling and applying pre- and post-emergence herbicides reduces the plant food available for pests and makes the seedlings more vulnerable to seedling pests, small birds and small mammals. A complex of pest attack seed and seedling (Table 14.3). Field mice dig up the seed, partridges, sparrows and skylarks graze seedlings on emergence. Woodpigeons strip the leaves of gappy crops and pheasants fell the

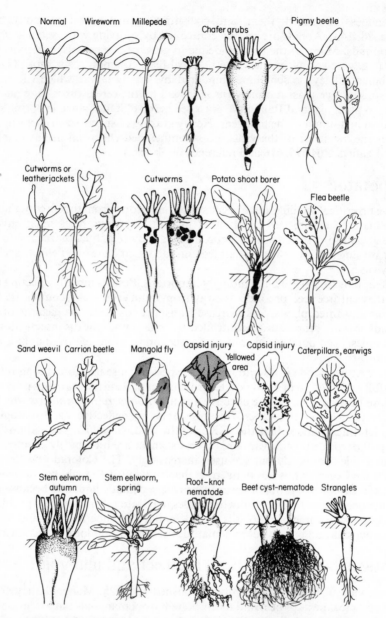

Fig. 14.1 Some examples of pest injury to sugar-beet. Strangles is not caused by insects.

plants. Like potatoes, however, once beyond the 6–8-leaf stage only the more insidious and prolific pests (aphids, cyst- and stem nematodes) cause serious injury. The long vegetative life of sugar-beet, mangold and fodder beet gives ample time for the dissipation of residues before tops or roots are fed to stock. No residue problems arise in sugar-beet roots because these are processed for sugar production.

Pests of beet and its allies are grouped in Table 14.3.

The importance of wireworms has declined since the introduction of insecticidal seed treatments, and crop failures from pigmy mangold beetle were few after crop rotation was

Table 14.3 Pests of beet and its allies.

A. *Seedling pests*		
Group 1	Wireworms Chafer grubs Millepedes Springtails Symphylids	Feed below ground only
Group 2	Field mice Pigmy mangold beetle Cutworms Leatherjackets	Feed above or below ground according to circumstances
Group 3	Small birds Mangold flea beetle Sand weevil Thrips	Feed above ground only
Group 4	Mangold fly larva	Mines the leaves
Group 5	Stem nematode	Invades growing points, petioles and leaves
B. *Pests of the mature crop*		
Group 6	Bean aphid Peach-potato aphid Capsids Beet leaf bug	Suck sap, aphids spread virus diseases
Group 7	Woodpigeon Tortoise beetles Silver Y moth larva	Consume leaves
Group 8	Beet cyst-nematode Stem nematode	Invades and destroys rootlets Causes crown canker in autumn

enforced but may increase now controls have gone. Chafer grubs and stem nematodes are not yet effectively controlled but serious attacks are infrequent. Other seedling pests, including mangold fly larvae within leaf-blister mines and pigmy beetle feeding above ground, are killed by a range of organophosphorus insecticides, but injury by birds has increased and is difficult to control. To kill mangold fly larvae in leaf blister mines and aphids on the leaves, organophosphorus compounds are used.

Attempts to assess insect injury to beet seedlings, by finding outbreaks and eliminating the insects on some plots with insecticides (Jones, 1953; Jones & Humphries, 1954) sometimes fail because the insecticides have unwanted side effects. To avoid these Jones, Dunning & Humphries (1955) defoliated plants artificially and hoed out some of them to give different plant populations thus simulating the attacks of seedling pests. Only complete defoliation lessened yield appreciably (Fig. 14.2). Increase in size of the remaining plants compensated losses in plant numbers. Even when three-quarters of the seedlings were removed, the yield was better than that from redrilling after destroying all the plants at the four-leaf stage (Fig. 14.3). Later experiments showed that complete defoliation in July had most effect, decreasing yield by 40% (Dunning & Winder, 1972) and recurrent defoliation (i.e. grazing) as by skylarks is also detrimental.

The peach-potato aphid and the bean aphid remain the most important pests of sugar-beet after establishment, the former being the principal vector of beet yellows viruses and the latter causing serious losses in dry, epidemic years. Mangold clamps are now uncommon and their role as overwintering sources of virus and aphids has been taken over by discarded beets at loading sites. These should be eliminated before April. By segregating steckling production, spraying steckling beds, or by sowing them under a cover crop, stecklings relatively free from

286

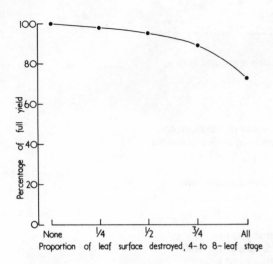

Fig. 14.2 The effect of artificial defoliation on yield of sugar-beet. (After Jones, Dunning & Humphries, 1955.)

virus can now be used to plant seed crops. In this way perhaps the most important source of virus, if not eliminated, has been greatly lessened. The same sprays can be used to control aphids and aphid vectors on seed and root crops. Aldicarb or thiofanox granules applied in the seed furrow control aphids up to the 6–8-leaf stage. Band spraying is also effective. British Sugar Plc operates an effective spray warning scheme. The dry summers of 1975–76 and the appearance of races of *Myzus persicae* resistant to organophosphorus insecticides led to a resurgence of yellows viruses in sugar-beet crops. Together with drought these greatly decrease yields of extractable sugar.

Beet cyst-nematode has spread in the Fens since it was first found in 1934 and has also appeared in many localities elsewhere. Crop rotation, previously enforced through factory contracts and the Beet Eelworm Orders (p. 231) has prevented serious losses in the Fens. Elsewhere infestations have been slow to develop possibly because 'highland' areas grow cereals more frequently, are usually infested with cereal cyst-nematode and are controlled by the fungi that keep that nematode in check. Stem nematodes rarely cause severe crown canker except in low-lying fields in river valleys.

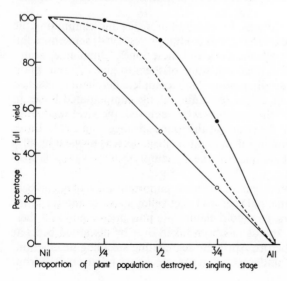

Fig. 14.3 The effect of plants killed upon yield of sugar-beet roots. The upper curve is based on six field trials in which plants were destroyed after singling. The lower curve shows what the relation between plant population and yield would be in the absence of compensation for gaps. The broken line shows the anticipated effect of irregular losses caused by pests. (After Jones, Dunning & Humphries, 1955.)

Pests of brassicas

Brassicas have small seeds that germinate quickly to produce delicate seedlings which grow rapidly. Many crops are sown directly (e.g. mustard, oilseed rape, kale, swedes, turnips and sometimes cabbages and Brussels sprouts). About 70% of all brassica crops excluding swedes and turnips are raised in compact seed beds or in peat/soil blocks or small containers for subsequent transplanting. Crops like oilseed rape and mustard have dense plant populations. Transplanted crops are more widely spaced and Brussels sprout may have as few as 10 000 plants ha^{-1} (one per m^2) which makes individual plants vulnerable to pest attack. Crops grown for seed or for human consumption are valuable and therefore better able to bear the cost of treatment; those grown for fodder, as cover crops or green manure, are rarely treated. Cabbages, cauliflower, calabrese and broccoli are damaged by the caterpillars. Plants near the coast in small plots and those in gardens inland suffer most from defoliation by caterpillars of the large white butterfly. Feeding skeletonizes the leaves and excreta makes the produce unsaleable. A choice of several sprays is available to rid brassicas of caterpillars (p. 88).

Most brassica crops withstand wireworms well but transplants are vulnerable to cutworms, leatherjackets and woodpigeons. Cruciferous flea beetles used to cause many failures of turnips, swedes and kale in southern and central England, but are now controlled by continued gamma-HCH thiram seed treatments which may need to be supplemented by a carbaryl or gamma-HCH spray when attacks are prolonged. Seed treatments also help to control minor pests that attack brassicas in seed beds; cabbage root fly, wireworms, cutworms, leatherjackets, gall and stem weevils as do carbofuran granules applied against cabbage root fly.

Cabbage root fly is a serious pest of early cabbage, early cauliflower and Brussels sprout in gardens, market gardens and wherever brassicas are grown often. In May, within 4 days of setting out, transplants from seed beds should receive routine treatment. Later plantings and transplantings also require protection. A range of treatments is available and the currently recommended insecticides persist for varying periods, persistence being important in relation to the anticipated harvest date (Table 14.4). Late-generation attacks on Brussels sprout buttons are difficult to control. Spraying with a short duration organophosphorus insecticide in a large volume of water is suggested.

The other serious pest of brassicas is cabbage aphid which transmits viruses (p. 34) and forms large colonies that smother, stunt and malform plants. Aphids cast skins and honey dew

Table 14.4 Insecticides and methods of controlling cabbage root fly on brassicas and time required for insecticide residues to dissipate.

	Granules				Drenches seedlings or trans- plants	Spray		
	broadcast before sowing	band appli- cation	spot appli- cation	peat blocks		†overall low volume	‡overall high volume	days to harvest
Carbofuran	+	+	+					42
Chlorfenvinfos	+	+	+		+	+		21
Chlorpyrifos	+	+	+		+	+		42
+disulfoton		+						42
Diazinon			+		+			14
Fonofos		+	+					42
+disulfoton		+	+					42
Iodofenphos							+	7
Trichlorphon							+	2

† before sowing   ‡ to Brussels sprout buttons

may soil cauliflowers and Brussels sprout, lowering their market value considerably. Organo-phosphorus insecticides effectively control the aphids but must not be applied to crops for human consumption or for fodder within a stated period before harvesting (Table 16.1) to avoid leaving toxic residues on or in the plants. They are best not applied after the end of October because slow growth and low temperatures lead to longer persistence.

Brassicas are hosts of the brassica cyst-nematode and beet cyst-nematode. The former is widespread and the latter multiplies freely on brassicas which tolerate large populations usually without showing collapse symptoms. Sugar-beet or mangolds planted after a run of brassica crops may be severely stunted. Therefore, in the areas where beet cyst-nematode is prevalent brassica and beet crops have to be considered as equivalent and only one or other but not both may be included in a four-course rotation intended to control beet cyst-nematode. The great expansion of the oilseed rape area is undoubtedly encouraging both species but the conse-quences will take several years to make themselves felt. Already however several severe attacks by the cabbage cyst-nematode have been reported from wide-spaced localities. One or two have been devastating.

Most of the pests that attack leaf brassicas (Brussels sprout, cabbage, calabrese, cauliflower, broccoli, turnips and swedes) also occur on mustard, oilseed rape and other brassica crops grown as seed but not all are economically important. The expansion of the oilseed rape area has increased their food supply and their potential range considerably. Some pests that tended to be local (e.g. cabbage stem flea beetle, cabbage stem winter weevil) are now found more generally. To the pests that attack the flowers and seed pods (blossom beetles, seed weevil, pod midge and cabbage aphid), spring rape is more susceptible than winter rape largely because the latter matures earlier. Seed weevil is more easily controlled in spring sown rape because adult weevils arrive in the green bud stage and sprays against them can be applied at green bud and yellow bud. Weevils invade winter rape in the early flowering stage and build up throughout the flowering period when spraying is precluded by possible harm to bees. After flowering the crops may be sprayed to control seed weevil and pod midge from the ground (triazophos or phosalone) or from the air (phosalone). White mustard is not attacked by seed weevil and mustard beetle is rare nowadays. Gamma-HCH seed treatments applied against ordinary flea beetles (*Phyllotreta* spp.) give some control of cabbage stem weevil but additional gamma-HCH sprays may be needed. Cabbage stem flea beetle on winter rape may be controlled by gamma-HCH sprays applied in October–November and in March–April when soil tempera-tures begin to rise, the object being to kill adults and young larvae before they penetrate deeply into stem and petiole tissue.

Woodpigeons feed on all brassicas including oilseed rape from mid-December to mid-March. They seem to prefer feeding from the ground and hence plants in and around patches of backward growth, alongside tracks and tramlines are more heavily grazed. Effective control measures against them have yet to be found. (For pests of vegetables see Booklet 2383 and for those of rape see Leaflet No. 780.)

Pests of leguminous crops

Leguminous crops fall into two groups: those with many small seeds to the ha (clover, trefoil, lucerne, sainfoin) and those with fewer larger seeds to the ha and thicker stems (peas, beans). The first group are grown for seed, and are of higher cash value. The seedling and flowering stages are critical.

Leguminous crops are unsuitable food for wireworms and are also somewhat 'resistant' on account of large plant populations or thick stems. All are attacked by pea and bean weevils (*Sitona* spp.) from shortly after germination onwards, especially during fine, sunny weather in early spring which activates the beetles while the plants are still slow-growing. *Apion* and *Sitona* spp. transmit broad bean stain and true mosaic viruses which greatly diminish yields and larvae of *Sitona* feed on the nitrogen-fixing nodules. Attacks on peas and field beans may be controlled by organophosphorus insecticides or by phorate or aldicarb granules applied in the seed furrow. To simulate injury by seedling pests of peas, George, Light & Gair (1962), cut segments from pea leaves and removed growing tops of plants. Removing 12.5% of the total leaf tissue when there were four expanded leaves decreased the yield by 8%. This amount of injury was more than is normally caused by pea and bean weevils and was not worth treatment, although greater injury is sometimes found around headlands. Removal of growing tops, which resembles bird damage, generally resulted in a slight increase in yield but sometimes delayed maturation of the crop.

Field beans suffer severely from bean aphid attacks in epidemic years, but these may now be effectively controlled by organophosphorus insecticides applied as sprays or as granules. Advances in the control of aphids, weevils and weeds make beans a more reliable and attractive crop than previously (McEwen *et al.*, 1981). Spraying at flowering time kills bees and should be avoided.

Pea moth is especially important in peas grown for freezing because even slight contamination may lead to produce being rejected. The incidence of pea moth is greatly influenced by peas grown for drying and canning, being greatest when a large proportion of the local pea crops are for harvesting dry. Pheromone traps have greatly simplified the timing of sprays against pea moth and also coming to a decision when spraying is desirable. Maturation delays caused by woodpigeons, weevils, thrips and aphids cause financial losses. Organophosphorus sprays such as azinphos-methyl kill pea aphid, pea moth, pea thrips, leaf miners and pea midge which is locally common. In sandy soils, pea early browning virus transmitted by *Trichodorus* spp. is sometimes a problem.

The pea cyst-nematode is a potential threat to the pea canning, packeting and quick freeze trade. Areas around the processing factories are favourably placed and tend to grow rather frequent crops. The nematode is present in most of these areas, e.g. Suffolk, where field beans (the other common host plant) have often been grown. Field populations and recorded infestations have begun to increase. To keep the pest in check long rotations of peas and field beans are desirable (1 year in 4 or longer). Oxamyl granules broadcast and thoroughly incorporated into the seed bed just before sowing prevent nematode damage. Nematode infestation inhibits nodule formation and heavily-infested plants run out of nitrogen in late June.

Stem nematode is a serious and devastating pest of red clover. On farms where it is prevalent even long rotations do not ensure protection, probably because weeds act as alternative hosts on which small populations survive. Varieties resistant to attack should be planted or a change made to lucerne or sainfoin. A separate race attacks white clover but is uncommon in Britain. Lucerne is attacked by yet another race of stem nematode which is mainly seed-borne. All seed should therefore be fumigated.

Where clover is grown for seed, clover seed-weevils may greatly reduce the yield. One of the problems in destroying the weevils is the great bulk of the clover crop which prevents insecticide from penetrating downwards between and amongst the stems and leaves down to soil level. Control measures are designed to overcome this (p. 121).

Most other pests of leguminous crops are minor or local.

Pests of umbelliferous crops

Carrots, parsnips and celery are grown on a field scale, especially in East Anglia. All have delicate seedlings but celery is raised in boxes and containers and transplanted, although direct sowing is sometimes practised. Carrot and parsnip seedlings are very susceptible to injury by wireworms, cutworms and leatherjackets. Carrot fly is an important pest of all three crops and large, endemic populations build up in intensive carrot growing areas. Of the two broods of flies (p. 159) the second is most abundant and is active over a long period in the late summer and autumn. Carrot seedlings are protected against larvae arising from the first brood (and from some incidental seedling pests) by seed treatments containing gamma-HCH, but not against larvae arising from the second brood. Disulfoton, triazophos or phorate granules applied to the soil as bands in front of the seed drills (bow-wave method) usually gives good protection. The possibilities of applying insecticides in other ways, such as deep placement alongside the rows are being explored. In Fen soils none of the currently available approved insecticides can be guaranteed to protect maincrop or large carrots from sowing time for more than about 20 weeks. The willow-carrot aphid is important because it spreads motley yellow dwarf virus which greatly lessens yield. In a bad year some direct injury is also done. As the aphid appears early in the year, seedlings need to be protected from the first true-leaf stage with an organophosphorus insecticide. Spraying has to be repeated 14 days later. Of a range of insecticides tried, demeton-S-methyl gave the best results in the 1961 epidemic (Dunn & Kirkley, 1962). The carrot cyst-nematode stunts carrots in a few localities where they are grown too frequently. On sandy soils carrots are sometimes attacked by root-knot nematodes or by root ectoparasitic nematodes. Celery transplants are crippled by lance nematodes in some peat soils. The celery leaf miner causes unsightly blisters on celery and parsnips but is rarely serious. It is easily controlled by a range of insecticidal sprays. Metaldehyde or methiocarb baits may be used against slugs on celery in gardens and allotments, but are rarely needed in fields.

Because carrots, parsnips and celery are grown for human consumption, insecticides used on them need to be chosen with care. Rates of application should be just adequate to give the desired kill and the last applications so timed that no harmful residues are left on or in the plants at harvest.

Pests of onions and leeks

Onions and leeks have delicate seedlings readily destroyed by wireworms, cutworms and leatherjackets. Onion fly is common in gardens and market gardens in some areas and can be controlled by carbofuran granules which also control wireworms and give a measure of protection against other seedling pests. It would be an advantage if all onion seed were fumigated to ensure that it is free from seed-borne stem nematodes. However, heavy attacks indicate that the nematode was already well established before sowing. Although dazomet, aldicarb, carbofuran and some organophosphorus nematicides protect onion seedlings from bloat, surviving larvae invade onion bulbs which rot in store after harvest (Whitehead, Tite & Fraser, 1973). Parcels of land where onions, beans or oats have been attacked are best avoided. Moist loam soils with an appreciable clay fraction are thought to favour the nematode which appears to maintain itself on weeds in the interim between susceptible crops. The peach-potato aphid and shallot aphid infest plants outdoors but are more common in stores; they may be vectors of viruses. Thrips sometimes cause injury during long periods of drought. Leek moth caterpillars mine leaves and stems in a few areas. Thrips are readily controlled by insecticides (e.g. malathion), but several sprays may be necessary to keep down the numbers of leek moths and prevent them laying eggs on leek plants. Where necessary aphids can be killed by organophosphorus insecticides. The chemical must be chosen to ensure that the minimum interval can be maintained between the last application and harvesting (see Table 16.1).

Pests of flax and linseed

Although once commonly grown, crops of flax and linseed are now uncommon. The dense plant populations and the unpalatability of the seedlings make both crops resistant to wireworm attack but they suffer from leatherjackets, possibly because they are often relegated to poorish wet land and grown after grass. Flea beetles (*Aphthona euphorbiae* and *Longitarsus parvulus*) also attack flax and linseed and may be controlled by spraying with gamma-HCH.

Perennial crops

Perennial crops occupy the land for a few years (e.g. strawberries) or for many years (e.g. fruit trees). Sometimes they are raised from seed but more often from runners or young plants propagated vegetatively. Ideally, planting stocks should be entirely free from pests and diseases, but complete freedom is almost impossible to arrange. After establishment, mobile pests migrate into the crop, less mobile pests are brought in and all begin to multiply, including those present at planting. If the perennial crop is of short duration, the stand can be destroyed and a period under other crops used to change the pest fauna as is done by crop rotation for annual crops. Tree crops are relatively permanent, however, and injurious pests must be controlled, especially those that cripple the trees rather than do temporary harm to foliage, flowers or fruit. Arthropods and nematodes that spread virus diseases are especially damaging because infected trees never recover. Trees that die for one reason or another must be replaced. Sometimes the old trees leave behind in the soil a complex of harmful organisms, bacteria, fungi and nematodes, which a young tree finds impossible to withstand. This is often referred to as a re-plant problem: it may arise in replacing single trees or whole orchards. The same kinds of organism may injure mature trees causing patches which decline and die. Sometimes soil sterilants, e.g. chloropicrin, can be used successfully to overcome re-plant and decline problems, but roots and the organisms associated with them may penetrate so deeply that they cannot be reached.

15

Pest Management

Chapter 1 describes how pest problems have arisen and shows that they are bound up with numbers. Subsequent chapters indicate that knowledge of each pest's life cycle and information about its enemies and competitors, i.e. its life-system, is essential if populations are to be managed intelligently. The basis of management is the planned manipulation of pest numbers, the object being to keep them continuously below those that inflict unacceptable damage to the host crop. Control is a term used loosely and often means no more than killing a percentage of the population sufficient to secure a satisfactory yield or maintain acceptable quality without regard to side effects.

Agricultural systems (agro-ecocystems) are man-made, usually consisting of one main plant species of uniform age and limited genetic variability, whereas natural systems contain many species of diverse ages and are self-perpetuating. Most arable crops are of short duration (often less than a year) and now rarely follow each other in set rotations. This makes a sustained policy of pest management difficult. The short vegetative life of the crop also means that there may be insufficient time for enemies to colonize pest populations before the crop is harmed. Changes in the species of crops grown may avoid some pests but it does little to help the build up of certain types of specific enemy. The term 'economic threshold', used in the first and subsequent chapters to indicate the population density above which economic injury occurs, is a convenient and helpful practical guide, but also somewhat arbitrary because the relationship between numbers and yield varies with the environmental stress to which the crop is subjected; the economics of treatment also change. It is not always realized that above ground damage results in a smaller root system because the supply of assimilates, amino acids and accessory substances fabricated by the shoots is diminished. The net effect, whether the plant is attacked above or below ground, is a root system less able to exploit the soil for water and nutrients. This may be important during drought or when the soil is marginally deficient in a particular nutrient.

Where the curve relating yield loss to population density has been studied over a wide range of densities, plant yield and height tend to diminish, not in direct proportion to the increase in population density, but according to some function of the population which gives a diminishing effect with population increase. In other words, yield decreases as numbers increase but more animals are needed to cause the same yield loss when numbers are great. An implication of this relationship is that kills of 80-90% are necessary to bring yield losses within acceptable limits and that the greatest benefits are from the last 10-20% of the animals killed.

The relationship between log number per plant, unit area or unit weight of soil and growth or yield is usually sigmoid (e.g. Seinhorst, 1965; Bardner & Fletcher, 1974). At small pest densities attacked plants may overcompensate, grow larger and yield better in the presence than in the absence of the pest. This phenomenon is well established for nematode pests in pot pests (see Fig. 15.1 A). Compensation for missing parts or missing plants is well known in the field (see Fig. 14.3). When pest numbers are excessive, yield either decreases asymptotically to zero or to a minimum. Seinhorst (1965) has modelled this situation for nematode pests. When the yield is zero at large densities $Y = z^{P-T}$, where Y is the observed yield, z is a constant slightly

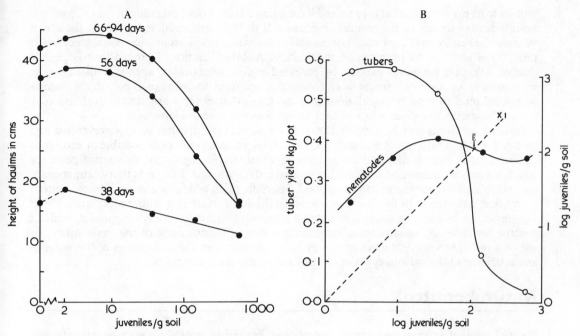

Fig. 15.1. A The relationship between the height of potato haulms measured 35, 56 and 66 days after planting and log population densities of potato cyst-nematode at planting. Heights did not change significantly between 66 and 94 days after planting. **B** Tuber yields and nematode population densities after harvest related to those at planting. The broken line (× 1) is the population maintenance line. Points above it indicate population increase and below it decrease. $\bar{E}$ is the equilibrium point at a little more than 100 eggs/g soil. At densities much greater than $\bar{E}$ nematode numbers after harvest are made up of eggs that did not hatch. Data from Peters (1961). Compare Fig. 1.3, p. 5, where unhatched eggs compose one third (0.33) of the population.

less than one representing the amount of plant tissue left after one nematode has fed, P the population density and T the density threshold after which the yield begins to decrease. Since T is notionally constant z^{-T} is also a constant c and so we may write $Y = cz^{P}$. When there is a minimum yield, this may be represented by Ym, a fraction of the full yield in the absence of nematodes, and the model becomes $(Y - Ym)/(1 - Ym) = Ym + (1 - Ym)cz^{P}$. The constant c has a value slightly greater than one and may be regarded as a measure of the rate of compensation, e.g. $c = 1.05$ would represent 5% compensation. Fitted to the data for haulm height 66–94 days after planting in Fig. 15.1 A, $Ym = 0.22$, $c = 1.05$, $z = 0.994$ and T, the threshold population density is 8 juveniles per g soil, i.e. $c = 0.994^{-8}$.

Although the model often describes field and laboratory data as well, in the field it is difficult to obtain yields in the absence of nematodes and so to estimate the value of c. As written the model takes no account of climatic or edaphic stress and probably the basic assumption that all individual nematodes exert the same effect regardless of where and when they attack is untrue.

Ferris (1978) showed how the yield loss curve for nematodes could be linked to the amount of nematicide needed to give the degree of control required at a given population density and so to achieve the best economic return.

A common feature of pest populations is that reproductive success increases as numbers decline. Therefore, successful control measures often lessen numbers to a point where reproduction is greatest, so that when control is relaxed or its effects wear off, the population soon returns to its previous level or may exceed it for a time. This applies generally to the microscopic nematodes at one end of the animal kingdom and to birds and small mammals at the other. Because the resilience of pest populations makes extermination virtually impossible, control is rarely final and has to be repeated indefinitely. Another feature of populations is genetic change; selection pressures exerted by repeated use of insecticides, agents of disease (e.g. myxomatosis on rabbits), frequent cultivation of resistant varieties (e.g. potato cyst-nematodes), all tend to modify populations in a direction that ensures greater survival and may render the insecticide, disease or resistant variety less effective or ineffective.

Control measures can be classified in various ways. One division is into preventive and curative, another into direct and indirect. Preventive measures are only possible or economic against serious pests that occur annually with considerable regularity, or against pests for which reasonably accurate forecasts can be made: these are few. For pests that occur sporadically only curative measures can be applied. Sporadic pests which feed in the concealment of plants (e.g. cabbage stem flea beetle) or beneath the soil surface (e.g. millepedes), are difficult to counter, for by the time injury is detected it is usually too late to do anything useful. Indirect control measures are usually those aimed at avoidance, modification of the environment or use of a resistant variety. Direct measures have, as their aim, the destruction of the noxious animal. Some of the various types of control are briefly discussed below.

Natural control

Natural control is that measure of population limitation brought about by climate and naturally occurring enemies; a limitation that occurs without direct human intervention. Our knowledge of the enemies of pests in Britain is insufficient though improving. For those animals that are pests, natural control is often inadequate or lags behind population increase and does not become effective until harm has been done. Occasionally however it is highly effective (e.g. the fungi that kill females and encysted eggs of the cereal cyst-nematode, p. 224). An important component in the limitation of arthropod pests is made up of other predatory or parasitic arthropods. The bulk of predatory arthropods in soil belong to the Coleoptera, and some, like the Carabidae and Staphylinidae, range widely and feed as young and adults on all stages of insects, while others feed on small slugs and snails. The effect of soil predators on the abundance of their prey and the unfortunate effects that persistent insecticides may have is well illustrated by work on cabbage root fly (Plate XXXIIIC–E and p. 166).

On plots where DDT, aldrin, and HCH had been broadcast and mixed into the soil, it was found that cabbages were more severely attacked by root fly than those grown on untreated plots. The concentrations of insecticides used were too small to kill the immature root flies but lessened the numbers of predatory beetles present for a period of 3 months. Hughes (1959), Hughes & Salter (1959) and Coaker & Worrall (1961) found that over thirty species of beetles attacked the immature stages of the cabbage root fly in soil. Egg predation accounted for about 60% of the total mortality to which carabids contributed about 24% and staphylinids 34%. The most important predatory species were *Bembidion lampros* (Herbst.) and *Trechus quadristriatus* (Schr.). In one series of experiments, egg survival was directly proportional to the number of *B. lampros* caught in pitfall traps and so prevented from attacking eggs. Both the activity of predators and the survival of eggs were affected by rainfall. Dry conditions in the soil surface favoured predation and increased egg mortality. Early work probably over-estimated predation, but this was later corrected by testing the gut contents of beetles collected against cabbage root fly anti-serum to show whether the food contained root fly or not.

In the 1960s it was observed that attacks of yellows viruses in sugar-beet sometimes increased following the use of DDT or trichlorphon to control flea beetles or mangold fly. Presumably this was due to the elimination of enemies present in the fields concerned at a time when

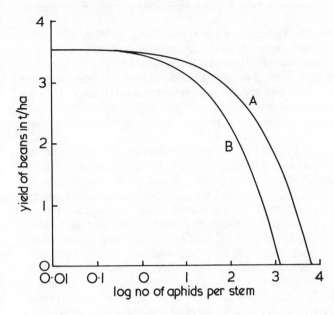

Fig. 15.2 Relationship between yield of field bean seed and log no of *Alphis fabae* per bean stem. **A**, peak numbers; **B**, average numbers between June 21 and July 11 (Data from Way *et al.*, 1958). Other data in the same source suggests that zero yield is approached asymptotically.

migrant aphids, especially *Myzus persicae*, were arriving. As specific aphid enemies were few, suspicions grew that these insecticides were killing ground predators, mostly carabids (p. 95). Since then much work has been done on ground beetles in cereal fields (Potts in Cherrett & Sagar, 1977). Significant inverse relationships have been established between aphid numbers and those of predatory arthropods. In experiments where polyphagous predators (e.g. ground beetles) were depleted with insecticides early in the season, a negative relationship was observed whereas the relationship between specific aphid predators was positive (numbers increased as aphid colonies expanded). Only after peak numbers did the relationship become negative indicating a marked depressing effect on aphid populations (Edwards *et al.*, 1979). This is not to say that they had no effect while aphid numbers were increasing; doubtless without specific enemies they would have increased faster. Although not often seen eating aphids in the field by day when most observations are made, dissections of gut contents implicated many species of ground beetle and a few staphylinide (Sunderland, 1975); most were caught on the plants at night when aphid predation seemed mostly to occur (Vickerman & Sunderland, 1975). In one series of tests, enzymes-linked immunosorbent assays indicated that 36% of spiders, 22% of staphylinid beetles and 16% of carabids fed on aphids when there were as few as one per cereal shoot. At very low aphid densities as in the very early stages of infestation (*c.* one per 36 shoots) the percentages were 23%, 11% and 3%. Evidently spiders and staphylinids play a part but how important that is depends on relative population densities of the three groups (Sunderland & Crook, 1982). Of the ground beetles, *Agonum dorsale* is thought to be especially important.

Above ground level, on and about plants, important arthropod predators belong mainly to the Hemiptera, Coleoptera and Diptera. Parasites mainly belong to the Hymenoptera and Diptera. In addition there are the larger, vertebrate predators, amongst which birds are probably most important. For most pests, the quantitative effects of parasitism and predation on numbers have not been studied, but abroad there are now many quoted instances of the harmful effects of persistent insecticides which have led to increases in pest numbers or the rise of new species to pest status.

Biological control

Biological control is the use of enemies, that is, predators, parasites, fungi, protozoa, bacteria and viruses to control a pest species. Some authors include natural control within biological control, as well as antibiotics and resistant varieties of plants (Sweetman, 1958). The definition could be enlarged further to cover the liberation of sterilized males, individuals bearing lethal genes or harmless animals that compete with pest species and replace them. Hinton (1957), on the other hand, suggested that biological control should be restricted to operations where permanent control is the objective and should exclude enemies used as living insecticides or needing to be replaced annually.

For historic reasons and probably also because of the great variations in climate and the long establishment of agriculture, Europe has more often been the donor of enemies to newly developed territories than the recipient (Franz, 1961). Although biological control has met with a measure of success in southern and south-eastern Europe, there has been little success in the more northerly parts of Europe except in glasshouses. In Britain few examples of successful biological control of outdoor pests can be quoted. Examples include the introduction of *Aphelinus mali* against the woolly apple aphid (p. 138), of *Encarsia formosa* against glasshouse whitefly (p. 69), of myxoma virus against the rabbit (p. 256) and the use of the predatory mite, *Phytoseiulus persimilis* to control red spider mites on cucumbers in glasshouses (p. 174).

Some of the most spectacular successes have been on oceanic or geographic islands in tropical or sub-tropical areas, but equally remarkable successes have been recorded on continental land masses. An example is the parasite *Aphidius salicis* (Hal.) introduced into Australia to control the carrot aphid. It spread to Tasmania and reduced populations so greatly that the carrot mottle-carrot red leaf viruses of which the aphid is the vector have virtually disappeared from carrot plantings in Australia and Tasmania. Few introductions are such outstanding successes and many fail. In the mainly temperate land mass of the U.S.A., up to 1956, some 485 agents were liberated against 77 species of pest. Ninety-five became established and exerted varying degrees of control against 22 pest species (Beirne, 1962). The scale of biological control operations is sometimes very great. In the campaign against the European corn borer (*Ostrinia nubilalis* (Hbn.)) after its discovery in the U.S.A. in 1917, between twenty and thirty million corn borer larvae were imported from Europe and elsewhere to breed parasites. In Canada, between 1910 and 1956, nearly one billion specimens of 220 species of parasites and predators were liberated against 68 pests.

Considerable care and much detailed background study is necessary before choosing an enemy for importation. There are many advantages in choosing an organism specific to the pest to be controlled, because then there is less risk of it transferring to a harmless or desirable species. But insistence on a high degree of specificity greatly limits the field of choice. Non-specific agents seem likely to be used increasingly because, against pests of annual crops, they are better able to survive temporary absence or scarcity of the host species. Successful 'permanent' biological control gives lasting benefits for relatively trifling cost and is especially useful in forests or over great tracts of crops. Temporary biological control requires the annual release of imported enemies or the release of native enemies in enhanced numbers and sometimes the release of the pest to maintain the enemies.

In 1892, the Australian ladybird, *Cryptolaimus montrouzieri* (Muls.), was introduced into California to control the citrus mealy bug. Although it became established, it did not survive the winter in numbers sufficient to prevent the build up of mealy bugs in spring. About 1917, it was found possible to breed ladybirds on potato sprouts infested with mealy bugs, making possible the annual release of millions of ladybirds at a cost less than that of spraying and controlling both citrus and citrophilous mealy bugs. In 1929 the introduction of two hymenopterous parasites from Australia replaced the release of ladybirds for the control of citrophilous mealy bugs but ladybirds were still used in California against citrus mealy bug and against mealy bugs in glasshouses. The ladybird has also been liberated in South Africa and elsewhere. In the above example a simple and cheap method of rearing made mass releases possible but, for many insect agents, the cost of the specialized food and rearing techniques make their use impracticable (Beirne, 1962). Ways of augmenting natural enemies are still being sought (Ridgeway & Vinson, 1977).

In recent years attempts have been made to use microbial agents such as nematodes, fungi, bacteria and viruses. The introduction of myxoma virus into Australia and Europe to control rabbits is an example in which spread was chiefly by insect vectors. Some microbial agents are sufficiently persistent to be sprayed onto crops like insecticides. Polyhedron and granule viruses of insects are especially promising against leaf-feeding lepidopterous and sawfly caterpillars. These viruses are often highly specific and restricted to one species (that from which they were isolated) or to closely related ones in the same genus. Isolates are being sought with somewhat wider host ranges which would infect a range of caterpillars, e.g. noctuids or those that attack cabbages (Burgess & Hussey, 1981). Before viruses can be used against insects much careful testing is necessary to ensure they are unlikely to attack humans or domestic animals. Fungi also attack insects and nematodes. Above ground attacks on insects depend very much on environmental conditions and heavy mortality is sometimes caused by naturally occurring fungi when conditions are favourable, e.g. *Entomophthora* spp., kill many adult wheat bulb flies. Spore suspensions of the fungus *Beauveria bassiana* (Bals.) and suspensions and dusts containing *Bacillus thuringiensis* (Berl.) have been used against insects particularly in the United States. A mixture of spores of *B. thuringiensis* and crystals of protoxin produced by it form a safe, easily prepared and stable microbial insecticide which is non-toxic to plants and virtually specific against larvae of Lepidoptera. In the insect's midgut, enzymes release potent toxins from the crystals that paralyse the gut and destroy its lining (Burges, 1970). *Bacillus popilliae* (Dutky) has been used in the U.S.A. to control the soil dwelling larvae of the Japanese beetle in turf. Various nematode trapping fungi have also been tried against nematode pests, so far without real success (see p. 241). Attempts have been made to use nematode parasites of insects (Welch, 1965; Burgess & Hussey, 1971; Gordon & Webster, 1974). Although some success has been claimed with *Neoaplectana* sp. the principle difficulties are that very large numbers must be cultured and that the nematodes need moisture after release otherwise they remain inactive. DD-136, an undescribed species of *Neoaplectana*, was claimed to be a vector of the bacterium *Achromobacter nematophilus* (Poinar and Thomas) which is responsible for killing the host insect, but this seems doubtful (Poinar & Thomas, 1966).

Control by disrupting reproduction

A means of control of which increasing use may be made is the development of substances that destroy gametes or the ability to produce them. Already ionizing radiations have been used to sterilize males of the screw worm (*Cochliomyia hominivorax* (Cqrl.)), a fly that lays eggs in open wounds on cattle after the manner of the sheep blowfly. Males sterilized by exposing pupae to gamma rays were liberated on the island of Curacao off the mainland of South America. They mated with wild females and lowered the fertility of eggs laid. After 6 months, fly populations were much smaller and only occasional batches of non-viable eggs were laid on wounded goats in test pens. After 12 months no screw worms were found on domestic animals and eradication seemed complete (Bushland, Linquist & Knipling, 1955). The method has since been used in an attempt to rid the south-eastern U.S.A. of the screw worm and on populations of Mediterranean fruit fly (*Ceratitis capitata* (Wied.)) on coffee and citrus in Nicaragua (Rhode *et al.*, 1971).

Chemical gameticides of the hormonal type might be used to achieve the same object by incorporation in baits or by spraying on to food plants. Substances that inhibit the development of the testes and ovaries of insects, birds and mammals have already been tried and show promise. 'Apholate', 'aphoxide', aziridine, tepa and meta-tepa are examples of alkylating agents producing irreversible sterility in insects when ingested or absorbed, without affecting mating behaviour or length of life. They are cheap, sterilize at small doses and are metabolized rapidly but are not specific. Sterilants can be used on field populations or for the mass release of artificially produced and sterilized insects but the results have been limited. Tepa sterilized Mexican fruit flies (*Anastrepha bidens* (Lw.)) provided a barrier to females entering California from Mexico, and house flies were reduced on an Italian island first by killing flies on a rubbish tip with Vapona strips and then liberating tepa-sterilized flies (Campion, 1972). Male sterilization has been attempted by liberating female house flies with pads impregnated with meta-tepa attached to their abdomens. Knipling (1962) points out that rendering an insect sterile is more effective than killing it, because it remains to compete with others of the same species and to prevent normal individuals increasing their breeding success. Triethylaminemelamine inhibited the testes of starlings in captivity when as little as 0.1 mg was added to the food for 3 days. Dosage could be regulated by feeding to obtain the required degree of sterility and so to regulate numbers. The chemical also had adverse effects on the reproductive organs of rats and foxes (Davis, 1962, 1970).

Mass releases of sterile screw worm flies were made until successful eradication was almost achieved. Mass releases of chemically sterilized animals, vertebrate and invertebrate, might also be made but the costs of rearing are great. Research to find new sterilants and ways of applying them to field populations is in progress.

Possibilities of insect control by novel methods, especially by genetic methods, is being considered increasingly. Whitten (1970) suggests that individuals with induced chromosome-inversions and without resistance to dieldrin could be released to displace fly populations that have become resistant to dieldrin. The suggestion is based on the genetical premiss that two forms cannot coexist if the hybrid between them is less fit than either, and in his example the hybrid is assumed to be sterile because of complications at meiosis. An advantage would be that once one strain with a multiple chromosome inversion had been developed, it could lead to a permanent system of control using females with resistance to dieldrin as booby-traps for non-resistant males. Whitten & Norris (1967) found that a dieldrin-resistant female whose thorax was coated with acetone-impregnated dieldrin could kill up to 100 susceptible males during attempted matings under laboratory conditions.

Booby-trap chemicals of the future seem unlikely to be conventional insecticides but will probably be similar to juvenile hormone and its analogues. Masner *et al.*, (1968) found that less than 1 μg of the hydrochloride of methyl-farnesoate would sterilize females of *Pyrrhocoris apterus* (L.) and that males could carry more than 1000 μg without affecting their behaviour adversely. The ratio of the amount that can be carried to the amount required to kill or sterilize the target sex determines effectiveness, and this ratio is much greater for methyl farnesoate than for dieldrin or other known insecticides. Moreover, the impact of sterile insects on control is greater than when the target sex is killed by an insecticide. Another advantage is that resistant insects are not required as carriers. Chemosterilants, if extensively used, are as likely to produce resistance as conventional insecticides but this resistance could also be circumvented by genetic displacement (Campion, 1972).

Other possible genetic methods of interfering with populations include the liberation of individuals with lethal genes of various kinds (conditional lethals) or other disadvantageous defects. Evidently population genetics offers a fruitful field of study from which future methods of pest management could arise.

Hormones

Hormones are chemical messengers and regulators within animals. Moulting and juvenile hormones of insects have been characterized and attempts made to use them as pesticides to disrupt development (Ellis *et al.*, 1970; Jennings, 1983). Moulting hormones and hormone mimics have also been synthesized and rich sources found in some plants, e.g. yew (*Taxus baccata* (L.)), where their presence may account for freedom from insect attack. Chitin antimetabolites (not hormones, Marks *et al.*, in Coats, 1982) have also been tested. A few commercial hormone and antimetabolite insecticides are available abroad, e.g. difluorbenzuron (Dimilin). None have found a practical use on field crops in the U.K. as yet.

Pheromones

Pheromones are chemical signals (semiochemicals, Nordlund *et al.*, 1981) that pass between individuals of the same or different species in the external environment and mediate their mutual behaviour or modify their physiology. By analogy, phytomones are key substances given out by plants that intergrate host finding, ovipositing, feeding or other aspects of behaviour in the animals that feed upon or parasitize them. Pheromones in the aerial environment are volatiles. Pheromones or phytomones in the soil environment are usually water-soluble compounds, e.g. the egg-hatching factors of cyst-nematodes that emanate from plant roots. Air currents are responsible for the pheromone plumes emitted, e.g. by female moths to attract their males. In the air or in the soil, pheromones are effective at great dilution and the sensoria of animals, whether highly organized arthropods or less elaborately organized nematodes, are ultra-sensitive detectors.

Pheromones are named and classified according to their function, e.g. sex pheromones, epideictic pheromones that influence spatial distribution and are thought to minimize competition, alarm and trail pheromones, and kairomones which aid insect parasites in the finding of their prey. Broadly, pheromones can be divided into attractants, arrestants, repellents and deterrents. Depending on population density and circumstances the same chemical may function in more than one way.

Pheromones are especially important in coordinating the colonial life of social insects. In almost all species of bisexual animals one of the sexes produces pheromones to attract the other. The chemical structures of almost 700 attractants have been determined. Most are from Lepidoptera (70%) and Coleoptera (12%). It has been found that the alarm pheromone of aphids (E-β-farnesene) is produced by one of two sets of glandular hairs on *Solanum berthaultii* (Hawkes), a wild potato (Gibson & Pickett, 1983). Allylisothiocyanate, a substance given out by cabbage and related cruciferae, attracts cabbage root flies. Female flies are attracted to expressed juice from swede plants, a favoured host, even when diluted with water to 1% (Holmes & Finch, 1983). Although some progress has been made in the characterization of cyst-nematode hatching factors, the structure of that responsible for hatching potato cyst-nematodes still eludes us. Some sex pheromones, e.g. that of the pea moth, are valuable aids in monitoring the appearance and abundance of pest species so making it possible to avoid unnecessary or untimely spraying. The sex attractant emitted by the pea moth female is (E,E)-8,10-dodecadien-1-yl acetate. The related (E)-10-dodecen-1-yl acetate is less effective but more stable and gives catches related to the density of the local field population. Traps last longer and are not occluded by excessive catches of males that arrive from distances possibly as great as 1 km when the pheromone itself is used (Greenway & Wall, 1981).

Although semiochemicals show much promise more work and sophisticated methods seem likely to be needed to capitalize the active work now being done on them. Knipling (in

Nordlund *et al.*, 1983) considers that primary reliance on natural enemies, plant resistance and non-chemical methods currently advocated and having recourse to insecticides only when pest densities reach economic thresholds, may have to be abandoned. Manipulating pest populations with semiochemicals will take time and be slow to achieve control. Satisfactory management programmes will have to operate over wide areas rather than be concerned with individual fields or farms. As yet there are few examples of successful control by pheromones used to disrupt mating or from annihilative trapping and none from Great Britain (Kydonieus & Beroza, 1982; Kydonieus *et al.*, 1982; Bartell, 1982; Wall, 1984).

Indirect control measures

Indirect measures include crop rotation, choice of crop, manipulation of sowing date, special cultivations and manuring to encourage vigorous growth. Rotation determines the frequency of annual crops, and the more frequently a given crop is grown, the greater the endemic populations of its pests tend to become. Rotation is most effective against immobile or relatively immobile pests such as cyst-nematodes, or insects overwintering in soil. It is ineffective against highly mobile pests (e.g. aphids, blossom beetles, or strong flying Diptera and Lepidoptera), against pests deriving their numbers from wild or semi-wild alternative hosts (e.g. frit fly with many grass hosts) or against pests with wide host ranges. The advantages of rotation on a particular field may be lost unless rotation is practised on neighbouring fields. For example, with pests like the carrot fly, an interval of 1 year is usually sufficient to rid the soil of larvae and pupae but, unless neighbouring land is free of carrots, flies may move in from surrounding fields. Even in an area where a four-course rotation in carrots is followed, on the average, each carrot field is flanked by an adjacent field that grew carrots in the previous year and from which flies will emerge in May. Intensive cropping on a short rotation leads, therefore, to serious pest problems of which the following are examples: pea moth in the Harlow area of Essex, cabbage aphid and cabbage root fly in market garden areas, e.g. Bedfordshire, potato cyst-nematode in Lincolnshire, beet cyst-nematode in the Isle of Ely. The cyst-nematodes are especially difficult to counter because of their persistence.

By changing the species or genus of a crop or altering its position in the rotation a measure of control is sometimes possible. Thus barley or wheat may be used as an alternative to oats on land prone to the oat race of stem nematode or infested with cereal cyst-nematode. Similarly sainfoin (a rare crop these days) on chalk land seems to encourage wireworms and may lead to so-called arable wireworm damage to cereals and roots, whereas lucerne, a possible alternative, does not.

Another indirect control measure is farm hygiene. This includes the eradication of harmful weed hosts, the timely destruction of crop residues and the removal of scrub and shelter.

Although indirect measures are helpful they are rarely entirely adequate and usually have a price. Following strict rotations of delaying planting dates may result in an inescapable financial loss. Economic pressures are such that farmers may find it best to plant the heaviest yielding variety when and where it is likely to perform well and to be prepared to use effective chemical control measures should they be necessary.

Resistant varieties

The appearance of the outstandingly successful organochlorine and organophosphorus insecticides in the late 1940s and early 1950s diverted attention from the breeding of crop varieties resistant to insects and mites. Effective fungicides and nematicides appeared later and compounds for use against viruses *per se* hardly exist even now. Consequently more progress was made in breeding for resistance to viruses, fungi and nematodes than to insects and mites. The physiological contact between most animals and the plants on which they feed is less intimate than that of viruses and fungi and their behaviour is more facultative. The intimacy of the relationship is, however, approached by some insects, e.g. gall-forming aphids, gall-wasps, cecidomyids, some eriophyid mites and some nematodes. Among the latter, cyst-forming, root-knot and other highly adapted nematode parasites induce and require specific tissue responses which result in the formation of transfer cells (Jones, M.G.K., 1981). Stem nematodes also need to be able to disrupt the middle lamella of cell walls. Such parasites, like fungi and viruses, must attain a working relationship with the tissues of their feeding sites which may be disrupted by single gene changes in the host possibly leading to a gene-for-gene relationship (see page 46, Table 15.1 and Jones *et al.*, 1981). Facultative feeders, capsids and root ectoparasitic nematodes, for example, may destroy the tissues on which they feed because their feeding is temporary and they are able to move to fresh sites. Breeding against such mobile pests is likely to be more difficult but once effected require greater genetic change for the pests to circumvent it.

Table 15.1 Presumed basis of the gene-for-gene relationship in potatoes and potato cyst-nematodes.

Resistance genes in potato	Products in sap of induced transfer cell	Products in saliva injected into transfer cell	Reaction between sap and saliva	Corresponding genes in nematode	Effect
h	Nil	Nil	None	nn	Female reproduces
h	Nil	Present	None	NN, Nn	Female reproduces
H	Present	Present	Incompatible	NN, Nn	Female dies
H	Present	Present	None	nn	Female reproduces

Note: Female nematodes with the gene *N* in the dominant condition cannot develop on plants with gene *H* in the dominant condition. Males feed little if at all and develop normally. Under natural conditions *N* genes probably moderated the rate of attack on hosts and were part of the *K* strategy of these highly adapted plant parasites. Resistance based in incompatibility is a form of hypersensitivity.

Resistance is of several types (Painter, 1951; Fowden *et al.*, 1972; Hedin, 1977). Sometimes attack is avoided because the activity of the pest is not properly synchronized with the vulnerable stages of the pest (*escape*).

Sometimes attack is tolerated (*tolerance*) because the host is able to replace lost parts or lost neighbours by compensatory growth, or by failing to react to toxins in host saliva which, in susceptible plants, cause necrotic lesions (*hypersensitivity*). Some hosts are less heavily attacked or avoided because of deficiencies in the train of stimuli that initiate and maintain feeding or trigger oviposition, or because they contain substances that deter or repel (*non-preference*). Again the quality of food provided may be inferior, contain nutrients in the wrong proportions or be unpalatable, or the cells may contain harmful toxins that interfere with metabolism, growth or reproduction (*antibiosis*). Excess fibre, morphological impediments, sticky hairs or other features may also aid resistance.

Escape is a precarious type of resistance depending for its use on the skill of the farmer and the vagaries of weather. Tolerance may allow large populations to breed which endanger susceptible cultivars. Antibiosis and non-preference are matters of degree. When weak only partial resistance results but a pest's capacity to increase may be diminished and controlled by chemicals or other means facilitated, i.e. resistant cultivars integrate well with other forms of control.

Fertilizers affect crop pests indirectly through their hosts; in field crops, nitrogen has the greatest effect on pest numbers, usually leading to an increase while potassium and phosphorus seem to have less effect. This is perhaps not surprising because nitrogen is the nutrient required in greatest quantity from soil. Evidence that organic fertilizers decrease pest numbers and increase plant resistance is equivocal. Although nitrogen may increase numbers and produce lush, nutritive growth, often by leading to more rapid growth, it helps crops to outgrow attacks or to compensate for losses of tissue or parts (Jones, 1976a).

In the sequence of events that begins with host finding and ends with feeding or oviposition, the pest receives a number of stimuli from the host in sequence. First, volatile substances (water soluble ones in soil) may be perceived that initiate searching (*excitants*). Reactions to the same or different compounds or different colours may direct it to its host, where *arrestants* ensure that it stays. *Incitants* initiate biting or piercing and sucking and *feeding stimulants* prolong feeding until appetite is satisfied. A similar sequence ends in oviposition. *Deterrents* and *repellents* may interfere with host finding, feeding and oviposition. Physical stimuli also play a part, including sometimes an appreciation of form, shape, rugosity, etc., and resistance to biting or piercing (by mouth parts or ovipositor). The plant microclimate and gross physical differences, e.g. in the shape and distribution of plant organs, may also affect the issue. For reviews of these factors see Hedin *et al.* in Maxwell & Harris (1974) and Hedin (1977).

It is conceivable that one volatile chemical might mediate all behavioural stages up to the initiation of feeding or oviposition but probably a galaxy of stimuli not all chemical is required to enable a pest with catholic tastes to find one or other of several hosts, and a more specialized feeder to find a few botanically-related hosts amongst many others. Moreover stimuli direct the animal to a particular part of its host. Since stimuli must have a measure of specificity they seem to come from secondary plant components, e.g. terpenoids, acetogens, alkaloids and phenylpropanes (Klun in Maxwell & Harris, 1974) which the host plant and its relatives have acquired during the evolution and to which the pest has become conditioned during its coevolution with the host.

Although plant breeders may lose resistance to pests and diseases in seeking heavy yields and other desirable properties for new cultivars, it is doubtful whether they can be blamed for the origin of new races of pests able to circumvent resistance. Most pest populations contain individuals with unusual genes which resistant cultivars select. Populations differ in different places and against some of these resistant cultivars they may be ineffective when first released. Sibling species and near relatives may also exist and replace the current population when plant resistance depletes or suppresses it. Previously the intruders may have been less fit than the resident population and unable to compete successfully.

The durability of resistance varies greatly. Against highly mobile *r*-strategists with short generation times resistance may not endure long. The best examples are rust-fungi of cereals and leafhoppers attacking rice. Amongst less mobile pests with longer generation times that incline to a *K* strategy, resistance is likely to endure longer. Here cyst-nematodes are a good example (see p. 218). There are, however, some examples of long-enduring resistance, e.g. vine rootstocks resistant to *Phylloxera*. So-called 'vertical' resistance based on major plant genes is more likely to be circumvented than 'horizontal' resistance based on polygenes. The latter are usually less likely to provide resistance as complete as that from monogenes regulating some intimate host-parasite relationship. (For a discussion see Robinson, 1976.) Their effects are usually less spectacular and more general resulting in diminished reproductive success, perhaps combined with a degree of tolerance of injury. This is sometimes referred to as field resistance.

Sources of heritable resistance are sought to establish gene pools of crop plants and their near relatives by collecting expeditions, often on an international basis. Although some preliminary progress has been made with resistance to carrot fly in carrots, to cruciferous pests in radish, calabrese and cauliflower and to onion fly in onions (Wheatley, 1983), there are few, if any, examples of commercially available field or vegetable crops with outstanding resistance to insects. Varieties of lettuce resistant to lettuce root aphid (p. 58) have been bred by the National Vegetable Research Station and swedes resistant to turnip root fly (p. 167) by the

Scottish Crop Research Institute. Cultivars of crops do vary somewhat in susceptibility to aphids which may be related to the way in which they feed. The stylets are inserted intercellularly and reach down to the phloem. The track of the stylets can be followed because around them a sheath of hardened saliva develops. If the stylets fail to contact the phloem they are partly withdrawn and re-inserted. The salivary sheath indicates the track taken and the number of attempts made. There is evidence that sensoria in the stylets respond to changes in the sucrose gradient, the source being the phloem (F. Kimmins, 1982; Pollard, 1971). Thus the track of the stylet is isolated from the surrounding tissues until the phloem is penetrated. The effect of feeding on the phloem seems not to have been studied but the possibility of a hypersensitive reaction and a gene-for-gene relationship exists.

A measure of resistance to the aphid *Sitobion avenae* occurs in some established winter wheat varieties, e.g. Kador, Bounty, Rapier, Virtue, Bouquet and Maris Ranger which is encouraging (Lowe, 1982). Resistance also occurs in some lines of the primitive wheat *Triticum monococcum* (L.) in which the degree of non-preference and antibiosis are correlated. The Moroccan carrot *Daucus capillifolius* (Gilli) is very resistant to carrot fly (p. 159) giving hybrids with cultivated carrot that retain much of that resistance. The hybrids are not resistant enough to be acceptable for sale in supermarkets but saleable roots can be obtained using about a third of the amount of insecticide normally required. Resistance seems to be correlated with the absence of concentrations of chlorogenic acid whereas resistance to aphids in some crop plants seems to be correlated with its presence. Resistance to *Aphis fabae* and *Acyrthosiphon pisum* occurs in relatives of *Vicia faba* (field bean, broad bean), e.g. in *V. narbonnensis* (L.) but has not been exploited. In wild *Vicia* spp., a thick cuticle may prevent penetration of first instar nymphs, hairiness at preferred feeding sites hinders settling and feeding and uncommon amino acids are toxic or deter feeding (Anon, 1983). Resistance to the cereal leaf beetle, *Oulema melanopa*, occurs in some unusual foreign wheats and is related to the degree of leaf-hairiness (Webster *et al.*, 1975).

The production of heavy yielding cultivars with good quality, desirable agronomic traits and an acceptable degree of resistance to a range of pests and diseases is an arduous and expensive task. Some of the properties desired may be mutually incompatible, e.g. the presence of glucosides to control pests in a wild species (e.g. in *Solanum chacoense* (Bitt.) (Sinden *et al.*, 1977) conferring resistance to Colorado bettle) or other undesirable traits (e.g. tuber shape in *S. bertaultii* (Hawkes) possessing aphid trapping and repelling hairs). Moreover the work is on-going as cultivars are out-classed or fail for various reasons including the depredations of races able to circumvent resistance or the appearance of new pests occupying the ecological niche freed by successful resistance. From initial crosses to the production of cultivars that have survived National Registration Trials rarely takes less than 15 years. Perhaps a comparison should be made between the costs of plant breeding and that of producing a new insecticide. Although the comparison would be difficult, an attempt should be made. As breeding for resistance increases and the newer techniques of genetic engineering, protoplast fusion and tissue culture take effect, the prospects of getting cultivars with resistance to major pests will improve. Using traditional plant breeding techniques does not necessarily require information about the mechanisms of resistance or their inheritance but, for some of the newer techniques to succeed, such information is essential.

Potatoes vary in susceptibility to attacks by slugs, e.g. Maris Piper and other ex *andigena* varieties are more susceptible than those that lack *andigena* genes. Commercially usable resistance against cyst-nematodes is found in oats and barley and to potato cyst-nematodes in potatoes and tomatoes, and resistance to races of stem nematodes in oats, clovers and lucerne. Usually resistance to nematodes is monogenetic (vertical) and its durability reasonable but not permanent. Polygenic resistance and tolerance to cyst-nematodes are being sought in potatoes. Considerable differences in tolerance have been observed (Table 15.2) but screening for tolerance is more difficult than screening for the lack of ability to reproduce (Evans, 1982) and

Table 15.2 Tolerance to nematode attack as shown by growth and yield of four potato cultivars in land lightly and heavily infected with *Globodera rostochiensis* Ro 1 (Evans, 1982).

Cultivar	6½ weeks		*Weeks after planting* 9 weeks		30 weeks
	Shoot weight g	Root weight g	Shoot weight g	Root weight g	Tuber yield t/ha
	Lightly infested, 5–12 eggs g⁻¹ soil				
'Dell'	297	36	698	56	46
'Peer'	230	26	471	47	25
'Crown'	239	28	657	48	56
'Cara'	293	24	1009	89	59
	Heavily infested, 95–131 eggs g⁻¹ soil (% uninfested)				
'Dell'	105 (35%)	27 (75%)	118 (17%)	15 (27%)	13 (28%)
'Peer'	48 (21%)	17 (65%)	145 (31%)	25 (53%)	11 (44%)
'Crown'	141 (59%)	37 (132%)	283 (43%)	63 (131%)	31 (55%)
'Cara'	208 (71%)	46 (191%)	652 (65%)	119 (134%)	55 (93%)

Note: 'Cara' is resistant and tolerant 'Crown' is susceptible and tolerant, 'Dell' is susceptible and intolerant. Resistance and tolerance seem to be inherited independently. Nematode attack increased the root weight of 'Crown' and 'Cara'.

may be complicated by resistance/tolerance of other pathogens, e.g. *Verticillium* (wilt.) (Evans *et al.*, 1982).

Attempts to introduce pest resistance into well-established cultivars almost always exacts a penalty expressed as diminished yield or poorer quality. The nematode-resistant potato cultivar Maris Piper has succeeded largely because it is an outstanding variety and is grown as such on land infested with *Globodera pallida* to which it is susceptible. Economic factors may force growers to plant desirable susceptible cultivars and give them protection rather than plant inferior cultivars with resistance. Housewives and supermarkets are also extremely conservative. In the Netherlands, consumer resistance has slowed the use of nematode-resistant potatoes in favour of the well-established, yellow-fleshed cultivar Bintje. In that country, in contrast, the potato starch manufacturers accept rapidly any new cultivar that improves starch yield and extraction, including ones resistant to nematodes.

Direct methods

Direct methods are those that repel, deter, trap or kill. Little success has been had with repellents or deterrents against birds or insects in Britain. The substance most commonly tried against insects has been naphthalene in one form or another. Deterrents such as scarecrows also soon become ineffective against birds. Physical methods include shooting, trapping and the use of heat, ultrasonics, electricity or radiation. Shooting and trapping are limited to birds and mammals. Insect traps (e.g. corrugated paper for codling moth pupation, grease bands for winter moth females, light traps for night flying moths) have not met with much success nor has the trap-cropping of cyst-forming nematodes by sowing a susceptible crop and destroying the plants before the immature females produce eggs. The very high energy applications required to kill nematodes or insects in soil make it unlikely that electrical or radiation methods will have practical value. Heat treatments are used against insects in seeds, or nematodes in bulbs and transplants. Often, however, the thermal death point of the pest is too close to that at which serious harm is done to the plant. Heavy rain often clears a crop of small insects such as aphids, so high-pressure sprays might be used to eradicate pests and irrigate at the same time (Beirne, 1970).

The greatest advance in direct control in the last two decades has been in the development of pesticides, that is chemical compounds used to kill animal pests, especially insects. Direct chemical control measures were first tried against the Colorado bettle around 1850 when it began to spread from the wild *Solanum rostratum* to cultivated potatoes but, until the 1940s, growers of field crops had to rely mainly on indirect control measures. Almost the only pesticides that could be applied to crops were arsenical compounds or plant extracts such as rotenone, pyrethrum, quassia or nicotine. The arsenical compounds had some persistence and gave protection against biting insects. The plant extracts were useful against insects with biting or piercing mouthparts but were ephemeral and killed only those insects hit by the dust or spray (rotenone, pyrethrum or quassia) or caught in transient pockets of toxic vapour (nicotine). Use of these materials was largely confined to fruit, to crops of high cash value or to field crops where drastic measures were essential to avoid total loss of crop.

After the discovery of the insecticidal properties of DDT in Switzerland in 1939, the picture changed rapidly. HCH soon followed and the organochlorine group of compounds was vigorously studied and exploited. Because of the persistence and activity of these compounds, they lessened cost and made possible the economic treatment of field crops of relatively low cash value. Costs were further cut by the introduction of low-volume spraying and by the introduction of seed treatments. Another important advance, pioneered in Germany, was the systemic organophosphorus compounds which came into use after 1946. Despite the toxic hazards, they had many advantages and recent trends have been towards compounds with lower mammalian toxicity, or less dangerous formulations such as granules and seed treatments. All these developments followed each other quickly and represented the greatest advance in pest control for half a century.

The rapid advances in insecticides were followed later by advances in compounds to control other pests. The oxime carbamates are effective nematicides but are non-specific and very toxic to mammals and therefore are usually formulated as granules for soil application. Warfarin greatly improved the control of rats and mice but the appearance of resistance in rats created a need for alternatives. Safe and effective means of killing or deterring noxious birds have still to be found. Control of mites is only partly sloved; compounds better than methiocarb, metaldehyde and copper sulphate are needed for use against slugs.

Selectivity of control measures.

One of the undesirable features of pesticides is that, besides the target species against which they are aimed, they kill many non-target species in the same environment. Cultural practices, light traps and juvenile hormones are usually also non-selective whereas attractant baits, pheromone traps, resistant cultivars and biological sprays are moderately selective. All these measures and some synthetic pheromones kill the same proportion of the population regardless of its density, i.e. they act as density-independent factors (p. 4), and although they would tend to diminish the rate of increase, they would contribute little to population stability. Genetic techniques and natural sex pheromones operate most effectively at low population densities. Pathogens, parasites and predators are more efficient at high population densities (see Table 15.3).

Phenological control

Manipulation of harvesting dates to curtail reproduction by pests has been termed phenological control. A good example is the early harvesting of early sown quick-maturing (first early)

Table 15.3 Control methods, their selectivity and relationship to population density. (Partly after Knipling 1972.)

Method	Selectivity			Density independent	Density at which most effective	
	Non selective	Moderately selective	Very selective		Low	High
Specific pathogens, parasites, predators			+			+
Generalized pathogens, parasites, predators		+				+
Genetic techniques			+		+	
Sex pheromones			+		+	
*Synthetic pheromones			+	+		
Biological sprays		+		+		
Attractant baits		+		+		
Resistant cultivars		+		+		
Juvenile hormones	+			+		
Light traps	+			+		
Pesticides	+			+		
Cultural practices	+			+		

* Used to block responses.

potato cultivars on land infested with potato cyst-nematodes (p. 223). The practice is possible only where the winter climate is mild and damaging frosts are rare, e.g. the Channel Islands, and coastal areas in Cornwall, S.W. Wales (Pembroke) and Strathclyde (Ayr). After planting, soil temperatures remain low (below 10°C) for a considerable time during which the potato plants establish root systems and escape attack. Encysted nematode eggs hatch and the juveniles begin to invade the roots when the soil temperature passes a threshold value of about 10°C. Thereafter the development of the females is well modelled by accumulated temperature in day degrees above this threshold (Jones, 1975). A certain number of day degrees are required to the peak of root invasion, more are required for the females to develop to sexual maturity and more still for the development of a full complement of eggs. Harvesting a fixed number of days after planting is unsatisfactory in seasons when temperatures are above average, especially in the last phase when fecundity is determined. A few extra days of egg production at the higher temperatures then prevailing may be critical (Webley & Jones, 1981). Commercial soil temperature 'intergrators' are available which would simplify the collection of accumulated temperatures in individual fields.

As with most other control measures, if practised long enough, populations may be selected that are better adapted to low temperatures (Hominick, 1979) or favour a related species already so adapted (Webley & Jones, 1981).

The use of accumulated temperatures to represent physiological rather than chronological age in computer simulation of the growth of host plants and their pests and in population growth of the latter is increasing. For pests it would be better to use temperature-development equations which can be fitted to experimental data (Jones, 1983). Unfortunately, for most pests such data is lacking. The usual relationship has the form shown in Fig. 15.3. The simpler temperature accumulation line can be fitted to the rising portion of the curve. The best fitting threshold temperature is where this line cuts the x-axis. Strictly there is also an upper threshold at the peak of the curve. The straight line is a poor fit near these thresholds and bears no relationship to development at temperatures beyond the peak. In the U.K. most pests operate below the temperature-development peak which can usually be ignored.

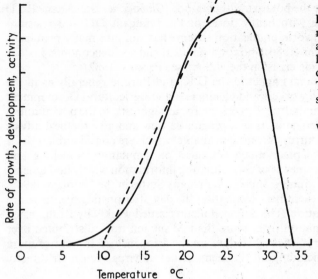

Fig. 15.3 The relationship between the rate of growth, development or activity and temperature. Solid line, form of relationship in the few species for which the data covers the full temperature range. The data can usually be fitted by a descriptive curve of the type:

$$R = A + (B + CT)/(1 + DT + ET^2)$$

where R is the rate T is the temperature and A, B, C, D and E are constants or coefficients (Jones, 1983).

Integrated pest control and pest monitoring

The idea of combining a number of control measures into a management scheme on a field or farm basis is not entirely new. However, increasing disenchantment with chemical control in the U.S.A. in the 1960s and early 1970s arising from the resurgence of old pests and the rise of secondary pests to primary status following the destruction of natural enemies and the appearance of resistance to insecticides as well as their other undesirable side effects, gave impetus to the development of new concepts in pest control. This was especially so for cotton, citrus and top fruits to which chemical treatments had been lavishly applied for a considerable period. Huffaker and his associates championed one new system and Knipling and his colleagues another. Both used the term 'integrated' in their descriptions and this has led to some confusion. Huffaker's approach was ecological, perhaps more correctly *synecological* taking into account many factors in the environment and has given birth to Integrated Pest Management (IPM) *sensu stricto*. Knipling's approach was *autecological*, dealing with one pest species only and based on a detailed understanding of its population dynamics. This is best termed Total Pest Management (TPM). The aim of IPM is to keep pest populations below their economic thresholds, preferably without the use of chemicals. The aim of TPM is eradication over a wide area using a combination of techniques including augmentation of natural enemies.

Guidelines for an IPM programme are in Table 15.4. Because each individual pest differs in its population strategy, no similar generalized guidelines can be given for TPM. The eradication programme for the screwworm fly (*Cochliomyia hominivorax* (Coquerel)) relied on the release of sterile males. That for the cotton boll weevil (*Anthonomus grandis* (Boheman)) also relied on the release of sterile males but combined with insecticides timed to kill individuals destined to overwinter in diapause and infest next year's crop (reproduction–diapause control or r–d control). Eradication is a difficult goal and, even if successful, re-introduction is a distinct possibility. Eradication may succeed against a small introduction of an alien pest near the limits of its climatic range (e.g. Colorado beetle in the U.K.) but is less likely to succeed against robust species whether endemic or naturalized. Though not part of a full-blooded TPM programme, the attempts to eradicate the potato cyst-nematode (*Globodera rostochiensis* Ro 1) from Long Island, New York State with heavy doses of the fumigant DD and resistant cultivars were unsuccessful. The bollworm eradication scheme has run into many problems.

For an account of IPM see Huffaker (1980) and for a political and socioenconomic history of IPM and TPM and of the 'insecticide crisis' in the U.S.A., see Perkins (1982).

IPM and TPM have not had the same impact in the U.K. and Europe generally as in the U.S.A. That is not to say that research workers and members of the Agricultural Development and Advisory Services (ADAS) are unmindful, disinterested or antagonistic to the potentialities of IPM. Those in contact with growers must take a pragmatic view and give the best advice they can with the knowledge and resources currently available, which are considerable.

Some IPM schemes in the U.S.A. place much emphasis on simulation modelling and computer-based 'delivery systems', i.e. means of disseminating information widely and quickly over large areas. The U.K. is a small country, smaller than many States in the Union, and 90% of the letters sent by first-class post reach their destination the day after posting, i.e. there are fewer pressures for faster communication. The Aphid Bulletins issued weekly by Rothamsted Experimental Station contain information from more than 20 suction traps distributed over the length and breadth of the country. These give data on the frequency of capture of more than 30 species of winged aphids, including all those important as virus vectors or directly

Table 15.4 Guidelines for an integrated pest control programme. (Adapted from Huffaker, 1980.)

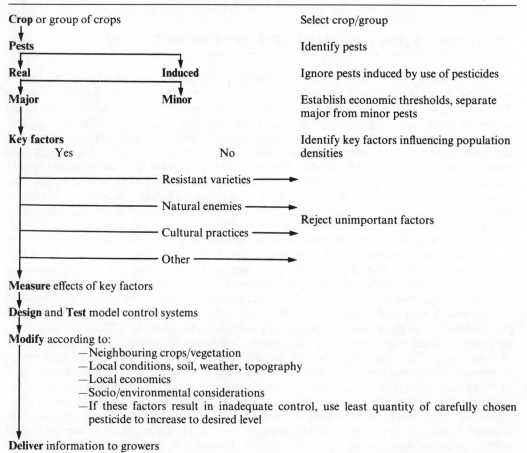

Crop or group of crops		Select crop/group
Pests		Identify pests
Real	**Induced**	Ignore pests induced by use of pesticides
Major	**Minor**	Establish economic thresholds, separate major from minor pests
Key factors		Identify key factors influencing population densities
Yes	No	
Resistant varieties →		
Natural enemies →		
Cultural practices →		Reject unimportant factors
Other →		

Measure effects of key factors

Design and **Test** model control systems

Modify according to:
- —Neighbouring crops/vegetation
- —Local conditions, soil, weather, topography
- —Local economics
- —Socio/environmental considerations
- —If these factors result in inadequate control, use least quantity of carefully chosen pesticide to increase to desired level

Deliver information to growers

injurious to field crops. On reaching recipients, the data is only 5–12 days old. During spring when winged aphids (e.g. *Myzus persicae* and *Aphis fabae*) are migrating to spring-sown crops, capture in suction traps usually precedes their detection in the fields. In summer and autumn, capture usually follows the appearance of alatae in crops and surrounding vegetation and supplies warning of the return migration to autumn-sown crops (e.g. *Rhopalosiphum padi*, vector of barley yellow dwarf virus to early-sown winter cereals). Rothamsted also co-ordinates the recording of large Lepidoptera caught in almost 300 well-dispersed light traps: these are operated by amateurs and began operation in 1933 (Taylor *et al.*, 1969–82, 1981–82). Monitoring schemes are also operated by ADAS for several pests, e.g. wheat bulb fly, pea moth, cutworms (*Agrotis segetum*), leatherjackets, and much general intelligence on other pests is collected at the Regional Centres. Access to this information, which is updated daily or as often as necessary, is available via a recorded telephone advisory service. When possible, forecasting systems are based on data obtained by monitoring coupled with the assessment of the economics of pest damage. Such systems can and do avoid unnecessary and untimely treatments, save farmers much money and are environmentally desirable (see foliage sprays, Table 4.2, page 63, compare years 1981 and 1982).

Information gathering and feedback from growers is an essential part of any control programme. Computers and modern electronic information transfer seem likely to play an increasing role in pest control.

To operate a thoroughgoing programme integrated control in any agricultural or horticultural system would demand much knowledge of the pest, its genetics and the behaviour of all the other organisms in the system. It is debatable whether the time and expense of acquiring all this knowledge, much of which may prove unusable or be rendered obsolete by new techniques, is useful. An alternative or complementary strategy, though less favoured, depends on the finding of selective pesticides or of using existing ones in ways that are selective, and knowing which species it is essential to preserve. Trends to overcome the undesirable properties of pesticides include shorter persistence, easy elimination from food chains and ready degradation to harmless end products (Price-Jones, 1968).

A general pesticide may be made to act selectively in various ways. Thus particles of insecticide may be coated with zein or encapsulated in gelatine so that it ceases to act as a contact poison and becomes exclusively a stomach poison, killing only phytophagous insects. A pesticide may be applied when the enemy to be preserved is absent or in a resistant stage. The pest's behaviour may also suggest ways of control less harmful than blanket treatments. In Switzerland, it was found that cockchafers, during the swarming flights in May, congregated about trees on the highest features of the landscape, and only these trees needed to be treated with persistent insecticide. Reducing the dosage of a pesticide is sometimes a means of increasing its selectivity and this may be further increased by placing the pesticide where it is most needed as in seed treatments, granules or baits.

A psychological factor against integrated control is the desire of the grower, suffering serious financial losses, to see an immediate and effective kill. Sometimes, also, unrealistic demands are made by the merchant or consumer. Thus, potatoes with even a few wireworm holes, carrots with an occasional carrot fly mine, or tinned produce containing occasional insects or insect parts such as legs or antennae, may be rejected out of hand. It might sometimes be better if these impediments were taken as signs of freedom from undesirable pesticide residue. Often, however, only powerful and adequately persistent pesticides are able to give the required kill of especially noxious pests, and against these perhaps the only satisfactory alternative is to breed for resistance.

A system incorporating crop rotation, soil fumigation and resistant varieties of potato but ignoring enemies was devised for the potato cyst-nematode and is being used in the Netherlands (Nollen & Mulder, 1969; Jones, 1973b). The basis of the system was to secure a 'kill' exceeding 90% of preplanting numbers. An increase rate of 100 times or more is then necessary for the

310

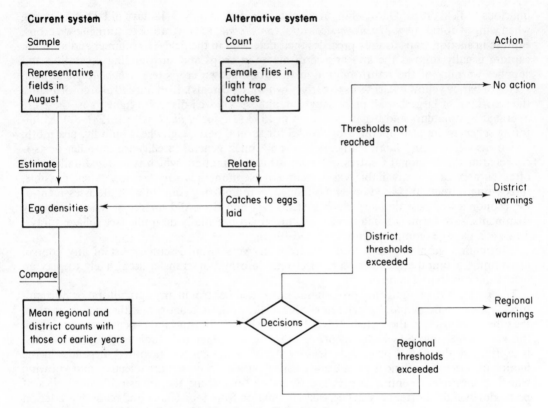

Fig. 15.4 Systems diagram for predicting wheat bulb fly attacks.

nematode to replace its numbers by harvest, but as the maximum crude increase rate of established infestations is rarely more than 25 to 50 times this replacement is not achieved. Selection by the resistant cultivars was ignored.

Simulation modelling and pest control

Modelling was mentioned in Chapter 1, p. 6. A simple model may be no more than an equation which adequately describes a particular stage of population growth, the relation between yield and population density or that between yield and financial return. More complex models usually contain a number of relationships, several variables which must be supplied (e.g. weather data, pest counts) and a number of parameters. The last are quasi constants used in equations the values of which must be determined by experimentation or deduced from data in the literature (e.g. rates of population growth or decline, rates of spread).

Models are made for various reasons: to understand how populations work, to indicate where research is needed, to improve ability to forecast effects of rotations, pesticides, resistant cultivars, enemies of juveniles and adults, and to understand and quantify the effects of competition within and between species. Model building usually begins with systems analysis. For example, the wheat bulb fly warning system (Fig. 15.4), or a more ambitious one taking in a major pest or pathogen and a range of interrelated factors (Fig. 15.5). Modelling invariably involves simplification and discarding all but essentials. It also requires much thought (Wit, 1970).

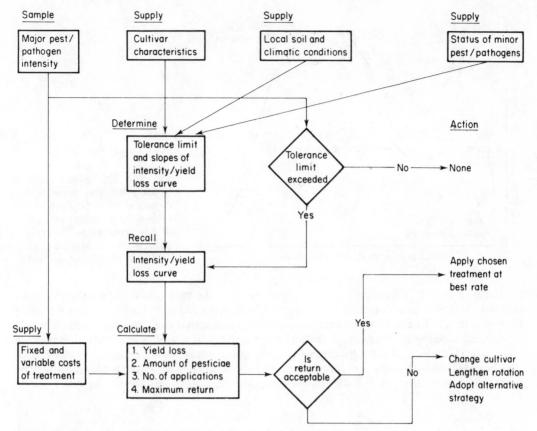

Fig. 15.5 Systems diagram for control of a major pest or pathogen.

Conceptual models

Conceptual models help understanding, a useful example being that of Southwood & Comins (1976). First consider Fig. 15.6, which generalizes the relationship between population density at time t (P_i) and time $t+1$ (P_f), i.e. between one generation and the next. The $\times 1$ line indicates an increase rate of 1. At values of P_i less than that corresponding to U on the $\times 1$ line, a bisexual species is *underpopulated*, its density too small for an adequate number of encounters between the sexes and, therefore, doomed to extinction. At S, the population is in equilibrium with its natural enemies. Above R, the so-called *release point*, its density is too great for natural enemies to prevent increase. The *crash point* is when numbers are just in balance with available environmental resources (food or space); this is also the *equilibrium point* about which populations oscillate. Populations greater than the equilibrium point must decrease. To understand Fig. 15.7 it is necessary to explain what is meant by r- and K- selection and their relationship to habitat stability. The growth rate of a population is r in the equation $dP/dt = rP$, where P is the number of individuals at any given time t, which may also be written $P_t = P_0 e^{rt}$, where P_0 is the number of individuals at $t=0$, P_t at time t and e is the base of Naperian (natural) logarithms. Hence r is the infinitesimal or innate capacity for increase and antilog e^r the finite increase rate (number of times a population increases in unit time). a' in Fig. 1.2, page 5, is e^{rm} or antilog$_e$ r_m where r_m is maximum value of r under ideal conditions.

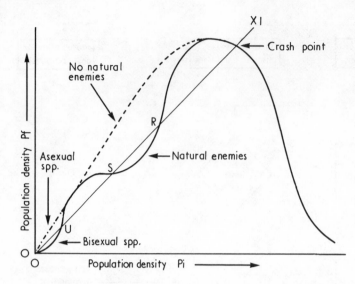

Fig. 15.6 The general relationship between population density at time t (P_i) and $t+1$ (P_f) at the beginning and end of one generation. The crash point is the equilibrium point E. For explanations see text.

Similarly, a in Fig. 1.2 = e^r or antilog$_e$ r_0 where r_0 is the rate observed in practice. As a population grows, it uses up resources and growth slows according to the logistic law $dP/dt = r$ $(K-P)$ where K is the equilibrium density (sometimes represented as E_l, the logistic equilibrium) or carrying capacity of the environment. The life strategy of a species has evolved in relation to its habitat of which, in plant feeding species, the host plant is a major component. An animal that has evolved in a stable, i.e. 'permanent' habitat tends not to overshoot its resources by more than a small margin. If it did the trait responsible, usually an excessive increase rate or

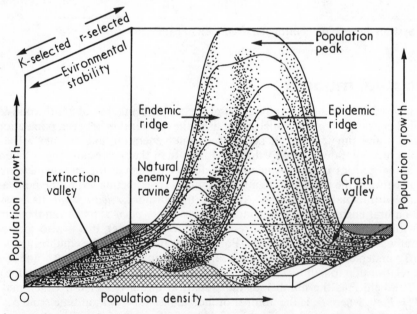

Fig. 15.7 A general or synoptic model to conceptualize the relationship between population density and population increase for populations occupying different positions in the r/K continuum. For explanations see text. Adapted from Southwood and Cumins (1976).

excessive damage to its host would be subject to negative selection. Hence, a conservative *K*-strategy is followed that minimizes population fluctuations. In animals living in an unstable, impermanent environment, rapid increase to exploit its food supply while it lasts is desirable and overshooting, if it occurs, has no effect on subsequent generations which develop elsewhere. These animals are *r*-selected. In fact *r*- and *K*-selection represent the two extremes of a continuous range of habitat types and pest species, and the degree of *r/k* selection exhibited by pests has a bearing on their exploitation of crops and the effects of natural enemies. Figure 15.7 is a general or synoptic model. Each curve of which it is composed is a variation of that in Fig. 15.6 simplified by omission of the effects of sexuality and of those due to the absence of natural enemies. It illustrates how populations of different *r/k* selection are likely to behave and at what population densities they can be expected to be endemic or epidemic, to become extinct or to overshoot resources with a resultant population crash. In *natural enemy ravine* numbers are held significantly below the carrying capacity of the environment which may or may not exceed the economic threshold. The aim of IPM is to use and deepen the ravine by encouraging, augmenting or supplementing the effects of natural enemies (Ridgeway & Vince 1977).

Population models

Many models are concerned with various aspects of population dynamics, population genetics and interspecific competition with the object of elucidating the best strategy or tactics to employ in controlling pests (policy models). Insect populations are often complex with overlapping generations and the effects of immigration and emigration (Gurney *et al.*, 1983). A simpler example is the model devised for potato and other cyst-nematodes which pass one generation a year (Jones & Perry, 1978; Jones & Kempton, 1978). The population segment of the model relates to changes in small areas of soil planted with a susceptible cultivar. The relationship $P_f = P_i$am' in Fig. 1.2, page 5, is replaced by $P_f = (1 - C_p) P_i$am'$ + C_p P_i$ where the parameter C_p is the fraction of eggs that fails to hatch and $(1 - C_p)$ the fraction that hatches and participates in population increase. m', the coefficient that modifies the rate of increase *a*, can be derived from the life cycle of the nematode (Jones & Kempton, 1978) and is in fact itself a derivative of the logistic law. It takes the form $m' = 1/(1 + (a-1)P_i)$ when population densities are expressed in terms of E_l the logistic equilibrium density, i.e. $P_i = \bar{P}_i/E_l$ and $P_f = \bar{P}_f/E_l$ where $\bar{P}_i$ and $\bar{P}_f$ are expressed as eggs per g of soil and P_i and P_f are dimensionless. When P_i is very small, $m' \to 1$ and *a* exerts its full effect. When $P_i = 1$, $\bar{P}_i = E_l$ and $m' = 1/a$, so the rate of increase $= a \times 1/a$ or 1. When P_i is larger than 1, $m' < 1/a$, hence the rate of increase falls below 1. To take account of the decrease in the food supply resulting from damage to the root system, the expression cz^{Pi} from Seinhorst's (1965) equation (p. 293) is incorporated into m' to give m (Fig. 1.2, p. 5). Now $cz^{\bar{E}}$ measures root damage at the real equilibrium point where $\bar{E}$ is the population density in the diminished root system and $\bar{E}/E_l$ measures the corresponding reduction in population density. As these two measures are equivalent we may write $cz^{\bar{E}} = \bar{E}/E_l = E$, hence $z^{Pi} = (\bar{E}/cE_l)^{Pi/E}$ or $(E/c)^{Pi/E}$ and m becomes $1/(1 + (a-1)P_i/c(E/c)^{Pi/E})$ which has the same properties relative to E as m' has to E_l.

The behaviour of the population at the equilibrium point can be investigated by differentiating the full equation, giving:

$$\Delta = \left[\frac{dP_f - P_i}{dP_i} \right]_{P_i = E} = (1 - C_p)(1 - 1/a)(\log_e(E/c) - 1)$$

May (1973) showed that values of Δ between -1 and 0 imply a steady approach to equilibrium without oscillations (Fig. 15.8A), values between -2 and -1 damped oscillations about the equilibrium (Fig. 15.8B) and values less than -2 large and unstable oscillations (Fig. 15.8C). In the equation above, $(1 - C_p)$ is a quasi constant, $(1 - 1/a)$ approaches 1 and the remaining

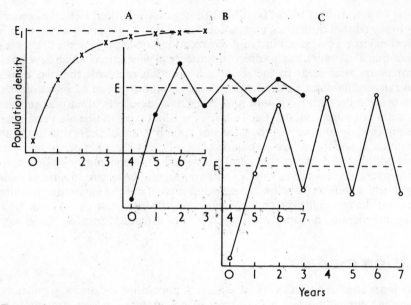

Fig. 15.8 Types of population oscillation. A, gradual approach to the logistic equilibrium E_l, root undamaged. B, damped oscillations about an equilibrium E, host suffers damage. C, large unstable oscillations about E, excessive multiplication, host damaged.

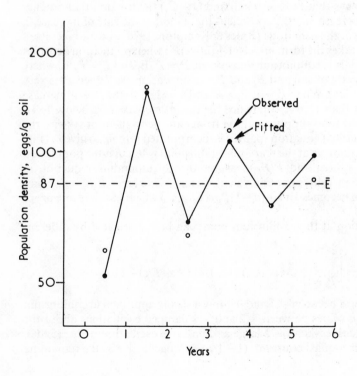

Fig. 15.9 Observed oscillations of *Globodera rostochiensis* in small plots planted with potatoes every year. Fitted oscillations derived from the population model compared with field data. $C_p = 0.33$, $\bar{E} = 87.3$, $\bar{E}/E_l = 0.19$, $\Delta = -1.75$, a is indeterminate (Jones & Parrott, 1969; Jones & Kempton, 1968). In large plots fluctuations are obscured because different parts of the plots may be in different phases of oscillation.

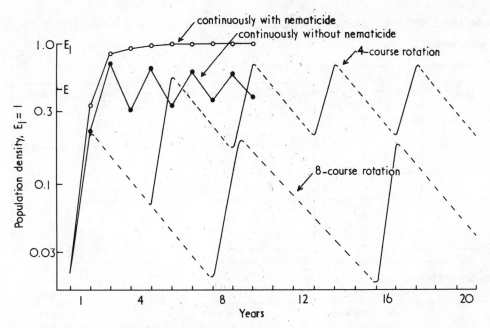

Fig. 15.10 Stylized projections from the cyst-nematode population model for *Globodera rostochiensis* or *G. pallida*. Continuous lines, susceptible cultivar grown; broken lines, non-susceptible crops grown. The population density scale is logarithmic in terms of $E_l = 1$, i.e. dimensionless.

term relates to the degree of damage to the root system. Evidently this last factor largely determines the nature of the equilibrium. The potato cyst nematode is strongly K-selected. Observed oscillations in small plots planted continuously with potatoes show damped oscillations (Fig. 15.9). The model fits these well with sensible parameters similar to those found in other experiments. The fitting to such data is part of the process of verifying this part of the model. Validation comes only when predictions from the whole model are in agreement with observed events over a longer time. It would be helpful to know what factors lead to the largest peak populations. Differentiating the model equation at the peak when $dP_f/dP_i = 0$ gives a complex answer indicating that several factors are involved of which a and $\bar{P}_i$ and $\bar{E}/E_l$ are important elements. When a resistant potato cultivar is grown a is virtually 0 and $P_f = C_p P_i$, i.e. there is no multiplication but the unhatched fraction of the population remains. When a non-host cultivar is grown (i.e. a crop other than potato or tomato), C_p is replaced by C_o which represents the larger fraction of the egg population that remains because non-hosts do not produce the egg-hatching factors that emanate from host roots (in Fig. 1.3, p. 5, C_p is 0.33 and C_o 0.67). Other modifications deal with resistant or susceptible potatoes and groundkeepers and with fumigant and oxime carbonate nematicides (see Fig. 15.10).

Complex models

Models that attempt to forecast the incidence of a range of pests and diseases of a particular crop or group of crops are more complex, contain several submodels and various inputs that influence calculations at different stages (Fig. 15.11). An example is EPIPRE (EPIdemics PREdiction and PREvention, Rabbinge & Rijsdijk, 1983). This model involves farmers and research workers and is modified as experience is gained (Table 15.5). The development of the submodel relating to cereal aphids is instructive. It ends with a simplification that relates exponential population growth to physiological rather than chronological time. Evidently,

Table 15.5 EPIPRE (EPIdemics PREdiction and PREvention) Rabbinge & Rijsdijk (1983).

Outline plan

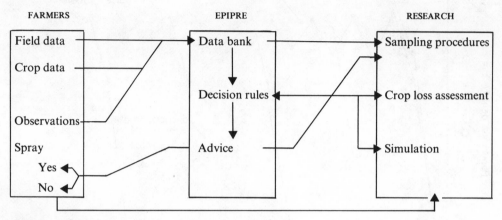

FARMERS   EPIPRE   RESEARCH

Growth of EPIPRE in use

Year	Pest	Disease	Fields
1978	*Puccinia striiformis*	Rust	400
1979	As above plus:		450
	Eresiphe graminis	Mildew	
	Sitobion avenae	Aphid	
1980	As above plus:		
	P. recondita	Rust	840
	Metapolophium dirhodum	Aphid	
	Rhopalosiphum padi	Aphid	
1981	As above plus:		
	Septoria spp.	Leaf spot/Glume blotch	1155

Development of a sub-model for cereal aphids

Objective	Explain and predict population dynamics
Limits	One field of winter wheat
Conceptualize system	Draw systems diagram with variables
Quantify relationships	Source, literature and experiments
Verify	Compare output with detailed data from best sites
Validate	Check predictions against reality, over several sites and years
Check sensitivity	Check relative effects of temperature, other weather data, enemies
Simplify	Exponential growth† of aphid numbers described by physiological time. The same for *S. avenae* and *M. dirhodum*
	Most environmental conditions and enemies have little effect on upsurge or peak

† See Fig. 4.6, page 43. Accumulated temperature is a measure of physiological time.

aphid epidemics once under way are beyond the release point *R* in Fig. 15.6 and are on the epidemic ridge shown in Fig. 15.7.

Some pest situations can only be studied initially by systems analysis and modelling. An example is the spread of nematodes within fields. It is virtually impossible to know when the first introductions were made and impossible by soil sampling to follow the course of the infestation until it is well established. It is easy to show that by their own unaided movements, nematodes would take very many years to colonize 1 ha (between 500 and 1414 host crop years for cyst-nematodes with one generation a year. See p. 227.) The harvesting of root crops with consequent spread of soil plays a dominant part in the early stages of spread which may be delayed initially until the primary infestation has multiplied to a stage when almost every particle of soil (say 1 ml) in the original focus contains one or more nematode cysts. Such models can only be verified and validated by comparing their predictions with well-documented case histories: these are few.

A dodge useful in modelling when, for example, the equilibrium population density is unknown or varies in different circumstances, is to set the unknown at 1 and to measure other population densities in relation to it as dimensionless numbers. The unknown can be dimensioned later supplying values appropriate for particular circumstances. In this way, the basic model is generalized but able to be made particular at will.

Legislative measures

Attempts to control pests by plant health regulations began with the Destructive Insects Act of 1877 which was prompted by fear of the introduction of Colorado beetle into Britain following the finding of a live beetle at Liverpool docks. The Act was revised and extended in 1907 and 1927, and all three Acts were consolidated, along with section 11 of the Agriculture (Miscellaneous Provisions) Act, 1949, by the Plant Health Act, 1967. The Plant Health Act (Northern Ireland), 1967, had the same effect in Northern Ireland. Under this Act the Minister of Agriculture may issue Statutory Orders conferring upon his authorized officers power to take such steps as seem necessary to prevent the introduction or spread of noxious organisms. The term organism includes insect, fungus, bacterium or any agent causing transmissible crop disease. Similarly, the term crop may include almost any kind of plant material. In some Orders, dangerous pests, e.g. the Colorado beetle, are made notifiable, that is, the owner or

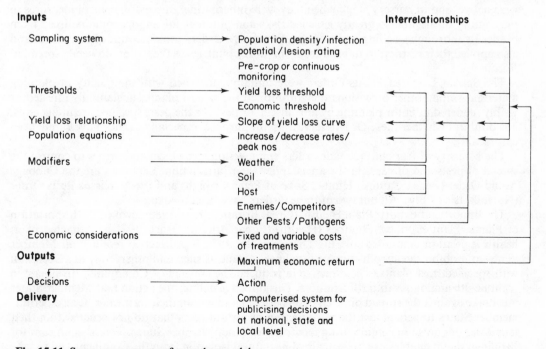

Fig. 15.11 Some components of complex models.

occupier of any land on which the pest is found must report the fact immediately to the Ministry of Agriculture. The Orders also specify the fines for non-compliance and the Acts authorize payment of compensation for crops removed or destroyed.

Orders are of three kinds: domestic, local and those concerning export and import. Domestic Orders concern the United Kingdom generally, although they are not necessarily the same for Scotland and Northern Ireland as for England and Wales. Examples are the various Colorado Beetle Orders, the Destructive Insects and Pests Order, the Beet Cyst Nematode Order, 1977, the various Destructive Pests and Diseases of Plants Order, 1965, the Potato Cyst Eelworm (Great Britain) Order, 1973 and the Sale of Diseased Plants Orders.

In 1953 the United Kingdom government ratified the International Convention for the Protection of Plants and Plant Products, Rome, December 1951 and has since adhered to subsequent amendments. The purpose of this Convention is to secure common and effective action to prevent the introduction and spread of pests and diseases of plants and plant products, and to promote measures for their control. Contracting governments, of which there are now more than 50, undertake to adopt legislative, technical and administrative measures specified in the Convention. Among other provisions, each contracting government undertakes as soon as possible to set up an official plant protection organization to inspect growing plants, areas under cultivation and so on, with the object of reporting the existence, outbreak and spready of diseases and pests and of controlling them. The organization is also expected to inspect consignments of plants and plant products moving in international traffic and to disinfest any consignments and their containers found to be infested and to issue certificates relating to the phytosanitary condition and origin of consignments.

The way in which British legislation has been guided by experience is well illustrated by the Colorado Beetle Orders. Early restrictions on the import of potatoes and produce likely to harbour beetles were severe. After a while it was recognized that potatoes and certain vegetables were less dangerous than leafy produce such as lettuce, and also that risks were related to season. Thus beetles were more likely to be found in produce at the times of the year when they were active, and in nursery stock when they were hibernating. Successful control measures in the country of origin also greatly lessened risks and potatoes for import could be freed from beetles by thorough washing and riddling. Successive Orders took account of these facts and also applied their restriction to produce from more countries as the Colorado beetle spread in Europe.

The Sale of Diseased Plants Orders were initially concerned with the quality of planting material. Among other provisions they prohibited the sale of plants substantially infested or visibly rendered unfit for planting by any 'insect' or pest. As the pests specified are indigenous, it is doubtful whether these Orders serve a useful purpose, especially as they have rarely been enforced.

The Ministry of Agriculture made orders empowering some local authorities to control the spread of pests and diseases in the areas under their jurisdiction. Examples are the Cabbage Aphid Order (Beds., Cambs., Hunts., Soke of Peterborough) and the Narcissus Pests Order (Norfolk, Isle of Ely). All but two of these Orders have been revoked.

The Import and Export (Plant Health) (Great Britain) Order, 1980, revoked the Importation of Plants, Plant Produce, Potatoes (Health) (Great Britain) Order, 1971 and enacted plant health legislation embodied in EEC Directive No. 197. The directive requires all Member States to prohibit the introduction of listed pests, some as such and others only in association with specified host plants. The directive introduced the concept of Community 'frontiers' in plant health analogous to tariff frontiers. This directive defines the action that Member States must take against the spread of listed pests between and within their territories. It also permits member States to act against the introduction of certain pests that do not occur within their territories but do so in neighbouring territories of other Member States. This is significant for Britain in that it enables use to be made of its natural isolation from the Continent and to take measures against such pests as the Colorado beetle. The 1980 Order contains for the first time measures against pests likely to be transported in soil and debris arriving with rooted trees and shrubs. Many serious pests in the New World and East Asia are also mentioned for the first time. Some important restrictions also apply to direct imports from countries outside the Community. The directive also places a statutory obligation on Great Britain to enforce the measures embodied in the directive and to issue phytosanitary certificates for exports to member States.

Under the 1980 Order, Plant Health Inspectors are empowered to enter, examine, mark and sample any material, plant pest or genetically manipulated material landed in Great Britain during the previous twelve months. Specifically prohibited pests include various fruit flies, tortrix moths, alien cutworms (*Spodoptora* spp.), some leaf mining Diptera, the Japanese beetle, Colorado beetle and potato cyst-nematodes. Tulip, narcissus and other flowers bulbs and some plants must be from a crop that has been previously inspected and found free and must themselves be free of stem nematodes (*Ditylenchus dipsaci*). Consignments of seed of onion, leek and chive must also be free from infestation with stem nematodes.

The Plant Pests (Great Britain) Order, 1980 applies specifically within Great Britain. It revokes the Sale of Diseased Plants Order, 1927 and the Sale of Diseased Plants Order (Scotland) 1927 and supersedes the Destructive Pests and Diseases of Plants Order, 1965 and the Destructive Pests and Diseases of Plants Order (Scotland) 1966. Its provisions are similar to those of the Import and Export (Plant Health) (Great Britain) Order, 1980 in that it requires notification of non-indigenous pests and of any subject to genetic manipulation, gives powers of entry, examination and sampling and enables remedial measures to be undertaken. Information may have to be divulged and offences and penalties are laid down. Similar Orders apply in Northern Ireland, e.g. the Import and Export of Plants and Plant Products (Plant Health) Order (Northern Ireland) 1981 and the Import and Export of Potatoes (Plant Health) Order (Northern Ireland) 1981. Both made under the Plant Health Act (Northern Ireland) 1967 and its amendments of 1967, 1972 and 1975.

All the above Orders are operated by the Plant Health Branch of the Ministry of Agriculture in England and Wales and the Departments of Agriculture for Scotland and Northern Ireland in those countries on whose Inspectors the arduous task of examining large consignments of home grown and imported produce devolves. Of all the material they have to scan, nursery stock is especially dangerous, for on it or in associated soil and debris pests are likely to be introduced with their hosts.

Another side of Inspectors' duties is the examination of produce for export so that it may comply with the phytosanitary regulations of the EEC and of the recipient countries. Apart from examination of consignments before shipment, more and more countries insist on inspection of plants during active growth. 'Freedom' from potato cyst-nematode, based on soil sampling, is now a common requirement. In many countries, certificates of freedom are demanded, although in fact absolute freedom cannot be guaranteed. What such a certificate means is that a sample of a certain size and collected in a specified manner has been examined and none of the organisms in question has been found therein (Jones, 1969). In many instances, previous examination of the crop, or knowledge of the environs from which the sample came, may make the certificate more trustworthy.

For informative articles on plant health see Ebbals and King, (1979) and for legislation affecting birds and small mammals see p. 245.

The various Orders, Rules and Regulations are made and operated with some subtlety. In recent Orders the word 'Pest' embraces harmful animals and disease organisms of various kinds. Similarly 'Plant' is broadly defined. Moreover Orders may not come into force if sampling is done unofficially for advisory or research purposes. As mentioned earlier Orders in Council confer a degree of flexibility which make workable rules that otherwise could be harsh and inflexible. Nevertheless the Orders can be strictly enforced if the need arises.

Economics

The economic consequences of pest attacks on crops include loss of revenue, increased costs of production, increased risks, losses to consumers, losses to society as a whole from wasted resources and finally a loss to the World community, especially those parts struggling against shortages of food and raw materials. When pests are properly controlled or managed the usual consequence is a larger, a better or more certain yield. Many attempts have been made to assess crop losses on a national and a global scale (Ordish, 1952; Anon, 1965, 1967, 1971; Cranmer, 1967; Bardner, 1973) but many estimates are no more than intelligent guesses, for assessing losses is difficult and beset by many pitfalls, yet it is on these assessments, however crude, subjective or inadequate, that research policy is based and funds and research facilities allocated. The farmer's objective is to maximize profit after making an intuitive judgement or a detailed calculation of costs and benefits of any new method. The result of an outstandingly successful new control measure is not always an increase in the farmer's margin of profit. If demand is inelastic the result is over-production and a fall in price. In such a situation the successful new measure will, in effect, free land for some other purpose and, in the long run, the community would benefit rather than the farmer provided the freed land were used sensibly. Because of this and because of variations in the price from season to season, the benefits from control are sometimes expressed in terms of the area of land freed rather than the cash generated by the yield increase. For a more detailed discussion of the economics of pest control consult Khan (1972).

16

Pesticides

Desirable properties of a pesticide

The term pesticide in this book refers to chemical compounds used to kill animal pests of field crops. The desirable properties of a pesticide include: (1) high toxicity to the pest to be controlled; (2) low toxicity to plants; (3) low toxicity to man and warm-blooded animals; (4) selectivity in use so that beneficial and harmless animals are not affected; and (5) leaving behind no undesirable residue. Because living processes in plants and animals have much in common and a measure of persistence may be necessary for successful use, it is almost impossible to satisfy the above conflicting requirements. Most pesticides are 'broad band' toxicants, and any selectivity they have is derived more from the manner in which they are used than from inherent properties. Although it is conceivable that highly selective poisons could be developed for individual pests and a few examples of relatively selective poisons are known (e.g. antu for rats), these would be costly to develop and uneconomic, because of the relatively small amounts that could be sold compared with less selective materials for wider use.

The desirable properties listed apply not only to the pesticide in its technically pure state but also to the various formulations in which it is applied. Other commercially desirable properties are: (1) low cost; (2) stability, including keeping without deterioration under extremes of temperature; (3) ease of manufacture; (4) availability of raw materials; (5) non-corrosive action against metal containers and parts of application machinery; (6) ready standardization by chemical or physical methods rather than bioassay; and (7) absence of patent rights restricting the use or manufacture of the material or its precursors.

Screening for pesticides

Most large chemical firms employ chemists to synthesize new classes of organic and inorganic compounds with the object of exploiting any that are useful. In screening them for biocidal activity, suitable dilutions are applied to cultures of test animals, e.g. locusts, aphids, grain insects, caterpillars, mosquitoes, houseflies, red spider mites, soil nematodes and molluscs. The dosing of exposed insects, mites or molluscs is relatively simple and percentage mortality easy to estimate. After dosage, insects are usually kept at a constant temperature and a high relative humidity to avoid complications caused by death from desiccation. The dosing of soil organisms is more difficult and the kill of small organisms such as nematodes often has to be assessed by indirect methods such as staining techniques which differentiate between living and dead eggs, juveniles or adults (Shepherd, 1962b), counts of juveniles that have invaded root systems, the number of root galls formed or the numbers of nematode cysts produced at the end of the season (Jones, 1961; Jones & Nirula, 1963; Southey, 1984). The yield of a crop is not always a satisfactory measure of nematicidal action in the field. Compounds of exceptionally high toxicity are studied further in glasshouse and field trials. Those that pass these trials successfully are tested for mammalian toxicity by feeding or injecting into test animals such as mice, rats or rabbits. Toxicity to animals, invertebrate or mammalian, is often expressed as the median lethal dose (LD_{50} or MLD), that is, the dose per kilo of body weight required to kill 50% of the test animals, because mortality is most conveniently and most accurately measured at that level. The hazards to operators applying the pesticide, to consumers from possible residues on treated crops, to live stock and to wild-life are also considered. Many compounds are discarded at one stage or another but a few are manufactured in test quantities and passed to research organizations for further trial.

The finding of marked biocidal activity in a chemical of a new type is usually followed by synthesis of homologues and substitution products. In these investigations, although some general principles have been established between molecular structure and insecticidal activity, especially in studying DDT and its homologues, it is almost impossible to predict whether a new derivative will have enhanced or diminished activity. Groupings known to confer toxicity include: (1) halogens, especially chlorine as in DDT and HCH; (2) isothiocyanate and cyanide; (3) nitro groups as in parathion; (4) unsaturated aliphatic chains as in dichloropropene; and (5) symmetrical molecular configuration as in DDT and TEPP. Changes in molecular structure may also be tried to improve formulation, increase or decrease vapour pressure, stability to weathering or sunlight, penetration of leaf or animal cuticle, or to decrease mammalian toxity.

Mechanisms of toxic action

Death by poisoning usually arises as a result of interfering with some vital process. The poisonous substance may react with groupings on a surface, interface, or on enzymes and prevent them from reacting with normal substrates, or so change the electron grouping that their reactivity is modified. Alternatively the poisonous substance may replace a growth substance or vitamin, or form a stable but ineffective enzyme complex. A wide range of enzymes contain manganese, copper or zinc, the ions of which function as coenzymes. Agents which combine with the metal are toxic. In toxic activity both chemical reactivity and spatial arrangement are concerned and often only one of several closely related isomers is highly toxic (e.g. the gamma isomer of HCH, *cis/trans* isomers of pyrethroids).

The action of organophosphorus compounds and some carbamates arises from the interference with an enzyme that splits acetylcholine into acetic acid and choline. In animals, the transmission of nervous impulses across nerve junctions (synapses), and from nerves to motor muscles, depends on the release of acetylcholine, which is immediately hydrolysed into acetic acid and choline so that the system may function again. Organophosphorus compounds and carbamates, absorbed or ingested, prevent hydrolysis so that acetylcholine accumulates, blocks nervous transmission and causes death. DDT, HCH, pyrethrum, rotenone and some other toxicants also act upon the nervous system, but the mechanisms are probably all different and little understood. Arsenical poisons affect oxidizing enzymes, hydrocyanic acid and cyanides inactivate enzymes that contain iron, preventing valency changes important in oxidation and reduction. The saturated hydrocarbon oils probably kill some insects and insect eggs (e.g. scale insects, lepidopterous eggs) by a physical, stifling process which deprives them of oxygen. The unsaturated, aromatic hydrocarbon oils, however, kill more by chemical action in which the percentage of phenols present is important (e.g. aphid and psyllid eggs) (Martin, 1973). Some inert dusts abrade the cuticle of insects, causing death by increased water loss in a dry atmosphere. Evidently the lethal action of an insecticide may be due to several effects involving the impairment of a number of life processes. For further information see Coates (1982).

Types of pesticide

Pesticides may be grouped according to the type of animal against which they are used (insecticides, acaricides, molluscicides, nematicides). Insecticides are further divided according to their mode of action: stomach poisons must be ingested, contact poisons gain entry through the integument, fumigants through the spiracles and possibly also through the integument, ovicides act against eggs, and systemic poisons are stomach poisons translocated by plants and ingested in plant sap by insects with biting or sucking mouth parts. Although this classification is useful, many insecticides have more than one mode of action.

In the following sections pesticides are considered in related groups, partly on origin and partly on chemical composition.

Plant derivatives and analogues

The plant derivatives toxic to pests vary greatly in chemical structure. They are all ephemeral, relatively harmless to beneficial insects on that account, and non-phytotoxic. Synthetic derivatives of some of them have greater stability.

Nicotine is a pungent brown liquid boiling at 247°C. It is obtained from tobacco waste, has the molecular formula $C_{10}H_{14}N_2$ and is a purine base which readily forms hydrochlorides and hydrosulphates. Nicotine used as a fumigant, or volatilized from dusts and sprays, enters the spiracles of insects and also penetrates their cuticle. It is especially valuable against insects with sucking mouth parts such as aphids, scales, suckers and thrips. The so-called fixed nicotines in which the base is combined to lessen volatility were also used as stomach poisons for caterpillars. Nicotine is very poisonous to man and vertebrates, but as it rapidly disappears, produce dusted or sprayed with it can be marketed within a relatively short time. Although once much in vogue, it has been replaced by organophosphates, carbamates and pyrethroids.

Rotenone is the chief insecticidal constituent of *Derris* and *Lonchocarpus* root. It is a complex organic compound of molecular formula $C_{23}H_{22}O_6$ forming colourless crystals melting at 163°C. Although less toxic to humans it is toxic to pigs and fish. Against insects it is one of the most highly toxic of all insecticides and is useful on fruit (e.g. against raspberry beetle) and vegetables.

Pyrethrum, from the flower heads of *Chrysanthemum cinerariaefolium* (Vis.) contains, as well as minor constituents (cinerins, jasmolins), pyrethrin I, which kills insects and pyrethrin II with knock-down action against flying insects. Pyrethrum is highly toxic to insects and safe to mammals, but unstable in light and therefore unsuitable for foliar application. It is toxic to fish.

Synthetic pyrethrinoids (see separate section below) have been developed to provide a new class of insecticides with a range of properties. Research first began on these at Rothamsted in the 1930s and now produces large returns in royalties to the National Research Development Corporation. Other insecticidal plant products include hellebore, ryania, sabadilla and quassia. Fundamental work into these products and others aimed at elucidating their chemistry, modes of action and the production of analogues, could be equally as rewarding as was the work on pyrethrum but is essentially very long-term. The insecticidal deterrent principles in the Indian neem tree and some other plants turn out to be antifeedants of complex structure of which polygodiol is currently thought most promising (Nakanishi, 1976).

Extracts from *Phytolacca dodecandra* (l'Héri.) and *Sapindus saponaria* (L.) kill molluscs. African marigolds (*Tagetes* spp.) and other Composites contain thienyls (unstable sulphur compounds), *Asparagus officinalis* (L.) contains asparagusic acid and plants such as *Sesame*, *Crotalaria*, *Polygonum hydropiper* and some Solanaceae contain other nematicidal or deterrent principles, namely alkaloids, phenolics, sesquiterpenes, diterpenes, polyacetylenes and sulphur containing structures. The thienyl compounds are interesting in that in the presence of oxygen and near ultraviolet light they generate singlet oxygen that attacks proteins and other cell components (Gommers, 1981).

Arsenical compounds

The arsenical insecticides are all stomach poisons. Because soluble arsenical compounds are phytotoxic, those used as dusts or sprays must be relatively insoluble until rendered soluble in the insect gut, where the pH of the digestive juices is an important consideration accounting for variations in toxicity between species. Arsenical compounds were much used against leaf-feeding insects before they were superseded by newer insecticides. They had a measure of persistence, but were toxic to mammals if ingested and, where repeatedly used, left undesirable accumulations of arsenic in the soil.

Paris green (copper acetoarsenite $(CH.COO)_3Cu.3Cu(AsO_2)_4$) was first used in 1867 against Colorado beetle, but, although an effective insecticide when used as dust or spray, it was often phytotoxic. For a long time it was the standard poison added to baits for use against locusts, cutworms, leatherjackets and slugs.

Lead arsenate ($PbHAsO_4$) is stable on exposed leaf surfaces and was much used as a spray against leaf-eating caterpillars on fruit trees and codling moth on apple.

Calcium arsenate ($CaHAsO_4$) was used as a dust on cotton and other field crops. Excess of calcium hydroxide in the dust probably helped to prevent the formation of soluble arsenical compounds and so avoided phytotoxicity.

Mercury compounds

Both mercuric ($HgCl_2$) and mercurous chloride or calomel (Hg_2Cl_2) have been used, the former mainly as a dip to protect the root systems of transplants against root-feeding insects (e.g. cabbage root fly). Calomel was used as a 4% dust against cabbage root fly and onion fly and, in the technically pure form, as a seed treatment against onion fly. Mercury vapour has insecticidal properties and has been used to protect stored grain. Its action is usually ovicidal. Mercury is expensive, its residues unacceptable and its usage restricted under the Pesticides Safety Precautions Scheme (Anon, 1984).

Organochlorine compounds

Organochlorine insecticides fall into two groups, the chlorinated hydrocarbons like DDT and HCH, and the cyclopentadiene derivatives like aldrin and dieldrin. With the exception of chlordane which is a liquid, all are crystalline or waxy solids soluble in organic solvents, insoluble in water and with relatively low vapour pressures. HCH and aldrin have the highest vapour pressures and are least persistent. They also have some fumigant action. All the members of the group, however, are powerful and persistent contact poisons and are also able to act as stomach poisons when ingested. Members of the group are neurotoxicants which penetrate insect cuticle rapidly, and sometimes it is sufficient merely for an insect to walk over a residual film to absorb a lethal dose. Because of these outstanding properties, the timing of applications is not critical and they can be used in a preventive or protective role.

Disadvantages arise from fat solubility, long persistence and a tendency to accumulate in predators at the end of food chains. Their usage is restricted under the Pesticides Safety Precautions Scheme. If frequently or excessively used residues accumulate in soil. Residues on edible plants are concentrated in the fatty tissues of herbivorous animals and are not readily excreted. Predatory animals also pick them up in contaminated prey. Another disadvantage is their harmful effects on beneficial insects if carelessly used.

DDT ($C_{14}H_9Cl_5$) was the first organochlorine compound exploited. It is remarkably stable and persistent and kills most insects except those that have become resistant to it. Except against plants of the family Cucurbitaceae, it is mainly non-phytotoxic, although some barley varieties are susceptible to injury from DDT sprays.

DDT

$Cl_3C—CH$

Cl

Cl

TDE

$H—C—CH$

Cl

Cl

Cl

Cl

TDE ($C_{14}H_{10}Cl_4$) is similar to DDT but has lower mammalian toxicity and can be used with greater safety on food plants, especially fruit trees. Dicofol is an acaricide.

Dicofol

$Cl_3C—COH$

Cl

Cl

DDE (inactive degradation product of DDT)

$Cl_2C=C$

Cl

Cl

HCH ($C_6H_6Cl_6$), is a crude product obtained by chlorinating benzene in ultraviolet light. It contains eight isomers of which gamma-HCH is the most active and comprises some 13% of the crude product. HCH is an effective soil insecticide but taints root crops such as carrots and potatoes unless the dosage is carefully regulated and distribution uniform. HCH is absorbed from the digestive tract by mammals and stored in fatty tissues and the liver. The fatal dose for man is 150 mg per kg of body weight, and growing animals are particularly susceptible (Morrison, 1972). Technically pure lindane contains 99% gamma-HCH, is more effective than HCH and less liable to cause taint but nevertheless should not be used on carrots or other produce for canning or processing.

gamma – HCH

Cl   Cl

Cl   Cl
Cl   Cl

Aldrin

Dieldrin

Heptachlor

Aldrin ($C_{12}H_8Cl_6$), *dieldrin* ($C_{12}H_8Cl_6O$) and *endrin*, a stereoisomer of dieldrin, are similar. Aldrin has little persistence on foliage but is a useful soil insecticide breaking down there to dieldrin. Dieldrin is persistent on surfaces and in soil. Both insecticides are relatively non-phytotoxic, do not taint, and are stable to alkalis; aldrin has often been incorporated in fertilizers. Against many insects, dieldrin is more toxic than DDT, HCH or aldrin. Like gamma-HCH, aldrin and dieldrin can be formulated as seed treatments and, because they are less phytotoxic, more concentrated treatments are possible. Those containing dieldrin have long persistence. Endrin is acaricidal.

Other organochlorine pesticides are chlordane ($C_{10}H_6Cl_8$), heptachlor ($C_{10}H_5Cl_7$) and campheclor (chlorinated camphene). Chlordane is used against earthworms in sports turf and campheclor is relatively innocuous to bees. For further details see Martin (1973) and West & Hardy (1961).

Organophosphorus compounds

A general formula that fits many organophosphorus compounds is:

$$
\begin{array}{c}
R_1 \qquad O \text{ or } S \\
\diagdown \quad \diagup \diagup \\
P \\
\diagup \quad \diagdown \\
R_2 \qquad X
\end{array}
$$

Where R_1 and R_2 may be alkyl, alkoxy, alkylamino or other groups and X may be an acid radicle, a reflection of the first half of the molecule with or without a bridging group between, or a great range of chain or ring structures.

The pioneer work on these compounds was done by Schrader in Germany before 1939. One of the earliest compounds was HETP (hexaethyl tetraphosphate), later found to be a mixture of esters of which TEPP (tetraethyl pyrophosphate) was the principal insecticidal ingredient. TEPP, now little used, is so rapidly hydrolysed that it has no persistence, acts only by contact and may be used as a nicotine substitute. Parathion, another of the early products, is stable and persistent on plant surfaces, kills a wide range of insects and behaves rather like DDT. Schradan and many other compounds that followed are remarkable because, after being sprayed on foliage or applied near the roots, they are absorbed and translocated (mainly upwards) in plants, and render them toxic to sucking insects, sometimes for long periods. Some have proved useful soil insecticides, e.g. chlorfenvinphos, which can be formulated as an emulsifiable concentrate, a seed treatment or a granule; many are acaricides: e.g. azinphos-methyl, dimethoate, malathion; and a few nematicides: e.g. ethoprophos, fenamiphos, thion-azin. It seems that diethyl organophosphorus compounds control cyst-nematodes better than dimethyl ones (Whitehead *et al.*, 1972). The early compounds were all highly toxic to man but research has produced several of relatively low mammalian toxicity. For example, fenchlorphos is an organophosphorus compound that can be sprayed onto or given orally to cattle to kill dipterous parasites.

Because of the many possible variations in chemical structure, the organophosphorus insecticides exhibit a wide range of physical and biological properties (Worthing & Walker, 1983). They are especially useful against sucking insects but sometimes kill a wide range of species. The characteristics of some of the more important compounds are in Table 16.1. The minimum time between application and harvesting is a rough guide to the persistence of the insecticide or its breakdown products on or in plants, but effective insecticidal action lasts for a much shorter time. Those that are systemic and persistent can be used as protectants, and with them timing of applications is not critical and a complete spray cover not essential.

Carbamates and oxime carbamates

Carbamates and oxime carbamates are cholinesterase inhibitors with the following general formulae.

Carbamate

Oxime carbamate

Carbamates (Table 16.2) have a relatively low mammalian toxicity, some systemic activity and persistence, and are rather selective in action. Carbaryl kills leaf-eating caterpillars and bees but is not very effective against aphids. It is harmless to fish, has some plant growth regulating activity and is recommended for the eradication of earthworms from sports turf. Pirimicarb is a selective aphicide, ometanate is an acaricide, propoxur is an insecticide and methiocarb is a non-systemic insecticide, acaricide and a powerful molluscicide (p. 184). Carbofuran is a systemic insecticide acaricide and nematicide.

Oxime carbamates (or more correctly carbamoyloximes) are powerful biocides extremely toxic to vertebrates and man. Aldicarb is a systemic insecticide, acaricide and nematicide used only in granules for soil application (Moss *et al.*, 1975). It is water soluble and, according to dosage and conditions, persists for 3 to 10 weeks. Oxamyl is similar but is sometimes used as a foliar spray and on certain plants is translocated downwards to kill nematodes attacking the roots (Peterson *et al.*, 1978). Thiofanox is used as a soil or seed treatment to control aphids and other foliage-feeding insects on sugar-beet and potatoes. When so used it assists in the control of nematode and aphid-borne viruses as does aldicarb. Aldicarb and thiofanox are listed as Part II substances under the Health and Safety (Agriculture) (Poisonous Substances) Regulations, 1975.

Synthetic pyrethroids

The structure of the pyrethroids is more complex than that of most other groups. Insecticidal activity of the natural pyrethrins, which depends on a particular spatial disposition of several structural features, can be maintained through a series of synthetic modifications so that the favourable characteristics of the parent compounds are preserved and intensified. The most active compounds can be directly related to pyrethrin I by successive isosteric replacements. Structurally, the pyrethroids have a relationship with prostaglandins (Elliott, 1980).

Pyrethrin I

Chrysanthemate   Rethrolone

R_1 = Me, CC$_2$ Me R_2 = Me, Et, Vinyl

Some established pyrethroids are in Table 16.3. Because of the complexities of structure there is much scope for *cis-trans* isomerism and different isomers may vary in potency. Bioallethrin which contains two out of eight isomers of allethrin is an improvement and (S)-bioallethrin containing one out of eight is better still. Similarly bioresmethrin is an improve-

Table 16.1 Some organophosphorus pesticides (see also Anon, 1984; Worthing & Walker, 1983).

Name	Formulation	Mode of action	Minimum interval between application and harvesting (days)	Classification in Schedule 2 of Health and Safety Regulations	Selectivity	Usage
† **Azinphos-methyl**, used only in mixtures with Demeton-S-methyl sulphone	*Sp, Ae	Contact and stomach poison. Sulphone is systemic	21	III	None	Against leaf-eating insects, with some action against mites
Carbophenothion	St	Contact – seed treatment only			None	As a seed treatment against wheat bulb fly. Long residual action
† **Chlorfenvinphos**	Gr, Sp, St	Contact and stomach poison	21	III	None	As a seed treatment against wheat bulb fly and to control cabbage root fly, carrot fly and frit fly
† **Chlorpyrifos**	Gr, Sp	Contact, ingestion, vapour	21	–	None	Granules to control cabbage root fly and leatherjackets in cereals. Sprays against aphids, caterpillars, frit fly and wheat bulb fly in seedlings
† **Demeton-S-methyl/Oxydemeton methyl**	Sp, Ae	Contact and systemic	21	III	None	Mainly against aphids and mites
Diazinon	Gr, Sp	Contact, penetrates locally	14	–	None	To control aphids, capsids, leaf miners, mites, springtails and thrips in glasshouses. Limited use agriculturally, spot treatment of cabbage root fly
Dimethoate	Sp, Ae	Systemic and contact	7	–	None	To control aphids, red spider mites and leaf miners. Against wheat bulb fly in seedlings
Disulfoton	Gr, Ae	Systemic	42	II	Some	As granules to control aphids, mites and carrot fly
Etrimphos	Sp	Contact	3	–	None	Caterpillars in brassicas, leatherjackets, late generation carrot fly
Fenitrothion	Sp, Ae	Contact	14	–	None	To control aphids, weevils, caterpillars, capsids, sawflies, midges, thrips, mites. On structures against grain pests in farm stores

Name	Formulation	Action	Days	Poison Rules	Predator effect	Notes
Fonofos	Gr	Contact	42	II	Some	As granules to control cabbage root fly, cabbage stem weevil and wheat bulb fly
Formothion	Sp, Ae	Contact and systemic	7	–	Some	To control aphids, red spider mites, mangold fly, and wheat bulb fly in seedlings
Heptenophos	Sp	Systemic	1	–	None	Aphids on wide range of crops
Iodofenphos	Sp	Stomach poison	7–14	–	None	Caterpillars in brassicas and cabbage root fly on Brussels sprout buttons
Malathion	Sp	Contact	1–7	–	None	Moderately powerful, broad band toxicant relatively safe for use in gardens and on food products
† Mevinphos	Sp	Systemic short persistence	3	II	Some	To control aphids, caterpillars and leaf miners where rapid kill is required close to harvest
Phorate	Gr, Ae	Systemic	42	II	Some – used selectively	Broad band toxicant to control aphids, carrot fly, frit fly, wireworm, capsids
Phosalone	Sp	Contact	21	–	None	Seedweevil and pod midge in brassica seed crops
Phosphamidon	Sp	Systemic	21	III	None	To control aphids, mites, leaf miners and Colorado beetle
Quinalphos	Gr, Sp	Contact and stomach poison	7–21	III	None	Control of caterpillars, celery fly, leatherjackets, late generation carrot fly
Thiometon	Sp, Ae	Systemic	21	III	Used selectively	To control aphids and mites
Thionazin	Sp, Gr	Systemic	56	II	None	Soil insecticide and nematicide
Triazophos	Sp	Contact and stomach poison, penetrates locally	21–28	III	None	Caterpillars, leatherjackets, wheat bulb fly, grass and cereal flies, seed weevil, pod midge, pea moth, late generation carrot fly
Trichlorphon	Sp	Contact and stomach insecticide	2	–	Used selectively	To control cabbage root fly, moth caterpillars, leaf miners and household pests

† Substances to which the Poison Rules apply (not Chlorpyrifos as granules).
* Gr = granule, Sp = spray, St = seed treatment, Ae = approved for aerial application to certain crops.

Table 16.2 Some carbamate pesticides (see also Anon, 1984; Worthing & Walker, 1983).

Name	Formulation	Mode of action	Minimum interval between application and harvesting (days)	Classification in Schedule 2 of Health and Safety Regulations	Selectivity	Usage
†**Aldicarb**	*Gr	Systemic	42–49	II	None	Aphids, soil and seedling pests, stem and cyst-nematodes. Assists control of insect and nematode-borne viruses
Carbaryl	Sp	Contact, slightly systemic	7	–	Some	Caterpillars, earwigs, pea moth, earthworms
Carbofuran	Gr	Systemic	42–49	II	None	Aphids, soil and seedling pests stem nematodes
Formetanate	Sp	Contact	?7	III	Some	Mainly an acaricide
Methiocarb	Pl, Ae	Contact	7	–	Some	Mainly a molluscicide
Oxamyl	Gr	Systemic	14–21	II	None	Used mainly as a nematicide and then controls some seedling pests
Pirimicarb	Sp, Ae	Systemic	3–7	–	Some	Mainly an aphicide
Propoxur	Sp, Gr	Contact	1–3	–	None	Jassids, bugs, aphids, millepedes, household pests. Knock down
†**Thiofanox**	St, Gr	Systemic	14–21	II	None	Aphids, seedling pests on sugar beet and potatoes. Partial control of virus spread

† Substances to which the Poison Rules apply.
* Gr = granule, Sp = spray, St = seed treatment, Pe = pellet, Ae = approved for aerial application.

Table 16.3 Some established synthetic pyrethroid insecticides (see also Anon, 1984; Worthing & Walker, 1983).

Name	Year of introduction	Stability	Knock down	LD_{50} mustard beetle
natural pyrethrins	before 1820		+	0.2
allethrin	1949		+	30.0
tetramethrin	1962		+	15.0
kadethrin	1976		+	0.6
pherothrin	1973			0.4
†resmethrin	1967			1.1
†bioresmethrin	1967			0.3
†permethrin	1973	+		0.25
†cypermethrin	1974	+		0.09
*†deltamethrin	1974	+		0.008
fenvalerate	1974	+		0.3

† Produced by M. Elliott, N. James and colleagues, Rothamsted Experimental Station.
* Deltamethrin is a Part III poison and subject to the Poison Rules.

ment on resmethrin and two of the isomers of cypermethrin are claimed to have greater insecticidal activity than the original mixture.

The synthetic pyrethroids preserve the favourable relative lack of toxicity to vertebrates. Bioresmethrin for example is 30 000 times more toxic to houseflies than to rats and so is outstandingly safe and especially suitable for the treatment of field crops immediately before harvesting. Resmethrin, bioresmethrin and subsequent pyrethroids are not readily metabolized by insects and do not need added synergists such as piperonyl butoxide often used with the natural product and some synthetic pyrethroids. The newer pyrethroids (Elliott, 1980) such as permethrin, cypermethrin decamethrin and fenralerate are much more stable on leaf surfaces (more than 2 weeks), have limited persistence in soils (2–4 weeks) and are rapidly metabolized by mammals. Because of their potency, rates of application are about one tenth that of other insecticides. Hence the newer pyrethroids are potential alternatives to some current insecticides no longer effective because resistant strains have developed or which have undesirable residues or are unduly persistent. Decamethrin is one of the most insecticidal substances yet discovered. Like the natural pyrethrins, synthetic ones are toxic to fish. Regrettably resistance to pyrethroids is not unknown.

Fumigants

Fumigants are used for the destruction of insects in confined spaces, such as buildings, silos, improvised or permanent chambers, tents, clamps and to kill nematodes in soil with or without a surface seal. They are a heterogeneous group, mostly of relatively simple compounds toxic to man, animals and plants. Because they must often penetrate between loosely packed products or along the labyrinth of soil pores, they usually have relatively small molecular weights and high vapour pressures at ordinary temperatures. Low water solubility and low fat solubility are also useful properties because less is then absorbed by produce, soil or other substrates and residues are smaller.

Compounds that have been used as fumigants include hydrocyanic acid gas, carbon disulphide, chloropicrin, methyl bromide, ethylene dibromide, ethylene dichloride, ethylene oxide, acrylonitrile, methallyl chloride and nicotine. Carbon disulphide was much used to rid French vineyard soils of *Phylloxera* before resistant rootstocks were found and was sometimes used to disinfest grain, but is dangerously inflammable and explosive. Hydrocyanic acid is excessively poisonous. Increasing use has been made of simple halogenated compounds of which methyl bromide boiling at 3.9°C is an excellent example. One of its disadvantages is lack of odour but

Table 16.4 The properties and application of some fumigant nematicides.

Name	Formulation	Boiling point °C	Vapour pressure (mmHg at 20°C)	Minimum soil temperature for application °C(°F)	Dosage rate kg/ha active ingredient	Method of application
*†Methyl bromide Bromomethane	Cylinders or canisters +2% chloropicrin	4	1380	5 (41)	500–1000	By pipeline from cylinder or canister, surface seal essential
Dichloropropane-dichloropropene	Liquid, often with other chlorinated hydrocarbon derivatives	106–111	18–25	7 (45)	250–500	Injected every 30 cm or less
†Chloropicrin Trichloronitromethane	Technically pure liquid	112	20	10 (50)	300–400	Injected every 30 cm or less
Ethylene dibromide Dibromoethane	In suitable solvent	132	8	16 (60)	50–100	Injected every 30 cm or less
§Dibromochloropropane	In suitable solvent or as granules	199	0.6	16 (60)	30–60	Injected every 30 cm or less, or rotavated in as granules
Metham-sodium Sodium methyl dithiocarbamate	32–35% in water	**Solid	–	10 (50)	300–400	Injected every 20 cm, or diluted and rotavated in
Dazomet Tetrahydrodimethyl thiadiazine thione	85% dust (prill)	**Solid	–	10 (50)	300–400	Applied as prill and rotavated in
Methyl isothiocyanate	40% in xylol	119	21	7 (45)	150–300	Injected every 20 cm or less

*Application by specialists only. **Nematicidal action due to breakdown to methyl isothiocyanate. †Part I substance: see Table 16.5. Approved products, see Anon (1984). §Rarely used in the U.K., now banned in U.S.A. because of evidence that it may be carcinogenic. §§No longer manufactured in U.S.A., sterilizes male workers in production plants. Has not found a use in U.K.

this can be overcome by admixture with a small percentage of chloropicrin. Leaks of halogenated compounds can also be detected with a halide lamp, a small blow-lamp which burns alcohol, has a copper-tipped burner and gives a blue flame when there are volatile halogen compounds in the atmosphere. Most of the other halogen compounds mentioned boil at much higher temperatures.

Fumigant nematicides

In Great Britain there are no fully satisfactory soil fumigants to kill nematodes. In the U.S.A. many thousands of acres of land are treated annually, mainly with dichloropropene fumigants, but the cash values of the crops concerned are greater and the soil temperatures are appreciably higher than in northern Europe.

Nematodes are almost invariably concealed below soil level and the few species that attack the aerial parts of plants are not controlled by systemic organophosphorus compounds. In the main therefore, nematode control involves treatment of the soil down to plough depth or deeper, which is far more difficult than treating plant surfaces, the soil surface or fumigating loosely packed grain. First there is the great mass of soil which has to be dosed amounting (down to 20 cm) to some 2500 t ha^{-1} and containing about 250–500 t of water. Its sheer bulk means that large quantities of toxicant are necessary. Then there is the problem of dispersing the toxicant evenly throughout the soil, and finally the soil is not inert; it adsorbs, inactivates and detoxifies many substances.

Successful soil fumigants usually possess an appreciable vapour pressure and are able to disperse themselves in the soil. Spot or strip injection is the commonest method of application but nematicides may also be applied as drenches, or as dust or granules mixed into the soil. Soil fumigation on a field scale is affected by many factors. Soil temperature influences vaporization and diffusion. Soil type is important: excessively fine particle size or much organic matter increases the adsorption of a toxicant and lowers its effective concentration. Particle size influences pore size and, along with moisture, determines porosity. In soils of low porosity, diffusion is slow, and excess of water increases the amount of nematicide held in solution. Finally, microorganisms in soil may hasten the breakdown of chemicals added.

During fumigation, gas tends to escape from the surface so that the top layer of soil is not effectively treated. On a field scale, the more volatile types of fumigant are expensive to use because they require a surface seal of tarpaulin, polythene sheeting or water-proof paper. Some of the less volatile nematicides would work better with a surface seal but usually the most that can be afforded is heavy rolling or, occasionally, a light watering to block surface pores temporarily.

At the time of fumigation the soil should be in fine seed-bed condition, drained to pF 1.8 and between 4.5° and 21°C, depending on the volatility of the compound used. After fumigation is complete, time must be allowed for the gas to escape before a crop is grown or injury may ensue. In Britain, the best time for application is early autumn when soil temperatures are high and soil moisture not excessive. However, this rarely finds favour with farmers, for in land left in seed-bed condition over winter the soil structure deteriorates and it is often difficult to produce a good seedbed next spring. Early spring, before planting, is less satisfactory for the soil is often too cold and too wet, and the gas may not dissipate rapidly enough after fumigation is finished.

Because nematicides are broadly toxic, they kill a range of soil organisms and their total effect is greater than from the kill of nematodes alone. Often unavailable nitrogen is set free and yield increased; sometimes weeds and pathogenic fungi are suppressed. Nitrifying bacteria are killed, so ammonia nitrogen accumulates and is much less readily leached from the upper layers of sand soils. This benefits small, slow growing seedlings. Most fumigant nematicides

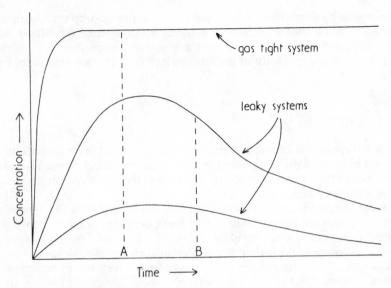

Fig. 16.1 Fumigation of closed and leaky systems. At A and B the concentration–time products of the closed and the less leaky system are equal. In the more leaky system the CTP takes longer to reach the same value or fails to reach it.

are slightly phytotoxic giving a small yield loss. The actual yield obtained after application is the algebraic sum of the various effects.

Table 16.3 lists some common fumigant nematicides and summarizes their properties. When infested materials are fumigated in gas-tight systems (e.g. a fumigation chamber), the dose required to achieve say a 95% kill (LD_{95}) is usually measured as a concentration-time product (CTP) which is approximately constant over a range of times and concentrations. In systems that leak and especially in soils, the CTP is represented by the area under the concentration-time curve (Fig. 16.1). Because the system leaks and because a proportion of gas is absorbed especially by organic matter the CTP may not be reached. When the fumigant is applied by spot or tine injection, the CTP may not be reached at points distant from the points of application. For some nematodes and nematicides contours of equal kill have been drawn; these show that the soil surface is always poorly fumigated. Attempts to improve kill are sometimes made by splitting the dose and applying the second half after the soil has been inverted, or by treating the soil surface with another nematicide, e.g. dazomet. Alternatively the fumigant may be applied by tines that move up the centre of shallow ridges; the apices of these ridges are skimmed off at planting and the seeds placed in effectively fumigated soil. In sandy soils prone to Docking disorder of sugar-beet the small amounts of D-D applied to the positions where the rows will run (p. 202) seem to disperse within a day: nevertheless fumigating and drilling in one operation sometimes results in thin stands. Very small doses of EDB give good control of cereal cyst-nematode on mallee soils in Victoria and South Australia (Gurner *et al.*, 1980).

Formulation of pesticides

With few exceptions, such as some fumigants, the technically pure pesticide is rarely used. Most are marketed only as products formulated ready for use, except that those intended for use as sprays have to be diluted with water, the softness or hardness of which can be important

(Ashworth & Crozier, 1973). Other common diluents include: inert dust, fertilizers, oils, organic solvents, water and air: water and air are the cheapest diluents. Dusts are used for foliage and soil application and for seed treatments. The chemical and physical properties of the diluents, which must be more or less inert, are important, for there must be chemical compatibility between the chemical and the diluent. Dust diluents include hydrated lime $(Ca(OH)_2)$ diatomite (SiO_2) and talc (hydrated magnesium or aluminium silicates). Some pesticides are unstable to alkaline diluents (see Table 16.1).

One of the most important physical properties of dusts is the diameter of the particles. The best range in the field is 10–30 μm, particles below 10 μm tend to drift instead of settling while those above 30 μm have a smaller relative surface to volume and make poor contact with plants or insects. Some dusts, such as diatomite, when below 10 μm are abrasive and therefore insecticidal in their own right, for they scratch the cuticle of insects and cause water loss. Abrasive action is of little use in the field but has been made use of in the protection of stored products such as grain. Other important physical properties include the specific gravity, the shape of the particle and bulk density. The more dense the particle is, the less its adhesive power, unless chemical forces are involved.

Any quantity of powder contains air spaces or voids and the proportion of voids is greatest in powders with very irregular-shaped particles, e.g. diatomite. More uniform particles have a greater bulk density. The shape also affects the adhesion; spheres allowing only point contact, cubes one-sixth of their surface area and plates, properly orientated, giving the greatest area of contact.

Electrostatic effects may arise when dry dusts are ejected from a duster, when the charges cause the particle to be attracted or repelled by the leaf surface. At a relative humidity of 75% or more, the charges disappear owing to conduction. The physical properties of dusts are therefore sometimes difficult to assess as they may act in conflicting ways and their effectiveness or otherwise is the resultant of the various tendencies.

Preparation of dusts

Liquid pesticides are sometimes mixed with absorbent fillers such as clays, while the more persistent crystalline compounds like DDT are diluted with crystalline minerals. There are two main methods of mixing: (1) the straight mix, in which the pesticide is ground to fine particle size and, if this tends to become pasty as a result of frictional heating, a small quantity of abrasive dust or solid carbon dioxide is added and the whole diluted with the main diluent; and (2) the impregnation method, in which the chemical is first dissolved in a light solvent and the resulting solution sprayed into the diluent in a mixer. Excess of solvent is recovered; that which remains may enhance the activity of the dust. By careful choice of diluent, it is possible to make a dust of the drifting type, one that settles at moderate speed or settles rapidly on soil or leaf surface. Dusts are now rarely used having been almost entirely replaced by sprays and granules.

Sprays

Water is usually the principal diluent of sprays and formulation depends on the solubility or otherwise of the pesticide. If it is soluble, it can be dissolved to the required concentration, e.g. sodium fluoride, nicotine; if insoluble, it is possible to make either suspensions or oil emulsions. Oil solutions are used in food storage practice.

Suspensions are produced by grinding the pesticide into small particles between 0.5 and 5.0 μm diam., as for dusts, and then adding wetting and stabilizing agents. This produces a dispersible paste or a dispersible powder concentrate (wettable powder). When these pastes or powders are mixed with water the particles form a suspension which does not readily settle.

Another method of forming a suspension is to dissolve the pesticide in a solvent miscible water so that, when excess of water is added to make up spray, the pesticide is precipitated in colloidal form.

Emulsions are suspensions of oil droplets in water. To make an emulsion, the pesticide is dissolved in an oily organic solvent, an emulsifying agent is added and the whole vigorously agitated or passed through a fine orifice. The molecules of the emulsifying agent (e.g. sulphite lye, bile salts, teepol, lissapol) have oil soluble and water soluble portions and act by orientating themselves at the oil/water interface. Fine droplets of oil become coated with a monomolecular film of emulsifying agent and no longer tend to run together and separate out. Emulsions used to be made up in the field, but now either stock emulsions, in which the water phase is reduced to a minimum, dispersible liquids or miscible oils are used. The thick, mayonnaise type of stock emulsion is less frost stable and more difficult to handle than the thin, liquid type. Stock emulsions or miscible liquids are poured into the spray tank where they are agitated by paddles or by recirculation through the spray pump.

The wetting, spreading and sticking of spray deposits can be improved or modified by various spray adjuvants. Wetting and spreading agents have properties similar to emulsifying agents, and one part of the molecule has an affinity for the waxy coverings of plant leaves and stems, the other is hydrophilic. Soaps are anionic surfactants and if mixed with cationic surfactants, they combine to produce a water-insoluble grease, and wetting properties are lost. With non-ionic surfactants this problem is avoided. The wetting properties of sprays are determined by the receding angle of contact which droplets make with the plant surface. A droplet with a contact angle approaching 180° tends to roll off, near 80° it sticks more readily and near 0° tends to stay on the surface and has good wetting properties. Stickers, such as casein or amine stearates, are used with arsenical or other insoluble pesticides.

The usual object in spraying is to obtain good coverage of the surfaces to be treated. Surfaces should not be overloaded to the point of run-off, emulsions should break on contact so that the oil phase separates on to the plant surface and wetting properties should be good enough to ensure sufficient surface contact but not to encourage run-off. Stickers and other spray adjuvants should help to provide a stable deposit not readily removed by weathering.

Oils and other organic solvents are especially useful as carriers. Many oils possess insecticidal properties of their own. They have a smothering effect owing to entry into the spiracles and may modify the lipoid layers of insect cuticle. They also possess affinities with plant and insect cuticles which give improved wetting and sticking properties to the sprays or dusts in which they are incorporated. Many oils and organic solvents are highly phytotoxic, therefore the grade and type must be chosen with care. This is particularly important for emulsions used in low-volume spraying where the concentration of oil in the spray may reach 5%. Oils are rarely used as the main diluents on crop sprays, but are sometimes used on waste land, marshes and scrub.

High-volume sprays are relatively dilute, contain much water and are applied at 700 or more litres per ha, medium-volume sprays are more concentrated and are applied at 200–700 l ha^{-1}, and low-volume sprays are more concentrated and are applied at 50–200 l ha^{-1}. Ultra-low-volume spraying applies oil-based pesticides by hand-held spraying machines incorporating a rotary atomiser at rates less than 50 l ha^{-1}. Droplet size is controlled to ensure efficient distribution on the target crop and good cover achieved. This method probably lessens environmental contamination as it distributes less toxicant per hectare. Early difficulties with low-volume sprays such as blockage of nozzles, toxicity to plants, spray drift and the harmful effects of drip when the sprayer remains stationary, have been largely overcome and low-volume sprays are now used extensively.

Sometimes concentrated sprays are fed into an air stream and blown into the crop through suitable nozzles. In these machines, the final diluent and propellant is air.

The efficiency of sprays depends much on the machinery used for the application. Reasonably distributed cover of plant surfaces is desirable, although this is less critical for persistent and systemic pesticides than for compounds with a transient life. Uniformity of deposit over the spray band applied is also important, a common fault being excess in the centre and too little at the margins. The kind of cover required varies to some extent with the pest to be controlled. Where only the tips of plants are infested, a tip cover is sufficient. Economies can sometimes be obtained in the spraying of seedlings if the pesticide can be confined to bands along the rows.

The durability of the spray deposit is also important. It should be stable in sunlight, not be readily removed by heavy rain but excessive persistence on or within plants is undesirable on produce for human or animal consumption. A problem associated with sprays is drift; apart from wind, drift particles may be taken up in air currents and 'fall out' on plants and animals many miles from the point of application.

Work is in progress on the control of spray droplet size and the application of a charge to attract droplets to crop surfaces. Effectively these surfaces are earthed, and therefore neutral, and charged particles whether positive or negative are attracted to them following curved paths that convey an ability to reach the undersides of leaves and other 'covered' surfaces. The useful range of droplet sizes produced by ordinary hydraulic spray nozzles is 100 to 250 μm diam. Particles with diameters less than 100 μm drift in wind and those greater than 250 μm tend to bounce off plants. Charged particles even when less than 100 μm diam. do not drift so readily and therefore reach their targets in greater numbers. When spraying tall dense crops (e.g. cereals or potatoes attacked by aphids, brassica seed crops attacked by pests feeding on the flower buds or seed pods) most of the spray is deposited on the upper parts where it is needed. The system works less well against pests operating below the canopy or at soil level. Advantages of the system are economic use of pesticide because much less is wasted and there is less drifting which is environmentally undesirable.

Three main systems of electrostatic spraying are presently being evaluated. In one type electrical forces both produce atomization of the spray liquid and charge the droplets, whereas in the others initial atomization is achieved either hydraulically or by means of a rotary atomizer, and the electrical charge is added afterwards. For further details and other possible developments see Walker (1980).

Granules

Granules and prills are similar to dusts except that the particles are larger, very much larger in the former. Clays, gypsum, sands, coal, sea shells, corncob grit and a variety of other materials are used as carriers and the usual method of incorporating the biocide is by impregnation. The dust from granules may be dangerous if inhaled and non-dusty materials are essential for the most highly toxic biocides, e.g. aldicarb. Additional dusts (not containing the biocide) and other materials are sometimes added to improve flowability in the ducts of applicators. Abrasive granule bases such as sand are undesirable as they wear out applicators reapidly.

Granules can be broadcast, applied as bands on the soil surface, placed beside or beneath seed rows or incorporated generally into the soil. Applied from the air or directly to plants by ground machinery, they are not subject to the problems associated with spray drift. Even distribution over soil or crops is desirable and, to control cyst-nematodes, thorough incorporation into the topsoil is essential for best results.

Granules are expensive to manufacture and bulky to transport and handle. Hence their use is usually limited to toxic biocides or where localized placement is required, e.g. in seed furrows. Because the ultimate despersion of the biocide they carry is in soil water, they somtimes fail to give the required degree of control if drought intervenes after application.

A parallel development of limited use is the encapsulation or microencapsulation of biocides in gelatin or other slow release materials. The main object is to extend persistence and to make the materials selective in action.

Seed treatments

Seed treatments are either powders, slurreys or steeps. In the powder type treatments, the most commonly used, the usual diluent is kaolin, to which is added a small amount of a non-phytotoxic oil to suppress dustiness and improve adhesion. The concentration of pesticide is great and is often in the range 20–60%. Sufficient powder dressing is added to the seed to coat individual grains with a layer approximately two particles deep, but the loading of the seed and the concentration of pesticide is determined by two factors, seed rate and the level of application to seed or row at which phytotoxicity occurs. Seed treatments as applied routinely by seed merchants often fail to achieve acceptable adhesion or sufficiently uniform distribution. Often less than half the expected dose is carried on the seed and whereas some seeds carry too much pesticide others scarcely have any. These facts help to explain the erratic behaviour of some seed treatments and the differences between results achieved in field experiments and farm practice.

Slurreys are thick, paste-type dressings mixed with the seed and allowed to dry on it. Steeps are solutions or suspensions similar in properties to sprays. Here also the concentration of pesticide is great and the more intimate association of seed and dressing raises problems of phytotoxicity. Steeps have been used to get seeds to carry or take up organophosphorus insecticides, but the concentration required makes some steeps hazardous to handle. On the whole, they have not yet found as much application as powder-type dressings. When seed is pelletted, the pesticide may be incorporated in the pelleting material or applied after pelletting.

Aerosols, smokes and fogs

Aerosols are formed by dissolving the pesticide in a solvent gaseous at ordinary temperatures so that the solution must be kept in a canister or cylinder under pressure. When the solution is released through a small orifice, the solution issues as a fine mist from which the solvent evaporates immediately, leaving fine particles of pesticide suspended in air. Smokes are made by combusting a mixture of pesticide and a suitable pyrotechnic. Some of the pesticide is destroyed but some is volatilized and dispersed as fine particles in the smoke. Fogs are similar to aerosols and smokes but contain both solid and liquid particles.

These formulations work best in confined spaces where the particles settle out and coat surfaces or are picked up by the bodies and wings of flying insects. Azobenzene smokes, for example, are used to control red spider mites in glasshouses. In the field, smokes and aerosols find little application because the drifting and settling of the toxic clouds cannot be controlled.

Aerial application of pesticides

In recent years there has been an increase in the number of crops treated from the air. Aerial treatment is especially convenient for large, compact areas of crop, for plantations, forests, tall, dense growing vegetation and difficult terrain.

Although aerial applications of pesticides are made in Britain, the small scattered fields and the convenience of ground machines makes the use of aircraft less attractive than elsewhere. Both helicopters and fixed-wing aircraft are used; the former have greater versatility and can operate from confined and restricted landing grounds, but they are more expensive to maintain and operate. Special equipment and formulations are required to cut down drift which, nevertheless, is an important disadvantage in a countryside of the British pattern. Where the helicopter is employed, use is made of the downward airstream produced by the rotor blades to improve the spray cover and crop penetration. Granules avoid some of these difficulties.

Compatibility and specifications

In formulating products or in combining formulated products to make multi-purpose sprays, great care is necessary to ensure that the various ingredients are mutually compatible otherwise changes may occur which render the individual pesticides inactive or cause harm to the crops sprayed. Phytotoxicity to crops, tainting of edible produce and the timing of spray applications are all important details. Pesticides must be manufactured in as pure a form as possible as the impurities, rather than the pesticide, may be phytotoxic, carcinogenic or cause taints and off flavours. To ensure that pesticides are satisfactory, FAO and WHO have published a series of specifications, e.g. FAO (1973).

Pesticide usage

Some idea of the quantities of crop protection chemicals used and the areas of different crops to which they are applied can be gained from Tables 16.5, 16.6 and 16.7. Herbicides make up the greatest weight of material applied and the greatest area treated. Approximately equal weights of fungicides and insecticides are applied but the area treated with insecticides is more

Table 16.5 Estimated annual usage of crop protection chemicals in field, fruit, vegetable and glasshouse crops, England and Wales, 1971–79 (Sly, 1981).

Type	†Hectares treated in 1000s	%	Weight applied tonnes	%
Organochlorine	147	1	149	1
Organophosphorus	910	6	482	2
Other	345	2	1097	5
Seed treatments	3736	26	578	2
Total insecticides	5138	36	2305	10
Fungicides	2074	15	2368	10
Herbicides	6936	49	17588	74
Other	142	1	1519	6
Total	*14290*	*100*	*23780*	*100*

† Some crops treated more than once, areas to which pesticides are applied are therefore somewhat smaller than those shown.

than twice that treated with fungicides. Although seed treatments are applied to a quarter of the treated area, the weight of material is only a small fraction of the total weight of chemicals applied. Organophosphorus insecticides are applied to 6% of the area but make up only 2% of the weight. Other materials are applied to 2% of the area but make up 5% of the weight. This group contains material applied to the soil of which greater amounts per application are required. Table 16.6 shows the weights of pest control chemicals used on vegetables and field crops in years for which the information is available. Contact organophosphates are applied mainly to vegetable crops mostly as chlorfenvinphos and triazaophos. Approximately equal amounts of systemic organophosphates are applied to vegetables, cereals and other arable crops. Of the carbamates almost equal amounts of pirimicarb and the carbamoyl oximes, aldicarb and oxamyl, are used but pirimicarb is mostly applied to cereals for aphid control and the other two are applied as granules mainly to field crops (especially sugar-beet and potatoes) to control seedling pests and nematodes. Dazomet is the principal soil sterilant used in the field. During the period covered, aldicarb, DDT, chlorfenvinphos, demeton–S–methyl and

Table 16.6 Usage of pest control chemicals on vegetables and field crops in certain years, tonnes (from Sly, 1981).

	Vegetables 1977	Cereals 1977	Other arable crops 1977	Fodder and forage crops 1979	Totals	%
Organochlorine						
Aldrin	2	<1	12	4	18	2
DDT	38	—	10	1	49	5
HCH	<1	1	29	1	32	3
All	*40*	*1*	*51*	*6*	*98*	*10*
Contact organophosphates						
Azinphos-methyl	5	—	1	<1	6	<1
Chorfenvinphos	50	—	—	1	51	5
Chlorpyrifos	4	3	—	1	8	1
Diazinon	5	—	—	—	5	<1
Fenithrothion	8	1	1	—	9	1
Malathion	3	—	—	—	4	<1
Triazophos	23	—	1	—	24	3
Trichlorphon	4	—	3	—	7	1
All	*102*	*4*	*6*	*2*	*114*	*12*
Systemic organophosphates						
Azinphos-methyl/demeton-S-methyl sulphone	3	—	—	<1	3	<1
Demephion	<1	5	—	—	5	<1
Demeton-S-methyl/oxydemeton methyl	30	27	37	<1	94	10
Dimethoate	8	61	26	<1	95	10
Disulfoton	21	—	32	—	53	7
Formothion	1	4	—	—	5	<1
Phorate	39	22	9	—	70	7
All	*102*	*119*	*104*	*<1*	*325*	*35*
Carbamates						
Aldicarb	6	—	77	<1	83	9
Carbofuran	18	—	3	—	21	2
Methiocarb	<1	6	<1	1	7	1
Oxamyl	10	<1	68	<1	78	8
Pirimicarb	1	67	9	<1	77	8
All	*35*	*73*	*157*	*1*	*266*	*28*
Soil sterilants						
Dazomet	71	—	46	—	117	12
Methylbromide	9	—	—	—	9	1
Molluscicide						
Metaldehyde	<1	3	3	5	11	1
Total	359	200	367	14	940	100

Table 16.7 Area in ha of principal field crops treated from the air in 1978 with the most commonly applied insecticides and molluscicides (from Sly, 1981).

	†Cereals	Potatoes	Sugar-beet	Peas	Beans	Oilseed rape	Brussels sprout
Demeton-S-methyl	14 245	9772	1309	6139	3503	95	240
Dimethoate	6570	5632	344	13 757	981	—	—
Pirimicarb	4739	654	168	1702	7084	14	32
Triazophos	10	13	—	14 370	868	15 656	62
Methiocarb	3852	5023	17	19	21	100	1127
Metaldehyde	1062	227	79	43	158	74	38

† mostly to wheat.

oxydemeton methyl, dazomet, dimethoate, disulfoton, phorate, oxamyl and pirimicarb were the chemicals used most commonly. The trend since then has been against DDT for which approval is due to be withdrawn on 1 October 1984 and in favour of pirimicarb because of the development of resistance to organophosphates in aphids, especially *Myzus persicae*. The newer and more stable pyrethroids are being used increasingly but approval is being given with caution because of their potency and relative persistence and because of possible side effects on non-target species. It must also be remembered that the usage of aphicides can vary greatly from year to year (see Table 4.2, p. 46).

Table 16.7 shows the areas of the principal crops treated from the air in 1978. Demeton-S-methyl, dimethoate and pirimicarb were employed mainly against aphids. Triazophos was used mainly to control pea moth and seed weevil and pod midge on oil seed rape. Methiocarb and metaldehyde were mostly to control slugs in potatoes and cereals.

Resistance to pesticides

Resistance to a pesticide may be said to occur when a pest can no longer be controlled by a treatment that was previously effective. The pest is not usually immune for it can still be killed if enough chemical is applied. For economic and environmental reasons this is undesirable and it is usual to turn to another chemical to which resistance is lacking.

Possibly the first evidence of resistance was that of the San Jose scale on apples in Washington State U.S.A. to lime sulphur in 1914. Resistance of scale insects on citrus in California to tent fumigation with HCN was reported about 1930 and in thrips to tartaremetic sucrose in 1942. Resistance of codling moth on apples in Colorado and Washington to lead arsenate appeared about 1943. The development or resistance in citrus and apple pest is perhaps not surprising as both crops had been intensively sprayed over many years.

The discovery of DDT and its intensive and extensive use in the 1940s turned what was a series of local curiosities into a matter of world importance because mosquitoes and lice, the carriers of human disease organisms, became resistant. Agriculturally, resistance in red spider mites and above all in the cotton boll weevil, were events of special importance. The failure of boll weevil control in the 1950s put some cotton-growing areas in the U.S.A., Mexico and Australia out of production. Resistance to pesticides had entered the political arena and the first efforts began to be made to find alternative methods. By the 1960s there were those who said that 'the golden age' of chemical control had passed and more recently to speak of 'superpests' and the 'pesticide crisis' (Perkins, 1982). Because of the increasing amount of time and money required for the development, testing and approval of new chemicals, the production of new materials has fallen behind the numbers of pest species in which resistance has been recorded (Fig. 16.2, Table 16.8).

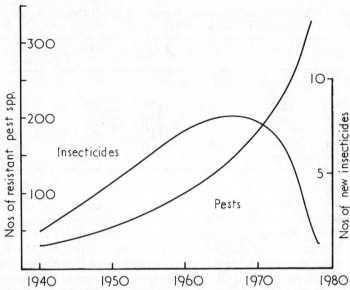

Fig. 16.2 The relationship between the production of new insecticides and the appearance of species with insecticide resistance (modified from Devonshire & Farnham, 1981).

Table 16.8 Occurrence of resistance in agricultural pests worldwide (Georghiou, 1980).

	No of species	%
Mites	38	15
Hemiptera Heteroptera	13	5
Homeoptera	43	16
Coleoptera	64	24
Lepidoptera	64	24
Diptera	27	10
Thysanoptera	7	3
Other	6	3
Total	*262*	*100*

Whatever the position may be in the U.S.A. or other countries, the problems of resistance to pesticides in field crops in the U.K. are slight at present. Much of the resistance recorded in aphids and red spider mites to a range of organophosphates, some organochlorines and to a few carbamates relates to crops under glass. Perhaps the best documented resistance outdoors is that of damson hop aphid (Muir, 1979) and potentially the most serious is the appearance of weak resistance in the peach potato aphid (*Myzus persicae*) to demeton-S-methyl in many areas but especially in East Anglia. Most aphids of this type are killed by a single application at the standard rate but some survive two or three times this dose probably because they do not come into direct contact with the spray when under the lowest leaves. However, their resistance protects them from the residual deposit and the systemic dose present in the plant. Stronger resistance occurs in *M. persicae* from glasshouses and occasionally in the field. Aphids with marked resistance to organophosphates also have resistance to carbamates such as pirimicarb. Aphids that are less resistant are readily controlled by pirimicarb but this insecticide acts more slowly and the delay might enable the aphids to transmit non persistent viruses before they die (Sawicki *et al.*, 1978). There is as yet no evidence for the occurrence of resistance in *Aphis fabae, Sitobion avenae, Metapolophium dirhodum* or in *Rhopalosiphum padi* (Stribley *et al.*, 1983).

Nature and management of pesticide resistance

In any numerous population individuals with unusual genes may exist or may arise by mutation. Some of these genes may influence the reactions of individuals possessing them towards chemical control agents. The situation is no different from that in relation to plants with resistance genes (p. 301) or indeed to any pest control practice. When genes, which initially may be somewhat deleterious, are selected by applying the chemical agent repeatedly, the individuals bearing those genes survive and the frequency of the genes increases in successive generations. Increase is slow at first especially if there is any mechanism which enables individuals with the old genes to persist (e.g. irregular emergence from diapause) or to immigrate from elsewhere. Eventually, the unusual genes become more frequent and the greater part of the population becomes resistant. Depending on the number of generations a year, on whether the genes are dominant or recessive, and on other factors resistance may be expected to appear within 10 to 15 years. According to Sawicki (1979), if the presence of a resistance gene doubles the survival chances of its carrier, then it takes 15 generations to change the gene frequency in the population from 1 in 10 000 to 1 in 30 but only seven more generations to change it to 1 to 1 (i.e. 50% of the population resistant).

Unusual genes operate in several ways. Occasionally they change behaviour so that the chemical is avoided. More usually ability to render the chemical harmless is enhanced by slowing penetration through the cuticle, improving the metabolic processes of detoxification or by changing slightly the structure of the nervous system to decrease its sensitivity, for most pesticides act upon nervous processes.

Delayed penetration of the cuticle confers resistance to many chemicals. Once inside the cuticle the process of detoxification varies with the type of chemical concerned but often includes degradation to less toxic and more soluble breakdown products that are more easily excreted. Early resistance may arise from a single major gene that confers resistance to a narrow range of related compounds. Some changes take in a wider range of chemicals and confer cross-resistance between groups of compounds. Thus, the peach potato aphid, *Myzus persicae*, in glasshouses have become resistant to many organophosphates, carbamate and pyrethroid insecticides by producing a single detoxifying enzyme effective against all three types. Because of the variety and complexity of resistance mechanisms it is difficult to decide on the best policy of pesticide usage to avoid the selection and intensification of resistance. Changing to another material may give temporary relief but equally it may select genes for new mechanisms whose combined effects may lead to extremely high levels of resistance to a wide variety of chemicals as has happened to some housefly populations.

As there is no longer a steady stream of new pesticides and because cross resistance to new materials may already exist, control strategies are necessary involving all available methods with the use of chemicals kept to the minimum to prolong their useful life as in IPM programmes (p. 307). (Sawicki, 1979; Devonshire & Farnham, 1981.)

The detailed study of insect resistance and the underlying biochemical mechanisms is important. Already it enables resistance in some insects to be detected by electrophoresis of enzymes which is more convenient and more rapid than by bioassays. It may also lead to the development of substances that suppress resistance by antagonising the mechanisms of detoxification. Some synergists (compounds that enhance the effect of pesticides) may operate in this way. Substances that are more toxic to resistant than to susceptible races may also be discovered and make possible the selection of susceptible populations so that the original pesticide may be used again.

Hazards from the use of pesticides

In 1950, a Working Party set up under the chairmanship of Professor (later Sir) S. Zuckerman by the Ministry of Agriculture and Fisheries considered in turn measures for the safety of human beings (Anon, 1951), residues in food (Anon, 1953) and risks to wild life (Anon, 1955). In 1960, because of renewed public disquiet, a Research Study Group was set up by the Ministers of Agriculture, of Health, and for Science, to enquire into the need for further research into the effects of toxic chemicals in agriculture and food (Anon, 1961). The Group's report covered: (*a*) risks to man and domestic animals from handling and applying poisonous chemicals; (*b*) the effect of residues in food; (*c*) the risks to wild life; and (*d*) the possible disturbance of the 'balance of nature'.

The use of pesticides in the United Kingdom is affected by several acts. Regulations made under the Poisons Act, 1972 make it compulsory to label substances containing scheduled poisons for sale with the name and concentration of the toxic agent. Persons using HCN for fumigation are protected by the Hydrogen Cyanide Fumigation Act, 1937 which is so worded that can include if necessary other dangerous fumigants such as ethylene oxide and methyl bromide. The latter is approved for use only by contractors with experienced staff. The Health and Safety at Work Act, 1974 and the Health and Safety (Agriculture) (Poisonous Substances) Regulations, 1975 made under it provide a comprehensive system of law covering those at work and the public at large in England, Scotland and Wales. Northern Ireland, the Channel Islands and the Isle of Man have similar regulations (Stell, in Scopes & Ledieu, 1980).

After the second Report of the Zuckerman Working Party a notification scheme was set up, at first voluntary but subsequently compulsory, under which a manufacturer importer or distributor is required to notify a new ingredient or a new formulation of an old material and to obtain clearance for safe use before putting on the market. The Pesticides Safety Precautions Scheme, as it is known, is a formally negotiated arrangement between the British Agrochemicals Association and the British Pest Control Association representing those seeking clearance for a product and the Government Departments and Agencies responsible for Agriculture and Health and Safety. Those seeking clearances must supply the evidence supporting safe use to an Advisory Committee on Pesticides and its Scientific Sub-Committee which make recommendations for the safe use of the chemical concerned. Application may then be made for the substance to be approved under the Agricultural Chemical Approvals Scheme. This is a statutory scheme operated on behalf of Agricultural Departments in the U.K. by the Agricultural Chemicals Approval Organization at the M.A.F.F. Harpenden Laboratory. This scheme is supported by the British Agrochemicals Association, the British Pest Control Association, the United Kingdom Supply Trade Association Limited, the National Unions and Associations of Farmers and Grocers in the U.K., and the National Association of Agricultural Contractors. The purpose of the scheme is to allow users to select, and advisors to recommend, efficient proprietary brands of pesticides and to discourage use of unsatisfactory products. A list of approved products is published annually (Anon, 1984, *et seq.*) and such products are marked by a distinctive label.

The Health and Safety (Agriculture) (Poisonous Substances) Regulations, 1975 contains a Second Schedule which contains three Parts (I, II and III) and specifies the protective clothing that must be worn appropriate to the pesticide being dispensed or sprayed. An abbreviated summary of these requirements is in Table 16.9: for more detailed information see Anon (1984) or Scopes & Ledieu (1980). Solid or granular applications are safer than sprays. Gaseous or atomized pesticides are most dangerous especially in confined spaces. Some relaxation of the protective clothing is permitted when Pt II or Pt III substances are used as granules which meet specifications laid down in the Regulations. Some of these concern freedom from toxic dust that might be inhaled by operatives. Many chemicals used in agriculture are poisons

Table 16.9 Protective clothing specified for pesticides listed in Parts I, II and III of the Second Schedule of the Health and Safety (Agriculture) (Poisonous Substances) Regulations, 1975.

Part I substances	Rubber gloves	Rubber boots	Overall and rubber apron, or mackintosh		Respirator for concentrates or toxic vapours
Part II substances	Rubber gloves	Rubber boots	As above, or a hood, or rubber coat and hood/ sou'wester	Face shield or dust mask	*Respirator
Part III substances	Rubber gloves	–	*Overall	Face shield	*Respirator, hood

* Required for applying aerosols or atomized fluids in confined spaces, for details see Anon (1984) and Poisonous Chemicals on the Farm, M.A.F.F. Booklet HS(G)2.

subject to the provisions of the Poisons Act, 1972 and the Lists and Rules made under it.

Labels on bottles, cartons, cans and other containers should show clearly the nature of the toxicant they contain, in which part of the second Schedule of the Health and Safety (Agriculture) (Poisonous Substances) Regulation, 1975 they are listed, and the protective clothing required. Any unusual risks should also be mentioned and what precautions are necessary to ensure no unacceptable residues remain on edible crops after harvest. Risks to bees, domestic animals, fish or other wild life should also be clearly listed. In addition the labels should show the dosage recommended for specific pests and crops.

Since about 1950, only two farm deaths have occurred arising from agricultural chemicals and there have been few cases of sickness. In the same period about 700 have died from accidents with tractors or other farm machinery. About 60 persons have been killed by falls; 20 by falls from ladders. Six were drowned in grain.

Toxic residues in soils

After applying the more persistent types of pesticides, such as arsenicals, organochlorine or organobromine compounds to plants or soils, residues may persist in the soil for months or years. The amounts added by successive treatments may exceed the rate of decay and cause accumulations which may harm root systems of sensitive plants, cause off-flavours in harvested produce and lead to small amounts being absorbed by plants which are later consumed by man or domestic animals.

Whether residues of pesticides or their breakdown products accumulate depends on the amounts applied, the frequency of application and the rate of loss. Often much is lost at the time of application by volatilization. Thus 20–40% of aldrin and possibly greater proportions of gamma-HCH sprayed on to crops disappears and never reaches the soil. Deposits of stable compounds of very low volatility such as dieldrin and DDT do find their way into soil, however, when the deposit weathers and is washed down by rain or dead foliage becomes incorporated in the soil surface. Losses from the soil are influenced by soil temperature, cultivations, soil type, site of application and the nature of the compound concerned. Sometimes it is the residues of breakdown products that accumulate rather than the pesticide itself.

The importance of residues in soil has increased with the increase in the use of pesticides. Although much attention is paid to the properties of pesticides in the developmental stages, the ultimate test of efficacy and freedom from undesirable side effects is not made until the material is on sale and is used extensively. Of the older pesticides, only those containing arsenic led to serious residue problems and these were found mainly in orchards sprayed repeatedly with lead arsenate. The outstanding success and widespread use of the organochlorine insecticides gave rise to residue problems. Their absolute volatility is low, which aids their persistance, but prevents them from spreading rapidly in soil as do some fumigants. Consequently more insecticide is probably used than would be necessary if it were thoroughly and uniformly distributed. Gamma-HCH and aldrin are the most volatile and least persistent. When worked into the surface soil or protected by a cover of vegetation, loss by volatilization is less than when applied to the exposed and undisturbed soil surface. Loss increases as soil temperature rises and is negligible when the soil surface is frozen. Once within the soil, organochlorine compounds are gradually redistributed by volatilization, diffusion, water percolation and cultivations until they occur more or less uniformly down to plough depth with traces in the sub-soil. Excess of moisture in soil containing aldrin appears to release it from sorption on soil particles and hasten volatilization which goes on slowly from the surface at all times. Degradation is another process which hastens loss. Breakdown products of gamma-HCH are detectable in soil within a few days of application; aldrin oxidizes slowly to dieldrin its epoxide, but dieldrin and DDT are less volatile and more stable than HCH and therefore persist longer. Earthworms and other organisms may play a part in the degradation of DDT to TDE (Edwards & Lofty, 1973).

With the possible exception of a few relatively persistent ones, organophosphorous, carbamate and pyrethroid compounds are probably short-lived in soil. The acid radicle containing the phosphorus atom is ultimately split off from organophosphates and becomes part of the phosphorus content of the soil. The various radicles (R_1, R_2 and X, p. 326) attached to the phosphorus atom also undergo degradation depending on what their particular properties are. The fate of oxime carbamates varies with their structure. In soils, plants, insects and vertebrates, aldicarb is rapidly oxidized to its sulphoxide, which is the principal active material, more slowly to sulphone, after which it is degraded to non-toxic products. Oxamyl is toxic *per se*, resists oxidation but is degraded to non-toxic products in other ways. It has a half life of about two weeks in soil. Crop uptake and volatilization are unimportant in the disappearance of oxime carbamates compared with degradation. Negligible amounts of aldicarb occur below 30 cm deep and so it is unlikely to contaminate drainage water except if used excessively. Most pyrethroids have little persistence but chemical modifications of the molecule have improved performance by increasing stability and persistence so that deltamethrin is sometimes thought of as a replacement for DDT.

Persistence of pesticides reflects the outcome of all processes modifying them in soil and important factors include the pesticide itself, its formulation, soil type, pH, moisture, temperature, amount applied, depth of placement and microbial activity. When pesticides are microbially degraded a second application usually disappears faster than the first. For further details consult Helling, Kearney & Alexander (1971).

Because of the risk from the use of pesticides many Government-sponsored reports have appeared since 1963 covering almost all aspects of pesticide usage.

Toxic residues in foods and on plants

All countries want pure food; the U.S.A. and Canada have laws determining the tolerance levels for those pesticides that leave residues on or in the crops to which they are applied. The largest amount which should occur in the food of a person per day is called the maximum acceptable daily intake and has been worked out for most foods and for pesticides in common use. Residues from chemicals, correctly used in an approved manner and at a rate that gives a satisfactory kill, are many times less than the dose that just causes obvious harm (the harmful dose) when fed continuously to experimental animals, such as the rat, guinea pig, mouse or rabbit. In Canada the maximum tolerance level for a particular combination of food and pesticide is calculated from (1) the harmful dose expressed in ppm in the diet; (2) a safety factor, usually 100 but less for those compounds for which there is adequate information about the effects on the human body; and (3) a food factor based on the proportion of the contami-

Table 16.10 Acceptable daily intakes and residue tolerances proposed for certain insecticides by FAO/WHO.

Pesticide	Acceptable daily intake in mg kg^{-1} of body weight	Residue tolerances on specific crops, in ppm by weight prior to consumption	
Aldrin and dieldrin	0.0001	aparagus, broccoli, Brussels sprouts, cabbage, cauliflower, cucumber, onions, parsnips, radishes	0.1
		fruit (other than citrus)	0.1
		citrus fruit	0.05
		rice (rough)	0.02
		potatoes	0.2
Chlordane	0.001	potatoes, turnips, parsnips, radishes	0.3
		asparagus, broccoli, Brussels sprouts, cabbage, celery, cauliflower, spinach, Swiss chard, lettuce	0.2
		beans, peas, tomatoes, collards	0.02
		wheat, rye, oats, rice (polished), maize	0.05
		cucumbers	0.1
		almonds, bananas, figs, filberts, passion-fruit, pineapples, strawberries, walnuts	0.1
		citrus, pome and stone fruits	0.02
DDT	0.005	apples, pears, peaches, apricots, small fruits (except strawberries), vegetables (except root vegetables), meat and poultry (on fat basis)	7
		nuts (shelled), strawberries, root vegetables	1
		cherries, plums, citrus and tropical fruit	3.5
Endrin	0.0002	apples, wheat, barley, rice	0.02
Heptachlor	0.0005	pineapple (edible portions)	0.01

Source: FAO, Ceres (1972) 5 (4), p. 7.

nated food consumed in an average diet. Thus, if the harmful dose is 10 ppm, the safety factor 100 and the food factor 0.2, the maximum tolerance is $10 \div (100 \times 0.2)$ or 0.5 ppm. The safety factor is intended to compensate for differences between the physiology of man and the test animals used, variation in susceptibility arising from age, health, individual idiosyncrasies and feeding habits. Apart from lead and arsenic, no tolerance levels for pesticides are enforced in Great Britain, but it is possible that tolerance levels may be determined and enforced in the future. Since 1964, FAO and WHO have published reports on toxic residues in foods, culminating in lists of daily intakes which, during an entire human lifetime, appear to be without appreciable risk on the basis of known facts. Table 16.10 is such a list for some common organochlorine pesticides. Analyses of pesticide residues in food stuffs in Great Britain published by the Laboratory of the Government Chemist and the M.A.F.F. Plant Pathology Laboratory, now known as the Harpenden Laboratory, show that home produced or imported produce contains residues which are, with few exceptions, well below the acceptable daily intakes recommended by the Codex Alimentarius Committee of FAO/WHO (Egan & Weston, 1977). Amounts ingested in total diets are trivial and, moreover, the last decade residues have declined.

In general, gamma-HCH is the main residue in cereals, fruits and preserves. DDT (as DDE, a breakdown product) is the main residue in meats. A range of residues is found in milk but at extremely low concentrations. Root vegetables contain scarcely any residues. Almost the only organophosphate residue found in food is malathion (the least toxic to mammals of all the organophosphorus pesticides) and in extremely small amounts. Out of 462 samples tested only single findings of demeton–S–methyl, disulfoton, fenchlorphos, fenithrothion and parathion were recorded, once more in trace amounts. It must be remembered that not all these residues originated in the U.K. Fenchlorphos and parathion are not included in Approved Products for Farmers and Growers (Anon, 1984).

The objectives of the voluntary Agricultural Chemicals Approved Scheme seem to have been achieved by a combination of methods namely rates of application of pesticides, rates of breakdown, minimum intervals between application and harvest, and the last date at which specified products can be applied to crops like Brussel sprout and leaf brassicas.

In studying systems of maximum pesticide residue levels, such as those used in the U.S.A., the Sanders Research Study Group pointed out that such levels were often wrongly misinterpreted as 'safety levels', when in fact they are the maximum amounts which result from approved use. The actual safeguard is correct application and the tolerance figure is only a means of detecting improper use. Further, the levels established must relate to the conditions in any particular country and cannot be established without adequate supporting data and sufficiently sensitive analytical methods.

At present, although tolerance levels (except for arsenic and lead) are not specified in Britain, the use of dangerous toxicants is limited in other ways. In the recommendation sheets, product labels and data sheets of approved chemicals used in agriculture and food storage, the maximum rate and frequency of application per crop and season are specified as well as the safe minimum intervals between application and harvest. These are drawn up so that, when the substance is used as specified, no undesirable residues will remain in or on the produce. Approved chemicals are listed under the Agricultural Chemicals Approval Scheme which deals with their biological efficiency (Anon, 1984, *et seq.*).

The main difficulties arose from the use of the persistent organochlorine compounds, especially DDT, dieldrin and HCH, which, being fat soluble, accumulate in body fat. Random samples of human body fat from 131 corpses were obtained from two centres in southern England, one semi-rural and the other urban. The average content of DDT at 2.2 ppm was much the same from both areas and in both, fat from males contained one and a half times as much DDT as that from females. The dieldrin content of fat was 0.2 ppm with no difference between the sexes. The DDT content was less than in West Germany and much less than in fat samples from Georgia and Washington, U.S.A. (11.7 ppm). American agricultural workers using DDT had 40–50 ppm in their body fat; and operatives from chemical works in the U.S.A., concentrations ranging from 264 to 1134 ppm, apparently without obvious harmful effects. Samples of human body fat collected in semi-rural areas of south-eastern England contained from 0.19 to 0.015 ppm of HCH. Between 1965 and 1967 the average from 237 samples collected in England was 0.31 ppm beta-HCH, which was well below danger limits (Morrison, 1972). The position has improved since then. The human body picks up many chemicals, natural and synthetic, and has great power of dealing effectively with most.

Hazards to livestock are due chiefly to heavy metals, e.g. lead, copper and arsenic, and not to pesticides. Provided instructions are obeyed and the movements of animals are controlled to prevent them feeding on sprayed crops, the immediate risks are small. But livestock, like human beings, pick up pesticide residues in their food and concentrate them in their body fat. Thus mutton fat used to contain small amounts of dieldrin, more being present in imported than home-produced meat. This is still the case but since the withdrawal of sheep dips containing dieldrin in 1966 amounts in mutton fat have decreased to 0.01 mg kg^{-1}. It is evident that a few commonly used pesticides may accumulate in the bodies of men and domestic animals, and on the long-term effects of this there is a need for intensified research and no cause for complacency.

Hazards to wild life

Between 1950 and 1960, as the use of pesticides increased, fears were expressed of their possible adverse effects on wild life. It soon became clear that schradan, which at that time was used to kill aphids in sugar-beet and Brussels sprout, caused many bird deaths, but complaints concerned deaths mainly of pheasants and partridges. As schradan was replaced by more selective organophosphorus compounds of lower toxicity to warm-blooded animals incidents of this kind decreased. In the springs of 1960 and 1961, following the widespread use of seed treatments containing aldrin, dieldrin and heptachlor on wheat to control wheat bulb fly, bird deaths increased alarmingly and caused a public outcry. At this time of the year, when other types of food are scarce, grain is often taken. Seed treatments containing a large percentage of insecticide and a small percentage or organomercury fungicide might almost be considered as poison bait for grain-eating birds. Notwithstanding the great concentration of insecticide in the dressings, the actual amounts carried by seeds are small and are decreased further after being buried in soil by drilling. In fact much buried grain would have to be consumed before a lethal dose was taken, but spilled grain or grain illegally scattered to kill pigeons carried enough poison for a lethal dose to be acquired in a single feed.

If only pigeons had been killed, there would have been less concern, but game birds and other desirable species were also killed and contaminated birds were eaten by predatory mammals and birds such as the peregrine falcon and Montagu harrier. Bird deaths have been greatly lessened by a voluntary agreement entered into by manufacturers and vendors of seed treatments, and by other interested parties. Seed treatments containing aldrin, dieldrin or heptachlor are no longer used. Seed treatments containing gamma-HCH are used for wireworm control as before. These have been widely used for many years and have not caused bird deaths.

It is regrettable that trouble was experienced with seed treatments largely because they are in many ways the most economic, convenient and appropriate method of applying insecticide, requiring no more than 150 g ha $^{-1}$ of active material. The withdrawal of dieldrin seed treatment for winter wheat has left a gap in the control of wheat bulb fly which has not been filled adequately by newer, non-organochlorine compounds.

The possible accumulation of persistent insecticides of the organochlorine group in the fatty tissues of animals and their progressive concentration up the food chain gave rise to much concern amongst naturalists and those interested in the preservation of rare or relatively rare species (Moriarty, 1975; Moore, 1972; Perring & Mellanby, 1977). Failure of some predatory birds to breed successfully was claimed to be caused by organochlorine compounds. After 1950 more egg breakages were recorded in the nests of peregrines, sparrow-hawks and golden eagles and the relative weight (largely thickness) of egg shells decreased from about 1945. According to Gunn (1972), these phenomena appear to have preceded the use of organochlorine pesticides and the greater use of organic chemical compounds, particularly organochlorines, for domestic, veterinary, agricultural and horticultural use began if anything after the thinning of egg shells first appeared. On the other hand the causal effect is accepted by most authorities (Ratcliffe, 1970; Hazeltine, 1972; Anon, 1972b; Cooke, 1973). Sensational claims that insecticides have caused the deaths of birds and other animals are often made in the press, on radio and television, often without objective evidence; condemnation precedes evidence because it supports the popular viewpoint, although what proportion of deaths can rightly be ascribed to pesticides is uncertain. Nevertheless, it cannot be denied that pesticides kill wild-life and laboratory tests show that some pesticides are toxic to fish (especially DDT and organochlorines), to mammals, birds, bees and also to many other invertebrates. There is great variation in susceptibility between species and within species, between sexes and age groups; young individuals succumb more rapidly than old.

The Minister of Agriculture's Advisory Committee on Poisonous Substances used in Agriculture and Food Storage under the chairmanship of Sir James Cook (Anon, 1964b) reported in March 1964, and concluded that 'There is insufficient evidence to justify a complete ban on any of the chemicals (persistent organochlorine pesticides) we have reviewed. There is, for instance, no basis for statements that these are severe liver poisons, nor is there any proof that DDT causes any injury while stored in the fat of human beings or animals. Similarly, DDT and dieldrin cannot be condemned as presenting a carcinogenic hazard to man. Nevertheless, we consider that the levels of dieldrin residues are undesirable and the evidence justifies a partial restriction of its use.' The Committee recommended that aldrin and dieldrin should cease to be added to fertilizers or sheep dips. Seed treatments containing aldrin, dieldrin and heptachlor are no longer approved. The use of aldrin is now restricted to the control of narcissus fly as a dip or spray for bulbs that will remain in the ground for 2 to 3 years and to the control of wireworms in potato after permanent pasture. All uses of DDT on farm crops will cease in 1984. Gamma-HCH, being less persistent, continues to be approved of a range of uses. Alternative organophosphorus, carbamate or pyrethroid materials to take the place of the organochlorine compounds are available although often these are more expensive, not always as effective or not available in satisfactory formulations. They are also more hazardous to those who have to apply them (Gair, 1971). In the Appendices of the 'Further Review' (Anon, 1969) are excellent surveys by Mr A.H. Strickland and others of the use of organo-chlorine pesticides, the needs for them and possible alternatives.

Although the use of pesticides in Britain has greatly increased in the last three decades, there have been few instances of 'blanket' or 'block' spraying of large areas of crop, forest or marsh from the ground or from the air, such as have occurred in the U.S.A., Canada and elsewhere. Most of the accounts of the misuse of pesticides sensationally publicized by Carson (1963) fortunately do not apply to Britain, and many of the fears expressed about the effects of pesticides on the so-called 'balance of nature' are based on emotion rather than fact. This is largely because there is much confusion in the public mind about the countryside, and many fail to realise that it is essentially man made and that man's impact on nature was greatest long ago (see Chapter 1). Because of the pressing need to produce more food for the world's increasing population, and to produce it more cheaply and efficiently in those countries that already have enough, pesticides have come to stay and will continue to find a place in IPM programmes. More research is needed to find selective pesticides, improved methods of application, techniques for delaying resistance to pesticides, and new ways of destroying pests (Graham-Bryce, 1983) so that undesirable side effects of all kinds can be minimized.

The alternatives to pesticides such as biological control and greater use of resistant cultivars pose their own problems. It cannot be assumed that biological control using bacteria, viruses and other micro-organisms will necessarily be possible, permanent or free from hazards. Experience with cultivars resistant to insects, fungi or nematodes shows only too clearly that races able to circumvent resistance already exist or are soon selected. Nor is there the time to garner all background knowledge before deciding what course of action to pursue when crops are at risk.

References

Note: Ministry of Agriculture Leaflets and Reference Books are mentioned at appropriate places in the text.

A'Brook, J. and Benigno, D. A. (1972). The transmission of cocksfoot mottle and phleum mottle viruses by *Oulema melanopa* and *O. lichenis*. *Ann. appl. Biol.* **72,** 105.

Acuna, A. Matarrita and Rojas, de L. (1958). Effectiveness of various organic phosphorus insecticides for control of thrips on onion. *Biologico*, **24,** 134.

Aitkenhead, P. (1981). Colorado beetle – recent work in preventing its establishment in Britain. *EPPO Bull.*, **11,** 225.

Andersen, S. and Andersen, K. (1982). Suggestions for determination of terminology of pathotypes and genes for resistance in cyst-forming nematodes especially *Heterodera avenae*. *EPPO Bull.*, **12,** 379.

Andrawes, N. R., Bagley, W. P. and Herrett, R. A. (1971a). Fate and carryover properties of Temik aldicarb pesticide [2-methyl-2-(methyl-thio) proprionaldehyde O-(methylcarbamoyl) oxime]. *J. Agric. Fd. Chem.* **19,** 727.

—— (1971b). Metabolism of 2-methyl-2-(methyl-thio) proprionaldehyde O-(methylcarbamoyl) oxime (Temik aldicarb pesticide) in potato plants. *J. Agric. Fd. Chem.* **19,** 731.

Anon. (1951). *Toxic chemicals in agriculture. Report of the working party on precautionary measures against toxic chemicals used in agriculture.* MAFF. London, HMSO 15 pp.

—— (1953). *Toxic chemicals in agriculture. Residues in Food. Report of the Working Party.* MAFF. London, HMSO 32 pp.

—— (1955). *Toxic chemicals in agriculture. Risks to wild life. Report of the Working Party.* MAFF. London, HMSO 30 pp.

—— (1961). *Toxic chemicals in agriculture and in food. Report of the Research Study Group.* MAFF. London, HMSO 65 pp.

—— (1962). *Infestation control.* MAFF. London, HMSO 61 pp.

—— (1964a). *Review of persistent organochlorine pesticides. Report of the Advisory Committee on Poisonous Substances used in Agriculture and Food Storage.* London, HMSO 68 pp.

—— (1964b). *Review of the persistent organochlorine pesticides. Supplementary Report by the Advisory Committee on Pesticides and other Toxic Chemicals.* London, HMSO 8 pp.

—— (1965a). *Losses in agriculture.* U.S. Dept. Agric. Washington, D.C.

—— (1965b). Chemical alternatives to organochlorine insecticides. *Plant Pathology*, **14,** Suppl. 36 pp.

—— (1967a). Chemical alternatives to organochlorine insecticides. *Plant Pathology*, **16,** Suppl. 44 pp.

—— (1967b). *Symposium on crop losses.* Rome, FAO 330 pp.

—— (1967c). Estimated crop losses due to plant-parasitic nematodes in the U.S.A. *J. Nematol.*, **1,** Suppl. 7 pp.

—— (1969). *Further review of certain persistent organochlorine pesticides used in Great Britain. Report of the Advisory Committee on Pesticides and Other Toxic Chemicals.* London, HMSO 148 pp.

—— (1970). Report of an international conference on rodents. Paris, EPPO Publications, Series A, No. 58.

—— (1972a). Plant parasitic nematode pathotypes. *Ann. appl. Biol.*, **71,** 263.

—— (1972b). Eggshell thinning and DDE. *Nature, Lond.*, **235,** 376, **239,** 411, **240,** 162 *et seq.*

—— (1980). *Factors affecting the application and use of nematicides in W. Europe.* Workshop Nematology Group Ass. Appl. Biologists, Rothamsted 1980. 116 pp + appendices.

—— (1983). *Plant breeding – an integrated discipline.* Abstracts of papers given at a meeting in Selwyn College, Cambridge, 1983. Wellesbourne Association of Applied Biologists. 97 pp.

Anon. (1984). *Approved products for farmers and growers.* ADAS, MAFF Reference Book 380 (84). London, HMSO 230 pp.

Anscombe, F. J. (1960). Soil sampling for potato root eelworm cysts. A report presented to the conference of Advisory Entomologists. *Ann. appl. Biol.*, **37**, 286.

Ashworth, R. de B. (1966). Certain aspects of the organisation of international crop protection. *PANS*, **12**, 179.

—— and Crozier, D. (1972). Standard waters: recommendations for the preparation of standard waters used for testing pesticidal and other formulations. Ed. G. R. Raw. Harpenden, Collaborative International Pesticides Council Ltd. 64 pp.

——, Heniet, J. and Lovatt, J. (1970). *Analysis of technical and formulated pesticides.* Vol. I. Harpenden, Collaborative International Pesticides Council Ltd. 1079 pp.

Atkinson, H. J. and Sykes, G. B. (1981). An analysis from aerial photography of spread of seed-borne *Ditylenchus dipsaci* in lucerne. *Nematologica*, **27**, 235.

Auclair, J. L. (1963). Aphid feeding and nutrition. *Ann. Rev. Ent.*, **8**, 439.

Baker, A. N. and Dunning, R. A. (1975). Association of populations of Onychiurid Collembola with damage to sugar beet seedlings. *Plant Pathology*, **24**, 150.

Balachowsky, A. and Mesnil, L. (1935/36). *Les insectes nuisible aux plantes cultivées.* 2 Vols. Paris, Busson. 1921 pp.

Banks, C. J. (1955). An ecological study of Coccinellidae (Col.) associated with *Aphis fabae* Scop. on *Vicia faba. Bull. ent. Res.*, **46**, 561.

Bardner, R., Edwards, C. A., Arnold, M. K. and Rogerson, J. P. (1971). The symptoms of attack by swede midge (*Contarinia nasturtii*) and effects on the yield of swedes. *Entomologia exp. appl.*, **14**, 223.

—— and Fletcher, K. E. (1979). Larvae of the pea and bean weevil, *Sitona lineatus*, and the yield of field beans. *J. agric. Sci., Camb.*, **92**, 109.

—— and —— (1974). Insect infestations and their effects on growth and yield of field crops. *Bull. ent. Res.*, **64**, 141.

Barnes, H. F. (1946–56). *Gall Midges.* Vols. 1–8. London, Crosby, Lockwood.

Bartell, R. J. (1982). Mechanisms of communication disruption by pheromone release in the control of Lepidoptera: a review. *Physiol. Ent.*, **7**, 353.

Bartlett, P. W. (1979). Preventing the establishment of Colorado beetle in England and Wales. In: *Plant health: the scientific basis of administrative control of plant diseases and pests.* Eds. D. L. Ebbels and J. E. King. Oxford, Blackwells. p. 248.

Bassett, P. (1978). Damage to winter cereals by *Zabrus tenebrioides* (Goeze.) (Coleoptera: Carabidae). *Plant Pathology*, **27**, 48.

Beament, J. W. L. (1961). The water relations of insect cuticle. *Biol. Rev.*, **36**, 281.

Becker, P. (1961). Leek moth occurring inland. *Plant Pathology*, **10**, 42.

Beirne, B. P. (1962). Trends in applied biological control of insects. *Ann. Rev. Ent.*, **7**, 387.

—— (1970). The practical feasibility of pest management Systems. In: *Concepts of Pest Management.* (Eds) R. Rabb and F. E. Gutherie. North Carolina State University, 158 pp.

—— (1972). The biological control attempt against European wheat stem sawfly, *Cephus pygmaeus* (Hym. Cephidae) in Ontario (Canada). *Can. Entomol.*, **104**, 987.

Bentley, E. W. (1964). A further loss of ground of *Rattus rattus* L. in the United Kingdom, during 1956–61. *J. anim. Ecol.*, **33**, 371.

—— and Larthe, Y. (1959). The comparative rodenticidal efficiency of five anti-coagulants. *J. Hyg. Camb.*, **56**, 19.

—— and Taylor, Eileen J. (1965). Growth of laboratory-reared ship rats (*Rattus rattus* L.). *Ann. appl. Biol.*, **55**, 193.

Bickerton, B. M. and Chapple, W. (1961). Starling roosts and their dispersal. *Agriculture, London*, **67**, 624.

Biddle, A. J. *et al.* (1983). Pheromone monitoring of the pea moth, *Cydia nigricana* (F). *Proc. 10th Int. Congr. Plant Protection, Brighton*, **1**, 161.

Bingham, J. and Lupton, F. G. H. (1958). Breeding spring oats for resistance to frit-fly attack. *Ann. appl. Biol.*, **46**, 493.

Blackman, R. L. (1974). *Invertebrate types: Aphids.* London, Ginn. 175 pp.

Blair, C. A. and Groves, J. R. (1952). Biology of the fruit tree red spider mite, *Metatetranychus ulmi* (Koch) in south-east England. *J. hort. Sci.*, **27**, 14.

Blake, C. D. (1962). Importance of osmotic potential as a component of the total potential of the soil water on the movement of nematodes. *Nature, Lond.*, **192**, 144.

Bombosch, S. (1955a). Beitrage zur Kenntnis des Moosknopfkäfers. *Zucker*, **8**, 46.

—— (1955b). Beitrage zur Kenntnis des Moosknopfkäfers. *Zucker*, **8**, 285.

Boudreaux, H. B. (1963). Biological aspects of some phytophagous mites. *Ann. Rev. Ent.*, **8**, 137.

Bowden, J., Cockrane, J., Emmett, B. J., Minall, T. E. and Sherlock, P. L. (1983). A survey of cut-worm attacks in England and Wales, and a description population model for *Agrotis segetum* (Lepidoptera: Noctnidae). *Ann. appl. Biol.*, **102**, 29.

—— and Jones, M. G. (1979). Monitoring wheat bulbfly. *Delia coarctata* (Fallen) (Diptera: Anthomyi-idae) with light traps. *Bull. ent. Res.*, **69**, 129.

Boyd, A. E. W. (1966). Sugar-beet strangles. *Edin. Sch. Agric. Tech. Bull.* **26**, 42 pp.

——, Erskine, D. S. C., Byford, W. J. and Webb, D. J. (1970). A herbicide induced abnormality in sugar beet. *Plant Pathology*, **19**, 163.

Brian, M. V. (1947). On the ecology of beetles of the genus *Agriotes* with special reference to *A. obscurus*. *J. anim. Ecol.*, **16**, 210.

Briggs, J. B. (1965). The distribution, abundance and genetic relationships of four strains of the rubus aphid (*Amphorophora rubi* (Kalt.)) in relation to raspberry breeding. *J. hort. Sci.*, **40**, 109.

Brindle, A. (1960). The larvae and pupae of the British Tipulinae (Diptera; Tipulidae). *Trans. soc. Brit. Ent.*, **14**, 63.

Bromilow, R. H. (1973). Breakdown and fate of oxime carbamate nematicides in crops and soil. *Ann. appl. Biol.*, **75**, 473.

Brown, E. B. (1969). Assessment of damage caused to potatoes by potato cyst-eelworm, *Heterodera rostochiensis* Woll. *Ann. appl. Biol.*, **63**, 493.

Brown, R. A. (1981). Gappiness, sugar beet yield loss and soil inhabiting pests. *Proc. 11th Brit. Insectic. Fungic. Conf.*, **3**, 803.

—— (1982). The ecology of soil inhabiting pests of sugar-beet, with special reference to *Onychiurus amatus*. PhD. Thesis, University of Newcastle-upon-Tyne.

Bry, R. E., Lang, J. H. and Bennis, N. M. (1971). Moth proofing woollen cloth with Gardona applied during padding. *J. econ. Ent.*, **64**, 990.

Burges, H. D. (1970). Control of insects by *Bacillus thuringiensis*. *Proc. 5th Br. Insectic. Fungic. Conf. Brighton 1969*, **2**, 405.

——, H. D. (Ed.) (1981). *Microbial control of pests and diseases, 1970–1980*. New York, Academic Press. 949 pp.

—— and Hussey, N. W. (Eds) (1971). *Microbial control of insects and mites*. New York, Academic Press. 861 pp.

Bushland, R. C., Lindquist, A. W. and Knipling, E. F. (1955). Eradication of screw worms through release of sterilized males. *Science*, **122**, 287.

Butler, C. G. (1962). *The world of the honey bee*. London, Collins, 226 pp.

—— (1970). Chemical communication in insects: behavioural and ecologic aspects. *Adv. Chemoreception*, **1**, 35.

Cammell, M. E. (1981). The black bean aphid, *Aphis fabae*. *Biologist*, **28**, 247.

—— and Way, M. J. (1977). Economics of forecasting for chemical control of the black bean aphid, *Aphis fabae*, on the field bean, *Vicia faba*. *Ann. appl. Biol.*, **85**, 333.

——, Way, M. J. and Heathcote, G. D. (1978). Distribution of eggs of the black bean aphid, *Aphis fabae* Scop. on the spindle bush, *Euonymus europaeus* L. with reference to forecasting infestations of the aphid on field beans. *Plant Pathology*, **27**, 68.

Campion, D. G. (1972). Insect chemosterilants: a review. *Bull. ent. Res.*, **61**, 577.

Carden, P. W. (1962). The cabbage leaf miner attacking Calabrese. *Plant Pathology*, **11**, 36.

Carson, R. (1963). *Silent spring*. London, Hamish Hamilton, 304 pp.

Carter, W. (1961). Ecological aspects of plant virus transmission. *Ann. Rev. Ent.*, **6**, 347.

Chapman, R. F. (1982). *The insects: structure and function*, 3rd ed. London, Hodder & Stoughton. 919 pp.

Chapman, R. N. (1931). *Animal ecology with especial reference to insects*. New York, McGraw-Hill. 464 pp.

Cherret, J. M. and Sagar, G. R. (Eds) (1977). *Origins of pest, parasite, disease and weed problems*. 18th Symposium of Brit. ecol. Soc. Oxford, Blackwell Scientific Publications. 413 pp.

Christie, J. R. (1929). Some observations on sex on the Mermithidae. *J. exp. Zool.*, **53**, 59.

Christie, J. R. (1959). *Plant nematodes, their bionomics and control.* Gainesville, Agric. Exp. Stn Univ. Fla. 256 pp.

Chua, T. H. (1977). Population studies of *Brevicoryne brassicae* (L.), its parasites and hyperparasites. *Res. popul. Ecol.*, **19**, 125.

Church, B. M., Jacob, F. H. and Thompson, H. V. (1953). Surveys of rabbit damage to wheat in England and Wales, 1950-52. *Plant Pathology*, **2**, 107.

——, Westmacott, M. H. and Jacob, F. H. (1956). Survey of rabbit damage to winter cereals, 1953-54. *Plant Pathology*, **5**, 66.

Clark, W. C. (1973). The ecological implications of parthenogenesis. In *Perspectives in aphid biology*. Ed. A. D. Low. Christchurch, Entomological Society of New Zealand. 103 pp.

Coaker, T. H. and Worrall, J. (1961). The biology and natural control of the cabbage root fly. *Rep. National Vegetable Res. Sta. Wellesbourne*, 1960. p. 42.

Coals, J. R. (Ed.) (1982). *Insecticide mode of action.* New York, Academic Press. 470 pp.

Cockbain, A. J. (1971). Aphids and virus diseases of pea. *Rep. Rothamsted exp. Stn* Pt 1 for 1970, p. 187.

—— and Gibbs, A. J. (1973). Host range and overwintering sources of bean leaf roll and pea enation mosaic viruses in England. *Ann. appl. Biol.*, **73**, 177.

Cockbill, G. F., Henderson, V. E. Ross, D. M. and Stapley, J. H. (1945). Wireworm populations in relation to crop production. 1. A large-scale flotation method for extracting wireworms from soil samples and results from a survey of 600 fields. *Ann. appl. Biol.*, **32**, 148.

Colquhoun, M. K. (1951). The wood pigeon in Britain. A.R.C. Report Series No. 10. London, HMSO 69 pp.

Conti, M. (1981). Wild plants and the ecology of hopper-borne viruses of grasses and cereals. In *Pests, pathogens and vegetation.* Ed. J. M. Thresh. London, Pitman. p. 109.

Conway, G. R. and Murdie, G. (1972). Population models as a basis for pest control. In *Mathematical models in ecology.* N. R. Jeffers (Ed.). London, Blackwell Scientific Publications. p. 195.

—— (Ed.) (1984). *Pest and pathogen control: strategy, tactics and policy models.* London, Wiley. (in press)

Cook, R. (1982). Cereal and grass hosts of some gramineous cyst-nematodes. *EPPO Bull.*, **12**, 399.

—— and York, P. A. (1982). Genetics of resistance to *Heterodera avenae* and *Meloidogyne naasi. Proc. IVth Int. Barley Genetics Symp. Edin. 1981.* p. 418.

Cooke, A. S. (1973). Shell thinning in avian eggs by environmental pollutants. *Environ. Pollut.*, **4**, 85.

Cooke, D. A. and Hull, R. (1972). The effect of soil fumigation with D-D on the yields of sugar beet and other crops. *Ann. appl. Biol.*, **71**, 59.

Cooke, R. C. (1962). The ecology of nematode-trapping fungi in soil. *Ann. appl. Biol.*, **50**, 507.

Cooper, J. I. (1971). The distribution in Scotland of tobacco rattle virus and its nematode vectors in relation to soil types. *Plant Pathology*, **20**, 51.

—— and Thomas, P. R. (1971). Potato viruses. *Rep. Scot. hort. Res. Inst.* 1970. p. 52.

Corbett, D. C. M. (1972). The effect of *Pratylenchus fallax* on wheat, barley and sugar beet roots. *Nematologica*, **18**, 303.

—— and Hide, G. A. (1971). Interactions between *Heterodera rostochiensis* Woll. and *Verticillium dahliae* Kleb on potatoes and the effect of CCC on both. *Ann. appl. Biol.*, **68**, 71.

Cranmer, H. H. (1967). *Plant protection and world crop production.* Leverkusen, Bayer Pflanzenschutz.

CSIRO (1970). *The insects of Australia: a textbook for students and research workers.* Melbourne University Press. 1029 pp.

Dainton, B. H. (1954a). The activity of slugs. I. The induction of activity by changing temperature. *J. exp. Biol.*, **31**, 165.

—— (1954b). The activity of slugs. II. The effect of light and air currents. *J. exp. Biol.*, **31**, 188.

Davis, D. E. (1962). Gross effects of triethylenemelamine on gonads of starlings. *Anatomical Record*, **142**, 353.

—— (1970). Principles for population control by gametocides. *Trans. Amer. Wildlife Nat. Res. Conf.*, **26**, 106.

Davis, R. A. (1970). *Control of rats and mice.* Bull. No. 181. MAFF, London. HMSO 28 pp.

—— (1963). Feral coypus in Britain. *Ann. appl. Biol.*, **51**, 345.

—— and Jenson, A. G. (1960). A note on the distribution of the coypu (*Myocastor coypus*) in Great Britain. *J. anim. Ecol.*, **29**, 397.

Day, M. F. and Venables, D. G. (1961). The transmission of cauliflower mosaic virus by aphids. *Aust. J. Biol. Sci.*, **14**, 187.

Devonshire, A. L. and Farnham, A. W. (1981). The nature and impact of insecticide resistance. *The Plantsman*, **3**, 14.

Dixon, A. F. G. (1971). Migration in aphids. *Science Progress, Oxf.*, **59**, 41.

—— (1977). Aphid ecology, life cycles, polymorphism and population regulation. *Ann. Rev. Ecol. Syst.*, **8**, 329.

Dobie, P. (1977). The contribution of the Tropical Stored Product Centre to the study of insect resistance in stored maize. *Trop. stored Prod. Inf.*, **34**, 7.

——, Greve, J. V., Khoti, K. and Kilminster, A. M. (1979). Inability of storage Bruchidae to infest winged beans (*Psophocarpus tetragonolobus*). *Ent. exp. & appl.*, **26**, 168.

—— and Kilminster, A. M. (1978). The susceptibility of triticale to post-harvest infestation by *Sitophilus zeamais* (Motschulsky) and *Sitophilus granarius* (L.). *J. stored Prod. Res.*, **14**, 87.

Doncaster, C. C. (1971). Feeding in plant parasitic nematodes: mechanisms and behaviour. In *Plant parasitic nematodes* (Eds) B. M. Zuckerman, W. F. Mai and R. A. Rohde. Vol. 2. p 137.

—— (1976). Feeding of the stem nematode *Ditylenchus dipsaci* on leaf tissue of field bean, *Vicia faba. J. Zool. Lond.*, **180**, 139.

Doncaster, J. P. (1943). The life history of *Aphis* (*Doralis*) *rhamni* B.d F. in eastern England. *Ann. appl. Biol.*, **30**, 101.

Dunn, E. (1949). Colorado beetle in the Channel Islands, 1947 and 1948. *Ann. appl. Biol.*, **36**, 525.

Dunn, J. A. (1952). The effect of temperature on the pea aphid–ladybird relationship. *2nd Rep. Nat. Veg. Res. Sta. Wellesbourne*, 1951. p. 21.

—— and Kirkley, J. (1962). Carrot-willow aphid. *Rep. nat. veg. Res. Stn Wellesbourne*, 1961. p. 53.

Dunning, R. A. (1957). Murid damage to seedling beet. *Plant Pathology*, **6**, 19.

—— and Winder, G. H. (1972). Some effects, especially on yield, of artificially defoliating sugar beet. *Ann. appl. Biol.*, **70**, 89.

Ebbels, D. L. and King, J. E. (1979). *Plant health: the scientific basis of administrative control of plant diseases and pests*. Oxford, Blackwells. p. 248.

Edwards, C. A. (1959). A revision of the British Symphyla. *Proc. zool. Soc. Lond.*, **132**, 403.

—— (1962). Springtail damage to bean seedlings. *Plant Pathology*, **11**, 67.

—— and Jeffs, K. (1974). Rate of uptake of DDT from soil by earthworms. *Nature, Lond.*, **247**, 157.

——, Sunderland, K. D. and George, K. S. (1979). Studies on polyphagous predators of cereal aphids. *J. appl. Ecol.*, **16**, 811.

Egan, H. and Weston, R. E. (1977). Pesticide residues: food surveys in the United Kingdom. *Pestic. Sci.*, **8**, 110.

El Khidir, I. (1963). The ecology of the cabbage whitefly (*Aleyrodes brassicae*, Walk). *Rep. Rothamst. exp. Sta.* 1962. p. 165.

—— (1972). Ecological Studies on *Aleyrodes brassicae* Walk. *Z. ang. Ent.*, **72**, 39.

Ellenby, C. (1952). Resistance to the potato root eelworm. *Nature, Lond.*, **170**, 1016.

Elliott, M. (1980). Established pyrethroid insecticides. *Pestic. Sci.*, **11**, 119.

Ellis, P. E., Morgan, E. D. and Woodbridge, A. P. (1970). Is there any hope for hormone mimics as pesticides? *PANS*, **16**, 434.

Emden, H. F. van and Williams, G. F. (1974). Insect stability and diversity in agro-ecosystem. *Ann. Rev. Entomol.*, **19**, 455.

Emmett, B. J. (1980). Key for the identification of lepidopterous larvae infesting brassica crops. *Plant Pathology*, **29**, 122.

Evans, G. O., Sheals, J. G. and Macfarlane, D. (1961). *The terrestrial Acari of the British Isles*. London, Brit. Mus. Nat. Hist. 219 pp.

Evans, K. (1982). Effects of infestation with *Globodera rostochiensis* (Wollenweber) Behrens Rol on the growth of four potato cultivars. *Crop Protection*, **1**, 169.

——, Greet, D. N. and Fatemy, F. (1982). Tolerance of potatoes to cyst-nematode attack. *Rep. Rothamsted expl. Stn for 1981, Pt 1*. p. 159.

—— and Stone, A. R. (1977). A review of the distribution and biology of the potato cyst-nematodes *Globodera rostochiensis* and *G. pallida. PANS*. **123**, 178.

FAO (1973). *Specifications for pesticides: organochlorine insecticides*. MAFF on behalf of FAO. London, HMSO 89 pp.

Feltwell, J. (1982). *Large white butterfly. The biology, biochemistry and physiology of* Pieris Brassicae *Linneaus*. The Hague, Junk. 535 pp.

Fenwick, D. W. (1961). Estimation of field populations of cyst-forming nematodes of the genus *Heterodera*. *J. Helminth. R. T. Leiper Suppl.* 63 pp.

Ferris, H. (1978). Nematode economic thresholds: derivation requirements and theoretical considerations. *J. Nematol.*, **10**, 341.

Finch, S. (1971). The fecundity of the cabbage root fly *Erioischia brassicae* under field conditions (Dipt., Anthomyiidae). *Entomologia exp. appl.*, **14**, 147.

Fisken, A. G. (1959a). Factors affecting the spread of aphid-borne viruses in potato in eastern Scotland. 1. Overwintering of potato aphids, particularly *Myzus persicae* (Sulzer). *Ann. appl. Biol.*, **47**, 264.

——(1959b). Factors affecting the spread of aphid-borne viruses in potato in eastern Scotland. 2. Infestation of the potato crop by potato aphids, particularly *Myzus persicae* (Sulzer). *Ann. appl. Biol.*, **47**, 274.

Fowden, G. M., Jenkins, J. N. and Parrott, W. L. (1972). Resistance of plants to insects. *Adv. Agronomy*. **24**, 187.

Franz, J. M. (1961). Biological control of pest insects in Europe. *Ann. Rev. Ent.*, **6**, 183.

Free, J. B. and Butler, C. G. (1959). *Bumble bees*. London, Collins. 208 pp.

Frisch, K. von (1950). *Bees, their vision, chemical senses and language*. Ithaca, New York, Cornell University Press. 119 pp.

Gair, R. (1971). Organochlorine alternatives—a review of the present position in the U.K. *Proc. 6th Br. Insectic. Fungic. Conf. Brighton*, **3**, 765.

George, K. S. and Gair, R. (1979). Crop loss assessment on winter wheat attacked by the grain aphid, *Sitobion avenae* (F.), 1974–77. *Plant Pathology*, **28**, 143.

——, Light, W. I. St. G. and Gair R. (1962). The effect of artificial defoliation of pea plants on the yield of shelled peas. *Plant Pathology*, **11**, 73.

Georghiou, C. P. (1980). Insecticide resistance and prospects for its management. *Residue Revs.*, **76**, 131.

Gerard, B. M. (1971). The effects on slugs of methiocarb, metaldehyde and aldrin used in potatoes. A.R.C. Slug Research Conference, Cardiff, 1971.

Gibson, R. W. (1971). Climatic factors restricting the distribution of the aphid *Rhopalosiphoninus latysiphon* to the subterranean parts of field potato plants. *Ann. appl. Biol.*, **69**, 89.

—— and Pickett, J. A. (1983). Wild potato repels aphids by release of aphid alarm pheromone. *Nature, Lond.*, **302**, 608.

Gilmore, S. K. (1970). Collembola predation on nematodes. *Search: Agriculture, Entomology, Limnology*, **1**, 1.

Gilsén, T. (1948). Aerial plankton and its conditions of life. *Biol. Rev.*, **23**, 109.

Golob, P. (1981). A practical appraisal of on-farm storage losses and loss assessment methods in Malawi. *Trop. stored Prod. Inf.*, **41**, 5.

Goodey, J. B. (1959). Gall-forming nematodes of grasses in Britain. *J. Sports Turf Res. Institute*, **10**, 1.

——, Franklin, M. T. and Hooper, D. J. (1965). *The nematode parasites of plants catalogued under their hosts*. Farnham Royal, Common Agricultural Bureaux. 214 pp.

Gommers, F. J. (1981). Biochemical interactions between nematodes of plants and their relevance to control. *Helm. Abstracts*, **50**, 9.

Gordon, R. and Webster, J. A. (1974). Biological control of insects by nematodes. *Helm. Abstracts*, **43**, 327.

Gough, H. C. (1947). A note on the occurrence in Yorkshire of *Celaena* (= *Apamea*) *secalis* L. (Lep. Canadrinidae), *Opomyza germinationis* L. (Dipt. Opomyzidae) and *Crepidodera ferruginea* Scop. (Col. Chrysomelidae) in winter wheat. *Ent. monthly Mag.*, **83**, 130.

—— and Dunnett, F. W. (1950). Rabbit damage to winter corn. *Agriculture, Lond.*, **57**, 374.

Gould, H. J. (1961). Observations on slug damage to winter wheat in East Anglia, 1957–59. *Plant Pathology*, **10**, 142.

——(1965). Observations on the susceptibility of maincrop potato varieties to slug damage. *Plant Pathology*, **14**, 109.

——(1970). Preliminary studies of an integrated control programme for cucumber pests and an evaluation of methods of introducing *Phytoseiulus persimilis* Athiars-Henriot for the control of *Tetranychus urticae* Koch. *Ann. appl. Biol.*, **66**, 505.

—— and Graham, C. W. (1970). Damage assessment of *Aphis fabae* on spring sown field beans. *Proc. 5th Brit. Fungic. Insectic. Conf. Brighton*. Vol. 2. p. 505.

Graham-Bryce, I. J. (1983). Novel chemical approaches to crop protection: needs and solutions. *Pestic. Sci.*, **14**, 261.

Greaves, J. H. (1970). Warfarin-resistant Rats. *Agriculture, Lond.*, **77**, 107.

—— and Ayres, P. (1967). Heritable resistance to warfarin in rats. *Nature, Lond.*, **215**, 877.

—— (1969). Some rodenticidal properties of coumatetralyl. *J. Hyg. Camb.*, **67**, 311.

Green, A. A. (1962). The trouble with cockroaches. *New Scientist*, **16**, 74.

Green, C. D. (1981). The effects of weeds and wild plants on the re-infestation of land by *Ditylenchus dipsaci* (stem and bulb nematode) and on the stability of its populations. In *Pests, pathogens and vegetation*. Ed. J. M. Thresh. London, Pitman. p. 217.

—— and Sime, S. (1979). The dispersal of *Ditylenchus dipsaci* with vegetable seeds. *Ann. appl. Biol.*, **92**, 263.

Green, R. E. (1978). Factors affecting the diet of skylarks, *Alauda arvensis* L. *J. anim. Ecol.*, **47**, 913.

—— (1979). The ecology of wood mice (*Apodemus sylvaticus*) on arable farm land. *J. Zool., Lond.*, **188**, 357.

—— (1980). Food selection by skylarks and grazing damage to sugar beet seedlings. *J. appl. Ecol.*, **17**, 613.

Greenway, A. R. and Wall, C. (1981). Attractant lures for males of the pea moth, *Cydia nigricana* (F.) containing (E)-10-Dodecen-1-yl acetate and (E, E)-8, 10-Dodecadien-1-yl acetate. *J. chem. Ecology*, **7**, 563.

Griffiths, D. A. (1960). Some field habitats of mites of stored food products. *Ann. appl. Biol.*, **48**, 132.

Gunn, D. L. (1972). Dilemmas in conservation for applied biologists. *Ann. appl. Biol.*, **72**, 105.

Gurner, P. S., Fisher, J. M. and Dubé (1980). Chemical control of cereal cyst-nematode (*Heterodera avenae*) on wheat by a new low-volume applicator. *Nematologica*, **26**, 448.

Gurney, W. S. C., Nisbet, R. M. and Lawton, J. H. (1983). The systematic formulation of tractable single-species population models incorporating age structure. *J. anim. Ecol.*, **52**, 451.

Harris, K. F. and Maramorosh, K. (Eds) (1977). *Aphids as virus vectors*. New York, Academic Press. 559 pp.

Hassell, M. P. (1978). *The dynamics of arthropod predator-prey systems*. Princeton University Press. 237 pp.

Hatchett, J. H. and Gallun, R. L. (1970). Genetics of the ability of the Hessian fly *Mayetiola destructor* to survive on wheats having different genes for resistance. *Ann. ent. Soc. Amer.*, **63**, 1400.

Hazeltine, W. (1972). Disagreement why brown pelican eggs are thin. *Nature, Lond.*, **239**, 410.

Heathcote, G. D. (1972). The beet leaf bug in East Anglia. *Suffolk Nat. Hist.*, **16**, 43.

—— and Gibbs, A. J. (1962). Virus diseases of British crops of field beans (*Vicia faba* L.). *Plant Pathology*, **11**, 69.

Hedin, P. A. (1977). *Host plant resistance to pests*. Washington DC, Amer. Chem. Soc. 286 pp.

——, Maxwell, F. G. and Jenkins, J. H. (1962). Insect plant attractants, feeding stimulants, repellents, deterrents and other related factors affecting insect behaviour. *Proc. Summer Inst. Biol. Control of Plant Insects and Diseases*. (Eds) F. G. Maxwell and F. A. Jackson U.S.A., Mississippi Univ. Press. 494 pp.

Heiznel, H., Fitter, R. and Parslow, J. (1972). *Birds of Britain and Europe with North Africa and the Middle East*. London, Collins. 336 pp.

Helling, C. S., Kearney, P. C. and Alexander, M. (1971). Behaviour of pesticides in soil. *Advances in Agronomy*, **23**, 147.

Hinton, H. E. (1957). Biological control of pests. Some considerations. *Science Progress*, **177**, 11.

Hollings, M. (1960). Aphid movement and virus spread in seed potato areas of England and Wales. *Plant Pathology*, **9**, 1.

Holmes, A. and Finch, S. (1983). Bioassay of volatile chemicals. *Rep. Nat. Veg. Res. Stn for 1982*, p. 34.

Homminick, W. M. (1979). Selection for hatching at low temperatures in *Globodera rostochiensis* by continuous cultivation of early potatoes. *Nematologica*, **25**, 322.

Howe, R. W. (1956a). The effect of temperature and humidity on the rate of development and mobility of *Tribolium castaneum* (Herbst) (Coleoptera, Tenebrionidae). *Ann. appl. Biol.*, **44**, 356.

—— (1956b). The biology of the two common storage species of *Oryzaephilus* (Coleoptera, Cucujidae). *Ann. appl. Biol.*, **44**, 341.

Huffaker, C. B. (Ed.) (1980). *New technology of pest control*. New York, John Wiley & Sons. 500 pp.

Hughes, A. M. (1976). *The mites of stored food.* Reference Book No. 409. 2nd Ed. MAFF. London, HMSO 400 pp.

Hughes, R. D. (1959). The natural mortality of *Erioischia brassicae* (Bouché) (Diptera, Anthomyiidae) during the egg stage of first generation. *J. anim. Ecol.*, **28,** 343.

—— (1960). A method of estimating the numbers of cabbage root fly pupae in the soil. *Plant Pathology*, **9,** 15.

—— and Salter, D. D. (1959). Natural mortality of *Erioischia brassicae* (Bouché) (Diptera, Anthomyiidae) during immature stages of first generation. *J. anim. Ecol.*, **28,** 231.

Hull, R. and Watson, M. (1946). Factors affecting the loss of yield of sugar been caused by beet yellows virus. *J. agric. Sci.*, **37,** part 4, 301.

Hunter, P. J. (1968). Studies of slugs in arable ground. *Malacologia*, **6,** 369.

—— and Symonds, B. V. (1970). The distribution of bait pellets for slug control. *Ann. appl. Biol.*, **65,** 1.

Hussey, N. W. (1979). Integrated control of glasshouse pests and diseases: its history and development. *Proc. Int. Symp. of IOBC/WPRS on Integrated Control in Agriculture and Forestry.* Vienna, I.O.B.C. p. 177.

Hutchinson, G. E. (1978). *An introduction to population ecology.* New Haven, Yale University Press. 260 pp.

Imms, A. D. (1977). *A general textbook of entomology* Vols 1 & 2. 10th Ed. revised by O. D. Richards and R. G. Davies. London, Chapman & Hall. 1354 pp.

—— (1978). *Outlines of entomology.* 6th Ed. revised by O. W. Richards and R. G. Davies. London, Chapman & Hall. 254 pp.

Jackson, D. J. (1920). Bionomics of weevils of the genus *Sitones* injurious to leguminous crops in Britain. *Ann. appl. Biol.*, **7,** 269.

Jennings, R. C. (1983). Insect hormones and growth regulators. *Pestic. Sci.*, **14,** 327.

Jepson, W. F. and Southwood, T. R. E. (1960). The spring oviposition peaks of frit fly and associated Diptera in young oats. *Plant Pathology*, **9,** 33.

Johnson, C. G. (1969). *Migration and dispersal of insects by flight.* London, Methuen. 763 pp.

Jones, D. P. (1945). Gall midges and grass seed production. *J. Minist. Agric.*, **52,** 248.

—— and Jones, F. G. W. (1947). Wireworms and the sugar beet crop: field trials and observations. *Ann. appl. Biol.*, **34,** 562.

Jones, F. G. W. (1951). The sugar beet eelworm order 1943. *Ann appl. Biol.*, **38,** 535.

—— (1953). The assessment of injury by seedling pests of sugar beet. *Ann. appl. Biol.*, **40,** 606.

—— (1955). Quantitative methods in nematology. *Ann. appl. Biol.*, **42,** 372.

—— (1960). Some observations and reflections on host finding by plant nematodes. *Meded. LandbHoogesch. Gent*, **25,** 1009.

—— (1961). The eelworm problem. *Proc. British Insecticide and Fungicide Conference*, 1961. Vol. 1. p. 17.

—— (1969a). Some reflections on distribution, quarantine and control. In *Nematodes of tropical crops.* Ed. J. E. Peachey, Farnham Royal, Comm. Agr. Bureaux. 354 pp.

—— (1969b). The control of the potato cyst-nematode. John Curtis 'Woodstock' Lecture. *J. R. Soc. Arts*, **118,** 179.

—— (1973a). Control of nematode pests, background and outlook for biological control and disease. In *Biology in pest and disease control.* (Eds) D. Price Jones and M. E. Solomon. *13th Symp. Br. ecol. Soc.* Oxford 1972. Oxford, Blackwell Scientific Pubs. 272 pp.

—— (1973b). Management of nematode populations in Great Britain, *Proc. 4th Tall Timbers Conference on Ecological Animal Control by Habitat Management, Tallahassee* 1972. p. 81.

—— (1975). The soil as an environment for nematodes. *Ann. appl. Biol.*, **79,** 113.

—— (1976a). Pests, resistance and fertilizers. *Proc. 12th Colloquium Int. Potash Inst. Izmir.* p. 233.

—— (1976b). *The golden nematode in Cyprus.* FAO Report. Rothamsted Experimental Station, Harpenden. 21 pp.

—— (1979). The problems of race-specificity in plant resistance breeding. *Proc. Brit. Crop Protection Conf. Brighton 1979*, Vol. **3,** 741.

—— (1980). Some aspects of the epidemiology of plant parasitic nematodes. In *Comparative epidemiology.* (Eds) J. Polti and J. Kranz. Wageningen, PUDOC. p. 71.

—— (1983). Weather and plant parasitic nematodes. *EPPO Bull.*, **13,** 103.

—— and Dunning, R. A. (1972). *Sugar beet pests.* Bull. No. 162. MAFF. London, HMSO, 3rd Ed. 113 pp.

Jones, F. G. W., Dunning, R. A. and Humphries, K. P. (1955). The effects of defoliation and loss of stand upon yield of sugar beet. *Ann. appl. Biol.*, **43**, 63.

—— and Humphries, K. P. (1954). The use of seed dressings containing γ-BHC in the establishment of sugar beet seedlings. *Ann. appl. Biol.*, **41**, 562.

——, Kempton, R. A. and Perry, J. N. (1978). Computer simulation and population models for cyst-nematodes (Heteroderidae: Nematoda). *Nematropica*, **8**, 36.

——, Larbey, D. W. and Parrott, D. M. (1969). The influence of soil structure and moisture on nematodes, especially *Xiphinema, Longidorus, Trichodorus* and *Heterodera* spp. *Soil Biol. Biochem.*, **1**, 153.

—— and Nirula, K. K. (1963). Hatching tests and counts of primary galls in the assessment of nematicides against *Meloidogyne* spp. *Plant Pathology*, **12**, 148.

—— and Parrott, D. M. (1969). Population fluctuations of *Heterodera rostochiensis* Woll. when susceptible potatoes are grown continuously. *Ann. appl. Biol.*, **63**, 175.

——, —— and Perry, J. N. (1981). The gene-for-gene relationship and its significance for potato cyst nematodes and their Solanaceous hosts. In *Plant parasitic nematodes* Vol. III. (Eds) B. M. Zuckerman and R. A. Rohde. New York, Academic Press. p. 23.

——, —— and Ross, G. J. S. (1967). The population genetics of the potato cyst-nematode, *Heterodera rostochiensis*: mathematical models to simulate the effects of growing eelworm-resistant potatoes bred from *Solanum tuberosum* ssp. *andigena. Ann. appl. Biol.*, **60**, 151.

—— and Perry, J. N. (1978). Modelling populations of cyst-nematodes (Nematoda, Heteroderidae). *J. appl. Ecol.*, **15**, 349.

Jones, M. G. (1942). A description of *Aphis (Doralis) rumicis* L., and comparison with *Aphis (Doralis) fabae* Scop. *Bull. ent. Res.*, **33**, 5.

—— (1969). Oviposition of frit fly (*Oscinella frit* L.) on oat seedlings and subsequent larval development. *J. appl. Ecol.*, **6**, 411.

—— (1971). Observations on changes in the female reproductive system of the wheat bulb fly, *Leptohylemyia cearitata* (Fall.). *Bull. ent. Res.*, **61**, 55.

—— (1972). Cereal aphids, their parasites and predators caught in cages over oat and winter wheat crops. *Ann. appl. Biol.*, **72**, 13.

—— (1976). The carabid and staphylinid fauna of winter wheat and fallow on a clay with flints soil. *J. appl. Ecol.*, **13**, 775.

—— (1979a). Abundance of aphids on cereals from before 1973 to 1977. *J. appl. Ecol.*, **16**, 1.

—— (1979b). Observations on primary and secondary parasites of cereal aphids. *Ent. mon. Mag.*, **115**, 61.

Jones, M. G. K. (1981). Host cell responses to endoparasitic nematode attack: structure and function of giant cells and synctia. *Ann. appl. Biol.*, **97**, 353.

Jones, R. A. C. (1981). The ecology of viruses affecting wild and cultivated potatoes in the Andean region of South America. In *Pests, pathogens and vegetation.* (Ed.) J. M. Thresh. London, Pitman. 89 pp.

——, Fribourg, C. E. and Slack, S. A. (1981). Set No. 2. Potato virus and virus-like diseases [with 140 slides]. In *Plant virus slide series* (Eds) O. W. Barnett and S. A. Tolin. South Carolina, Clemson University, 59 pp.

Jones, T. (1954). The external morphology of *Chirothrips hamatus* (Trybom) (Thysanoptera). *Trans. R. ent. Soc. Lond.*, **105**, 163.

Jørgensen, J. (1955). The onion fly *Hylemyia antiqua* (Meig). Investigations on bionomics and experiments on carrots. *Tidsskr. Planteavl.*, **59**, 252.

Kennedy, J. S. and Booth, C. O. (1951). Host alternation in *Aphis fabae* Scop. Feeding preferences and fecundity in relation to the age and kind of leaves. *Ann. appl. Biol.*, **38**, 25.

——, Day, M. F. and Eastop, V. F. (1962). *A conspectus of aphids as vectors of plant viruses.* London, Comm. Inst. Ent. 114 pp.

Kershaw, J. (1957). The beet leaf bug and beet crinkle disease. *Sugar beet pest investigations*, p. 6. Committee paper 444, MAFF.

Kerry, B. R., Crump, D. H. and Mullen, L. A. (1982). Natural control of the cereal cyst nematode, *Heterodera avenae* Woll. by soil fungi at three sites. *Crop Protection*, **1**, 99.

Khan, H. H. (1972). Economic aspects of crop losses and disease control. In *Economic nematology* (Ed.) J. M. Webster. London, Academic Press. p. 1.

Kimmins, F. (1982). The probing behaviour of *Rhopalosiphum maidis*. *Proc. 5th Int. Symp. Insect–Plant Relationships.* Wageningen, PUDOC. p. 411.

Knipling, E. F. (1962). Potentialities and progress in the development of chemosterilants for insect control. *J. econ. Ent.*, **55**, 782.

—— (1972). Entomology and the management of man's environment. *J. Aust. Ent. Soc.*, **11**, 4.

—— and Gilmore, J. E. (1971). *Population density relationships between hymenopterous parasites and their aphid hosts—a theoretical study*. Tech. Bull. ARS, USA Dept. Agric. No. 1428, 34 pp.

Kydonieus, A. F. and Beroza, M. (Eds) (1982). *Insect suppression by controlled release pheromone systems*. Vol. I. Boca Raton Fla, CRC Press. 274 pp.

——, —— and Zweig, G. (Eds) (1982). *Insect suppression by controlled release pheromone systems* Vol. II. Boca Raton Fla, CRC Press. 312 pp.

Lack, D. (1954). *The natural regulation of animal numbers*. Oxford, Clarendon Press. 343 pp.

Lambers, D., Hille, Ris. (1966). Polymorphosm in Aphididae. *A. Rev. Ent.*, **11**, 47.

Laveglia, J. and Dahm, P. A. (1977). Degradation of organophosphorus and carbamate insecticides in the soil and by microorganisms. *Ann. Rev. Ent.*, **22**, 483.

Lee, D. L. and Atkinson, H. J. (1976). *Physiology of nematodes*. London, Macmillan Press. 215 pp.

Lees, A. D. (1955). *The physiology of diapause in arthropods*. Cambridge Univ. Press. 151 pp.

—— (1962). Phenological aspects of diapause. *Ann. appl. Biol.*, **50**, 596.

—— (1966). The control of polymorphism in aphids. *Adv. Ins. Physiol.*, **3**, 207.

—— (1967). The production of the apterous and alate forms in the aphid *Megoura viciae* Buckton, with special reference to the role of crowding. *J. insect Physiol.*, **13**, 289.

Lewis, T. (1968). Windbreaks, shelter and insect distribution. *Span*, **11**, 186.

—— (1973). *Thrips, their biology, ecology and economic importance*. London, Academic Press. 349 pp.

Lloyd, H. G. (1962). The distribution of squirrels in England and Wales, 1959. *J. anim. Ecol.*, **31**, 157.

—— (1963). Spring traps and their development. *Ann. appl. Biol.*, **51**, 329.

—— (1970a). Variation and adaptation in reproductive performance. *Symp. zool. Soc. Lond.*, **26**, 165.

—— (1970b). Post-myxomatosis rabbit populations in England and Wales. *EPPO Publ. Series A*, **58**, 197.

Lowe, A. D. (Ed.) (1973). *Perspectives in aphid biology*. Christchurch, Entomological Society of New Zealand. p. 123.

Lowe, H. J. B. (1982). Some observations on susceptibility and resistance of winter wheat to the aphid *Sitobion avenae* (F.) in Britain. *Crop Protection*, **1**, 431.

MacArthur, R. H. and Wilson, E. O. (1967). *The theory of island biogeography*. Princeton, University Press. 203 pp.

McEwen, F. L., Schroeder, W. T. and Davis, A. C. (1957). Host range and transmission of the pea enation mosaic virus. *J. econ. Ent.*, **50**, 770.

McEwen, J. *et al.* (1983). The effects of irrigation, nitrogen fertilizer and control of pests and pathogens on spring sown field beans (*Vicia faba* L.) and residual effects on two following winter wheat crops. *J. agric. Sci., Camb.*, **96**, 129.

McMillan, J. H. (1975). Interspecific and seasonal analysis of the gut contents of three Collembola (Family: Onychiuridae). *Rev. Ecol. Biol. du Sol.*, **12**, 449.

Maas, P. W. T. and Heijbroek, W. (1983). Biology and pathogenicity of the yellow beet cyst-nematode, a host race of *Heterodera trifolii* in the Netherlands. *Nematologica*, **28**, 77.

Mabbott, T. W. (1960). Observations on the development of potato root eelworm, *Heterodera rostochiensis* Woll. on the potato tuber and the importance of such development in the spread of this nematode on washed tubers. *Europ. Potato J.*, **3**, 236.

Mai, W. F. and Lyon, H. H. (1975). *Pictorial key to genera of plant nematodes*. 4th ed., revised Ithaca, Plant Pathology Dept., Cornell Univ. 61 pp.

Marshall, G. M. (1960). Seed-borne diseases in commercial samples, IV and V. *Ann. appl. Biol.*, **48**, 34.

Martin, H. (1984). *The scientific principles of crop protection*, 7th ed. London, Edward Arnold. 596 pp.

Masamune, T., Anetai, M., Takasugi, M. and Katsui, N. (1982). Isolation of a natural hatching stimulus, glycinoeclepin A. for the soya bean cyst nematode. *Nature Lond.*, **297**, 495.

Maskell, F. E. (1970). Further seed dressing trials against wheat bulb fly (*Leptohylemyia coarctata* (Fall.)) in East Anglia, 1966–68. *Plant Pathology*, **19**, 42.

Masner, P., Slama, K. and Landa, V. (1968). Sexually spread insect sterility induced by analogues of juvenile hormone. *Nature, Lond.*, **219**, 395.

Massee, A. M. (1954). *Pests of fruit and hops*, 3rd ed. London, Crosby Lockwood. 338 pp.

Matthews, L. H. (1960). *British Mammals*. New Naturalist series. London, Collins Ltd. 410 pp.

Maxwell, F. G. and Harris, F. A. (1974). Proceedings of the Summer Institute on biological control of plant insects and diseases. Jackson, USA, University of Mississippi Press. 647 pp.

May, R. M. (1973). Stability in randomly fluctuating versus deterministic environments. *Amer. Naturalist*, **107**, 621.

—— (1981). *Theoretical ecology. Principles and applications*, 2nd Ed. Oxford, Blackwells Scientific Publications. 489 pp.

Mead-Briggs, A. R. (1966). Rabbits and myxomatosis. *Agriculture, Lond.*, **73**, 196.

Moore, N. W. (1972). Toxic chemicals and wildlife section. *Rep. Monks Wood exp. Stn* for 1969–71, p. 24.

Moriarty, F. (1975). *Pollutants and animals: a factual perspective*. London, Allen and Unwin. 140 pp.

Moss, S. R., Crump, D. and Whitehead, A. G. (1975). Control of potato cyst-nematodes, *Heterodera rostochiensis* and *H. pallida*, in a sandy, peaty and silt loam soils by oximecarbamate and organophosphate nematicides. *Ann. appl. Biol.*, **81**, 359.

Morrison, F. O. (1972). Lindane and stored products. *Residue Reviews*, **41**, 113.

Moss, J. E. (1933). The natural control of the cabbage caterpillars, *Pieris* spp. *J. anim. Ecol.*, **2**, 210.

Muir, R. C. (1979). Insecticide resistance in damson-hop aphid, *Phorodon humuli* in commercial hop gardens in Kent. *Ann. appl. Biol.*, **92**, 1.

Murphy, P. W. and Doncaster, C. C. (1957). A culture method for soil meiofauna, and its application to the study of nematode predators. *Nematologica*, **2**, 202.

Murton, R. K. (1965). *The wood pigeon*. New Naturalist series. London, Collins. 256 pp.

—— (1971). Why do some birds feed in flocks? *Ibis*, **113**, 534.

—— and Isaacson, A. J. (1962). The functional basis of some behaviour in the wood-pigeon, *Columba palumbus. Ibis*, **104**, 503.

—— and Ridpath, M. G. (1962). The autumn movements of the wood-pigeon. *Bird study*, **9**, 7.

—— and Wright, E. N. (1968). *The problems of birds as pests*. Institute of Biology Symposium No. 17, London, 254.

Nakanishi, K. (1976). Insect growth regulators. In *Semaine d'Etude sur le Thème Produits Naturels et la Protection de Plantes*. (Ed.) G. B. Marioni-Bettolo. *Pont. Acad. Sci. Scri. Varia*, **41**, (8) II, 7, 14 pp.

Nault, L. R., Bowers, W. S. and Edwards, L. J. (1973). Aphid alarm pheromones. *2nd Int. Congr. Plant Pathology, Minneapolis. Abstr. No.* 0312.

Newell, P. F. (1966). The nocturnal behaviour of slugs. *Med. biol. Illustr.*, **16**, 146.

Newton, H. C. F. (1928). The biology of the flea beetles (Phyllotreta) attacking cultivated Cruciferae. *J. S. E. agric. Col. Wye*, No. 25, 90.

Nollen, H. M. and Mulder, A. (1969). A practical method for economic control of the potato cyst-nematode. *Proc. 5th Br. Insectic. Fungic. Conf. Brighton 1969*, **3**, 671.

Nordlund, D. A., Jones, R. L. and Lewis, W. J. (Eds) (1981). *Semiochemicals, their role in pest control.* New York, Wiley and Sons. 306 pp.

Norton, D. C. (1978). *Ecology of plant parasitic nematodes*. New York, Wiley & Sons. 268 pp.

Oakley, J. N. (1980). Damage to barley germ by *Limothrips* spp. *Plant Pathology*, **29**, 99.

Ordish, G. (1952). *Untaken harvest. Man's loss of crops from pest, weed and disease*. London, Constable. 171 pp.

Oostenbrink, M. (1971). Quantitative aspects of plant-nematode relationships. *Indian J. Nematol.*, **1**, 68.

Owen, M. (1977). The role of wildfowl refuges on agricultural land in lessening the conflict between farmers and geese in Britain. *Biological Conservation*, **11**, 209.

Painter, R. H. (1951). *Insect resistance in crop plants*. New York, Macmillan & Co. Ltd. 520 pp.

Paramonov, A. A. (1968). *Plant-parasitic nematodes*. Vol. I. (Ed.) K. Skrjabin. Jerusalem, Israel Program for Scientific Translations. 390 pp.

Parr, W. J. and Hussey, N. W. (1967). Biological control of red spider mites on cucumbers: effects of different predator densities at introduction. *Rep. Glasshouse Crops Res. Inst. 1966*. p. 135.

Perkins, J. H. (1982). *Insects, experts and the insecticide crisis. The quest for new post-management strategies*. New York, Plenum Press. 304 pp.

Perrins, C. (1974). *Birds*. Countryside Series. London, Collins. 176 pp.

Perring, F. H. and Mellanby, K. (Eds) (1977). *Ecological effects of pesticides*. London, Academic Press. 193 pp.

Perry, J. N., Macaulay, E. D. M. and Emmett, B. J. (1981). Phenological and geographical relationships between catches of pea moth in sex-attractant traps. *Ann. appl. Biol.*, **97**, 17.

Perry, R. N. and Clarke, A. J. (1981). Hatching mechanisms of nematodes. *Parasitology*, **83**, 435.

—, Wharton, D. A. and Clarke, A. J. (1982). The structure of the egg-shell of *Globodera rostochiensis* (Nematoda: Tylenchida). *Int. J. Parasitol.*, **12**, 481.

Peters, B. G. (1961). *Heterodera rostochiensis* population density in relation to potato growth. *J. Helminth.*, *R.T. Leiper Suppl.*, p. 141.

Peterson, C. A., Wilt, P. P. Q. de and Edgington. L. V. (1978). A rationale for the ambimobile translocation of the nematicide oxamyl in plants. *Pestic. Biochem. Physiol.*, **8**, 1.

Peterson, R., Mountfort, G. and Hollom, P. A. D. (1974). *A field guide to birds of Britain and Europe*. London, Collins. 344 pp.

Petherbridge, F. R. and Jones, F. G. W. (1944). Beet eelworm (*Heterodera schachtii* Schm.) in East Anglia, 1934–1943. *Ann. appl. Biol.*, **31**, 320.

— and Stapley, J. H. (1944). Two important wheat pests. *Agriculture, Lond.*, **51**, 320.

Pitkin, B. R. (1980). Variation in some British material of the *Onychiurus armatus* group (Collembola). *Systematic Entomol.*, **5**, 405.

Plumb, R. T. (1971). The control of insect-transmitted viruses of cereals. *Proc. 6th Br. Insectic. Fungic. Conf. Brighton*, **1**, p. 307.

—, Lennon, E. A. and Gutteridge, R. A. (1982). Barley yellow dwarf virus. *Rep. Rothamsted expl. Stn for 1981*. p. 195.

Poinar, G. O. and van der Laan, P. A. (1972). Morphology and life history of *Sphaerularia bombi*. *Nematologica*, **18**, 239.

— and Thomas, G. M. (1966). Significance of *Achromobacter nematophilus*. Poinar and Thomas (Achromobacteraceae: Enbacteriales) in the development of the nematode, DD-136 (Neoaplectana sp. Steinernematidae). *Parasitology*, **56**, 385.

Pollard, D. G. (1971). Some aspects of plant penetration by *Myzus persicae* (Sulz.) nymphs (Homoptera, Aphididae). *Bull. ent. Res.*, **61**, 315.

Powell, W. (1982). The identification of hymenopterous parasitoids attacking cereal aphids in Britain. *Systematic Entomology*, **7**, 465.

Price Jones, D. (1968). Integrated control of pests. *PANS*, **14**, 514.

Proeseler, G. (1982). Ein bisher unbekannte Krankheit der Beta-Rüber in der DDR. *Nachrichtenblatt für den Pflazenschutz in der DDR*, **36**, 20.

Quick, H. E. (1960). British slugs (*Pulmonata; Testacellidae, Arionidae, Limacidae*). *Bull. British Museum (Nat. Hist.) Zoology*, **6**, (3).

Rabbinge, R. and Rijsdijk, F. H. (1983). EPIPRE: a disease and pest management system for winter wheat taking account of micrometeorological factors. *EPPO Bull.*, **13**, 297.

Raski, D. J. and Hewitt, W. R. (1963). Symposium on interrelationships between nematodes and other agents causing plant diseases. Plant nematodes as vectors of plant viruses. *Phytopathology*, **53**, 39.

Ratcliffe, D. A. (1970). Changes attributable to pesticides in egg breakage frequency and eggshell thickness in some British birds. *J. appl. Ecol.*, **7**, 67.

Raw, F. (1951). The ecology of the garden chafer *Phyllopertha horticola* (L.) with preliminary observations on control measures. *Bull. ent. Res.*, **42**, 605.

Rennison, B. D., Hammond, L. E. and Jones, G. L. (1968). A comparative trial of norbormide and zinc phosphide against *Rattus norvegicus* on farms. *J. Hyg. Camb.*, **66**, 147.

Rhode, R. H., Simon, J., Perdomo, A., Gutiérrez, J., Dowling, Jr. C. F. and Linquist, D. A. (1971). Application of sterile-insect-release technique in Mediterranean fruit fly suppression. *J. econ. Ent.*, **64**, 708.

Ribbands, C. R. (1953). *The behaviour and social life of honey bees*. London, Bee Research Association Ltd. 352 pp.

Ridgway, R. L. and Vinson, S. B. (Eds) (1977). *Biological control by augmentation of natural enemies*. New York, Plenum Press. 480 pp.

Robinson, R. A. (1976). *Plant pathosystems*. Berlin, Springer-Verlag. 184 pp.

Roesler, R. (1937). Drehherzkrankheit und Herzlosigkeit bei Kohl. *Kranke Pflanze*, **14**, 124.

Ross, J. (1972). Myxomatosis and the rabbit. *Br. vet. J.*, **128**, 172.

Runham, N. W. and Hunter, P. J. (1970). *Terrestrial slugs*. London, Hutchinson University Library. 184 pp.

Salt, G. and Hollick, F. S. J. (1944). Studies of wireworm populations. 1. A census of wireworms in pasture. *Ann. appl. Biol.*, **31**, 52.

Santos, M. S. N. de A. (1972). Production of male *Meloidogyne* spp and attraction to their females. *Nematologica*, **18**, 291.

Sawicki, R. M. (1979). Resistance to pesticides. 1. Resistance of insects to insecticides. *Span*, **22**, 50.

——, Devonshire, A. L., Rice, A. D., Moores, G. M., Petzing, S. M. and Cameron, A. (1978). The detection and distribution of organophosphorus and carbamate insecticide-resistant *Myzus persicae* (Sulz.) in Britain in 1976. *Pestic. Sci.*, **9**, 189.

Schofield, R. K. and Costa, J. V. B. da (1935). The determination of the pF at permanent wilting and at the moisture equivalent by the freezing point method. *3rd Int. Congr. Soil Sci. Oxford*, **1**, 6.

Scopes, N. E. A. and Biggerstaff, S. M. (1971). The production, handling and distribution of the white fly *Trialeurodes vaporariorum* and its parasite *Encarsia formosa* for use in Biological Control programme in glasshouses. *Plant Pathology*, **20**, 111.

—— and Ledieu, M. (1983). *Pest and disease handbook.* 2nd Ed. Croydon, BCPC Publications. 672 pp.

Seinhorst, J. W. (1956). Population studies on stem eelworms. (*Ditylenchus dipsaci*). *Nematologica*, **1**, 159.

—— (1957). Some aspects of the biology and ecology of stem eelworms. *Nematologica*, **2**, Suppl., 355.

—— (1965). The relations between nematode density and damage to plants. *Nematologica*, **11**, 137.

Seymour, P. R. (1979). *Invertebrates of economic importance in Britain: common and scientific names.* MAFF Tech. Bull. No. 6 GD/2. London, HMSO 132 pp.

Sharma, R. D. (1971). Studies on the plant parasitic nematode *Tylenchorhynchus dubius. Meded Landb-Hoogesch. Wageningen*, **71**, 1.

Shaw, M. J. P. (1970a). Effects of population density on alienicolae of *Aphis fabae* Scop. (1) The effect of crowding on the production of alatae in the laboratory. *Ann. appl. Biol.*, **65**, 191.

—— (1970b). Effects of population density on alienicolae of *Aphis fabae* Scop. (2) The effects of crowding on the expression of migratory urge among alatae in the laboratory. *Ann. appl. Biol.*, **65**, 197.

—— (1970c). Effects of population density on aliencolae of *Aphis fabae* Scop. (3) The effect of isolation on the development of form and behaviour of alatae in a laboratory clone. *Ann. appl. Biol.*, **65**, 205.

Shepherd, A. M. (1962a). *The emergence of larvae from cysts in the genus* Heterodera. Farnham Royal, Comm. Agric. Bur. 90 pp.

—— (1962b). New blue R, a stain that differentiates between living and dead nematodes. *Nematologica*, **8**, 201.

—— and Clark, S. A. (1978). Cuticle structure and 'cement' formation at the anterior end of female cyst-nematodes of the genera *Heterodera* and *Globodera* (Heteroderidae: Tylenchida). *Nematologica*, **24**, 201.

——, —— and Dart, P. J. (1972). Cuticle structure in the genus *Heterodera. Nematologica*, **18**, 1.

——, —— and Hooper, D. J. (1980). Structure of the anterior alimentary tract of *Aphelenchoides blastophthorus* (Nematoda: Tylenchida, Aphelenchina). *Nematologica*, **26**, 313.

Sherlock, P. L. (1983). Natural incidence of disease in cutworms, especially *Agrotis segetum* in England and Wales. *Ann. appl. Biol.*, **102**, 49.

Shorten, M. (1946). A survey of the distribution of the American grey squirrel (*Sciurus carolinensis*) and the British red squirrel (*S. vulgaris leucourus*) in England and Wales in 1944–5. *J. anim. Ecol.*, **15**, 82.

—— (1954). *Squirrels.* The New Naturalist Series. London, Collins. 212 pp.

Simmonds, S. P. (1971). Observations of the possible control of *Tetranychus urticae* on strawberries by *Phytoseinulus persimilis. Pl. Path.*, **20**, 117.

Sinden, S. L., Sanford, L. L. and Osman, S. F. (1979). Glycoalkaloids and resistance to the Colorado beetle in *Solanum chacoense* Bitt. *Am. Pot. J.*, **56**, 479.

Sly, J. M. A. (1981). *Pesticide usage.* Ref. Book No. 523. MAFF, London, H.M.S.O. 94 pp.

Smart, L. E. and Stevenson, J. H. (1982). Laboratory estimation of toxicity of pyrethroid insecticides to honeybees: relevance to hazard in the field. *Bee World*, **63**, 150.

Smith, K. M. (1948). *A textbook of agricultural entomology.* Cambridge, University Press. 286 pp.

Solomon, M. E. (1949). The natural control of animal populations. *J. anim. Ecol.*, **18**, 1.

Sorauer, P. (1953). *Hanbuch der Pflanzenkrankheiten. Tierishe Schädlinger an Nutzflanzen. Erster Teil.* Berlin, Paul Parey. 518 pp.

South, A. (1965). Biology and ecology of *Agriolimax reticulatus* (Müll) and other slugs: spatial distribution. *J. anim. Ecol.*, **34**, 403.

Southey, J. F. (1972). *Anguina tritia.* In *Descriptions of plant parasitic nematodes*, Set 1. No. 13. St Albans, Comm. Inst. Helmanthol. 4 pp.

—— (Ed.) (1978). *Plant nematology*, 3rd Ed. ADAS, MAAF, GDI Ref. Book No. 407. London, HMSO 440 pp.

—— (Ed.) (1985). *Laboratory methods for work with plant and soil nematodes.* ADAS. MAFF Reference Book No. 40. London, HMSO (in preparation).

——, Alphey, T. J. W., Cooke, D. A., Whiteway, J. A. and Mathias, P. L. (1982). Surveys of beet cyst nematode in England 1977-80. *Plant Pathology*, **31,** 163.

Southwood, T. R. E. (1978). *Ecological methods: with particular reference to the study of insect populations*, 2nd Ed. London, Chapman & Hall. 524 pp.

—— (1981). Bionomic strategies and population parameters. In *Theoretical Ecology*, 2nd Ed. (Ed.) R. M. May. Oxford, Blackwell Scientific Publications. p. 30.

—— and Comins, H. N. (1976). A synoptic population model. *J. anim. Ecol.*, **45,** 949.

—— and Jepson, W. F. (1961). The frit fly—a denizen of grassland and a pest of oats. *Ann. appl. Biol.*, **49,** 556.

Spaull, A. M. (1982). *Helicotylenchus vulgaris* and its association with damage to sugar beet. *Ann. appl. Biol.*, **100,** 501.

Speyer, E. (1928). The red spider mite (*Tetranychus telarius* L.). *Jour. Pom. and Hort. Sci.*, **7,** 161.

Stǎry, P. (1976). Parasite spectrum and relative abundance of parasites of cereal aphids in Czechoslovakia (Hymenoptera, Aphidiidae). *Acta entomol. bohemoslov.*, **73,** 216.

Stevenson, J. H., Needham, P. H. and Walker, J. (1978). Poisoning of honeybees by pesticides: investigations of the changing pattern over 20 years. *Rep. Rothamsted. expl. Stn for 1977 Pt 2*, p. 55.

—— and Smart, L. E. (1979). Oil seed rape and honeybee poisoning—1978, 1979. *Proc. Brit. Crop Protection Conf. Brighton 1979*, **1,** p. 117.

Stephenson, J. W. (1967). The distribution of slugs in a potato crop. *J. appl. Ecol.*, **4,** 129.

—— (1971). Slugs, formulation of facts and of a molluscicide in gelatin. *Rep. Rothamsted exp. Stn for 1970, Pt 1*, p. 198.

—— and Knutson, L. V. (1966). A résumé of recent studies of invertebrates associated with slugs. *J. econ. Ent.*, **59,** 356.

Stone, A. R. (1973). *Heterodera pallida* n. sp. (Nematoda: Heteroderidae) a second species of potato cyst-nematode. *Nematologica*, **18,** 591.

——, Platt, H. M. and Khalil, L. F. (Eds) (1983). *Concepts in nematode systematics.* London, Academic Press. 388 pp.

Stribley, M. F., Moores, G. D., Devonshire, A. L. and Sawicki, R. M. (1983). Application of the FAO-recommended method for detecting insecticide resistance in *Aphis fabae* Scop., *Sitobion avenae* (F.), *Metopolophium dirhodum* (Walk.) and *Rhopalosiphum padi* (L.). *Bull. ent. Res.*, **73,** 107.

Strickland, A. H. (1962). Recent advances in applied entomology. *Span*, **5,** 69.

—— (1965). Pest control and productivity in British Agriculture. *J. R. Soc. Arts*, **113,** 62.

Sturhan, D. (1970). *Ditylenchus dipsaci*—doch ein Artenkomplex? *Nematologica*, **16,** 327.

Sunderland, K. D. (1975). The diet of some predatory arthropods in cereal crop. *J. appl. Ecol.*, **12,** 507.

—— and Crook, N. E. (1982). Serological detection of aphid predators. *Rep. Glasshouse Crops Res. Stn for 1981*, p. 112.

Sylvester, E. S. (1980). Cumulative and propagative transmission by aphids. *Ann. Res. Ent.*, **25,** 257.

Sweetman, H. L. (1958). *The principles of biological control.* Dubuque, Iowa, W. C. Brown & Co. 560 pp.

Swenson, K. G. (1973). Insects as vectors of plant viruses and virus-like organisms. In *Perspectives in aphid biology*. (Ed.) A. D. Lowe. Christchurch, Entomological Society of New Zealand. p. 92.

Taylor, C. E. (1972). Nematode transmission of plant ciruses. *PANS*, **18,** 269.

—— and Gordon, S. C. (1970). A comparison of four nematicides for the control of *Longidorus elongatus* and *Xiphinema diversicaudatum* and the viruses they transmit. *Hort. Res.*, **10,** 133.

Taylor, K. D. (1963). Some aspects of grey squirrel control. *Ann. appl. Biol.*, **51,** 334.

Taylor, L. R. (1965). Flight behaviour and aphid migration. *Proc. N. cent. Brch. ent. Soc. Am.*, **20,** 9.

Taylor, L. R. *et al.* (1969-82). Rothamsted insect survey, 1st-13th *Rep. Rothamsted expl. Stn for 1968-1981, Pt 2*, pagination varies.

—— —— (1981-1982). Synoptic monitoring for migrant insect pests in Great Britain and Western Europe. *Rep. Rothamsted expl. Stn for 1980 and 81 Pt 2*, p. 41 and p. 23.

Thompson, A. L. (Ed.) (1964). *A new dictionary of birds*. London, Nelson. 928 pp.

Thompson, H. V. (1963). The limitations of control measures. *Ann. appl. Biol.*, **51**, 326.

—— and Worden, A. N. (1956). *The Rabbit*. The New Naturalist series. London, Collins. 240 pp.

Thorne, G. (1961). *Principles of nematology*. New York, McGraw-Hill. 543 pp.

Thorpe, W. H. (1931). The biology, post embryonic development and economic importance of *Cryptochaetum iceryae*. *Proc. zool. Soc. Lond.*, **104**, 929.

Thresh, J. M. (1982). Cropping practices and virus spread. *Ann. Rev. Phytopathol.*, **20**, 193.

Torrance, L. and Jones, R. A. C. (1982). Increased sensitivity of detection of plant viruses obtained by using a fluorogenic substrate in enzyme-linked immunosorbent assay. *Ann. appl. Biol.*, **101**, 501.

Triantaphyllou, A. C. (1971). Genetics and cytology. In *Plant parasitic nematodes* (Eds) B. M. Zuckerman, W. F. Mai and R. A. Rohde, **2**, 1.

Trudgill, D. L. (1967). The effect of environment on sex determination in *Heterodera rostochiensis*. *Nematologica*, **13**, 263.

Ubler, B. (1978). Zur Frage Schädlickheit subterraner Collembolen in Zückerrubenständen. *Z. Pflkrankh. PflSchutz.*, **85**, 594.

—— (1980). Untersuchungen zur Nahrungswahl von *Onychiurus fimatus* Gisin (Onychiuridae, Collembola) einem Aufgangsschädling der Zuckerrübe. *Z. angew. Ent.*, **90**, 333.

Varley, G. and Gradwell, R. G. (1960). Key factors in population studies. *J. anim. Ecol.*, **29**, 399.

Vickerman, G. P. and Sunderland, K. D. (1975). Arthropods in cereal crops nocturnal activity, vertical distribution and aphid predation. *J. appl. Ecol.*, **12**, 755.

——, Wratten, S. D. (1979). The biology and pest status of cereal aphids in Europe: a review. *Bull. ent. Res.*, **69**, 1.

Walker, J. O. (1980). *Spraying systems for the 1980s*. Monograph No. 24. *Proc. Symp. Royal Holloway Coll. UK.* 1980. London BPCP Publications. 319 pp.

Wall, C. (1984). Exploitation of insect communications by man—fact or fantasy? *Proc. 12th Symp. R. ent. Soc., Lond.* (in press).

Wallace, H. R. (1961a). The bionomics of the free-living stages of zoo-parasitic and phyto-parasitic nematodes—a critical survey. *Helm. Abstr.*, **30**, 1.

—— (1961b). The nature of resistance in chrysanthemum varieties to *Aphelenchoides ritzemabosi*. *Nematologica*, **6**, 49.

—— (1963). *The biology of plant parasitic nematodes*. London, Edward Arnold. 280 pp.

Wardle, R. A. (1929). *The problems of applied entomology*. Manchester, University Press. 587 pp.

Watson, M. A. and Plumb, R. T. (1972). Flight behaviour of aphids. *Ann. Rev. Ent.*, **17**, 425.

Way, M. J. (1962). Definition of diapause. *Ann. appl. Biol.*, **50**, 595.

—— (1963). Mutualism between ants and honeydew-producing Homoptera. *Ann. rev. Ent.*, **8**, 307.

—— (1967). The nature and causes of annual fluctuations in numbers of *Aphis fabae* Scop. on field beans *Vicia faba*. *Ann. appl. Biol.*, **59**, 175.

——, Bardner, R., Baer, R. van, and Aitkenhead, P. (1958). A comparison of high and low volume sprays for control of the bean aphid, *Aphis fabae* Scop. on field beans. *Ann. appl. Biol.*, **46**, 399.

Webley, D. (1962). Experiments with slug baits in South Wales. *Ann. appl. Biol.*, **50**, 129.

—— (1964). Slug activity in relation to weather. *Ann. appl. Biol.*, **53**, 407.

—— and Jones, F. G. W. (1981). Observations on *Globodera pallida* and *G. rostochiensis* on early potatoes. *Plant Pathology*, **30**, 217.

Webster, J. A. and Smith, D. H. (1971). Seedlings used to evaluate resistance to the cereal leaf beetle (*Oulema melanopus*: Col., Chrysomelidae). *J. Econ. Ent.*, **64**, 925.

——, ——, Rathke, E. and Cress, C. E. (1975). Resistance to cereal leaf beetle in wheat: density and length of leaf-surface pubescence in four wheat lines. *Crop Science*, **15**, 199.

Welch, H. E. (1965). Entomophilic nematodes. *Ann. Rev. Ent.*, **10**, 275.

West, T. F. and Hardy, J. E. (1961). *Chemical control of insects*. London, Chapman & Hall. 206 pp.

Wheatley, G. A. (1970). The problem of carrot fly control on carrots. *Proc. 5th Br. Insectic. Fungic. Conf.* 1969, **1**, 248.

—— (1983). Entomology. *Rep. National Vegetable Res. Stn for 1982*, p. 25.

Williams, J. J. W. and Carden, P. W. (1961). Cabbage stem flea beetle in East Anglia. *Plant Pathology*, **10**, 85.

Whitehead, A. G. (1973). Control of cyst-nematodes (*Heterodera* spp.) by organophosphates, oxime-carbamates and soil fumigants. *Ann. appl. Biol.*, **75**, 439.

Whitehead, A. G. (1975). Chemical control of potato cyst-nematode. *ARC Research Rev.*, **1**, 17.

——, Dunning, R. A. and Cooke, D. A. (1971). Docking disorder and root ectoparasitic nematodes of sugar beet. *Rep. Rothamsted exp. Stn for 1970* Pt 2, 219.

—— and Fraser, J. E. (1972). Injury to field beans (*Vicia faba* L.) by *Tylenchorhynchus dubius*. *Plant Pathology*, **21**, 112.

——, Fraser, J. E. and Storey, G. (1972). Chemical control of potato cyst-nematode in sandy clay soil. *Ann. appl. Biol.*, **72**, 81–88.

——, Tite, D. J. and Fraser, J. E. (1973). Chemical control of stem nematode. *Rep. Rothamsted exp. Stn.* Pt 1 1972. p. 166.

Whitten, M. J. (1970). Genetics of pests in their management. In *Concepts of Pest Management*. (Eds) R. L. Rabb and F. E. Guthrie. Raleigh, North Carolina State University, p. 119.

—— and Norris, K. K. (1967). 'Booby-trapping' as an alternative to sterile males for insect control. *Nature, Lond.*, **216**, 1136.

Whiteway, J. A., Alphey, T. J. W., Mathias, P. L. and Southey, J. F. (1982). Computer mapping of records of beet cyst-nematode. *Plant Pathology*, **31**, 157.

Wigglesworth, V. B. (1972). *The principles of insect physiology*, 7th Ed. London, Methuen. 827 pp.

Wilde, J. de (1962). Photoperiodism in insects and mites. *Ann. Rev. Ent.*, **7**, 1.

Winfield, A. L. (1961). Field observations on the control of blossom beetles (*Meligethes aeneus* F.) and cabbage seed weevils (*Ceuthorhynchus assimilis* Payk.) on mustard-seed crops in East Anglia. *Ann. appl. Biol.*, **49**, 539.

—— (1965). Potato root eelworm in Holland, Lincolnshire. *NAAS Quart. Rev.*, **67**, 110.

——, Wardlow, L. R. and Smith, B. F. (1967). Further observations on the susceptibility of maincrop potato cultivars to slug damage. *Plant Pathology*, **16**, 136.

Winslow, R. D. (1954). Provisional lists of some host plants of some root eelworms (*Heterodera* spp.). *Ann. appl. Biol.*, **41**, 591.

—— (1955). The hatching responses of some root eelworms of the genus *Heterodera*. *Ann. appl. Biol.*, **43**, 19.

—— (1960). Some aspects of the ecology of free-living and plant-parasitic nematodes. In: *Nematology, fundamentals and recent advances with emphasis on plant parasitic and soil forms.* (Eds) J. N. Sasser and W. R. Jenkins. Chapel Hill, Univ. N. Carolina Press, pp. 341–415.

—— (1964). Soil nematode population studies. 1. Migratory root-Tylenchida and other nematodes of the Rothamsted and Woburn six-course rotations. *Paedobiologia*, **4**, 65.

Wit, C. T. de (1970). Dynamic concepts in biology. In *The use of models in agricultural and biological research.* (Ed.) J. G. W. Jones. Maidenhead, Grassland Research Institute. p. 9.

Wood, F. H. and Foot, M. A. (1975). Treatment of potato tubers to destroy cysts of potato cyst nematode: a note. *N.Z. J. expl. Agric.*, **3**, 349.

Woodville, H. C. (1970). Results of a three year survey of saddle-gall midge (*Haplodiplosis equestris* (Wagn.)) on spring barley. *Plant Pathology*, **19**, 141.

Worthing, C. R. and Walker, S. B. (1983). *The pesticide manual*, 7th Ed. Malvern, BCPC Publications. 700 pp.

Wratten, S. D. (1978). Effects of feeding position of the aphids *Sitobion avenae* and *Metopolophium dirhodum* on wheat yield and quality. *Ann. appl. Biol.*, **90**, 11.

Wright, D. W. and Coaker, T. H. (1968). Development of dieldrin resistance in carrot fly in England. *Plant Pathology*, **17**, 178.

——, Geering, Q. A. and Dunn, J. A. (1951). Varietal differences in the susceptibility of peas to attack by the pea moth, *Laspeyresia nigricana* (Steph.). *Bull. ent. Res.*, **41**, 663.

——, Hughes, R. D., Salter, D. D. and Worrall, J. (1960). The biology and natural control of the cabbage root fly. *Rep. Nat. Veg. Sta. Wellesbourne*, 1959, 35.

——, Hughes, R. D. and Worrall, J. (1960). The effect of certain predators on the numbers of cabbage root fly (*Erioischia brassicae* (Bouché)) and on the subsequent damage caused by the pest. *Ann. appl. Biol.*, **48**, 756.

—— and Kirkley, J. (1963). Carrot-willow aphid. *Rep. Nat. Veg. Res. Sta. Wellesbourne*, 1962, 52.

Wright, E. N., Inglis, I. R. and Feare, C. J. (Eds) (1980). *Bird problems in agriculture*. London, BPCP Publications. 210 pp.

York, P. A. (1980). Relationship between cereal and root-knot-nematode *Meloidogyne naasi* and the growth and grain yield of spring barley. *Nematologica*, **17**, 575.

Yuksel, H. S. (1960). Observations on the life cycle of *Ditylenchus dipsaci* on onion seedlings. *Nematologica*, **5,** 289.

Zuckermann, B. M., Mai, W. F. and Rhode, R. A. (Eds) (1972). *Plant parasitic nematodes*. New York. Academic Press. Vol. I, 345 pp., Vol. 2, 347 pp.

—— and Rhode, R. A. (Eds) (1981). *Plant parasitic nematodes*. Academic Press, New York. Vol. 3, 508 pp.

Index: Latin names of animals and plants

Note: Principal page references are in bold type.

Index